Nematodes as Environmental Indicators: From Theory to Practice

2nd Edition

This volume is dedicated to the memories of our colleagues
Gregor Yeates, Diana Wall and Brian Darby,
with all our appreciation of their legacies and contributions to the science of nematology.

Nematodes as Environmental Indicators: From Theory to Practice

2nd Edition

Edited by

Thomais Kakouli-Duarte

South East Technological University, Ireland

Gerard Korthals

Wageningen University and Research, The Netherlands

Sara Sánchez-Moreno

National Museum of Natural Sciences, Spain

Gerhard du Preez

North-West University, South Africa

and

Ron G.M. de Goede

Wageningen University and Research, The Netherlands

CABI

CABI is a trading name of CAB International

CABI
Nosworthy Way
Wallingford
Oxfordshire OX10 8DE
UK

Tel: +44 (0)1491 832111
E-mail: info@cabi.org
Website: www.cabi.org

CABI
200 Portland Street
Boston
MA 02114
USA

Tel: +1 (617)682-9015
E-mail: cabi-nao@cabi.org

A catalogue record for this book is available from the British Library, London, UK.

ISBN-13: 9781800624207 (hardback)
9781800624214 (ePDF)
9781800624221 (ePub)

DOI: 10.1079/9781800624221.0000

Commissioning Editor: Rebecca Stubbs
Editorial Assistant: Emma McCann
Production Editor: Shankari Wilford

Typeset by Straive, Pondicherry, India
Printed and bound in the UK by CPI Group (UK) Ltd, Croydon, CR0 4YY

Contents

Contributors vii

Foreword xi

Preface xiii

Part A – Nematode Communities as Part of Ecosystems and Food Webs

1 Roles of Nematodes in Ecosystems: Their Biology and Ecology 1
H. Ferris, W.H. Van der Putten, G.W. Yeates and T. Moens

2 Nematode Diversity in Marine, Freshwater and Terrestrial Ecosystems 37
N. Majdi, M. Schratzberger, D.L. Porazinska, W. Traunspurger, J.H. Schmidt, J. Ingels, S. Geisen and M. Hodda

3 The Evolution of Nematode Diversity 66
S. Sánchez-Moreno, D.L. Porazinska and L. Robertson

Part B – Describing Nematode Communities

4 General Community Indices for the Analysis of Nematode Assemblages 77
D.A. Neher and B.J. Darby

5 Indices Developed Specifically for Analysing Nematode Assemblages 91
J.H. Schmidt, G. du Preez, S. Sánchez-Moreno and H. Ferris

6 Nematode-Based Pollution Indices for Aquatic Ecosystems 105
S. Höss, W. Traunspurger, F. Semprucci and M. Schratzberger

7 Nematode Contributions to Ecosystem Functioning and Bioindication Under Different Land Uses 119
G. du Preez, R. Campos-Herrera, M. Daneel, G. Korthals, J.H. Schmidt and C.D. Sprunger

Part C – Practical Implications on the Application of Nematodes as Bioindicators

8 Case Studies Using Nematode Assemblage Analysis in Terrestrial Habitats 132
P. Nagy, T. Kakouli-Duarte, S. Costa, L. Ruess and M. Argyropoulou

9 **Case Studies Using Nematode Assemblage Analysis in Aquatic Habitats** 150
C. Gambi, S. Höss, F. Ape, R. Danovaro, S. Mirto, W. Traunspurger and A. Zullini

Part D – Threats Affecting the Global Distribution of Nematodes

10 **Nematodes in Extreme Environments** 167
X. Xue, D.H. Wall and B.J. Adams

11 **Nematodes as Indicators of Climate Change and Environmental Restoration** 182
Q. Li, W. Liang and S. Sánchez-Moreno

12 **Soil Nematode Communities Across Large Spatial Scales** 192
S. Geisen, R.A. Wilschut, M. Liu, D.H. Wall and S. Sánchez-Moreno

Part E – Methods for Studying Nematodes in the Laboratory and International Standards

13 **Ecotoxicity Testing with Nematodes** 206
S. Höss and P.L. Williams

14 ***Caenorhabditis elegans* and Ecotoxicogenomics** 222
C. Lagido, C. Zhang, L. Atira, A. Vieyra-Ramirez, T. Kakouli-Duarte and S.R. Stürzenbaum

15 **Protocols for Nematode Sampling, Extraction, Morphological and Molecular Identification, Index Calculation and Toxicity Testing** 250
R.G.M. de Goede, G. du Preez, S. Geisen, S. Höss and T. Kakouli-Duarte

Part F – Policy, Education, Communication and Commercialization

16 **Environmental Policy and Nematodes** 273
T. Kakouli-Duarte, L. Atira, L. Da Costa Monteiro, A. Godjo, C. Jansen van Rensburg, A. Karpinska, A. Pulavarty, F. Semprucci, A. Vieyra-Ramírez and M. Viketoft

17 **Education and Environmental Nematology** 292
H. Fourie, I. Dehennin, R.G.M. de Goede, L. Cortada, G. Korthals, D.A. Neher, M. Hodda and W. Bert

18 **Commercial Aspects of the Use of Nematodes as Bioindicators** 312
S.J. Forster, C.P. Villenave, C. Chauvin, B. Calvo and M.W. Trett

Index 343

Contributors

Byron J. Adams, Department of Biology, 4102 Life Sciences Building (LSB), Brigham Young University, and Monte L. Bean Life Science Museum, Provo, UT 84602, USA. Email: bjadams@byu.edu.

Lilian Atira, Molecular Ecology and Nematode Research Group (MENRG), enviroCORE, Department of Applied Science, South East Technological University, Kilkenny Road Campus, Carlow R93 V960, Ireland. Email: Lilian.Atira@postgrad.setu.ie

Francesca Ape, Institute of Marine Sciences, National Research Council (ISMAR-CNR), Via Gobetti 101, 40129 Bologna, Italy. Email: francesca.ape@cnr.it

Maria Argyropoulou, Department of Zoology, School of Biology, Aristotle University of Thessaloniki, GR-54124 Thessaloniki, Greece. Email: margyrop@bio.auth.gr

Wim Bert, Nematology Research Unit, Faculty of Sciences, Ghent University, K.L. Ledeganckstraat 35, B-9000 Ghent, Belgium. Email: wim.bert@ugent.be

Beatriz Calvo, Department of Pathology and Population Sciences, Royal Veterinary College, Hawkshead Lane, Hatfield, Hertfordshire, AL9 7TA, UK. Email: bcalvou@gmail.com

Raquel Campos-Herrera, Instituto de Ciencias de la Vid y del Vino (Gobierno de La Rioja, CSIC, Universidad de La Rioja), Finca La Grajera Ctra, Burgos Km. 6 Salida 13 Lo-20, 26007 Logroño, Spain. Email: raquel.campos@icvv.es

Camille Chauvin, ELISOL Environnement, ZA des Tourels, 10 Avenue du Midi, 30 111 Congénies, France. Email: camille.chauvin@elisol-environnement.fr

Laura Cortada, Nematology Research Unit, Faculty of Sciences, Ghent University, K.L. Ledeganckstraat 35, B-9000 Ghent, Belgium. Email: Laura.CortadaGonzalez@UGent.be

Sofia R. Costa, Centre of Molecular and Environmental Biology (CBMA), Department of Biology, University of Minho, Campus de Gualtar, 4710-057 Braga, Portugal. Email: sofia.costa@bio.uminho.pt

Luana Da Costa Monteiro, Marine Biology Research Group, Biology Department, Faculty of Sciences, University of Ghent, Krijgslaan 281, Building S8, Ground Floor, B-9000 Ghent, Belgium. Email: Luana.DaCostaMonteiro@UGent.be

Mieke Daneel, Unit for Environmental Sciences and Management, North-West University, Private Bag X6001, Potchefstroom 2520, South Africa; and Agricultural Research Council–Tropical and Subtropical Crops, Private Bag X11208, Nelspruit 1200, South Africa. Email: mieke@arc.agric.za

Roberto Danovaro, Department of Life and Environmental Sciences, Polytechnic University of Marche, Via Brecce Bianche, 60131 Ancona, Italy. Email: r.danovaro@univpm.it

Brian J. Darby†, formerly Department of Biology, University of North Dakota, 10 Cornell St. Stop 9019, Grand Forks, ND 58202, USA.

Inge Dehennin, Nematology Research Unit, Faculty of Sciences, Ghent University, K.L. Ledeganckstraat 35, B-9000 Ghent, Belgium. Email: inge.dehennin@ugent.be

Ron G.M. de Goede, Soil Biology Group, Wageningen University and Research, PO Box 47, 6700 AA Wageningen, The Netherlands. Email: Ron.deGoede@gmail.com

Gerhard du Preez, Unit for Environmental Sciences and Management, North-West University, Private Bag X6001, Potchefstroom 2520, South Africa. Email: Gerhard.DuPreez@nwu.ac.za

Howard Ferris, Department of Nematology, University of California, Davis, Davis, CA 95616, USA. Email: hferris@ucdavis.edu

Simon J. Forster, Physalia Associates, Little Pook Hill, Heathfield Road, Burwash Weald, East Sussex, TN19 7LA, UK. Email: simon.forster@physalia.uk

Hendrika Fourie, Unit for Environmental Sciences and Management, North-West University, Private Bag X6001, Potchefstroom 2520, South Africa. Email: Driekie.Fourie@syngenta.com

Cristina Gambi, Department of Life and Environmental Sciences, Polytechnic University of Marche, Via Brecce Bianche, 60131 Ancona, Italy. Email: c.gambi@univpm.it

Stefan Geisen, Laboratory of Nematology, Wageningen University and Research, 6708 PB Wageningen, The Netherlands. Email: stefan.geisen@wur.nl

Anique Godjo, Molecular Ecology and Nematode Research Group (MENRG), enviroCORE, Department of Applied Science, South East Technological University, Kilkenny Road Campus, Carlow R93 V960, Ireland. Email: godjoanique@gmail.com

Mike Hodda, National Research Collections Australia, CSIRO, GPO Box 1700, Canberra, Australia. Email: mike.hodda@csiro.au

Sebastian Höss, Ecossa, Giselastr. 6, 82319 Starnberg, Germany. Email: hoess@ecossa.de

Jeroen Ingels, Earth Sciences New Zealand, 301 Evans Bay Parade, Hataitai, Wellington 6021, New Zealand. Email: jeroen.ingels@niwa.co.nz

Candice Jansen van Rensburg, Department Zoology and Entomology, University of the Free State, PO Box 339, 205 Nelson Mandela Drive, Park West, Bloemfontein 9301, South Africa. Email: JvRensC@ufs.ac.za

Thomais Kakouli-Duarte, Molecular Ecology and Nematode Research Group (MENRG), enviroCORE, Department of Applied Science, South East Technological University, Kilkenny Road Campus, Carlow R93 V960, Ireland. Email: thomae.kakouli@setu.ie

Anna Karpinska, Molecular Ecology and Nematode Research Group (MENRG), enviroCORE, Department of Applied Science, South East Technological University, Kilkenny Road Campus, Carlow R93 V960, Ireland. Email: annakarpinska44@gmail.com

Gerard Korthals, Wageningen Plant Research, BU Biointeractions and Plant Health, PO Box 16, 6700 AA Wageningen, The Netherlands. Email: Gerard.Korthals@wur.nl

Cristina Lagido, Department of Chemistry and Bioscience, University of Aalborg, Fredrik Bajers Vej 7H, 9220 Aalborg, Denmark. Email: cmr@bio.aau.dk

Qi Li, CAS Key Laboratory of Forest Ecology and Silviculture, Institute of Applied Ecology, Chinese Academy of Sciences, 72 Wenhua Road, Shenyang, Liaoning 110016, China. Email: liq@iae.ac.cn

Wenju Liang, Institute of Applied Ecology, Chinese Academy of Sciences, 72 Wenhua Road, Shenyang, Liaoning 110016, China. Email: liangwj@iae.ac.cn

Manqiang Liu, Centre for Grassland Microbiome, State Key Laboratory of Grassland Agro-Ecosystems, College of Pastoral Agriculture Science and Technology, Lanzhou University, Lanzhou 730000, China. Email: liumq@njau.edu.cn

Nabil Majdi, Réserve Naturelle Nationale de la Forêt de la Massane, Laboratoire Arago, Jardin Méditerranéen du Mas de la Serre, Mas Reig, Route des Crêtes, 66650 Banyuls-sur-Mer, France. Email: n.majdi@hotmail.fr

Simone Mirto, Institute of Anthropic Impacts and Sustainability in Marine Environment, National Research Council (IAS-CNR), 90146 Palermo, Italy. Email: simone.mirto@ias.cnr.it

Tom Moens, Marine Biology Section, Department of Biology, Ghent University, Krijgslaan 281/S8, B-9000 Ghent, Belgium. Email: Tom.Moens@ugent.be

Péter I. Nagy, Department of Zoology and Ecology, Institute for Wildlife Management and Nature Conservation, Hungarian University of Agriculture and Life Sciences, H-2100 Gödöllő 1 Páter K. str., Hungary. Email: Nagy.Peter.Istvan@uni-mate.hu

Deborah A. Neher, Department of Agriculture Landscape and Environment, University of Vermont, 63 Carrigan Drive, Burlington, VT 05405, USA. Email: deborah.neher@uvm.edu

Dorota L. Porazinska, Department of Entomology and Nematology, University of Florida, 1881 Natural Area Drive, Gainesville, FL 32611, USA. Email: dorotalp@ufl.edu

Anusha Pulavarty, Molecular Ecology and Nematode Research Group (MENRG), enviroCORE, Department of Applied Science, South East Technological University, Kilkenny Road Campus, Carlow R93 V960, Ireland. Email: anushapulavarty@gmail.com

Lee Robertson, Forestal Pathology Group, Center of Forest Sciences, National Institute for Agricultural and Food Research and Technology (INIA-CSIC), Crta. Coruña km 7.5, 28040 Madrid, Spain. Email: robertson.lee@inia.csic.es

Liliane Ruess, Institute of Biology, Ecology Group, Humboldt-Universität zu Berlin, Unter den Linden 6, 10099 Berlin, Germany. Email: liliane.ruess@biologie.hu-berlin.de

Sara Sánchez-Moreno, Department of Biodiversity and Evolutionary Biology, National Museum of Natural Sciences (MNCN-CSIC), José Gutiérrez Abascal 2, 28006 Madrid, Spain. Email: sarasm@inia.csic.es

Jan Henrik Schmidt, Institute for Epidemiology and Pathogen Diagnostics, Julius Kühn Institute (JKI) – Federal Research Centre for Cultivated Plants, Messeweg 11/12, 38104 Braunschweig, Germany. Email: jan-henrik.schmidt@julius-kuehn.de

Michaela Schratzberger, Centre for Environment, Fisheries and Aquaculture Science, Pakefield Road, Lowestoft, NR33 0HT, UK. Email: michaela.schratzberger@cefas.gov.uk

Federica Semprucci, Department of Biomolecular Sciences (DiSB), University of Urbino, Via Ca' Le Suore, 2-4, 61029 Urbino (PU), Italy. Email: federica.semprucci@uniurb.it

Christine Sprunger, Department of Plant, Soil, and Microbial Sciences, Michigan State University, Plant and Soil Sciences Building, 1066 Bogue St. Room A286, East Lansing, MI 48824, USA; and W.K. Kellogg Biological Station, 3700 East Gull Lake Dr., Hickory Corners, MI 49060, USA. Email: sprunge5@msu.edu

Stephen R. Stürzenbaum, Analytical, Environmental and Forensic Sciences Department, Institute of Pharmaceutical Sciences, School of Cancer & Pharmaceutical Sciences, Faculty of Life Sciences & Medicine, King's College London, London, SE1 9NH, UK. Email: stephen.sturzenbaum@kcl.ac.uk

Walter Traunspurger, Animal Ecology, Bielefeld University, Konsequenz 45, 33615 Bielefeld, Germany. Email: traunspurger@uni-bielefeld.de

Marcus W. Trett, Physalia Associates, Little Pook Hill, Heathfield Road, Burwash Weald, East Sussex, TN19 7LA, UK Email: marcus.trett@physalia.uk

Wim H. Van der Putten, NIOO-KNAW, Netherlands Institute of Ecology, PO Box 50, 6700 AB Wageningen, The Netherlands. Email: w.vanderputten@nioo.knaw.nl

Alejandra Vieyra-Ramírez, Molecular Ecology and Nematode Research Group (MENRG), enviroCORE, Department of Applied Science, South East Technological University, Kilkenny Road Campus, Carlow R93 V960, Ireland. Email: C00304800@setu.ie

Maria Viketoft, Department of Ecology, Swedish University of Agricultural Sciences, PO Box 7044, SE-750 07 Uppsala, Sweden. Email: Maria.Viketoft@slu.se

Cécile P. Villenave, ELISOL Environnement, ZA des Tourels, 10 Avenue du Midi, 30 111 Congénies, France. Email: cecile.villenave@elisol-environnement.fr

Diana H. Wall†, formerly Department of Biology and School of Global Environmental Sustainability, 108 Johnson Hall, Fort Collins, CO 80523-1036, USA.

Phillip L. Williams, College of Public Health, University of Georgia, Rhodes Hall, Health Sciences Campus, Athens, GA 30602, USA. Email: pwilliam@uga.edu

Rutger A. Wilschut, Laboratory of Nematology, Wageningen University and Research, 6708 PB Wageningen, The Netherlands. Email: rutgerwilschut@gmail.com

Xia Xue, The Fifth Affiliated Hospital of Zhengzhou University, 3 Kangfu Front St., Erqi District, Zhengzhou, Henan 450052, China. Email: xue-xiasophia@hotmail.com

Gregor W. Yeates†, formerly PO Box 1758, Palmerston North 4440, New Zealand.

Chubin Zhang, Analytical, Environmental and Forensic Sciences Department, Institute of Pharmaceutical Sciences, School of Cancer & Pharmaceutical Sciences, Faculty of Life Sciences & Medicine, King's College London, London, SE1 9NH, UK. Email: chubin.zhang@kcl.ac.uk

Aldo Zullini, Dipartimento di Biotecnologie & Bioscienze, Università di Milano-Bicocca, Piazza della Scienza 2, 20126 Milano, Italy. Email: aldo.zullini@unimib.it

Foreword

Research into free-living nematodes of soil, marine and freshwater habitats is a major component of studies on ecology, pollution monitoring and remedial soil recovery, among others. Free-living nematodes form a crucial tool in such studies because they occur in all types of environments, are abundant, fill all trophic levels in diverse food webs and are known to show well-defined responses to environmental challenges. Free-living nematodes also enhance nutrient cycling and stimulate plant yield. The use of species of free-living nematodes as bioindicators is a widely applied approach to analyse and evaluate ecological conditions of any biotope. It has gained increasing importance in the last decades with the visible evidence of the ubiquitous impact of human activities and resulting climate change on the environment. The worldwide environmental problems and their impact on food security and economic and social aspects, including the awareness to 'save our planet', are reflected in the extended and more integrated approach to describe and understand nematode ecological strategies and functioning. This underpins analysis of nematode sensitivity to environmental disturbance and how to mitigate ecological threats, as well as the use of indicator species in assessment of remedial work on damaged ecological habitats. The study of nematodes at various levels, be it molecular, cellular, organismal, community and as part of ecosystems, helps to address relevant issues and problems with environmental change and biodiversity crisis. Biosensor strains of *Caenorhabditis elegans* determine whether environmental samples can activate mechanisms implicated in human disease.

An update of the successful first edition, entitled *Nematodes as Environmental Indicators* (2009) edited by M.J. Wilson and T. Kakouli-Duarte, is necessary to illustrate the more recent developments and to expand the coverage as a practical guide that includes relevant protocols and case studies. This second edition puts nematodes as bioindicators in a broader picture and increases their visibility commercially as a tool for industry and policy makers. In the light of this widened and more practical approach, Thomais Kakouli-Duarte is joined by four new co-editors, Gerhard du Preez, Gerard Korthals, Sara Sánchez-Moreno and Ron de Goede, each being responsible for a specific topic that includes two to four chapters written by international experts.

This second edition, *Nematodes as Environmental Indicators: From Theory to Practice*, comprises six parts, starting with up-to-date information and recent developments in environmental nematology in Part A (Chapters 1–3) entitled 'Nematode Communities as Part of Ecosystems and Food Webs', with a new chapter on evolution of nematode diversity. Part B (Chapters 4–7), 'Describing Nematode Communities', has two new chapters on pollution indices for aquatic ecosystems and ecosystem functioning, and nematode contributions to bioindication under different land uses. Part C (Chapters 8 and 9) on 'Practical Implications on the Application of Nematodes as Bioindicators' introduces new and experienced researchers to the practical aspect with case studies of work on nematodes as bioindicators. Part D (Chapters 10–12) is new for this second edition and in 'Threats Affecting the Global Distribution of Nematodes' it provides information on pertinent environmental issues and discusses how this important knowledge can be utilized to alleviate or counteract the effects of such universal threats. Part E (Chapters 13–15) includes up-to-date information on 'Methods for Studying Nematodes in the Laboratory and International Standards' and describes the rapid evolution in technologies and automation in the use of biosensors and ecotoxicogenetics. This part also shows how the best-studied nematode model organism, *Caenorhabditis elegans*, provides a powerful tool in the search for suitable biosensors and stimulates evolution in ecotoxicogenomics. The final section of the book, Part F (Chapters 16–18) entitled 'Policy, Education, Communication and Commercialization', explores and stimulates approaches to make education in environmental nematology more holistic, attractive and assessable as the basis for detection and problem solving for all stakeholders.

Ecology of free-living nematodes is essentially the study of relationships between nematodes and all aspects of their environment. Such knowledge underpins the use of nematodes as bioindicators and biosensors, aspects that are increasingly indispensable for environmental biomonitoring and remediation. This book focuses on these aspects, while also making an important and significant contribution to the overall study of the ecology of free-living nematodes. It is a timely volume that is especially enhanced by the use of practical approaches and case studies, and the editors are to be congratulated on assembling such a comprehensive coverage. The book is relevant to researchers, policy makers and industry, and makes a vital contribution to efforts to alleviate ecological damage as well as the monitoring and prevention of future damage.

Wilfrida Decraemer
Ghent University, Belgium

Roland N. Perry
University of Hertfordshire, UK

March 2025

Preface

Do we need another nematology book? We always need another nematology book! Do we need the second edition of the volume *Nematodes as Environmental Indicators?* Yes, and now more than ever.

This project was started nearly 20 years ago; we refer to the first edition that became a reality due to the progressive and collegial thinking and approach of our colleague, Dr Mike Wilson. Mike was the inspirator of the first edition and played a catalytic role for the collective and international awareness of environmental nematology two decades ago, within the wider international nematology community. Now, Mike has left science and enjoys playing music and the sunshine and culture of Greece, and this is not a bad thing to do after all! But he left a legacy and a foundation that the current co-editors and authors of the present volume built upon to create *Nematodes as Environmental Indicators: From Theory to Practice*.

Thomais Kakouli-Duarte had been thinking for at least 7–8 years now about this second edition. The decision to commit to one was taken when she met and chatted about it with our omnipresent and evergreen colleague, Professor Rolo Perry, at the 33rd Symposium of the European Society of Nematologists in Ghent, back in September 2018. Rolo not only encouraged and helped her to make up her mind to start the second edition but also reviewed it and wrote the foreword to it together with our esteemed colleague, Professor Frieda Decraemer. Although 7 years have elapsed since that meeting, with one pandemic and one International Congress of Nematology later, the second edition volume is now a reality! The 7th International Congress of Nematology in May 2022 in Antibes provided the exact necessary fertile ground where the current co-editors met, brainstormed, debated and decided to join the effort. In addition, many of the authors in the current volume were also present at the congress and that gave an excellent opportunity to connect and get involved. Special thanks are therefore due to Pierre Abad and his team, and to the International Federation of Nematology Societies (IFNS), for providing us such a fantastic collective nematology experience that spearheaded the start of this edition.

Indeed, we have tried to make this edition a truly collective and collaborative effort, with contributors spanning all continents and extending from academic and research arenas to commercial sectors. Our world is facing many challenges, with environmental degradation being a major problem and geopolitical uncertainty intensifying it. The need for collective action in environmental nematology was never greater. At the same time, advancements in the area can facilitate this action and this volume not only provides the state-of-the-art but also goes one more important step forward: it provides a more hands-on approach to truly become the environmental nematology textbook of favour in environmental science and practice. These advancements are the result of the work of colleagues who have authored parts of this book, and we are ever so thankful to them for their contributions.

Mentioning our colleagues, we would like to take a moment and remember our colleagues Gregor Yeates, Diana Wall and Brian Darby. Gregor authored a chapter in the first edition, Diana has authored a chapter in this book and Brian authored chapters in both editions. This volume is dedicated to their memory and is a token of honour to their valuable contributions to nematology and nematode ecology.

It has been an amazing experience and journey to see this volume materializing. We have learned many things about nematodes, about each other and about ourselves on the way, and we are thankful. We would like to thank all the contributors to the present volume because without them there would be no volume! Many thanks also to the various reviewers of the chapters who helped us to produce a high-quality volume and, of course, our thanks and gratitude go to Frieda and Rolo. CABI has been a pleasure to work with, and we thank them for their guidance and patience and the opportunity to publish this volume. We hope that this second edition will be a valuable resource and tool to all new starting and more mature nematologists and environmental scientists for the years to come, and until the need for the third edition becomes a reality. We are confident that we have paved the way towards that prospect.

Collegially,

Thomais Kakouli-Duarte, Gerard Korthals, Sara Sánchez-Moreno, Gerhard du Preez and Ron de Goede

May 2025

1 Roles of Nematodes in Ecosystems: Their Biology and Ecology

Howard Ferris[1]*, Wim H. Van der Putten[2], Gregor W. Yeates[†3] and Tom Moens[4]
[1]*Department of Nematology, University of California Davis, California, USA;*
[2]*NIOO-KNAW, Netherlands Institute of Ecology, Wageningen, The Netherlands;*
[3]*Formerly PO Box 1758, Palmerston North, New Zealand;* [4]*Marine Biology Section, Department of Biology, Ghent University, Belgium*

Abstract

The organisms of soil and aquatic food webs perform ecological functions and services that are essential for life on Earth. Nematodes, the most abundant multicellular animals on the planet, are prevalent in terrestrial, marine and freshwater ecosystems. Humans have likely been aware of the indicator characteristics of nematodes since before recorded history, first as human and animal parasites by their presence in faeces, skin lesions and sputum, and progressively, with improvements in microscopy, as free-living soil and aquatic forms. The abundance and diversity of nematodes with different feeding habits and ecological functions in soil systems, for example, indicate resource availability and the favourability of environmental conditions. Because of their diversity and ubiquity, nematodes are convenient indicators of similar functions performed by other organisms in their environment. In this chapter we provide insights into the biological and ecological properties of nematodes that underlie their usefulness as environmental indicators.

1.1 Introduction

Human society and, in fact, life on the planet are entirely dependent on a variety of ecosystem services, many of them provided by soil biodiversity (Wall, 2004; Millennium Ecosystem Assessment, 2005). Nematodes play a major role in most soil-borne ecosystem services. They influence the provisioning of food, fibre, clean water and air, and regulation of pests and diseases. Nematodes participate in the transformation of organic matter into mineral and organic nutrients that provide plant nutrition and crop productivity and are available to other organisms in soil and sediment food webs (Ingham *et al.*, 1985; Ferris *et al.*, 1998, 2004c). Nematode feeding activities on soil microbes and other organisms contribute to food web stability through a variety of complex metabolic and behavioural interactions. However, their feeding activities may also destabilize food webs, leading to ecosystem transitions. Some nematodes cause plant, animal or human diseases, while others may influence pest damage to plants by regulating insect abundance (Viglierchio, 1991). In natural ecosystems, nematodes contribute to spatial and temporal diversity in plant communities and, therefore, to the diversity of plant-associated communities of microbes and invertebrates, both above- and below-ground (De Deyn and Van der Putten, 2005). Most effects in

*Email: hferris@ucdavis.edu

 Nematodes as Environmental Indicators: From Theory to Practice, 2nd Edition (T. Kakouli-Duarte *et al.* eds)
DOI: 10.1079/9781800624221.0001

which nematodes are involved may be due to the combined influences of nematodes with many other soil organisms (Kardol *et al.*, 2005; Sánchez-Moreno *et al.*, 2008).

1.1.1 Indicator value

Because nematodes are so omnipresent and abundant in ecosystems, they may serve as relatively easily obtainable and informative indicators of environmental disturbance (Bongers 1990; Ferris *et al.*, 2001; Yeates, 2003; Höss *et al.*, 2004; Schratzberger *et al.*, 2006; Heininger *et al.*, 2007; Ridall and Ingels, 2021; Du Preez *et al.*, 2022). In fact, a recent review of the literature documented 672 instances of the use of nematodes as bioindicators (Du Preez *et al.*, 2022). One nematode species, *Caenorhabditis elegans* (Rhabditina), has become a central model for genomic studies that aim to relate gene expression to the development and functioning of organisms. Indeed, *C. elegans* was the first multicellular organism whose genome was sequenced (CESC, 1998). Studies on *C. elegans* underpin many of the advances in molecular biology. With advances in technology, several other nematode genomes have been sequenced, leading to greater understanding of the effects of environmental cues on nematode development, moulting, ageing, longevity, behaviour and other life-history characteristics.

There are estimates of between 40,000 and 10,000,000 species in the phylum Nematoda (Blaxter, 1998; Yeates and Boag, 2006). Clearly, continued exploration activities are necessary to verify these and other estimates. For example, one provocative speculation is that there might be as many as 100,000,000 nematode species, even before considering the cryptic diversity among morphologically indistinguishable taxa (Lambshead, 1993; Nadler and Perez-Ponce de Leon, 2011). Molecular techniques have improved understanding of the phylogenetic relationships of nematodes with other animal groups. There has been considerable debate about their associations relative to the ecdysozoan theory based on the occurrence of the *Hox* gene in various animal groups (De Rosa *et al.*, 1999; Gutierrez and Sommer, 2004). However, that theory is disputed by studies that included a larger number of genes in the analysis (Blair *et al.*, 2002). Nematodes are now generally classified as the phylum Nematoda in the Acoelomata of the Metazoa (see outline classification in Appendix 1).

1.1.2 Body structure

Nematodes are built on a simple plan, with the digestive tract and body wall being concentric tubes. The typical nematode has a mouth (= oral aperture) leading to a stoma (= buccal cavity), connected to the lumen of a muscular and/or glandular pharynx (= oesophagus) that leads to a tubular intestine, a rectum and posterior anus. The lack of a complete mesodermal lining of the body cavity led to the characterization of nematodes as pseudocoelomate (Hyman, 1951). The body is cylindrical or spindle-shaped and the tail beyond the anus varies enormously in length. Adult females range in length from 0.3 mm to 8 m. A vast range of head structures that facilitate food ingestion parallels the great diversity of nematodes and can be useful indicators of feeding habits (Fig. 1.1). The tubular gonads lie in the cavity between the digestive tract and body wall, and they, together with the associated genital structures, are further sources of species differentiation. The typical life history of nematodes includes an egg and, usually, four juvenile stages (= 'larvae') preceding the adults. The morphology of the juvenile stages converges with that of the adult; there is no metamorphosis.

1.1.3 History

There is a long history to our knowledge of nematodes. They were recorded as human parasites by the Chinese around 2700 BC and as plant parasites by Shakespeare in 1594 (Viglierchio, 1991) and more formally by Needham (1744). Although our attention has been drawn to nematode species that, through parasitism, adversely affect the quality of life of humans, domestic animals and food sources, the vast majority of nematode taxa are 'free-living'; they feed on bacteria, algae, fungi, diatoms, microinvertebrates and the epistrate deposits of aquatic sediments (Yeates *et al.*, 1993; Moens and Vincx, 1997; Traunspurger *et al.*, 1997; Schratzberger *et al.*, 2019; Hodda, 2022). With growing awareness

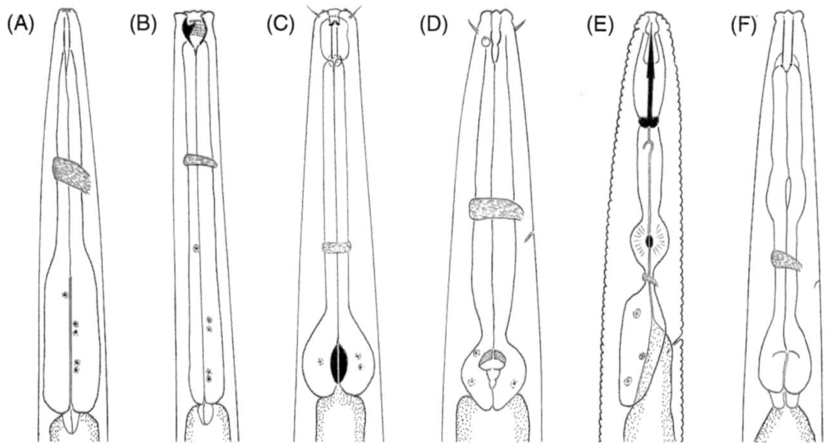

Fig. 1.1. Head and pharyngeal regions of six nematodes found in soil and water environments. (A) Dorylaimida: Dorylaimidae, *Dorylaimus*; (B) Mononchida, Mylonchulidae, *Mylonchulus*; (C) Chromadorida, Ethmolaimidae, *Ethmolaimus*; (D) Plectida, Plectidae, *Plectus*; (E) Tylenchina, Hoplolaimidae, *Rotylenchus*; (F) Rhabditina, Rhabditidae, *Rhabditis*. In most, the location of the circum-pharyngeal nerve ring, pharyngeal gland nuclei and amphid are shown. The diameter of the stylet lumen is ~5 μm in (A) and ~0.1 μm in (E). Diagrammatic and not to scale. (Author's own image compiled from various sources and observations.)

of the potential extinction of yet undescribed species – of both nematodes and other phyla – our ignorance of nematode diversity, habitat range and ecological amplitude continues to be challenged. In the latter part of the 20th century, the roles of nematodes in biological turnover in soil came to be appreciated (Yeates, 1984; Ingham *et al.*, 1985; Ferris *et al.*, 1997, 2004c). Awareness of the impact of nematode activity on populations of other organisms, and thus on ecosystem processes, developed concomitantly with awareness of the importance of biodiversity. Understanding of the possibilities of interpreting the abundance and diversity of nematode assemblages in various habitats has led to the development of information-rich 'indices' that may be used to characterize disturbance and other environmental conditions (Bongers, 1990; Ferris *et al.*, 2001; Yeates, 2003; Du Preez *et al.*, 2022).

1.2 Nematode Biology – Individuals and Populations – An Outline

1.2.1 Feeding by nematodes

Most nematodes utilize the energy fixed by plant photosynthesis. They may feed directly on primary producers, such as higher plants (e.g. *Aphelenchoides* on foliar parts, *Ditylenchus* on stems, *Pratylenchus* and *Meloidogyne* on roots) and unicellular algae (e.g. *Chromadorita*, *Daptonema* and *Pareudiplogaster* on diatoms), or on microbes associated with decomposing plant material (e.g. *Aphelenchus*, *Filenchus* on fungal hyphae; *Rhabditis*, *Plectus*, *Leptolaimus* on bacteria). Soil organic matter, litter, marine detritus, animal faeces and cadavers are important resources for microbes and thus for microbial- and epistrate-feeding nematodes. Higher trophic levels feed as predators of nematodes and other microinvertebrates (e.g. *Mononchus*, *Nygolaimus*, *Enoploides*, *Sphaerolaimus*) or as parasites of invertebrate and vertebrate animals (e.g. *Strongyloides*, *Thelastoma*, *Ascaris*). Therefore, except for herbivores in the second trophic level, most nematodes are third (bacterivores, fungivores) or fourth (carnivores) trophic-level organisms. Omnivores may predate on all trophic levels, depending on the trophic position of the resources that they feed on. A few marine species in low-oxygen environments obtain energy from chemical bonds in their environment or from symbiotic sulfur-oxidizing bacteria (e.g. *Astomomema*) (Polz *et al.*, 2000; Van Gaever *et al.*, 2006; Blaxter and Koutsovoulos, 2015).

1.2.2 Physical environments occupied by nematodes

Nematodes are influenced by the nature of their physical environment (e.g. soil or sediment texture, water chemistry) and by gradients within it (e.g. redox potential, plant root distribution, temperature, moisture). There is a clear distinction between the physical environment occupied by a nematode and the food resource it uses, and each may vary with the nematode life stage. In typical bacterial-feeding, hyphal-feeding and predacious nematodes, all post-hatching stages, while feeding, occupy soils, sediments or decaying vegetation. Plant-feeding nematodes such as *Pratylenchus* and *Radopholus* feed solely on plant cell contents but migrate between plants and soil. The cyst nematodes, *Heterodera* and *Globodera*, and the root-knot nematode *Meloidogyne*, hatch as second-stage juveniles that migrate, without feeding, into roots, where they develop into saccate females feeding on plant resources via highly specialized transfer cells.

In terrestrial habitats, eggs of ascarids (e.g. *Ascaris*) do not hatch outside the invertebrate or vertebrate host and they complete their life histories within the living host; adults inhabit the stomach and intestine of the definitive host and primarily consume food ingested by the host. In contrast, intestinal parasites of vertebrate hosts (e.g. *Ancylostoma*, *Strongylus*, *Trichostrongylus*, *Haemonchus*) feed in the intestinal mucosa, commonly evidenced by blood in host intestines and faecal matter. Their eggs hatch in faeces and grow as bacterial-feeding stages in a non-living substrate before reinfecting a new vertebrate host. Other nematode parasites of humans and other vertebrates (e.g. *Dracunculus*) have complex life cycles with invertebrates as secondary hosts that are consumed by definitive hosts in which the nematodes complete their life cycle. Nematodes that feed on bacteria, fungi, and on other component organisms of decomposition food chains, aggregate where their food sources are plentiful, such as the plant rhizosphere or near the soil surface where decay of leaf litter and other organic debris is occurring. Of course, their abundance in any environment will depend on favourable conditions of temperature, moisture and aeration, and the activity of predators.

1.2.3 Nematode response to stimuli

Behavioural responses to stimuli are essential for the success of nematodes in locating food resources, hosts and mates. While thermotaxis, phototaxis and geotropism are responses to basic environmental cues, they are important in host location in taxa as diverse as *Ancylostoma*, *Stephanurus*, *Anguina*, *Mermis* and *Trichostrongylus*. Chemotaxis is commonly involved in food finding and food selection by free-living nematodes. Responses to different kinds of chemicals and/or bacterial food tend to be highly species-specific and potentially contribute to the small-scale patchiness of nematode populations (Venette and Ferris, 1998; Moens *et al.*, 1999; Höckelmann *et al.*, 2004; Wilschut *et al.*, 2017). There is an increasing knowledge base regarding semiochemical signalling between plant roots and both parasitic and beneficial nematodes (Rasmann *et al.*, 2012; Ali and Davidson-Lowe, 2015; Pineda *et al.*, 2015). However, although plant-parasitic nematodes are highly efficient in locating plant roots, only a few plant signals that allow root location have been identified and characterized (Kulkarni *et al.*, 2018). For example, CO_2 attracts many species of plant-parasitic nematodes (McCallum and Dusenbery, 1992; Robinson, 1995), ethylene attracts *Heterodera* juveniles (Wubben *et al.*, 2001), and onion-root exudates attract *Ditylenchus* juveniles (Spiegel *et al.*, 2003). Rühm *et al.* (2003) found that an unknown kairomone in *Sinapsis alba* exudates served as an attractant to *Heterodera schachtii* juveniles and a range of root metabolites that impact the behaviour of soil nematodes has been documented (Sikder and Vestergard, 2020). An interesting example of molecular interactions between host and parasite is that of an ascaroside pheromone released by nematodes that is metabolically edited by plant cells to generate chemical signals that repel infective stages (Manohar *et al.*, 2020). Clearly, to successfully invade their host plants, plant-parasitic nematodes must evolve molecular mechanisms to overcome plant defence responses (Ali *et al.*, 2018).

Some of the responses of nematodes to stimuli seem to be species-specific; root exudates from favourable host plants increase egg hatching of cyst nematodes of the Heteroderidae (Wang *et al.*, 1997; Zasada *et al.*, 2013) and exudates of cotton roots increase egg hatching of *Rotylenchulus reniformis* (Sankaralingam and McGawley, 1994). Among the early work on sex attractants was that of Green and Plumb (1970) with *Heterodera* spp. and *Globodera* spp.; other work has included studies on *Panagrolaimus*, *Rhabditis*, *Panagrellus*, *Ditylenchus*, *Belonolaimus* and Trichostrongylidae. The sensitivity of nematode response to chemical stimuli in their environment is demonstrated by the ability of *C. elegans* to detect and distinguish between odour compounds at picomolar (10^{-12}) concentrations (Bargmann and Mori, 1997) and the attraction to plant roots of bacterial-feeding nematodes with adhering rhizobium bacteria (Horiuchi *et al.*, 2005).

1.2.4 Abundance and diversity of nematodes

A square metre of soil, forest litter or aquatic sediment may contain millions of individual nematodes belonging to over 400 species. Nematodes may also be abundant and diverse within living substrates. N.A. Cobb (1915) recorded 40,000 individual nematodes from the stomach of a wallaby. In New Zealand, domestic cattle and sheep can be infected by as many as 27 and 29 nematode species, respectively (McKenna, 1997). As a result, identifying the taxa in each nematode assemblage, and assessing the absolute and relative contribution of each taxon to the nematode assemblage, would be a prodigious task. Unless a study is restricted to a particular group (e.g. comparing pre-planting and post-harvest populations of plant-feeding nematodes or assessing the abundance of infective juveniles of Trichostrongylidae on pasture herbage), it is useful to present results in information-rich indices. Such indices may be either those widely used in ecological studies, such as the Shannon–Weiner index of diversity (H′), Margalef index of richness (D or SR) and Pielou's evenness (J′) (Pielou 1975; Yeates, 1984; Magurran, 1988; see Chapter 4, this volume), or indices specifically developed for nematodes (Bongers, 1990; Ferris *et al.*, 2001; see Chapter 5, this volume).

Nematode grazing on microbes can significantly increase nutrient cycling and plant responses in localized patches (Ingham *et al.*, 1985; Ferris *et al.*, 1998; Zhang *et al.*, 2017). Although there have been various studies, there is a need for understanding the factors governing the distribution of patches, when scaling up from a roughly homogeneous patch (e.g. a uniform agricultural field) to the landscape and regional scales for policy development (Coleman *et al.*, 1992; Müller and Lenz, 2006; Sánchez-Moreno *et al.*, 2008). The organizational structure of ecosystems is mediated by the energy available to develop network designs that maximize the energy fluxes through them that are compatible with the constraints of the environment. The eco-exergy of a system includes not only the free energy of matter, but also the available energy of embodied information utilized for interactions and the performance of ecosystem processes (Jørgensen and Mejer, 1979; Odum, 1988).

1.2.5 Effects of nematodes on their resources

Nematodes may affect the growth rates, health and yield of plants or animals (e.g. infection of soybeans or potatoes with cyst nematodes or of grazing mammals with trichostrongylids). In contrast, grazing and bioturbation by microbial feeding nematodes may stimulate nutrient cycling and plant yield (Ingham *et al.*, 1985; Alkemade *et al.*, 1992; Aller and Aller, 1992; Fu *et al.*, 2005). By its very nature, parasitism involves mutual adjustment between host and parasite to permit coexistence without serious harm to either component of the interaction. Indeed, some studies suggested that low burdens of plant-feeding nematodes may stimulate plant growth under favourable conditions, perhaps by causing or stimulating release of root exudates that result in microbial activity and plant growth promotion that more than offsets the damaging effects of parasitism (Bardgett *et al.*, 1999). On the other hand, there is loss of production, or fitness, when the parasite burden increases (De Rooij-Van der Goes, 1995) or when the host is additionally stressed.

An important effect of nematodes on their resources is what might be considered resource farming. A substantial proportion of the bacteria consumed by bacterial-feeding nematodes survives passage through the nematode digestive tract (Yeates, 1969a,b). Additionally, bacteria adhere to the nematode cuticle. As the nematodes move through the soil in response to semiochemical signals from other resources, they distribute the internal and external burden of bacteria to those resources and new colonies are established (Fu *et al.*, 2005; Horiuchi *et al.*, 2005). Similarly, entomopathogenic nematode species carry *Photorhabdus* and *Xenorhabdus* bacteria into the bodies of insects where they are released and multiply. In well-documented cases, nematodes are the phoretic passengers in distribution to their food sources; for example, the phoretic relationship between nematode of the genus *Bursaphelenchus* and insect vectors. Nematodes of the family Chambersiellidae (e.g. *Diastolaimus*, *Macrolaimus*) are assumed to be transported by insects between lichen colonies on above-ground parts of plants. The nematodes may also benefit from necromeny in that they have access to bacteria involved in decomposition of the insect cadaver after its death (Cid del Prado-Vera *et al.*, 2023).

For both soil-inhabiting and aquatic nematodes, physical conditions can be more important than seasonal changes or management practices in determining the structure of the nematode assemblage (Yeates, 1984; Heip *et al.*, 1985; Griffiths *et al.*, 2003; Bardgett, 2005). Feeding rates of the predacious marine nematode *Enoploides longispiculosus*, which may exert a strong top-down control over nematode and ciliate prey communities, are strongly reduced or even completely impeded by subtle shifts in silt content, mean grain size and water content of intertidal flats (Gallucci *et al.*, 2005). Stressful abiotic factors may interact; for example, a classic experiment by Griffiths *et al.* (2000) demonstrated that, because of their effects on soil biodiversity, 'stress on stress' combinations of adverse environmental conditions strongly influenced the stability of ecosystem services. Similarly, damaging effects of plant-feeding nematodes may depend on external conditions, such as drought (Haverkort *et al.*, 1992) or the presence of other pathogens or symbionts, which may lead to synergistic (De Rooij-Van der Goes, 1995) or antagonistic (Brinkman *et al.*, 2005; Hol *et al.*, 2007) interactions.

Nematodes may differentially stress one plant species, thereby indirectly benefiting other plant species. For example, damage to clover roots by cyst nematodes may result in the leakage of nitrogen compounds, which benefit grasses that are not directly affected by the nematodes (Bardgett *et al.*, 1999). Damage to roots by *Meloidogyne incognita* reduces plant vigour and provides a competitive advantage to *Cyperus* weed species, which are damaged less by the nematodes (Thomas *et al.*, 2005).

In many cases, nematologists have identified individual nematode species as primary causes of crop losses. However, ecologists point to the importance of the spatial and temporal patterns in communities of organisms in soils and sediments (Blanchard, 1990; Traunspurger, 2000; Ettema and Wardle, 2002; Ettema and Yeates, 2003; Fisher, 2003; Yeates, 2003; Van Gaever *et al.*, 2004; Bardgett *et al.*, 2005; Michiels and Traunspurger, 2005). Even in agricultural fields, which are considered to have relatively homogeneous soils and to be uniformly managed, there is considerable variability in the spatial and temporal composition and abundance of plant-parasitic nematode species (Quist *et al.*, 2019). A variety of approaches has been used in efforts to understand this heterogeneity (Goodell and Ferris, 1980; Ferris *et al.*, 1990; Prot and Ferris, 1992; Robertson and Freckman, 1995; Quist *et al.*, 2019). Spatial and temporal patterns and variation also are important when considering the roles of nematodes in natural ecosystems, which are far less homogeneous than agricultural soils, and when using nematodes as environmental indicators. In such conditions, it is important to understand the determinants and consequences of nematode diversity within samples (alpha diversity), between samples within fields (beta diversity) and between fields within landscapes (gamma diversity). Further, it is important to consider that, as conditions change and resources become depleted, successional changes will lead to altered species dominance. Such compositional developments are relevant factors to be included in included in diversity assessments.

With respect to effects of the diversity of nematodes on resources, Laakso and Setälä (1999) regarded diversity among functional groups as more important than diversity within them. On the other hand, a diversity of sizes and activities of microbial-feeding nematodes may provide access of predators to a greater range of soil structural channels and to the microbes that are at the bases of these energy channels. When the parasitic nature of nematodes is considered, and when parasitism is host-species- or even variety-specific, such as that of potato cyst nematodes (*Globodera* spp.), resolution at the functional group level is inadequate. In such cases, identification at the species or even pathotype level is necessary, for example when designing nematode control strategies (Folkertsma *et al.*, 2001). Similar examples may also exist in food web interactions. For example, taxonomically very closely related bacterial-feeding nematode species may have differential effects on phytodetritus decomposition and on the activity and community composition of the associated bacteria (Venette and Ferris, 1998; De Mesel *et al.*, 2003, 2004; Fu *et al.*, 2005). Therefore, the necessary levels of identification for functional interpretation depend on the specific nature of the ecosystem functions, the nematodes involved in performing them and the environmental conditions in which they are being performed. Considering the time-consuming processes and uncertainties involved in morphological and morphometric identification of nematodes species and their functions, molecular diagnostic approaches will increasingly play key roles in soil ecology studies in the future (Geisen *et al.*, 2018).

1.3 Nematode Feeding and its Ecosystem Consequences

1.3.1 Food resources of nematodes

Many of the ecosystem functions and services provided by nematodes are direct consequences of their feeding activity and of the physiological processes of digestion and metabolism. Across the diversity of the phylum are representatives that ingest a vast array of resources to drive their metabolic processes. We might divide them into general categories of *grazers* or *browsers*, which feed on food resources that continue to live, and *predators* whose feeding results in the death of their prey. Within these two general categories we can separate nematodes according to the nature of the food resource: (i) *herbivores* feeding on living tissues of higher plants; (ii) *carnivores* feeding on animal tissues, vertebrate (as parasites) or invertebrate (as predators or parasites); (iii) *fungivores* feeding on fungi; (iv) *bacterivores* feeding on prokaryotic organisms; and (v) *unicellular eukaryote feeders*, feeding on ciliates, other protozoans or diatoms and unicellular algae (Yeates *et al.*, 1993). Elsewhere in this chapter we have used terms such as *plant feeders* and *microbial feeders*, but it is useful to recognize that more precise terminology can be applied. Some nematodes (e.g. in the families Ascarididae and Thelastomatidae) that inhabit the gastrointestinal tract of vertebrates or invertebrates may actually be grazing on intestinal microbes, or on food ingested by the host, rather than on tissues of the host. In some cases, it is difficult to link a specific food source with the nematode feeding habit. For example, a soil-dwelling member of the family Rhabditidae, siphoning the soil solution to ingest bacteria, may also be obtaining other resource molecules from the solution. Similarly, large-mouthed nematodes in freshwater or marine sediments may be ingesting an assortment of organisms and material besides diatoms; consequently, they are conveniently classified as *epistrate feeders* (Moens and Vincx, 1997; Traunspurger, 2000).

In each case, and as detailed below, there are nematodes that are *specialists* in their feeding habits with stoma structures, behavioural attributes or specific biochemical requirements (e.g. transfer cells of *Heterodera* and *Meloidogyne*) that are adaptations to high-intake feeding on a narrow range of food sources; others are *generalists* and capable of obtaining resources from a wider range of sources. Some of the latter are omnivores, crossing feeding-type boundaries and feeding at more than one trophic level in the food pyramid. For example, some may be carnivorous as well as bacterial-feeding (e.g. *Pristionchus*), or bacterial-feeding as well as feeding on unicellular eukaryotes (e.g. *Daptonema*, *Thalassomonhystera*). Omnivory appears to be quite common in some nematodes, and food sources from different trophic levels may be utilized simultaneously (e.g. many Dorylaimida), in different life stages (e.g. bacterial-feeding juveniles of hookworms whose adults parasitize the gastrointestinal tract of mammals), or follow temporal or environmental fluctuations in the availability of different resources (e.g. the marine nematode *Enoplus brevis* feeding on cyanobacteria, diatoms, oligochaetes, nematodes and rotifers) (Hellwig-Armonies *et al.*, 1991).

Of course, in a group with enormous diversity of habitat, there are exceptions to the simplistic categorizations of feeding types. Some nematodes appear to be able to transport, or allow diffusion, of dissolved organic molecules across the cuticle, as an alternative to, or perhaps in addition to, stomal ingestion. Cuticular microvilli are used in the insect-inhabiting *Bradynema*; *C. elegans* has been cultured on chemically defined media, without other living organisms (Vanfleteren, 1980); dissolved organic matter may be ingested via the stoma (Chia and Warwick, 1969). Several marine nematodes, including some newly discovered species of the anoxic oceanic abyss, lack any orifice for the digestive system and derive their resources either directly or via symbiotic bacteria (e.g. *Astomonema* and several Stilbonematinae) (Hentschel *et al.*, 1999; Dover, 2000).

The characterization of such unusual feeding strategies, and the challenges of modelling the roles and functions of nematodes in a food web context, may benefit from stable isotope approaches, using both natural and experimentally enriched abundances of, mainly, ^{13}C and ^{15}N (Moens *et al.*, 2002, 2005a). Chemosynthetically produced carbon, for instance, is typically very depleted in ^{13}C, while ^{15}N fractionates significantly with trophic level and is thus a useful tracer of trophic position. The technology has been used successfully in tracking the fate of carbon from plant residues into the microbial biomass and into nematodes (Minoshima *et al.*, 2007), and from microalgae and bacteria of intertidal regions into nematodes (Moens *et al.*, 2002; Van Oevelen *et al.*, 2006).

Stoma morphology in the phylum Nematoda ranges from simple apertures of fixed diameter, which apparently limit the size of the ingested material, to permanently cavernous features or structures that can be opened to enormous size for ingestion of large prey. Even simple tubular stomata adapted for ingestion of bacteria (e.g. in the Rhabditidae) or diatoms (e.g. in *Gonionchus*, *Praeacanthonchus*) often have small teeth or denticles, presumably to abrade, crush or filter ingested particles. Additionally, there may be other cuticular structures in the stoma and/or pharynx that serve to abrade or rupture food before it enters the intestine; for example, the grinder in the bulbous posterior region of the pharynx of many rhabditid nematodes.

Intestinal parasites of vertebrates are often equipped with teeth that allow tearing of mucosa to provide access to tissue and blood (e.g. *Ancylostoma*). Many other nematodes are equipped with fixed or moveable teeth (Chromadoridae, Diplogastridae, Enoplidae, Mononchida), or with hollow spears or stylets (Dorylaimida, Tylenchida) for piercing a resource and withdrawing contents. In most Tylenchida, and the plant-feeding Longidoridae, the stylet lumen is very narrow (~0.1 and 0.5 μm in diameter, respectively), significantly limiting type and size of ingested material. Many omnivorous and predatory Dorylaimida have a wide stylet lumen (≥ 6 μm) or just a single mural tooth (e.g. *Nygolaimus*), allowing ingestion of larger food sources. The stoma and its armature are, however, not a definitive indicator of food resource; some nematodes with narrow stylet lumens are effective predators. For example, *Seinura* spp. ingest their food resources from large-bodied prey through a <0.5 μm stylet aperture, while other predators (e.g. *Anatonchus tridentatus*), with a wide stoma, may ingest whole organisms and are frequently observed to have intact nematode prey in their intestine (Small, 1987). Others (e.g. *Tripylina* and other Tripylidae) are able to open an apparently narrow stoma very wide to ingest nematode prey intact (Cid del Prado-Vera *et al.*, 2012).

1.3.2 Microbial feeding in soils and sediments

Nematode activities in soils and sediments include the redistribution of resources so that they are more available to other consumers. Some nematodes, such as bacterivores, may also transport adherent microbes to other locations where they gain access to new resources (Fu *et al.*, 2005). Their ability to locate suitable feeding patches from a distance, for example by sensing cues of decomposition-associated end products such as CO_2 (Klingler, 1965; Pline and Dusenbery, 1987; Riemann and Schrage, 1988) or responding to semiochemical signals (Horiuchi *et al.*, 2005; Ali *et al.*, 2011), contributes to their phoretic efficiency (e.g. Jatala *et al.*, 1974). Generalist bacterivores include those considered 'enrichment opportunists' (*sensu* Ettema and Bongers, 1993) which appear to simply imbibe aqueous suspensions of their food, with larger particles restricted only by stomatal diameters of 2 μm or less. These nematodes apparently have little other restriction on the types of bacteria ingested (Venette and Ferris, 1998; Salinas *et al.*, 2007) and it may be the stomatal dimensions of the feeding stage that limits food size for maintenance of populations. Other bacterivore nematodes feed with activities involving simple to elaborate head probolae, muscular contractions to open a closed stoma during ingestion, sweeping motions of the head, or thrashing motions of the whole body. Such behaviours provide access to food by stirring sediments or disrupting adherence of bacteria to surfaces (Moens *et al.*, 2004).

Several important ecosystem functions and services are well documented for bacterial-feeding nematodes in soils and aquatic systems. Nematode grazing on microbes may result in greater metabolic activity in their prey populations, essentially keeping in a growth phase and reproductively active. Besides redistribution of bacteria adherent on the nematode cuticle, around 30% of the bacteria ingested by bacterial-feeding nematodes are not digested and assimilated; in fact, they may still be alive when defecated (Yeates, 1969a,b; Ingham *et al.*, 1985; Bird and Ryder, 1993; Fu *et al.*, 2005; Ghafouri and McGhee, 2007). The proportion of ingested bacteria that survives passage through the nematode intestine may vary with food availability and bacterial density (Moens *et al.*, 2006). In effect, the redistribution of bacteria is a form of 'resource farming', to be compared with ants that manage aphids or fungi for their own food provisioning: it results in the increase of resource provisioning for bacterivorous nematodes; however, this predator–prey dynamics may be distorted by overgrazing (Fu *et al.*, 2005). Another example of such nematode-aided bacterial farming is the migration of entomopathogenic nematodes into arthropods where they release toxic bacteria that kill the arthropod. The bacteria multiply in the cadaver and provide food for the nematode (Boemare, 2002).

It has been suggested that nematodes can discriminate among soil microbes. Rodger *et al.* (2004) cultured bacterial-feeding nematodes (*C. elegans*, *Coarctadera cystilarva*; Rhabditidae) on four species of bacteria and measured nematode attraction to those bacteria. On agar, there seemed to be a 'substrate legacy' affecting the subsequent ability of these bacterial-feeding nematodes to locate the food on which they had been cultured. Venette and Ferris (1998) observed differences among population growth rates of bacterial-feeding nematode species on different bacterial colonies. This area of food selectivity within nematode feeding types needs further exploration in order to understand its ecological implication.

1.3.3 Excretion by bacterial-feeding soil nematodes

Another important ecosystem service emerges from the nature of the substrate ingested and subsequently assimilated across the intestinal wall. Bacterial-feeding soil nematodes may excrete materials assimilated that are in excess of their needs, in forms that are available to other organisms. A familiar example of the mineralization of digested organic molecules is the participation of most organisms in the carbon cycle. In liberating energy from ingested materials, nematodes have been calculated to release, across the cuticle, about 40% of the ingested carbon in the form of CO_2 (Klekowski *et al.*, 1972; Ferris *et al.*, 1995). CO_2 returns to the atmosphere and is available to plants for photosynthetic fixation of complex molecules. But the ingested molecules from which the respired carbon is derived may contain minerals in excess of the needs of the nematode for maintenance, growth and reproduction. Such excesses are excreted in mineral forms that can be taken up by plants and are a key mechanism explaining how nematodes contribute to primary production.

The best-studied example is the excretion of excess nitrogen in the form of ammonium, which is then available for uptake by plants or for bacterial transformation to nitrates and, subsequently, to atmospheric nitrogen, including the potent greenhouse gas N_2O. The mineralization service resulting from nematode digestion of organic molecules may be enhanced by differences in the molecular ratios of the food and the consumer. The C:N ratio of the biomass of many bacteria is in the range of 4:1 while that of their nematode predators is around 5.9:1 (Ferris *et al.*, 1997). The C:N ratio of some fungi averaged 8.5:1 while that of their nematode grazers was 9.1:1 (Chen and Ferris, 1999). Ratios of C:N in these ranges suggest that for every atom of

C ingested, the bacterial-feeding nematode ingests about 8% more than its body requirement for N. After respiratory mineralization of around 40% of the ingested C, the nematode has, in total, about 18% more N than required for body structure and reproductive output. The excretions of bacterial-feeding nematodes alone may enhance available mineral N in the soil by 20% or more (Ferris *et al.*, 1998). Even though the C:N ratios of fungal-feeding nematodes are apparently similar to those of their food substrate, the N associated with respired C (around 5% of N intake) is excreted as excess, varying with fungal substrate (Chen and Ferris, 1999; Okada and Ferris, 2001). The combined excretions of all consumer organisms in the soil food web may account for 80% of total mineralized N (Sánchez-Moreno *et al.*, 2008). This point is often overlooked in microbial ecology that is classically excluding the role of soil fauna in ecosystem functioning and ecosystem productivity.

1.3.4 Phoresy or vectoring by and of plant-feeding nematodes

Feeding activities of nematodes, in addition to facilitating nitrogen mineralization, contribute to other ecosystem services. Stylet-bearing nematodes may provide avenues of ingress into host or prey tissues; for example, access of bacteria and fungi through the migration and feeding activities of endoparasitic herbivores in roots and plant storage organs. This may be associated with weakened physical barriers and defence mechanisms allowing fungal and bacterial infections of plant tissue (e.g. *Fusarium*, *Phytophthora parasitica* and *Pseudomonas solanacearum* in tobacco) (Powell, 1971). Also, stylet-bearing nematodes may provide phoretic transport of bacteria or viruses into hosts (Hewitt *et al.*, 1958; Brown *et al.*, 1993). The reported instances seem to provide no obvious advantage to the nematode; in fact it may be to the detriment of the nematode, as it appears to enable exploitation of nematode behaviour and biology by other organisms. For example, the transport of *Clavibacter* bacteria by *Anguina* spp. to their developing seed galls in Gramineae, where the nematode may be outcompeted for the resource by the bacteria (Bird and Stynes, 1977).

Plant viruses vectored by *Longidorus* spp. and *Xiphinema* spp. may weaken the host and render it a poorer resource for the nematode (Brown *et al.*, 1993, 1995). While there might be reciprocal, but hitherto uncovered benefits for both organisms, a clearer example of mutually beneficial phoresy is the transport of *Bursaphelenchus xylophilus* by long-horned beetles (Coleoptera: Cerambycidae: *Monochamus*) to healthy pine trees where they enter through feeding wounds on young twigs. Invasion of the nematodes into the pine tree, and the damage they cause in resin canals, renders the tree less vigorous and more favourable for oviposition by the beetles, followed by larval development and pupation. The emerging adults become contaminated by nematodes nictating in the pupal chamber and transfer them to healthy trees (Togashi and Shigesada, 2006).

1.4 Effects of Nematodes as Regulators of Populations and on Ecosystem Production and Succession

Nematode feeding has the potential to control the magnitude of available resources and consequently impact ecosystem structure and function. Some general examples are shown in Fig. 1.2 and more specific observations are provided in the following sections.

1.4.1 Effects of nematode feeding on plant community composition

In addition to the much-studied effects of plant-feeding nematodes on plant performance and crop yield, the reduction in the growth rate and fitness of higher plants potentially decreases the exclusionary competitiveness of those plants and confers relatively greater fitness on their competitors. Consequently, ecosystem succession progresses, and plant diversity is increased (Van der Putten, 2003). The result in natural systems is that the susceptible and less competitive plants are reduced or even eliminated from the plant community (Wilschut and Geisen, 2021). More constrained examples are provided in agricultural systems where aggressive strains

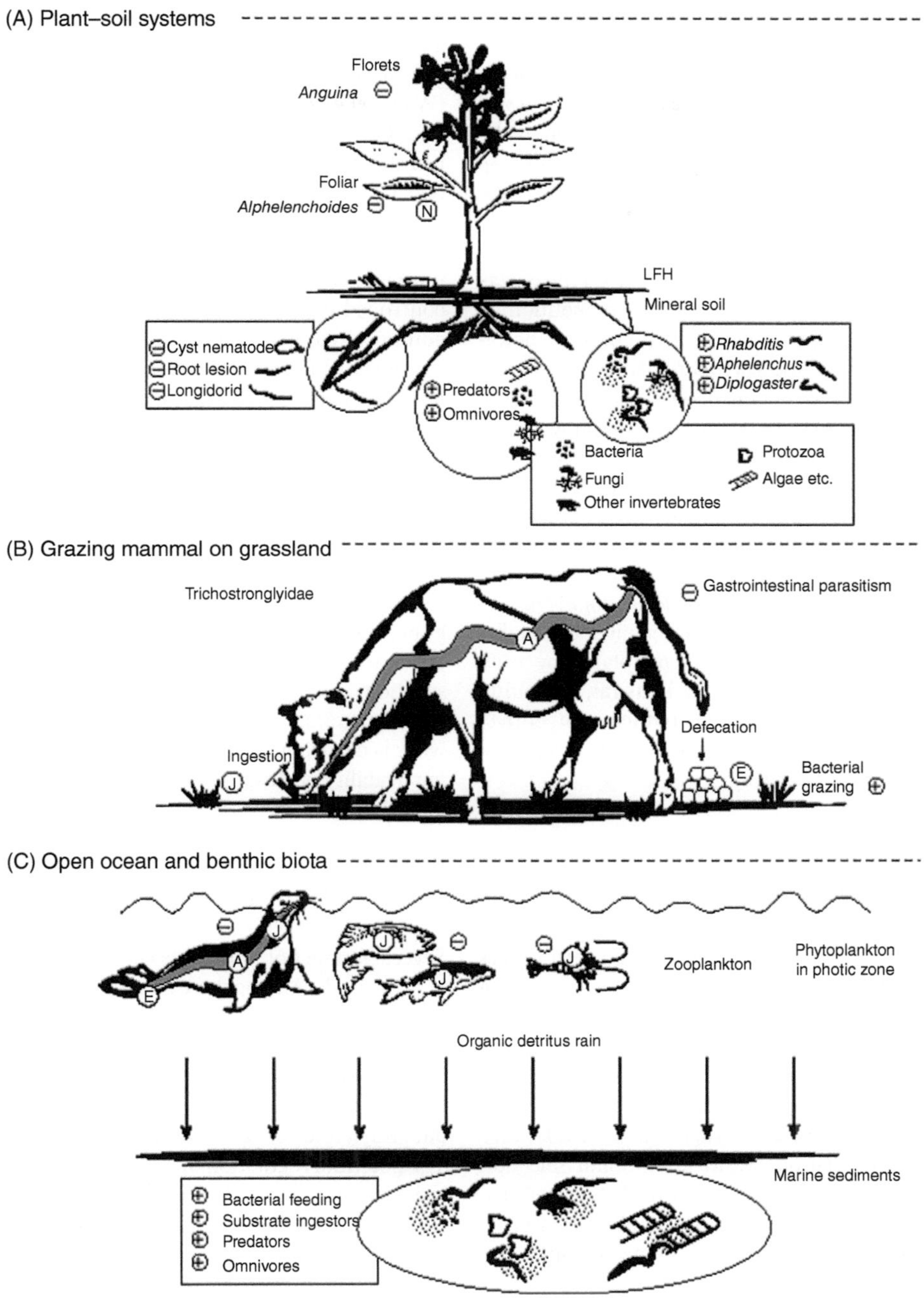

Fig. 1.2. Potential consequences of nematode feeding on food resources (+, positive effects of grazing on microbes and nutrient availability; –, potential loss of plant or animal production) in (A) plant–soil systems, (B) grassland with grazing mammals and (C) open ocean pelagic and benthic biota; LFH indicates the organic soil horizons: L (litter), F (partially decomposed), H (humus). Nematode metabolic activities and excretion are not specifically shown. For simplicity only selected nematode groups are included. Sites of nematode activity are labelled E, J, A or N to represent occurrence of egg, juveniles, adults or the species as a whole; in soils and sediments developmental stages are not separated. (Inspiration from Wardle *et al*., 2004.)

of plant-feeding nematodes have often been introduced with their susceptible, but agronomically desirable, hosts (Ferris *et al.*, 2003). A seeding rate designed for maximizing production will provide a less vigorous stand of the crop in the presence of nematode herbivores, opening up the canopy and reducing competitiveness with weeds, resulting into yield losses (Alston *et al.*, 1991; Schroeder *et al.*, 2005). Interestingly, applied ecologists have seldom exploited the effects of nematode herbivory on plant competition by designing cropping systems to enhance herbivore nematode species that render the weeds less competitive. For example, seed- and bud-feeding nematodes of the family Anguinidae may reduce the fitness of certain weeds, which clearly affects the plant community. The effects of *Mesoanguina amsinckiae* on the growth of coast fiddleneck (*Amsinckia intermedia*: Boraginaceae) in wheat was modelled by Pantone *et al.* (1989a,b), and in a field trial in Texas *Ditylenchus phyllobius* reduced the density of the target silverleaf nightshade (*Solanum elaeagnifolium*) plants by 66% (Northam and Orr, 1982). Thus far, exploiting root-feeding nematodes to reduce competitiveness of weed species has been relatively unsuccessful (Thomas *et al.*, 2005).

1.4.2 Nematode parasites of vertebrates

Similar examples of reduced competitiveness certainly occur among vertebrates, including humans, when parasitized by nematodes. Individuals with genetic characteristics or behaviour patterns that render them more susceptible to nematode infections may be selected against, resulting in greater average fitness or tolerance in the population. Multigenerational selection by farmers of the sheep in a flock that perform better under uniform parasite challenge can lead to sheep more tolerant of nematode infection (Bisset and Morris, 1996; Morris *et al.*, 2000).

The shift from mixed cropping and mixed grazing to more intensive livestock farming has often led to significant problems with nematode parasites of grazing animals for which anthelmintic drenches provided relief. However, apart from perceived non-target effects, these drenches have resulted in genetic selection for anthelmintic-resistant nematodes. Alternatively, the use of the nematode-trapping fungus *Duddingtonia flagrans* has reduced nematode parasitism of sheep with acceptable economic return (Waller *et al.*, 2004).

The functional impact of nematode parasitism in human populations is to stimulate societal responses of sanitation and other public health measures and the search for appropriate parasite management tools. River blindness in humans, caused by *Onchocerca volvulus*, can be controlled by annual treatment with the orally administered microfilaricidal drug, Mectizan® (ivermectin) (Cupp and Cupp, 2005). Elephantiasis or lymphatic filariasis, caused by *Wuchereria bancrofti*, can be locally eliminated using two drugs: ivermectin (Mectizan®) and albendazole in sub-Saharan Africa, or diethylcarbamazine and albendazole elsewhere (Richard-Lenoble *et al.*, 2003). However, for dracunculiasis, caused by Guinea worm (*Dracunculus medinensis*), no vaccine or medication is available, and public health programmes are designed to reduce exposure to infective juveniles. Nature-based solutions for nematode control might provide more sustainable control measures, but more work is needed in order to test the feasibility and applicability of this principle.

1.4.3 Nematode regulation of populations of other invertebrates

Several families of nematodes include associates of arthropods. Among those most studied, because of their potential as biocontrol agents, are the families Allantonematidae, Heterorhabditidae, Mermithidae, Phaenopsitylenchidae, Sphaerulariidae, Steinernematidae and Tetradonematidae (Kaya and Stock, 1997; Lacey *et al.*, 2001). The potential regulation of arthropods by nematodes has been most clearly demonstrated by the exploitation of heterorhabditid and steinernematid nematodes that carry toxic bacteria, *Photorhabdus* and *Xenorhabdus*, respectively, into their hosts. Mass production of these nematodes on artificial diets has fostered an industry dedicated to biological approaches in management of those insects that have a life stage in the soil (Ehlers and Shapiro-Ilan, 2005). There have also been attempts to capitalize on the direct parasitism and feeding of *Romanomermis culicivorax* on mosquito larvae (Petersen, 1985) and *Deladenus siricidicola* has been used successfully in biocontrol of the wood wasp *Sirex* in New Zealand and Australia (Bedding, 1993).

Other examples of regulation of invertebrates are the parasitism of slugs and snails by nematodes in the families Agfidae, Alloionematidae, Angiostomatidae and Rhabditidae. Of these, species of *Phasmarhabditis* have been developed as biological control agents for slugs (Wilson and Grewal, 2005).

1.4.4 Nematode influence on aquatic bacterial community diversity

Perhaps less spectacular, but with potentially important consequences for instance in decomposition processes, are observations that mucus secretions by aquatic nematodes may selectively favour settlement of specific strains of bacteria (Moens *et al.*, 2005b). Even low grazing rates by nematodes may significantly affect bacterial community composition, while high grazing rates may depress bacterial community diversity (De Mesel *et al.*, 2004).

1.4.5 Nematode suppression of mycelial growth

The feeding of aphelenchid nematodes (Riffle, 1967; Sutherland and Fortin, 1968; Ruess and Dighton, 1996) (Table 1.1) may reduce the rate of spread of saprophytic and mycorrhizal fungi in Petri dishes. Under standard conditions, feeding on fungi affects both the populations and morphometrics of particular Aphelenchidae, Tylenchidae and Tylencholaimidae (Faulkner and Darling, 1961; Townshend and Blackith, 1975; Ruess and Dighton, 1996; Okada and Kadota, 2003). Nematode populations may reduce, as well as be reduced by, fungi. For example, *Pleurotus ostreatus* (Basidiomycetes) supported population increase of *Filenchus misellus* and *Tylencholaimus parvus*, whereas *Aphelenchus avenae* populations were reduced as a result of predation by *P. ostreatus* (Okada and Kadota, 2003; Okada *et al.*, 2005). Feeding by Tylenchidae and Tylencholaimidae on decomposer fungi in mineral and organic soil horizons may enhance mineralization of nitrogen in those horizons (Okada and Ferris, 2001).

Nematode grazing on fungal mycelia may play other roles in complex interactions in soils. Laboratory studies suggest that feeding by *Aphelenchoides* sp. on the biocontrol fungus *Trichoderma harzianum* may constrain its efficacy as a biocontrol agent (Bae and Knudsen, 2001) and nematode feeding on mycorrhizal fungi could reduce exploration of the soil matrix by their mycelia (Table 1.1). *Aphelenchoides* spp. suppress colonization of *Pinus ponderosa* seedlings by the ectomycorrhiza *Suillus granulatus* (Riffle, 1975). In a pot study, Hussey and Roncadori (1981) demonstrated that grazing by *A. avenae* on the vesicular-arbuscular endomycorrhizae *Glomus margarita* and *Glomus etunicatus* could retard shoot and root growth of cotton. However, they concluded that the large population levels of *A. avenae* required probably precluded any significant interactions under field conditions.

There are many studies of fungi as biological control agents of nematodes (e.g. Timper, 2014). Often, the interactions are complex. Moreover,

Table 1.1. Colony size of two ectotrophic mycorrhizal fungi grown for 3 weeks at 25°C after addition of five initial densities of *Aphelenchus avenae* (Aphelenchidae). (After Sutherland and Fortin, 1968.)

	Amanita rubescens		*Suillus granulatus*	
Initial nematode population	Colony diameter (mm)	Final nematode population	Colony diameter (mm)	Final nematode population
0	49[a]	0[a]	40[a]	0[a]
5	47[a]	18,166[b]	34[b]	5,322[b]
10	43[a]	57,055[c]	26[c]	18,011[b,c]
25	34[b]	57,555[c]	21[d]	34,811[c,d]
50	25[c]	81,360[d]	16[e]	37,011[d]

[a–e]In each column, values not followed by the same superscript letter differ significantly at $P < 0.01$.

while *Pochonia chlamydosporia* is being assessed as a potential biocontrol agent for economically important plant-feeding nematodes, such as *Globodera* spp., *Heterodera* spp. and *Meloidogyne* spp., its intraspecific variants differ in their host preference. Further, its persistence in soil requires saprophytic activity (Mauchline *et al.*, 2004), raising questions about its functional activities in the soil food web. Therefore, while being spectacular to observe, fungal biocontrol of nematodes needs more work in order to result in predictable applications.

1.4.6 Nematode predation and omnivory

Undisturbed soils often have abundant specialist predators of nematodes, for example Diplogastrida and Mononchida, as well as an abundance of generalist predators, mainly Dorylaimida. Observational evidence of such systems supports the hypothesis that the predators, in combination with organisms of other taxa that benefit from similar environmental conditions, have regulatory, or even suppressive, effects on the abundance and temporal stability of nematodes occupying lower trophic levels in the food web, including herbivores, fungivores and bacterivores (e.g. Wardle *et al.*, 1995; Jackson *et al.*, 2019). Specialist and generalist predators are, however, quite sensitive to soil disturbance and chemical amendments (Korthals *et al.*, 1996; Berkelmans *et al.*, 2003; Tenuta and Ferris, 2004). For those reasons, they are at relatively low abundance in agricultural soils, where their potential prey exhibit unregulated population increase when provided with suitable resources and environmental conditions. Interestingly, conversion of such disturbed systems to reduced tillage and organic production does not immediately result in a more structured soil food web with greater connectance to higher trophic levels. This is probably because predators may be slow colonizers, have longer life cycles and lower productivity. Colonization and regulatory balance in the soil food web may require considerable time, often being decades rather than years, including recovery of soil structure (Korthals *et al.*, 1996; Yeates *et al.*, 1999a; Sánchez-Moreno *et al.*, 2006; Morriën *et al.*, 2017).

In some marine sediments, predacious nematodes may reach high abundances and dominate biomass and densities of other nematodes in the assemblage. Both laboratory and field evidence indicate that nematodes of the genus *Enoploides* exert substantial top-down control over prey density and community composition (Moens *et al.*, 2000; Hamels *et al.*, 2001; Gallucci *et al.*, 2005). Many other aquatic nematodes are presumed (based on stoma morphology) to be predators, but the lack of empirical evidence on their actual food sources and feeding rates hampers proper assessment of the importance of top-down regulation within the meiobenthic component of aquatic food webs. However, a recent compilation of nematode feeding habits based on observations of diets and on stomatal and pharyngeal morphology of the entire phylum Nematoda (Hodda, 2022) provides a useful basis for ecological analyses where direct observations are lacking.

1.5 Nematodes in a Community and Ecosystem Context

1.5.1 Economic crop loss due to plant-feeding nematodes

Galled wheat florets ('cockles') (Needham, 1744), galls on cucumber roots (Berkeley, 1855) and patches of stunted plants in sugarbeet fields (beet tiredness) (Schacht, 1859) were among the first-documented signs of nematode damage to plants. Over the past 150 years, there have been many investigations of the management of the pathogenic effects of plant-feeding nematodes. There are examples in books edited by Perry and Moens (2006), Subbotin and Chitambar (2018) and Sikora *et al.* (2021).

1.5.2 Management of microbial-feeding nematode function in an agricultural context

Nematode management activities in agricultural fields are usually aimed at reductions in abundance of plant-parasitic nematodes, often without consideration of the concomitant effects on microbial-feeding or predator nematodes. In the development of more sustainable agricultural systems, focus on the nurture of other components of the soil food web may be important. As an example, in the Mediterranean climate of the central valley of Northern California, annual crops are grown during the spring and summer months with the aid of irrigation. Following harvest in August or September, fields are left fallow, irrigation ceases and the soil becomes very dry prior to the start of winter rainfall in November. During the September to November fallow period, soil temperatures would be conducive to biological activity if water was available. Cover crops, when used, are usually planted with the first rain in late November. In a study of the transition from conventional to organic farming practices, crops planted in the spring following incorporation of a winter-grown legume cover crop but without application of mineral fertilizers exhibited symptoms of nitrogen deficiency. Microbial biomass was at high levels following cover crop incorporation, but population levels of bacterial-feeding nematodes were very low (Fig. 1.3). About six weeks after planting the summer crop, bacterial-feeding and fungal-feeding nematodes had increased on the newly available resources and the nitrogen-deficiency symptoms in the crop disappeared, with crop yield and mineral N concentrations being strongly correlated (Fig. 1.4).

In microcosm experiments, soil mineral N levels could be increased by 20% or more when bacteria were grazed upon by bacterial-feeding nematodes (Ferris *et al.*, 1998) and fungal-feeding nematodes may also contribute to nitrogen mineralization (Chen and Ferris, 1999; Zhang *et al.*, 2017). The abundance of bacterial-feeding nematodes was considered an indicator of concomitant abundance of other bacterial grazers,

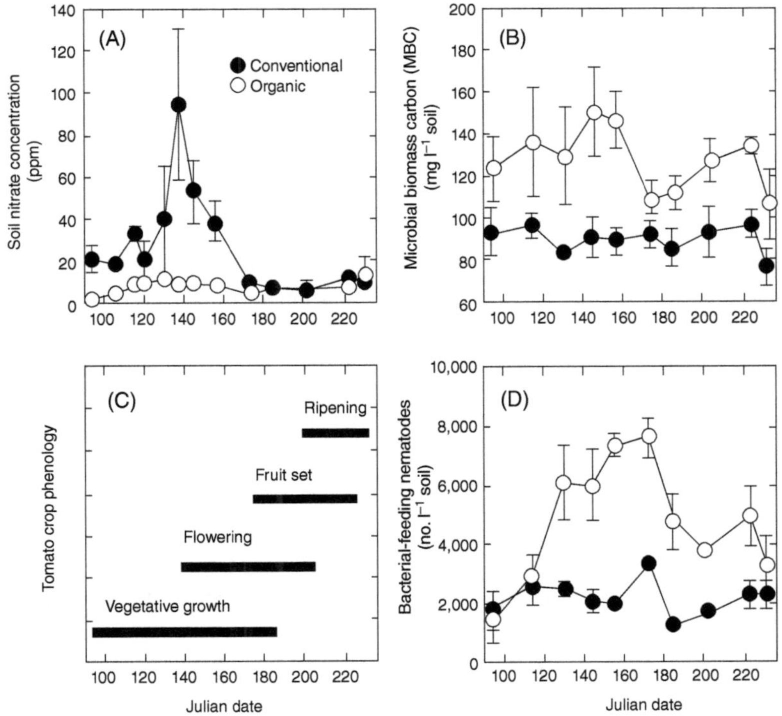

Fig. 1.3. Relationship of soil nitrate availability, microbial biomass and bacterial-feeding nematode abundance in soils under organic and conventional farming systems to tomato crop phenology. Data are means and standard errors across four replicates; Julian days run from 1 January. (A) Extractable soil nitrate concentration (from Temple, 1993). (B) Microbial biomass expressed as microbial biomass carbon (data from Gunapala and Scow, 1998). (C) Tomato crop phenology (data from Flint, 1985). (D) Abundance of bacterial-feeding nematodes (redrawn from Ferris *et al.*, 1996).

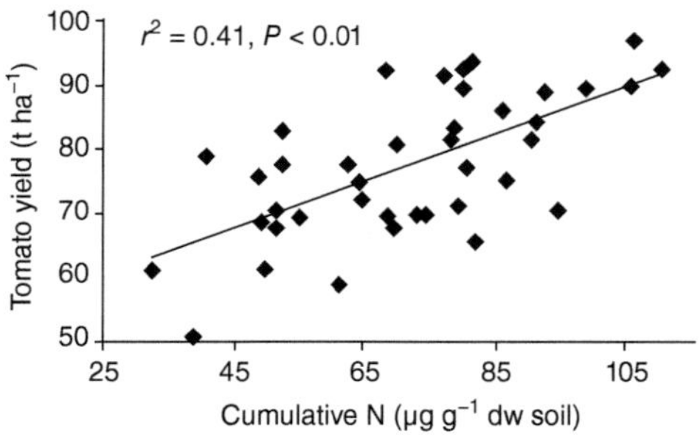

Fig. 1.4. Tomato crop yield in August 1996 in relation to cumulative soil N measurements over five sampling dates in April and May, prior to planting. dw soil = dry weight of soil. (Redrawn from Ferris *et al.*, 1996.)

especially protozoa. The window of opportunity for increasing the abundance of bacterial- and fungal-feeding nematodes and other organisms in field soil in the spring, at the time of cover crop incorporation, was during the warm soil temperature period of September to November. After that period, winter soil temperatures were too cool for nematode reproduction and biological activity.

In field plots where the soil was irrigated during the September to November period, there were greater abundances of bacterial-feeding nematodes in the spring and greater nitrogen availability at the time of establishment of the new crop. Nitrogen-deficiency symptoms were not seen in those plots.

Such field and laboratory studies support the conclusion that designed and holistic management of the soil food web, and indeed the entire soil ecosystem, is a major component of sustainable agricultural production systems. As a caveat, it should be noted that there is a downside to the late summer irrigation. The costs of water and pumping are additional financial overheads in the operation and weeds that would otherwise be dormant at that time may grow actively and require management (Ferris *et al.*, 2004c).

1.5.3 Coastal sand dune nematodes controlled by many factors

Vegetation succession in coastal foredunes is, in part, driven by plant species-specific negative feedbacks with the soil community (Van der Putten *et al.*, 1993). Initially, it was assumed that nematodes play a major role in the soil pathogen complexes, because of the specific occurrence of specialized root feeders in association with the different dominant plant species (Van der Putten and Van der Stoel, 1998). On other continents, the pioneer species marram grass (*Ammophila arenaria*) has been introduced in the absence of the nematode species and became invasive (Van der Putten *et al.*, 2005). This would all point at root-feeding nematodes controlling plant abundance, and specialized nematodes driving succession. However, in the native range nematodes turned out to be controlled by species-specific factors: competition (Brinkman *et al.*, 2005), arbuscular mycorrhizal fungi (De la Peña *et al.*, 2006), endophytes (Hol *et al.*, 2007), soil microorganisms (Piśkiewicz *et al.*, 2007) or by the plants themselves, such that they did not have the opportunity to control plant abundance (Fig. 1.5). Therefore, the current opinion is that, in the root zone of marram grass, damaging levels of plant-feeding nematodes are regulated by multifactorial pressures to levels that are too low to control plant performance. In European dunes, where *A. arenaria* naturally occurs, plant control therefore appears to be the result of the whole assemblage of soil biota. How such assemblages may be plant-species-specific (Van der Putten *et al.*, 1993) is still a question to be resolved. In spite of the uncertainties, these phenomena, such as plants controlling cyst nematodes to sublethal densities, may stimulate new approaches to sustainable nematode management in agricultural systems (Van der Putten *et al.*, 2006).

1.5.4 Intertidal nematode populations governed by environmental modulation of top-down regulation

Nematodes inhabiting intertidal habitats often migrate vertically in relation to the tidal cycle. The genus *Enoploides*, abundant in many fine to medium sandy sediments along coasts and estuaries in north-western Europe, is a voracious and selective predator of, for example, oligochaete, nematode and ciliate prey, and potentially exerts significant top-down control over its prey communities (Moens *et al.*, 2000; Hamels *et al.*, 2001).

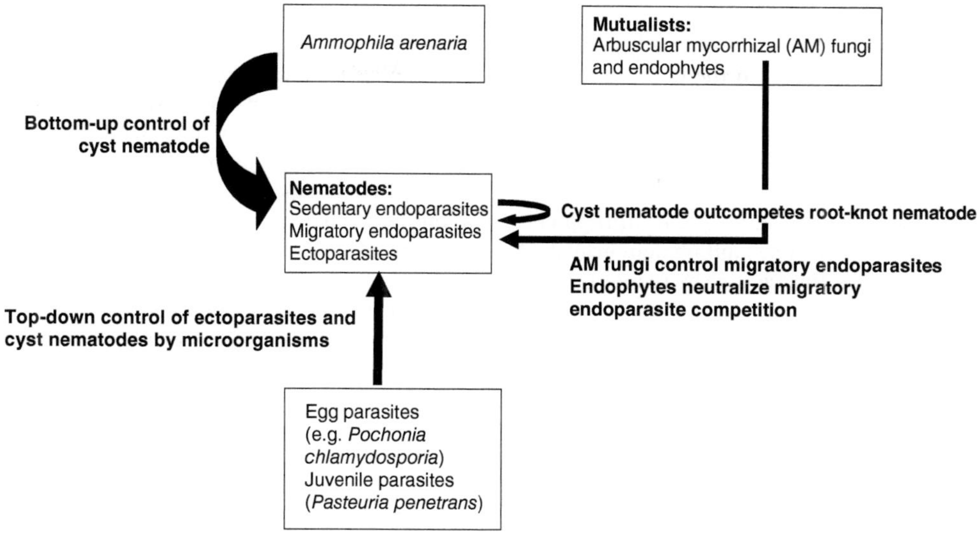

Fig. 1.5. Schematic of multiple controls on root-feeding nematodes on marram grass (*Ammophila arenaria*) in coastal sand dunes in north-western Europe. (After Van der Putten, 2003. Reproduced with permission from John Wiley and Sons.)

Enoploides is generally restricted to the upper 3 cm of the sediment, with clear density peaks in the upper 2 cm. Total abundance of all other nematode species in a fine to medium sandy sediment on the Molenplaat, Schelde Estuary (south-western Netherlands), peaks at a depth of 4–5 cm. These sediments are porous, and groundwater rapidly drains at low tide, leaving the upper sediment relatively dry after a few hours of exposure.

Laboratory experiments have demonstrated that the predation efficiency of *Enoploides* on nematode prey is strongly impaired by even mild sediment desiccation (Gallucci *et al.*, 2005). Since this and similar intertidal sites are exposed for 8 to 18 hours per day, optimal foraging by *Enoploides* must be restricted to relatively short periods during, and shortly after, inundation. Probably as an avoidance of sediment drying, *Enoploides* migrate a short distance into the sediment during low tide. Remarkably, some of its most abundant prey species at the Molenplaat exhibit the opposite vertical migration behaviour, peaking at or near the sediment surface only upon low tide exposure (Steyaert *et al.*, 2001). Many of these prey species rely on diatoms as their principal food. Since diatom production on intertidal flats is largely restricted to the very surface of the sediment (i.e. the photic zone), the upward migration of prey nematodes during low tide at the Molenplaat is likely a strategy to optimize feeding. In contrast, their downward migration upon inundation may be interpreted as an avoidance of predation by *Enoploides*. *Enoploides* distribution, in turn, appears to be largely controlled by sediment effects on the activity of the predator (Steyaert *et al.*, 2001; Gallucci *et al.*, 2005).

1.5.5 Multifactorial control over nematode abundances in a salt marsh benthic food web

Experimentally imposed diesel contamination of a *Spartina alterniflora* salt marsh caused high mortality of meiobenthic harpacticoid copepods and concomitant (transient) increases in both microphytobenthos biomass (MPB) and nematode abundance, and in meiofaunal grazing rates on the MPB. Total meiobenthic (copepods + nematodes) grazing was lower in diesel-contaminated than in unaffected plots, suggesting: (i) that MPB is top-down controlled by meiobenthic (here mainly copepod) grazing; and (ii) that harpacticoid copepods and nematodes compete for this limiting resource (Carman *et al.*, 1997; Fleeger *et al.*, 2006). The system is yet more complex, however, in that inclusion of the naked goby, *Gobiosoma bosc*, a burrowing fish predacious on meiofauna, affects both MPB and meiofauna in a multifactorial manner. Specifically, presence of the naked goby reversed the diesel-induced increase in nematode and abundance of MPB, while at the same time enhancing abundance of cyanobacteria (Fleeger *et al.*, 2006). Bioturbation of the sediment by *G. bosc* physically disrupts the sediment surface and MPB patches, while at the same time limiting light availability to the benthos through an increased turbidity. Hence, the diesel-induced release of the MPB from grazing control by harpacticoid copepods is counteracted by a fish-induced light limitation (which in turn improves the competitive ability of cyanobacteria over diatoms), and nematodes are again bottom-up controlled by a limited availability of MPB.

1.5.6 Nematodes of lakes and oceans

In intertidal and shallow sediments, local primary production by MPB is often the predominant carbon source fuelling nematode assemblages (Riera *et al.*, 1996; Moens *et al.*, 2002, 2005a). Nematodes in deeper (or light-limited) environments are primarily dependent on the rain of organic debris from the photic zone above, as well as input from rivers and streams. In both shallow and deep environments, the relative importance of direct grazing on primary production and associated bacteria and on heterotrophic debris of protozoa and other organisms is somewhat unclear. *Rhabditis marina* is almost uniquely associated with seaweed wrack on beaches (Sudhaus, 1976) and decomposition of dead fish (Gerlach, 1977) or marine mammals (Debenham *et al.*, 2004) has quantifiable impacts on nematode assemblages. Littoral macrophytes, and their associated periphyton, provide limited *in situ* resources for nematodes (Peters and Traunspurger, 2005). However, in sediment-settling studies, nematodes from an intertidal salt marsh settled at four to ten times the densities on to the macrophyte *Fucus* than on other substrates. In addition, nematodes from the deep ocean settled at greater densities on a sulfide medium than on other substrates. The experiments suggest that nematodes and other meiofaunal organisms can settle selectively when suspended in water. That may affect their dispersal in relation to resources (Mevenkamp *et al.*, 2016).

As in soils, meiobenthic nematodes contribute to ecological processes (Alkemade *et al.*, 1992; Aller and Aller, 1992; Montagna 1995; Traunspurger *et al.*, 1997). Energy availability is generally negatively correlated with the depth through which phytodetrital food sinks (Suess, 1980). Tietjen *et al.* (1989) found strong correlations between deep-sea meiofaunal abundance (dominated by nematodes) and fluxes of both organic carbon (r^2 = 0.98) and nitrogen (r^2 = 0.97). Nematode species richness in the deep sea of the North Atlantic Ocean increases with latitude, in contrast to patterns of molluscan and isopod diversity. This is likely related to the increase in primary productivity with latitude in that area (Lambshead *et al.*, 2000), a correlation which was also confirmed at the equator in the Pacific Ocean (Lambshead *et al.*, 2002).

In the deep-sea hydrothermal vents and seeps, as well as in some anoxic coastal sediments (such as subsurface sediments in mangroves) with an abundance of sulfur-reducing bacteria, chemoautotrophic production may replace sedimentation of particulate food from the euphotic zone as the primary carbon and energy source for some nematode genera or assemblages (Dover, 2000). Chemoautotrophic bacteria may be directly grazed upon, as in the case of *Halomonhystera disjuncta* feeding in mats of the sulfide-oxidizing *Beggiatoa* at a 1280 m deep Arctic mud volcano (Van Gaever *et al.*, 2006). However, more specialized relationships between chemoautotrophic bacteria and nematodes have evolved. For example, the adaptation of *Stilbonema* and *Laxus* to sulfide-rich sediments involves bacterial ectosymbionts capable of respiratory reduction of nitrate to nitrite (Hentschel *et al.*, 1999); the nematodes act as vectors for the bacteria, but probably also graze upon them (Polz *et al.*, 2000; Murfin *et al.*, 2012). The unrelated mouthless nematode *Astomonema* lacks a functional stoma and probably derives at least part of its nutrition from bacterial endosymbionts (Giere *et al.*, 1995). These endosymbionts are Gammaproteobacteria which appear closely related to the bacterial ectosymbionts on stilbonematinid nematodes and in gutless oligochaetes (Musat *et al.*, 2007). The marine nematode *Oncholaimus campylocercoides* is adapted to sulfidic sediments by the development of polysulfur chains and S-8 rings in the epidermis; these disappear on return to normal oxygen levels (Thiermann *et al.*,

2000). While populations of deep-sea mussels have been used as bioindicators of habitat condition (Jones *et al.*, 2006), it may be some time before there is sufficient data so that nematodes can be used similarly.

1.5.7 Species diversity within vertebrate hosts

Intra-host speciation of nematodes in the fore stomach of the Macropodinae (kangaroos and wallabies) has led to a flock of 36 genera and 256 species of the subfamily Cloacininae (Chabertiidae) (Cobb, 1915; Beveridge *et al.*, 2002; Chilton *et al.*, 2017). Current molecular evidence supports the hypothesis that early radiation of the strongyloid genera in marsupials may have occurred in the marsupial wombats (Marsupialia: Vombatidae) (Sukee *et al.*, 2022). A cladistic analysis of nematode species suggested that the aggregations within a given host are polyphyletic and probably evolved by host switching or colonization rather than by co-speciation. This radiation differs from that in ruminants where overlapping distribution of hosts and human management has confounded host distributions. There are, however, species flocks of strongylid nematodes in equids (Bucknell *et al.*, 1996).

1.6 Nematodes as Environmental Modifiers

1.6.1 Sediment agitation by nematodes

Bioturbation is physical disturbance of the mineral and organic particles comprising the substrate. Bioturbation by nematodes is of major importance in the direct physical enhancement of fluxes of oxygen and nutrients, and indirectly through the stimulatory effect of such enhanced fluxes on microbial activity (Alkemade *et al.*, 1992; Aller and Aller, 1992). Alkemade *et al.* (1992) demonstrated that a 30% enhancement of saltmarsh grass (*Spartina anglica*) decomposition in sediments with an abundant *Diplolaimella dievengatensis* population was almost entirely due to bacterial stimulation through increased oxygen availability as a result of bioturbation by nematodes. Riemann and Schrage (1978) observed that aquatic nematodes rapidly aggregate both sediment and organic matter particles when dispersed in water. They hypothesized an underlying trophic relationship, where mucus secreted by the nematodes stimulates growth of bacteria, which could in turn be grazed upon.

This hypothesis of trophic relationship was later modified into the 'enzyme sharing' concept: nematodes may produce exoenzymes that start the decomposition of complex molecules and promote establishment and growth of bacteria, which then take over the organic matter decomposition. Both nematodes and bacteria then feed on the 'soup' of dissolved organic matter released from this shared use of enzymes (Riemann and Helmke, 2002). Several chromadorid nematodes from intertidal mudflats (e.g. the widespread and often abundant *Ptycholaimellus ponticus*) build tubes by agglutinating sediment and small organic particles through mucus produced by the ventral gland. This affects sediment stability, enlarges the surface available to decomposers (the total inner surface area of the tube may be up to five times greater than the sediment surface), and probably offers the nematodes shelter against the effects of water currents as well as predation (Nehring *et al.*, 1990; Nehring, 1993).

1.6.2 Nematodes influence ecological succession in grasslands

Effects of nematodes on ecological succession have been studied mainly in terrestrial grassland systems (e.g. Van der Putten *et al.*, 2006). Very few studies have related the occurrence of plant-parasitic nematodes to the performance of aquatic macrophytes (plants) (Fritz *et al.*, 2004) and some studies have investigated the role of nematodes in the decline of wild nitrogen-fixing shrubs (Oremus and Otten, 1981; Maas *et al.*, 1983; Zoon *et al.*, 1993). As already detailed, plant-feeding nematode populations in the root zone of the foredune marram grass (*A. arenaria*) all appear to be regulated by a range of factors. However, in secondary grassland succession following removal of intensive farming practices, nematodes may influence plant performance (Verschoor *et al.*, 2001) and plant succession (De Deyn *et al.*, 2003), although cessation of agricultural production reduces the numbers of root-feeding nematodes to levels that may be too low to reduce growth of the plants (Wubs *et al.*, 2019). Nevertheless, while usually abundance of plant-feeding nematode species is relatively low, local aggregates of high abundance might still result in substantial effects on plant productivity (Verschoor, 2002). Moreover, host preference of some nematodes for dominant, fast-growing plant species may explain their contribution to plant community composition (Van Ruijven *et al.*, 2005) or succession (De Deyn *et al.*, 2003).

Biodiversity–ecosystem functioning studies have greatly fuelled thinking about species diversity versus the presence of individual species. For example, while plant species can influence the composition of nematode assemblages in soil, the actual plant species present can be more important for nematode diversity than the diversity of the plant community per se (De Deyn *et al.*, 2004; Viketoft *et al.*, 2005). The presence of certain plant-feeding nematodes may also enhance biotic resistance of soils against introduced exotic plant species (Van Ruijven *et al.*, 2005). Most studies mentioned above have involved inoculation experiments. Studies in prairie ecosystems in the 1970s concluded that nematodes use up to one-quarter of the net primary productivity (Stanton, 1988). However, that conclusion was based experiments with selective biocides and the results have not been verified by inoculation studies with known nematode abundances.

1.6.3 Nematodes may influence succession of aquatic microbial communities

The transport of, and grazing on, bacteria by nematodes may affect settlement (Moens *et al.*, 2005b), community composition (De Mesel *et al.*, 2004), densities and activity (Traunspurger *et al.*, 1997) of bacteria, and hence likely their contribution to important ecosystem processes such as organic matter mineralization. Microcosm studies using an estuarine sample of enrichment opportunists (e.g. *R. marina*, *Panagrolaimus paetzoldi*) and general opportunists (e.g. several species of Monhysteridae) indicate that the succession often observed from the former group to the latter does not passively follow the organic matter and bacterial dynamics. Instead, overgrazing of bacteria by rhabditid nematodes may facilitate monhysterids by suppressing bacterial densities to levels at which their feeding behaviour provides them with a competitive advantage (Moens *et al.*, 2000; dos Santos *et al.*, 2008).

1.6.4 Nematodes may predispose hosts to other organisms

A classic case of nematode effects on plant physiology that increase susceptibility to another pathogen is that of *Mesocriconema xenoplax*, an ectoparasite, on *Prunus* spp. Feeding by the nematode on the roots increases infection of the above-ground parts of the tree to *Pseudomonas syringae* (Lownsbery *et al.*, 1973; Ferris *et al.*, 2004a). Nematodes may predispose hosts to other nematodes, leading to competitive or facilitative effects. In general, the more complex the relationship between root-feeding nematodes and their host plants, the more competitive they are within and among species (Eisenback, 1993; Umesh *et al.*, 1994). On the other hand, nematodes may avoid direct competition by feeding on different cell layers in the root cortex (Bongers and Bongers, 1998; Siddiqi, 2000). Another issue of competition or facilitation, which has attracted far less attention, is that between root-feeding nematodes and other soil invertebrates. These interactions may lead to contrasting patterns, such as the induction of antagonistic effects by earthworms against cyst nematodes (Blouin *et al.*, 2005), to the facilitative effects of wire worms (Elateridae) on root-knot nematodes by indirect promotion of *Meloidogyne* abundance in mixed plant communities (De Deyn *et al.*, 2007).

Interactions may also be inhibitory; such interactions demonstrated between congeneric species of *Diplolaimelloides* do not appear to be due to direct competition for food or space (De Mesel *et al.*, 2006; dos Santos *et al.*, 2008, 2009). Similarly, inhibition among soil bacterial-feeding nematode species (Bongers *et al.*, 2001; Postma-Blaauw *et al.*, 2005) may not be completely due to competitive interactions but could conceivably be mediated through bacterial defence signals (Venette and Ferris, 1998; Phillips *et al.*, 2003). Probably, when explored more intensively, such indirect effects among nematodes and between nematodes and other organisms will be recognized as important determinants of community interactions.

1.7 Synoptic Integration to System, Landscape and Biosphere Levels

While literature dealing with nematode faunae as soil (health) indicators in different farming and natural systems is abundant (Du Preez *et al.*, 2022), only a few studies have dealt with the complexities of regional/landscape/ocean-floor zone/global distribution patterns of nematode faunae (e.g. Proctor, 1984, 1990; Song *et al.*, 2017; Van den Hoogen *et al.*, 2019). To be comprehensive, such analyses generally require the aggregation of data sets from a wide range of researchers and then development of a strategy for assembly and reporting the conclusions. It can be a formidable undertaking, for example as has been carried out by Van den Hoogen *et al.* (2019). While there are some similar studies for aquatic environments, for example Schratzberger *et al.* (2007) and Van den Hoogen *et al.* (2019), there is an underlying problem that causes of high local diversity in deep-sea sediments and is very poorly understood. Thus, the use of nematode diversity per se as an indicator in deep-sea sediments is not yet well established. However, for freshwater rivers and lakes, systems have been developed as indicators of pollution that are centred on the abundance of nematode 'species at risk' which can be used in combination with DNA metabarcoding (Höss *et al.*, 2011, 2016; Schenk *et al.*, 2020).

In marine systems, there has been a strong emphasis on comparisons of nematode assemblages at beta and gamma scales, but there is a lack of process-oriented or mechanistic understanding of drivers of the patterns. Most studies on nematode assemblages in soil are based on field plots or single-crop farm fields (e.g. Wardle *et al.*, 1995; Berkelmans *et al.*, 2003; Ferris and Matute, 2003). However, feedback loops among above-ground and below-ground biota are important ecological drivers in terrestrial ecosystems (Sánchez-Moreno *et al.*, 2008). The spatial patterns of soil biota have important above-ground consequences on both plant community structure and on individual plants, and vice versa (Ettema and Wardle, 2002). Studies indicate that nematodes can enhance the control of above-ground pests through both bottom-up and top-down control activities (Bezemer *et al.*, 2005). This linking of above- and below-ground subsystems is a new and promising development indicating that proper soil management is an essential foundation for sustainable crop protection against even above-ground pests.

A simple example of the problems of scale and resolution in considering nematode roles and services is in their function as herbivores and entry points for carbon and energy into the soil food web (e.g. Yeates *et al.*, 1999b). Individual nematodes and their activities represent the minimum patch, and the population of a species represents the more general local aggregate of patches. In agriculture, such patches of plant-feeding nematodes may cause local loss of yield, with field-scale loss being determined by the field-scale nematode population level and distribution, which reflects the age of the infestation, edaphic conditions, crop history and management practices. Similarly, strong aggregations of nematodes and other organisms in the root zones of uprooted orchard trees can result in death or poor growth of trees in a subsequent orchard planting (e.g. Ferris *et al.*, 2003, 2004a). As with most bioindicator systems, the aggregation of organisms into patches generates problems in measuring abundance and diversity at local and larger scales. The problem is compounded when the patches are not visible and the assessment of environmental condition is based on a sample, or multiple samples, of soil or water without knowledge of patch distribution (Ferris *et al.*, 1990; Ettema and Wardle, 2002; Ettema and Yeates, 2003). A more extensive spatial sampling study has revealed that various nematode species show individually different abundance relationships with space (Quist *et al.*, 2019).

In another example of functional linkages between above-ground and below-ground biota, an abundance of bacterial-feeding nematodes provides important services in well-managed organic farms, whereas in conventional production systems, where nutrients are supplied from external sources, a similar abundance of nematodes may contribute little benefit to crop growth (Ettema and Bongers, 1993; Ferris *et al.*, 1998). While these effects occur at the ped/microsite level, scaling up by using Geographic Information Systems and multilayer mapping techniques will allow a more comprehensive understanding of the functions and services of nematodes at the landscape level.

1.8 Humans, Nematodes and Ecosystem Management

Nematodes are intrinsic components of natural ecosystems. The early focus of helminthology and nematology on nematodes as pests and parasites of humans, plants and animals has developed into a realization that parasitism plays a role in natural ecosystems in regulating species abundance and maintaining species diversity. Moreover, the awareness of the role of nematodes in decomposition processes as well as their potential role in controlling outbreaks of insect pests and their use as environmental indicators has led to a paradigm shift from consideration of nematodes only as undesirable pests to a broader recognition of their contributions in ecosystem services (e.g. Bongers, 1990; Ferris *et al.*, 2004b; Wilschut and Geisen, 2021). There is increasing awareness that nematode populations may be optimized by managing nature rather than utilizing broad-spectrum control measures with unknown ecological consequences.

Plant-feeding nematodes may contribute to coexistence and succession in natural plant communities (De Deyn *et al.*, 2003) and the same principles drive the need for crop rotation in agriculture (e.g. Ferris *et al.*, 1996; Van der Putten *et al.* 2006). In the past, when wide crop rotations were applied, nematodes were less of a problem. However, the globalization of world food and commodity markets and the development of expensive specialized machinery have led to the need for specialization among growers. Rotations have become so narrow that it is difficult to exploit natural population dynamics to reduce parasite population levels. The side effects of chemical nematicides on soil life and neighbouring aquatic ecosystems, and on groundwater quality, have resulted in awareness that these chemicals threaten the sustainability of crop production and have highlighted the need for biological control and more eco-friendly pesticides (Tenuta and Ferris, 2004). However, biological control, because it represents a series of interlocking biological populations, and their respective environments, is much more variable and less predictable, limiting the generalization of findings in one crop at specific site conditions (Timper, 2014; Van der Putten *et al.*, 2006). This leaves considerable challenges for nematologists, agronomists, entomologists and ecologists to develop farming systems that maximize nematode control while fulfilling the high demands of global trade and economy on cropping systems.

When humans modify natural selection, competition and species replacement, through crop breeding for high yield, disease resistance, weed control and other features, the need for understanding impacts throughout the food chain is paramount. When pests are not subjected to natural regulatory pressures of competition, predation and genetic barriers, massive outbreaks may result, similar to those seen after the application of broad-spectrum pesticides. Biological and organic agriculture each aim for conditions that more or less restore biodiversity and habitat complexity, avoiding chemical crop protection and mineral fertilizers (Tenuta and Ferris, 2004). The question of whether maximal yields are achievable with natural equilibria of plant pests and their antagonists, non-chemical fertilizers and sustainable soil tillage is still open, and nematode management is one of the key issues to be solved in order to reach that state (Temple, 1993). Sustainable crop production will probably require a blending of cropping practices applied in the past and development of new farming systems. Almost certainly, it will include the use of cultivars that have not been selected for maximum yield under the umbrella of synthetic pesticides and mineral fertilizers.

1.9 Suitability of Nematodes as Environmental Indicators

The development of nematodes as bioindicators in soil and aquatic systems has required determination of appropriate ways to assess and quantify their contribution to ecological processes, and the validation of their ability as indicators of environmental condition. Several unique characteristics of nematodes facilitated those developments. In summary, nematodes occur in all soil and aquatic systems: in acidified forest soils, in heavily polluted soil and water, in heavy clay soil, in deep-sea sediments, in rotting plant material, in composts, and in any habitat in which organic material is decomposed. Different nematode taxa exhibit specificities of food sources and changes in the food web are mirrored in shifts among feeding groups. Species in many families within the Tylenchina feed exclusively on the roots of higher plants but never on bacteria. Species in the Cephalobidae and Plectidae feed on bacteria but not on higher plants or fungi. Species in the Mononchidae and Anatonchidae are specialist predators of other nematodes and do not feed on higher plants or fungi. The transparent nature of the nematode body allows easy observation of mouth and pharyngeal structures, which allows inference of feeding habit (e.g. Hodda, 2022). Robust techniques for extracting nematodes from soil and other substrates have been developed and can be applied to all taxonomic and functional groups. Most importantly, nematode species vary in their responses to stress factors and resource enrichment; some species are extremely sensitive to pollutants and others are extremely tolerant (Korthals *et al.*, 1996; Ferris *et al.*, 2004b; Tenuta and Ferris, 2004). They vary in lifespan: some species with a generation time of days, others of months or even a year. The matrix of feeding habit, generation time and sensitivity to environmental disturbance allows the designation of functional guilds of different nematode taxa with similar response characteristics (Bongers and Bongers, 1998; Ferris and Bongers, 2006).

In this chapter, we have hinted at the opportunities of using nematodes as biological indicators of the condition of their environment. This issue is still a wide-open research question, also now the European Union has adopted the law on soil monitoring, which requires insights into the indicator value of soil organisms, as not every individual species may be monitored. However, we have also mentioned the challenges involved in assessing the diversity and abundance of nematode assemblages, let alone of assemblages of all soil biota. Other important considerations are the issues of succession within assemblages of organisms and the timing of population assessments in relation to an environmental perturbation or seasonal event. In an agricultural system, for example, a faunal analysis conducted six weeks after a disturbance or introduction of resources may look quite different than would have been the case two weeks after the disruption. Resources will have changed in abundance and nature as they are utilized and the species with different feeding habits and life-history strategies will have changed accordingly; succession and/or recovery will have occurred in the nematode fauna. In many cases, a single assessment of the nematode assemblage may be inadequate for determination of the short-, medium- and long-term impacts of an environmental event, the video equivalent of a series of samples at appropriate time intervals will be more revealing than the snapshot of a single assessment (Berkelmans *et al.*, 2003; Ferris and Matute, 2003, Ferris *et al.*, 2004b).

Finally, we need to remain cognizant that our planet is always in a state of flux and that the amplitude and frequency of the fluxes are not constant. They are impacted by climate change, geopolitical disturbances, marketing opportunities and pressures, availability of fossil fuels and natural resources, and myriad other forces. While we focus on the bioindicator potential of nematodes with regard to ecosystem services, we need to reflect that the very nature of environmental conditions changes, currently faster than in almost any moment in time. Climate change, biodiversity, land-use intensity and contamination are altering at an alarming rate; for example, the accumulation of microplastics in the ocean and contamination from the effluent of aquaculture systems and human habitation. Therefore, using nematodes as bioindicators appears an open challenge, with targets that are often moving faster than research progresses.

1.10 Concluding Points

1. The nematode body plan and life history, although apparently limiting, are sufficiently adaptable to allow species of the phylum Nematoda to occupy a wide range of physical environments, utilize a wide range of resources, become extremely diverse and achieve large populations. Nematodes are the most abundant multicellular animals on the planet.

2. Nematodes have key impacts on populations of many microbes, plants and animals, and on ecosystem processes. They play key roles in many ecosystem processes and services.

3. The abundance and diversity of nematode species vary with ecosystem, physical environment and management.

4. Nematodes are valuable environmental bioindicators, whether in relation to diverse ecosystem services, plant diseases, management of parasites in grazing mammals, human health or insect control, etc. Some such indicators have been used for decades. The challenge is to use nematodes also as indicators for ecosystem health.

5. There are opportunities for further studies on the use of nematodes as indicators for other ecosystem functions, such as the state of restoration in semi-natural ecosystems and the capacity of soils to sustain diverse plant communities. However, indicators must be tailored for each question and environment, with the response variable(s) and indicator being context-specific.

6. Anthelminthic drenches developed for nematodes of livestock relieve suffering when applied to humans. Nematode-trapping fungi, extensively studied in relation to plant-feeding nematodes, have also been successfully deployed to manage gastrointestinal nematodes of livestock. These cases clearly demonstrate the transferability of knowledge of the relationships between nematodes and their total environment. Their usefulness in soils still needs to be further studied.

7. Root-feeding nematodes influence diversity and succession in natural vegetation, although they may operate in assemblages of many soil-inhabiting species. While their absence correlates with plant invasiveness, especially when plants are introduced into new habitats without their original root feeders, the role in plant abundance control needs further studies. Similarly, changes in environmental and climatic conditions, and availability of resources, influence succession in nematode assemblages, but consequences for ecosystem flows and fluxes need to be further established.

8. In agricultural and natural systems, nematodes and other soil biota influence plant–pest interactions above ground; healthy vegetation and above-ground biodiversity, therefore, depend on the composition of the soil community, and probably also on a healthy soil.

Further information on nematode families, genera and species referred to as examples in this chapter is available in Nemaplex (https://nemaplex.ucdavis.edu, accessed July 2024).

Author Note

The original lead author of Chapter 1 in the first edition of this book, the world-renowned soil biologist, nematologist, ecologist and systematist, Dr Gregor Yeates, died on 6 August 2012. As his co-authors, in revising and updating the material in the chapter, we recognize that each of us made substantial contributions in the first edition to the biology, ecology and insights detailed herein. But it is clear to us that the central thread and organization of the material, and the breadth and depth of the supporting citations, reflect Gregor's energy, drive, passion and synoptic view of nematodes in soil and aquatic ecosystems.

References

Ali, J.G. and Davidson-Lowe, E. (2015) Chapter 8 – Plant cues and factors influencing the behavior of beneficial nematodes as a belowground indirect defense. *Advances in Botanical Research* 75, 191–214.

Ali, J.G., Alborn, H.T., Stelinski, L.L. (2011) Constitutive and induced subterranean plant volatiles attract both entomopathogenic and plant parasitic nematodes. *Journal of Ecology* 99, 26–35.

Ali, M.A., Anjam, M.S., Nawaz, M.A., Lam, H. and Chung, C. (2018) Signal transduction in plant–nematode interactions. *International Journal of Molecular Sciences* 19, 1648. DOI: 10.3390/ijms19061648.

Alkemade, R., Wielemaker, A., de Jong, S.A. and Sandee, A.J.J. (1992) Experimental evidence for the role of bioturbation by the marine nematode *Diplolaimella dievengatensis* in stimulating the mineralization of *Spartina anglica* detritus. *Marine Ecology Progress Series* 90, 149–155.

Aller, R.C. and Aller, J.Y. (1992) Meiofauna and solute transport in marine muds. *Limnology and Oceanography* 37, 1018–1033.

Alston, D.G., Bradley, J.R., Coble, H.D. and Schmitt, D.P. (1991) Impact of population density of *Heterodera glycines* on soybean canopy growth and weed competition. *Plant Disease* 75, 1016–1018.

Bae, Y.S. and Knudsen, G.R. (2001) Influence of a fungus-feeding nematode on growth and biocontrol efficacy of *Trichoderma harzianum*. *Phytopathology* 91, 310–306.

Bardgett, R.D. (2005) *The Biology of Soil: A Community and Ecosystem Approach*. Oxford University Press, Oxford.

Bardgett, R.D., Cook, R., Yeates, G.W. and Denton, C.S. (1999) The influence of nematodes on below-ground processes in grassland ecosystems. *Plant and Soil* 212, 23–33.

Bardgett, R.D., Bowman, W.D, Kaufmann, R. and Schmidt, S.K. (2005) A temporal approach to linking aboveground and belowground ecology. *Trends in Ecology and Evolution* 20, 634–641.

Bargmann, C.I. and Mori, I. (1997) Chemotaxis and thermotaxis. In: Riddle, D.L., Blumenthal, T., Meyer, B.J. and Priess, J.R. (eds) *C. elegans II*. Cold Spring Harbor Laboratory Press, Woodbury, New York, pp. 717–737.

Bedding, R.A. (1993) Biological control of *Sirex noctilio* using the nematode *Deladenus siricidicola*. In: Bedding, R., Akhurst, R and Kaya, H. (eds) *Nematodes and the Biological Control of Insect Pests*. CSIRO Publications, East Melbourne, Australia, pp. 11–20.

Berkeley, M.J. (1855) Vibrio forming cysts on the roots of cucumbers. *Gardeners' Chronicle*, 7 April, 220.

Berkelmans, R., Ferris, H., Tenuta, M. and van Bruggen, A.H.C. (2003) Effects of long-term crop management on nematode trophic levels other than plant feeders disappear after one year of disruptive soil management. *Applied Soil Ecology* 23, 223–235.

Beveridge, I., Chiton, N.B. and Spratt, D.M. (2002) The occurrence of species flocks in the nematode genus *Cloacina* (Strongyloidea: Cloacininae), parasitic in the stomach of kangaroos and wallabies. *Australian Journal of Zoology* 50, 597–620.

Bezemer, T.M., De Deyn, G.B., Bossinga, T.M., van Dam, N.M., Harvey, J.A. *et al.* (2005) Soil community composition drives aboveground plant–herbivore–parasitoid interactions. *Ecology Letters* 8, 652–661.

Bird, A.F. and Ryder, M.H. (1993) Feeding of the nematode *Acrobeloides nanus* on bacteria. *Journal of Nematology* 25, 493–499.

Bird, A.F. and Stynes, B.A. (1977) The morphology of a *Corynebacterium* sp. parasitic on annual ryegrass. *Phytopathology* 67, 828–830.

Bisset, S.A. and Morris, C.A. (1996) Feasibility and implications of breeding sheep for resilience to nematode challenge. *International Journal for Parasitology* 26, 857–868.

Blair, J.E., Ikeo, K., Gojobori, T. and Hedges, S.B. (2002) The evolutionary position of nematodes *BMC Evolutionary Biology* 2, 7.

Blanchard, G.F. (1990) Overlapping microscale dispersion patterns of meiofauna and microphytobenthos. *Marine Ecology Progress Series* 68, 101–111.

Blaxter, M.L. (1998) *Caenorhabitis elegans* is a nematode. *Science* 282, 2041–2046.

Blaxter, M. and Koutsovoulos, G. (2015) The evolution of parasitism in Nematoda. *Parasitology* 142, S26–S39.

Blaxter, M.L., De Ley, P., Garey, J.R., Liu, L.X., Scheldeman, P. *et al.* (1998) A molecular evolutionary framework for the phylum Nematoda. *Nature* 92, 710–75.

Blouin, M., Zuily-Fodil, Y., Pham-Thi, A.T., Laffray, D., Reversat, G. *et al.* (2005) Belowground organism activities affect plant aboveground phenotype, inducing plant tolerance to parasites. *Ecology Letters* 8, 202–208.

Boemare, N. (2002) Biology, taxonomy, and systematics of *Photorhabdus* and *Xenorhabdus*. In: Gaugler, R. (ed.) *Entomopathogenic Nematology*. CAB International, Wallingford, UK, pp. 57–78.

Bongers, T. (1990) The maturity index: an ecological measure of environmental disturbance based on nematode species composition. *Oecologia* 83, 14–19.

Bongers, T. and Bongers, M. (1998) Functional diversity of nematodes. *Applied Soil Ecology* 10, 239–251.

Bongers, T., Ilieva-Makulec, K. and Ekschmitt, K. (2001) Acute sensitivity of nematode taxa to $CuSO_4$ and relationships with feeding-type and life-history classification. *Environmental Toxicology and Chemistry* 20, 1511–1516.

Brinkman, E.P., Duyts, H. and Van der Putten, W.H. (2005) Consequences of variation in species diversity in a community of root-feeding herbivores for nematode dynamics and host plant biomass. *Oikos* 110, 417–427.

Brown, D.J.F., Halbrendt, J.M., Robbins, R.T. and Vrain, T.C. (1993) Transmission of nepoviruses by *Xiphinema americanum*-group nematodes. *Journal of Nematology* 25, 349–354.

Brown, D.J.F., Robertson, W.M. and Trudgill, D.L. (1995) Transmission of viruses by plant nematodes. *Annual Review of Plant Pathology* 33, 223–249.

Bucknell, D., Hoste, H., Gasser, R.B. and Beveridge, I. (1996) The structure of the community of strongyloid nematodes in domestic equids. *Journal of Helminthology* 70, 185–192.

Carman, K.R., Fleeger, J.W. and Pomarico, S.M. (1997) Response of a benthic food web to hydrocarbon contamination. *Limnology and Oceanography* 42, 561–571.

CESC (*C. elegans* Sequencing Consortium) (1998) Genome sequence of the nematode *C. elegans*: a platform for investigating biology. *Science* 282, 2012–2018.

Chen, J. and Ferris, H. (1999) The effects of nematode grazing on nitrogen mineralization during fungal decomposition of organic matter. *Soil Biology and Biochemistry* 31, 1265–1279.

Chia, F.S. and Warwick, R.M. (1969) Assimilation of labelled glucose from seawater by marine nematodes. *Nature* 224, 720–721.

Chilton, N.B., Shuttleworth, M.A., Huby-Chilton, F., Koehler, A.V., Jabbar, A. *et al.* (2017) Speciation in the genus *Cloacina* (Nematode: Strongylida): species flocks and intra-host speciation. *Parasitology* 144, 1828–1840. DOI: 10.1017/S0031182017001238.

Cid del Prado-Vera, I., Ferris, H., Nadler, S.A. and Lamothe Argumedo, R. (2012) Four new species of *Tripylina* Brzeski, 1963 (Enoplida: Tripylidae) from México, with an emended diagnosis of the genus. *Journal of Nematode Morphology and Systematics* 15, 71–86.

Cid del Prado-Vera, I., Ferris, H. and Subbotin, S.A. (2023) A new species of the genus *Diastolaimus* and redescription of *D. mexicanus* Cid del Prado, 2012 (Rhabditida: Chambersiellidae) from Mexico. *Nematropica* 53, 30–40.

Cobb, N.A. (1915) Nematodes and their relationships. *Yearbook of the United States Department of Agriculture* 1914, 457–490.

Coleman, D.C., Odum, E.P. and Crossley, D.A. (1992) Soil biology, soil ecology and global change. *Biology and Fertility of Soils* 14, 104–111.

Cupp, E.W. and Cupp, M.A. (2005) Impact of ivermectin community-level treatments on elimination of adult *Onchocerca volvulus* when individuals receive multiple treatments each year. *The American Journal of Tropical Medicine and Hygiene* 73, 1159–1161.

De Deyn, G.B. and Van der Putten, W.H. (2005) Linking above- and belowground biodiversity. *Trends in Ecology and Evolution* 20, 625–633.

De Deyn, G.B., Raaijmakers, C.E., Zoomer, H.R., Berg, M.P., De Ruiter, P.C. *et al.* (2003) Soil invertebrate fauna enhances grassland succession and diversity. *Nature* 422, 711–713.

De Deyn, G.B., Raaijmakers, C.E., van Ruijven, J., Berendse, F. and Van der Putten, W.H. (2004) Plant species identity and diversity effects on different trophic levels of nematodes in the soil food web. *Oikos* 106, 576–586.

De Deyn, G.B., Van Ruijven, J., Raaijmakers, C.E., De Ruiter, P.C. and Van der Putten, W.H. (2007) Above- and belowground insect herbivores differentially alter nematode communities by plant diversity shifts. *Oikos* 116, 923–930.

De la Peña, E., Rodríguez Echeverría, S., Van der Putten, W.H., Freitas, H. and Moens, M. (2006) Mechanism of control of root-feeding nematodes by mycorrhizal fungi in the dune grass *Ammophila arenaria*. *New Phytologist* 169, 829–840.

De Ley, P. and Blaxter, M. (2002) Systematic position and phylogeny. In: Lee, D.L. (ed.) *The Biology of Nematodes*. Taylor & Francis, London, pp. 1–30.

De Ley, P. and Blaxter, M.L. (2004) A new system for Nematoda: combining morphological characters with molecular trees, and translating clades into ranks and taxa. *Nematology Monographs and Perspectives* 2, 633–653.

De Mesel, I., Derycke, S., Swings, J., Vincx, M. and Moens, T. (2003) Influence of bacterivorous nematodes on the decomposition of cordgrass. *Journal of Experimental Marine Biology and Ecology* 296, 227–242.

De Mesel, I., Derycke, S., Moens, T., Van der Gucht, K., Vincx, M. *et al.* (2004) Top-down impact of bacterivorous nematodes on the bacterial community structure: a microcosm study. *Environmental Microbiology* 6, 733–744.

De Mesel, I., Derycke, S., Swings, J., Vincx, M. and Moens, T. (2006) Role of nematodes in decomposition processes: does within-trophic group diversity matter? *Marine Ecology Progress Series* 321, 157–166.

De Rooij-Van der Goes, P.C.E.M. (1995) The role of plant-parasitic nematodes and soil-borne fungi in the decline of *Ammophila arenaria* (L.) Link. *New Phytologist* 129, 661–669.

De Rosa, R., Grenier, J.K., Andreeva, T., Cook, C.E., Adoutte, A. *et al.* (1999) Hox genes in brachiopods and priapulids and protostome evolution. *Nature* 399, 772–776.

Debenham, N.J., Lambshead, P.J.D., Ferrero, T.J. and Smith, C.R. (2004) The impact of whale falls on nematode abundance in the deep sea. *Deep-sea Research Part 1 – Oceanographic Research Papers* 51, 701–706.

dos Santos, G.A.P., Derycke, S., Genevois, V.G.F., Coelho, L.C.B.B., Correia, M.T.S. *et al.* (2008) Differential effects of food availability on population development and fitness of three species of estuarine, bacterial-feeding nematodes. *Journal of Experimental Marine Biology and Ecology* 355, 27–40.

dos Santos, G.A.P., Derycke, S., Genevois, V.G.F., Coelho, L.C.B.B., Correia, M.T.S. *et al.* (2009) Interactions among bacterial-feeding nematode species at different levels of food availability. *Marine Biology* 156, 629–640.

Dover, C.L.V. (2000) *The Ecology of Deep-Sea Hydrothermal Vents*. Princeton University Press, Princeton, New Jersey.

Du Preez, G., Daneel, M., De Goede, R., Du Toit, M.J., Ferris, H. *et al.* (2022) Nematode-based indices in soil ecology: application, utility, and future directions. *Soil Biology and Biochemistry* 169, 108640.

Ehlers, R.-U. and Shapiro-Ilan, D.I. (2005) Mass production. In: Grewal, P.S., Ehlers, R.-U. and Shapiro-Ilan, D.I. (eds) *Nematodes as Biocontrol Agents*. CAB International, Wallingford, UK, pp. 65–78.

Eisenback, J.D. (1993) Interactions between nematodes in cohabitance. In: Khan, M.W. (ed.) *Nematode Interactions*. Chapman & Hall, London, pp. 134–174.

Ettema, C.H. and Bongers, T. (1993) Characterization of nematode colonization and succession in disturbed soil using the Maturity Index. *Biology and Fertility of Soils* 16, 79–85.

Ettema, C.H. and Wardle, D.A. (2002) Spatial soil ecology. *Trends in Ecology and Evolution* 17, 177–183.

Ettema, C.H. and Yeates, G.W. (2003) Nested spatial biodiversity patterns in a New Zealand forest and pasture soil. *Soil Biology and Biochemistry* 35, 339–342.

Faulkner, L.R. and Darling, H.H. (1961) Pathological histology, hosts and culture of the potato rot nematode. *Phytopathology* 51, 778–786.

Ferris, H. and Bongers, T. (2006) Nematode indicators of organic enrichment. *Journal of Nematology* 38, 3–12.

Ferris, H. and Matute, M. (2003) Structural and functional succession in the nematode fauna of a soil food web. *Applied Soil Ecology* 23, 93–110.

Ferris, H., Mullens, T.A. and Foord, K.E. (1990) Stability and characteristics of spatial description parameters for nematode populations. *Journal of Nematology* 22, 427–439.

Ferris, H., Lau, S. and Venette, R. (1995) Population energetics of bacterial-feeding nematodes: respiration and metabolic rates based on carbon dioxide production. *Soil Biology and Biochemistry* 27, 319–330.

Ferris, H., Venette, R.C. and Lau, S.S. (1996) Dynamics of nematode communities in tomatoes grown in conventional and organic farming systems, and their impact on soil fertility. *Applied Soil Ecology* 3, 161–175.

Ferris, H., Venette, R.C. and Lau, S.S. (1997) Population energetics of bacterial-feeding nematodes: carbon and nitrogen budgets. *Soil Biology and Biochemistry* 29, 1183–1194.

Ferris, H., Venette, R.C., van der Meulen, H.R. and Lau, S.S. (1998) Nitrogen mineralization by bacterial-feeding nematodes: verification and measurement. *Plant and Soil* 203, 159–171.

Ferris, H., Bongers, T. and de Goede, R.G.M. (2001) A framework for soil food web diagnostics: extension of the nematode faunal analysis concept. *Applied Soil Ecology* 18, 13–29.

Ferris, H., Jetter, K.M., Zasada, I.A., Chitambar, J.J., Venette, R.C. *et al.* (2003) Risk assessment of plant-parasitic nematodes. In: Sumner, D. (ed.) *Exotic Pests and Diseases: Biology and Economics for Biosecurity*. Iowa State University Press, Ames, Iowa, pp. 99–119.

Ferris, H., McKenry, M.V., Jaffee, B.A., Anderson, C.E. and Juurma, A. (2004a) Population characteristics and dosage trajectory analysis for *Mesocriconema xenoplax* in California *Prunus* orchards. *Journal of Nematology* 36, 505–516.

Ferris, H., Bongers, T. and de Goede, R.G.M. (2004b) Nematode faunal analysis to assess food web enrichment and connectance. *Nematology Monographs and Perspectives* 2, 503–510.

Ferris, H., Venette, R.C. and Scow, K.M. (2004c) Soil management to enhance bacterivore and fungivore nematode populations and their nitrogen mineralization function. *Applied Soil Ecology* 24, 19–35.

Fisher, R. (2003) Spatial and temporal variations in nematode assemblages in tropical seagrass sediments. *Hydrobiologia* 493, 43–63.

Fleeger, J.W., Tita, G., Carman, K.R., Millward, R.N., Moser, E.B. *et al.* (2006) Does bioturbation by a benthic fish modify the effects of sediment contamination on saltmarsh benthic microalgae and meiofauna? *Journal of Experimental Marine Biology and Ecology* 330, 180–194.

Flint, M.L. (ed.) (1985) *Integrated Pest Management for Tomatoes*, 3rd edn. Publication No. 3274. University of California Statewide IPM Project, Division of Agriculture and Natural Resources, Oakland, California.

Folkertsma, R.T., van Koert, P., Rouppe van der Voort, J.N.A.M., de Groot, K.E., Kaemmenga, J.E. *et al.* (2001) The effects of founding events and agricultural practices on the genetic structure of three metapopulations of *Globodera pallida*. *Phytopathology* 91, 753–758.

Fritz, K.M., Evans, M.A. and Feminella, J.W. (2004) Factors affecting biomass allocation in the riverine macrophyte *Justicia americana*. *Aquatic Botany* 78, 279–288.

Fu, S., Ferris, H., Brown, D. and Plant, R. (2005) Does the positive feedback effect of nematodes on the biomass and activity of their bacteria prey vary with nematode species and population size? *Soil Biology and Biochemistry* 37, 1979–1987.

Gallucci, F., Steyaert, M. and Moens, T. (2005) Can field distributions of marine predacious nematodes be explained by sediment constraints on their foraging success? *Marine Ecology Progress Series* 304, 167–178.

Geisen, S., Snoek, L.B., ten Hooven, F.C., Duyts, H., Kostenko, O. *et al.* (2018) Integrating quantitative morphological and qualitative molecular methods to analyse soil nematode community responses to plant range expansion. *Methods in Ecology and Evolution* 9, 1366–1378.

Gerlach, S.A. (1977) Attraction to decaying organisms as a possible cause for patchy distribution of nematodes in a Bermuda beach. *Ophelia* 16, 151–165.

Ghafouri, S. and McGhee, J.D. (2007) Bacterial residence time in the intestine of *Caenorhabditis elegans*. *Nematology* 9, 87–91.

Giere, O., Windoffer, R. and Southward, E.C. (1995) The bacterial endosymbiosis of the gutless nematode, *Astomonema southwardorum* – structural aspects. *Journal of the Marine Biological Association of the United Kingdom* 75, 153–164.

Goodell, P. and Ferris, H. (1980) Plant-parasitic nematode distributions in an alfalfa field. *Journal of Nematology* 12, 136–141.

Green, C.D. and Plumb, S.C. (1970) The interrelationships of some *Heterodera* spp. indicated by the specificity of the males attractants emitted by their females. *Nematologica* 16, 39–46.

Griffiths, B.S., Ritz, K., Bardgett, R.D., Cook, R., Christensen, S. *et al.* (2000) Ecosystem response of pasture soil communities to fumigation-induced microbial diversity reductions: an examination of the biodiversity–ecosystem function relationship. *Oikos* 90, 279–294.

Griffiths, B.S., Neilson, R. and Bengough, A.G. (2003) Soil factors determined nematode community composition in a two year pot experiment. *Nematology* 5, 889–897.

Gunapala, N. and Scow, K.M. (1998) Dynamics of soil microbial biomass and activity in conventional and organic farming systems. *Soil Biology and Biochemistry* 30, 805–816.

Gutierrez, A. and Sommer, R.J. (2004) Evolution of dnmt-2 and mbd2-like genes in the free-living nematodes *Pristionchus pacificus*, *Caenorhabditis elegans* and *Caenorhabditis briggsae*. *Nucleic Acids Research* 323, 6388–6396.

Hamels, I., Moens, T., Muylaert, K. and Vyverman, W. (2001) Trophic interactions between ciliates and nematodes from an intertidal flat. *Aquatic Microbial Ecology* 26, 61–72.

Haverkort, A.J., Boerma, M., Velma, R. and Van de Waart, M. (1992) The influence of drought and cyst nematodes on potato growth. 4. Effects on crop growth under field conditions of 4 cultivars differing in tolerance. *Netherlands Journal of Plant Pathology* 98, 179–191.

Heininger, P., Höss, S., Claus, E., Pelzer, J. and Traunspurger, W. (2007) Nematode communities in contaminated river sediments. *Environmental Pollution* 146, 64–76.

Heip, C., Vincx, M. and Vranken, G. (1985) The ecology of marine nematodes. *Oceanography and Marine Biology: An Annual Review* 23, 399–489.

Hellwig-Armonies, M., Armonies, W. and Lorenzen, S. (1991) The diet of *Enoplus brevis* (Nematoda) in a supralittoral salt marsh of the North Sea. *Helgoländer Meeresuntersuchungen* 45, 357–372.

Hentschel, U., Berger, E.C., Bright, M., Felbeck, H. and Ott, J.R. (1999) Metabolism of nitrogen and sulfur in ectosymbiotic bacteria of marine nematodes (Nematoda, Stilbonematinae). *Marine Ecology Progress Series* 183, 149–158.

Hewitt, W.B., Raski, D.J. and Goheen, A.C. (1958) Nematode vector of soil-borne fanleaf virus of grapevines. *Phytopathology* 48, 586–595.

Höckelmann, C., Moens, T. and Jüttner, F. (2004) Odor compounds from cyanobacterial biofilms acting as attractants and repellents for free-living nematodes. *Limnology and Oceanography* 49, 1809–1819.

Hodda, M. (2022) Phylum Nematoda: feeding habits for all valid genera using a new, universal scheme encompassing the entire phylum, with descriptions of morphological characteristics of the stoma, a key, and discussion of the evidence for trophic relationships. *Zootaxa* 5114, 318–451. DOI: 10.11646/ZOOTAXA.5114.1.3.

Hol, W.H.G., de la Peña, E., Moens, M. and Cook, R. (2007) Interaction between a fungal endophyte and root herbivores of *Ammophila arenaria*. *Basic and Applied Ecology* 8, 500–509.

Horiuchi, J., Pritiviraj, B., Bais, H.P., Kimball, P.A. and Vivanco, J.M. (2005) Soil nematodes mediate positive interactions between legume plants and rhizobium bacteria. *Planta* 222, 848–857.

Höss, S., Traunspurger, W., Severin, G.W., Juttner, I., Pfister, G. *et al.* (2004) Influence of 4-nonylphenol on the structure of nematode communities in freshwater microcosms. *Environmental Toxicology and Chemistry* 23, 1268–1275.

Höss, S., Claus, E., Von der Ohe, P.C., Brinke, M., Güde, H. *et al.* (2011) Nematode species at risk – a metric to assess pollution in soft sediments of freshwaters. *Environment International* 37, 940–949. DOI: 10.1016/j.envint.2011.03.013.

Höss, S., Heininger, P., Claus, E., Möhlenkamp, C., Brinke, M. *et al.* (2016) Validating the Nema-SPEAR[%]-index for assessing sediment quality regarding chemical-induced effects on benthic communities in rivers. *Ecological Indicators* 73, 52–60. DOI: 10.1016/j.ecolind.2016.09.022.

Hussey, R.S. and Roncadori, R.W. (1981) Influence of *Aphelenchus avenae* on vesicular-arbuscular endomycorrhizal growth response in cotton. *Journal of Nematology* 13, 48–52.

Hyman, L.H. (1951) *The Invertebrates.* Volume III: *Acanthocephala, Aschelminthes, and Entoprocta. The Pseudocoelomate Bilateria*. McGraw-Hill, New York.

Ingham, R.E., Trofymow, J.A., Ingham, E.R. and Coleman, D.C. (1985) Interactions of bacteria, fungi and their nematode grazers on nutrient cycling and plant growth. *Ecological Monographs* 55, 119–140.

Jackson, L.E., Bowles, T.M., Ferris, H., Margenot, A., Hollander, A. *et al.* (2019) Plant and soil biodiversity across the borders between arable and forest ecosystems in a Mediterranean landscape. *Applied Soil Ecology* 136, 122–138.

Jatala, P., Jensen, H.J. and Russell, S.A. (1974) *Pristionchus lheritieri* as a carrier of *Rhizobium japonicum*. *Journal of Nematology* 6, 130–131.

Jones, W.J., Won, Y.-J., Maas, P.A.Y., Smith, P.J., Lutz, R.A. *et al.* (2006) Evolution of habitat use by deep-sea mussels. *Marine Biology* 148, 841–851.

Jørgensen, S.E. and Mejer, H. (1979) A holistic approach to ecological modelling. *Ecological Modelling* 7, 169–189.

Kardol, P., Bezemer, T.M., van der Wal, A. and Van der Putten, W.H. (2005) Successional trajectories of soil nematode and plant communities in a chronosequence of ex-arable lands. *Biological Conservation* 126, 317–327.

Kaya, H.K. and Stock, S.P. (1997) Techniques in insect nematology. In: Lacey, L.A. (ed.) *Manual of Techniques in Insect Pathology*. Academic Press, London, pp. 281–324.

Klekowski, R.Z., Wasilewska, L. and Paplinska, E. (1972) Oxygen consumption by soil-inhabiting nematodes. *Nematologica* 18, 391–403.

Klingler, J. (1965) On the orientation of plant nematodes and of some other soil animals. *Nematologica* 11, 4–18.

Korthals, G.W., Van de Ende, A., Van Megen, H., Lexmond, T.M., Kammenga, J.E. *et al.* (1996) Short-term effects of cadmium, copper, nickel, and zinc on soil nematodes from different feeding and life-history strategy groups. *Applied Soil Ecology* 4, 107–117.

Kulkarni, P., Wilschut, R.A., Verhoeven, J.F.K., Van der Putten, H.W. and Garbeva, P. (2018) LAESI mass spectrometry imaging as a tool to differentiate the root metabolome of native and range-expanding plant species. *Planta* 248, 1515–1523. DOI: 10.1101/322867, DOI: 10.1007/s00425-018-2989-4.

Laakso, J. and Setälä, H. (1999) Sensitivity of primary production to changes in the architecture of belowground food webs. *Oikos* 87, 57–64.

Lacey, L.A., Frutos, R., Kaya, H.K. and Vail, P. (2001) Insect pathogens as biological control agents: do they have a future? *Biological Control* 21, 230–248.

Lambshead, P.J.D. (1993) Recent developments in marine benthic biodiversity research. *Oceanis* 19, 5–24.

Lambshead, P.J.D., Tietjen, J., Ferrero, T. and Jensen, P. (2000) Latitudinal diversity gradients in the deep sea with special reference to North Atlantic nematodes. *Marine Ecology Progress Series* 194, 159–167.

Lambshead, P.J.D., Brown, C.J., Ferrero, T.J., Mitchell, N.J., Smith, C.R. *et al.* (2002) Latitudinal diversity patterns of deep-sea marine nematodes and organic fluxes: a test from the central equatorial Pacific. *Marine Ecology Progress Series* 236, 129–135.

Lownsbery, B.F., English, H., Moody, E.H. and Schick, F.I. (1973) *Criconemoides xenoplax* experimentally associated with a disease of peach. *Phytopathology* 63, 994–997.

Maas, P.W.Th., Oremus, P.A.I. and Otten, H. (1983) Nematodes (*Longidorus* sp. and *Tylenchorhynchus microphasmis* Loof) in growth and nodulation of sea buckthorn (*Hippophaë rhamnoides* L.). *Plant and Soil* 73, 141–147.

Magurran, A.E. (1988) *Ecological Diversity and Its Measurement*. Cambridge University Press, Cambridge.

Manohar, M., Tenjo-Castano, F., Chen, S., Zhang, Y.K., Kumari, A. *et al.* (2020) Plant metabolism of nematode pheromones mediates plant–nematode interactions. *Nature Communications* 11, 208. DOI: 10.1038/s41467-019-14104-2.

Mauchline, T.H., Kerry, B.R. and Hirsch, P.R. (2004) The biocontrol fungus *Pochonia chlamydosporia* shows nematode host preference at the intraspecific level. *Mycological Research* 108, 161–169.

McCallum, M.E. and Dusenbery, D.B. (1992) Computer tracking as a behavioral GC detector: nematode responses to vapor of host roots. *Journal of Chemical Ecology* 18, 585–592.

McKenna, P.B. (1997) Checklist of helminth parasites of terrestrial mammals in New Zealand. *New Zealand Journal of Zoology* 24, 277–290.

Mevenkamp, L., Van Campenhout, J. and Vanreusel, A. (2016) Experimental evidence for selective settlement of meiofauna from two distinct environments after sediment suspension. *Journal of Experimental Marine Biology and Ecology* 474, 195–203.

Michiels, I.C. and Traunspurger, W. (2005) Seasonal variation of biodiversity and assemblage structure in freshwater nematodes. *Archiv für Hydrobiologie* 163, 183–194.

Millennium Ecosystem Assessment (2005) *Ecosystems and Human Well-being: Current State and Trends. Findings of the Condition and Trends Working Group*. Island Press, Washington, DC.

Minoshima, H., Jackson, L.E., Cavagnaro, T.R. and Ferris, H. (2007) Short-term fates of ^{13}C-depleted cowpea shoots in no-till and standard tillage soils. *Soil Science Society of America Journal* 71, 1859–1866.

Moens, T. and Vincx, M. (1997) Observations on the feeding ecology of estuarine nematodes. *Journal of the Marine Biological Association of the United Kingdom* 77, 211–227.

Moens, T., Verbeeck, L., De Maeyer, A., Swings, J. and Vincx, M. (1999) Selective attraction of marine bacterivorous nematodes to their bacterial food. *Marine Ecology Progress Series* 176, 165–178.

Moens, T., Herman, P.M.J., Verbeeck, L., Steyaert, M. and Vincx, M. (2000) Predation rates and prey selectivity in two predacious estuarine nematodes. *Marine Ecology Progress Series* 205, 185–193.

Moens, T., Luyten, C., Middelburg, J.J., Herman, P.M.J. and Vincx, M. (2002) Tracing organic matter sources of estuarine tidal flat nematodes with stable carbon isotopes. *Marine Ecology Progress Series* 234, 127–137.

Moens, T., Yeates, G.W. and De Ley, P. (2004) Use of carbon and energy sources by nematodes. *Nematology Monographs and Perspectives* 2, 529–545.

Moens, T., Bouillon, S. and Gallucci, F. (2005a) Dual stable isotope abundances unravel trophic position of estuarine nematodes. *Journal of the Marine Biological Association of the United Kingdom* 85, 1401–1407.

Moens, T., dos Santos, G.A.P., Thompson, F., Swings, J., Fonsêca-Genevois, V. *et al.* (2005b) Do nematode mucus secretions affect bacterial growth? *Aquatic Microbial Ecology* 40, 77–83.

Moens, T., Bergtold, M. and Traunspurger, W. (2006) Feeding ecology of free-living benthic nematodes. In: Eyualem, A., Andrássy, I. and Traunspurger, W. (eds) *Freshwater Nematodes: Ecology and Taxonomy*. CAB International, Wallingford, UK, pp. 105–131.

Montagna, P.A. (1995) Rates of metazoan meiofaunal microbivory: a review. *Vie at Milieu* 45, 1–9.

Morriën, E., Hannula, E.E., Snoek, L.B., Helmsing, N., Zweers, H. *et al.* (2017) Composition of soil biodiversity networks, functional changes in nutrient dynamics, and consequences for vegetation succession. *Nature Communications* 8, 14349. DOI: 10.1038/ncomms14349.

Morris, C.A., Vlassoff, A., Bisset, S.A., Baker, R.L., Watson, T.G. *et al.* (2000) Continued selection of Romney sheep for resistance or susceptibility to nematode infection: estimates of direct and correlated responses. *Animal Science* 70, 17–27.

Müller, F. and Lenz, R. (2006) Ecological indicators: theoretical fundamentals of consistent application in environmental management. *Ecological Indicators* 6, 1–5.

Murfin, K.E., Dillman, A.R., Foster, J.M., Bulgheresi, S., Slatko, B.E. *et al.* (2012) Nematode–bacterium symbioses – cooperation and conflict revealed in the Omics age. *Biological Bulletin* 223, 85–102.

Musat, N., Giere, O., Gieseke, A., Thiermann, F., Amann, R. *et al.* (2007) Molecular and morphological characterization of the association between bacterial endosymbionts and the marine nematode *Astomonema* sp. from the Bahamas. *Environmental Microbiology* 9, 1345–1353.

Nadler, S.A. and Perez-Ponce de Leon, G. (2011) Integrating molecular and morphological approaches for characterizing parasite cryptic species: implications for parasitology. *Parasitology* 138(Sp. Iss. 13), 1688–1709.

Needham, T. (1744) A letter concerning chalky tubulous concretions, called malm: with some microscopical observations on the farina of the red lily, and of worms discovered in smutty corn. *Philosophical Transactions of the Royal Society of London* 42, 634–641.

Nehring, S. (1993) Tube-dwelling meiofauna in marine sediments. *Internationale Revue der Gesamten Hydriobiologie* 78, 521–534.

Nehring, S., Jensen, P. and Lorenzen, S. (1990) Tube-dwelling nematodes: tube construction and possible ecological effects on sediment–water interfaces. *Marine Ecology Progress Series* 64, 123–128.

Northam, F.E. and Orr, C.C. (1982) Effects of a nematode on biomass and density of silverleaf nightshade. *Journal of Range Management* 35, 536–537.
Odum, H.T. (1988) Self organization, transformity, and information. *Science* 242, 1132–1139.
Okada, H. and Ferris, H. (2001) Temperature effects on growth and nitrogen mineralization of fungi and fungal-feeding nematodes. *Plant and Soil* 234, 253–262.
Okada, H. and Kadota, I. (2003) Host status of 10 fungal isolates for two nematode species, *Filenchus misellus* and *Aphelenchus avenae*. *Soil Biology and Biochemistry* 35, 1601–1607.
Okada, H., Harada, H., Tsukiboshi, T. and Araki, M. (2005) Characteristics of *Tylencholaimus parvus* (Nematoda: Dorylaimida) as a fungivorus nematode. *Nematology* 7, 843–849.
Oremus, P.A.I. and Otten, H. (1981) Factors affecting growth and nodulation of *Hippophaë rhamnoides* L. ssp. *rhamnoides* in soils from two successional stages of dune formation. *Plant and Soil* 63, 317–331.
Pantone, D.J., Williams, W.A. and Maggenti, A.R. (1989a) An alternative approach for evaluating the efficacy of potential biocontrol agents of weeds. 1. Inverse linear model. *Weed Science* 37, 771–777.
Pantone, D.J., Williams, W.A. and Maggenti, A.R. (1989b) An alternative approach for evaluating the efficacy of potential biocontrol agents of weeds. 2. Path analysis. *Weed Science* 37, 778–783.
Perry, R.N. and Moens, M. (eds) (2006) *Plant Nematology*. CAB International, Wallingford, UK.
Peters, L. and Traunspurger, W. (2005) Species distribution of free-living nematodes and other meiofauna in littoral periphyton communities of lakes. *Nematology* 7, 267–280.
Petersen, J.J. (1985) Nematode parasites. *American Mosquito Control Association Bulletin* 6, 110–122.
Phillips, D.A., Ferris, H., Cook, D.R. and Strong, D.R. (2003) Molecular control points in rhizosphere food webs. *Ecology* 84, 816–826.
Pielou, E.C. (1975) *Ecological Diversity*. Wiley, New York.
Pineda, A., Soler, R., Pozo, M.J., Rasmann, S and Turlings, T.C.I. (2015) Above–belowground interactions involving plants, microbes and insects. *Frontiers in Plant Science* 6, 318.
Piśkiewicz, A.M., Duyts, H., Berg, M.P., Costa, S.R. and Van der Putten, W.H. (2007) Soil microorganisms control plant ectoparasitic nematodes in natural coastal foredunes. *Oecologia* 152, 505–514.
Pline, M. and Dusenbery, D.B. (1987) Responses of the plant-parasitic nematode *Meloidogyne incognita* to carbon dioxide determined by video camera-computer tracking. *Journal of Chemical Ecology* 13, 1617–1624.
Polz, M.F., Ott, J.A., Bright, M. and Cavanaugh, C.M. (2000) When bacteria hitch a ride. *American Society of Microbiology News* 66, 531–539.
Postma-Blaauw, M.B., de Vries, F.T., de Goede, R.G.M., Bloem, J., Faber, J.H. *et al.* (2005) Within-trophic group interactions of bacterivorous nematode species and their effects on the bacterial community and nitrogen mineralization. *Oecologia* 142, 428–439.
Powell, N.T. (1971) Interactions between nematodes and fungi in disease complexes. *Annual Review of Phytopathology* 9, 253–273.
Proctor, D.L.C. (1984) Towards a biogeography of free-living soil nematodes. I. Changing species richness, diversity and densities with changing latitude. *Journal of Biogeography* 11, 103–117.
Proctor, D.L.C. (1990) Global overview of the functional roles of soil-living nematodes in terrestrial communities and ecosystems. *Journal of Nematology* 22, 1–7.
Prot, J.C. and Ferris, H. (1992) Sampling approaches for extensive surveys in nematology. *Journal of Nematology* 24, 757–764.
Quist, C.W., Gort, G., Mooijman, P., Brus, D.J., van den Elsen, S. *et al.* (2019) Spatial distribution of soil nematodes relates to soil organic matter and life strategy. *Soil Biology and Biochemistry* 136, 107542. DOI: 10.1016/j.soilbio.2019.107542.
Rasmann, S., Ali, J.G., Helder, J. and Van der Putten, W.H. (2012) Ecology and evolution of soil nematode chemotaxis. *Journal of Chemical Ecology* 38, 615–628. DOI: 10.1007/s10886-012-0118-6.
Richard-Lenoble, D., Chandenier, J. and Gaxotte, P. (2003) Ivermectin and filiariasis. *Fundamental and Clinical Pharmacology* 17, 199–203.
Ridall, A and Ingels, J. (2021) Suitability of free-living marine nematodes as bioindicators: status and future considerations. *Frontiers in Marine Science* 8, 685327. DOI: 10.3389/fmars.2021.685327.
Riemann, F. and Helmke, E. (2002) Symbiotic relations of sediment-agglutinating nematodes and bacteria in detrital habitats: the enzyme-sharing concept. *Marine Ecology* 23, 93–113.
Riemann, F. and Schrage, M. (1978) The mucus-trap hypothesis on feeding of aquatic nematodes and implications for biodegradation and sediment texture. *Oecologia* 34, 75–88.
Riemann, F. and Schrage, M. (1988) Carbon dioxide as an attractant for the free-living marine nematode *Adoncholaimus thalassophygas*. *Marine Biology* 98, 81–85.

Riera, P., Richard, P., Grémare, A. and Blanchard, G. (1996) Food source of intertidal nematodes in the Bay of Marennes-Oléron (France), as determined by dual stable isotope analysis. *Marine Ecology Progress Series* 142, 303–309.

Riffle, J.W. (1967) Effect of an *Aphelenchoides* species on the growth of a mycorrhizal and pseudomycorrhizal fungus. *Phytopathology* 57, 541–544.

Riffle, J.W. (1975) Two *Aphelenchoides* species suppress formation of *Suillus granulatus* ectomycorrhizae with *Pinus ponderosa* seedlings. *Plant Disease Reporter* 59, 951–955.

Robertson, G.P. and Freckman, D.W. (1995) The spatial distribution of nematode trophic groups across a cultivated ecosystem. *Ecology* 76, 1425–1433.

Robinson, A.F. (1995) Optimal release rates for attracting *Meloidogyne incognita*, *Rotylenchulus reniformis*, and other nematodes to carbon-dioxide in sand. *Journal of Nematology* 27, 42–50.

Rodger, S., Griffiths, B.S., McNicol, J.W., Wheatley, R.W. and Young, I.M. (2004) The impact of bacterial diet on the migration and navigation of *Caenorhabditis elegans*. *Microbial Ecology* 28, 358–365.

Ruess, L. and, Dighton, J. (1996) Cultural studies on soil nematodes and their fungal hosts. *Nematologica* 42, 330–346.

Rühm, R., Dietsche, E., Harloff, H.J., Lieb, M., Franke, S. *et al.* (2003) Characterization and partial purification of a white mustard kairomone that attracts the beet cyst nematode, *Heterodera schachtii*. *Nematology* 5, 17–22.

Salinas, K.A., Edenborn, S.L., Sextone, A.J. and Kotcon, J.B. (2007) Bacterial preferences of the bacterivorous soil nematode *Cephalobus brevicauda* (Cephalobidae): effect of bacterial size and type. *Pedobiologia* 51, 55–64.

Sánchez-Moreno, S., Minoshima, H., Ferris, H. and Jackson, L.E. (2006) Linking soil properties and nematode community composition: effects of soil management on soil food webs. *Nematology* 8, 703–715.

Sánchez-Moreno, S., Smukler, S., Ferris, H., O'Green, A.T. and Jackson, L.E. (2008) Nematode diversity, food web condition, and chemical and physical properties in different soil habitats of an organic farm. *Biology and Fertility of Soils* 44, 727–744.

Sankaralingam, A. and McGawley, E.C. (1994) Influence of *Rhizoctonia solani* on egg hatching and infectivity of *Rotylenchus reniformis*. *Journal of Nematology* 26, 486–491.

Schacht, H. (1859) Über einige Feinde der Rübenfelder. *Zeitschrift Vereines Rübenzucker – Industrie Zollverein* 9, 175–179.

Schenk, J., Höss, S., Brinke, M., Kleinbölting, N., Brüchner-Hüttemann, H. *et al.* (2020) Nematodes as bioindicators of polluted sediments using metabarcoding and microscopic taxonomy. *Environment International* 143, 105922. DOI: 10.1016/j.envint.2020.105922.

Schratzberger, M., Bolam, S., Whomersley, P. and Warr, K. (2006) Differential response of nematode colonist communities to the intertidal placement of dredged material. *Journal of Experimental Marine Biology and Ecology* 334, 244–255.

Schratzberger, M., Warr, K. and Rogers, S.I. (2007) Functional diversity of nematode communities in the southwestern North Sea. *Marine Environmental Research* 63, 368–389.

Schratzberger, M., Holterman, M., Van Oevelen, D. and Helter, J. (2019) A worm's world: ecological flexibility pays off for free-living nematodes in sediments and soils. *BioScience* 69, 867–876.

Schroeder, J., Thomas, S.H. and Murray, L.W. (2005) Impacts of crop pests on weeds and weed–crop interactions. *Weed Science* 53, 918–922.

Siddiqi, M.R. (2000) *Tylenchida: Parasites of Plants and Insects*, 2nd edn. CAB International, Wallingford, UK.

Sikder, M.M. and Vestergard, M. (2020) Impacts of root metabolites on soil nematodes. *Frontiers in Plant Science* 10, 1792. DOI: 10.3389/fpls.2019.01792.

Sikora, R.W., Desaeger, J. and Molendijk, L. (eds) (2021) *Integrated Nematode Management: State-of-the-Art and Visions for the Future*. CAB International, Wallingford, UK.

Small, R.W. (1987) A review of the prey of predatory soil nematodes. *Pedobiologia* 30, 179–206.

Song, D., Pan, K., Tariq, A., Sun, F., Li, Z. *et al.* (2017) Large-scale patterns of distribution and diversity of terrestrial nematodes. *Applied Soil Ecology* 114, 161–169.

Spiegel, Y., Burrows, P.M. and Bar-Eyal, M. (2003) A chemo attractant in onion root exudates recognized by *Ditylenchus dipsaci* in laboratory bioassay. *Phytopathology* 93, 127–132.

Stanton, N.L. (1988) The underground in grasslands. *Annual Review of Ecology and Systematics* 19, 573–589.

Steyaert, M., Herman, P.M.J., Moens, T., Widdows, J. and Vincx, M. (2001) Tidal migration of nematodes on an estuarine tidal flat (The Molenplaat, Schelde Estuary, SW Netherlands). *Marine Ecology Progress Series* 224, 299–304.

Subbotin, S.A. and Chitambar, J.J. (eds) (2018) *Plant Parasitic Nematodes in Sustainable Agriculture of North America* (2 volumes). Springer, Cham, Switzerland.

Sudhaus, W. (1976) Vergleichende Untersuchungen zur Phylogenie, Systematik. Ökologie, Biologie und Ethologie der Rhabditidae (Nematoda). *Zoologia* 125, 1–229.

Suess, E. (1980) Particulate organic carbon flux in the oceans – surface productivity and oxygen utilization. *Nature* 288, 260–263.

Sukee, T., Beveridge, I., Koehler, A.V., Hall, R.S., Gasser, R.B. *et al.* (2022) Phylogenetic relationships of the strongyloid nematodes of Australasian marsupials based on mitochondrial protein sequences. *Animals* 12, 2900. DOI: 10.3390/ani12212900.

Sutherland, J.R. and Fortin, J.A. (1968) Effect of the nematode *Aphelenchus avenae* on some ectotrophic, mycorrhizal fungi and on a red pine mycorrhizal relationship. *Phytopathology* 58, 519–523.

Temple, S.R. (1993) *A Comparison of Conventional, Low Input and Organic Farming Systems: The Transition Phase and Long-Term Viability*. Progress Report on Project LW-89-18.1 to USDA-SARE. United States Department of Agriculture – Sustainable Agriculture Research and Education, College Park, Maryland.

Tenuta, M. and Ferris, H. (2004) Relationship between nematode life-history classification and sensitivity to stressors: ionic and osmotic effects of nitrogenous solutions. *Journal of Nematology* 36, 85–94.

Thiermann, F., Vismann, B. and Giere, O. (2000) Sulphide tolerance of the marine nematode *Oncholaimus campylocercus* – a result of internal sulphur formation. *Marine Ecology Progress Series* 193, 251–259.

Thomas, S.H., Schroeder, J. and Murray, L.W. (2005) The role of weeds in nematode management. *Weed Science* 53, 923–928.

Tietjen, J.H., Deming, J.W., Rowe, G.T., Macko, S. and Wilke, R.J. (1989) Meiobenthos of the Hatteras Abyssal Plain and Puerto Rico Trench: abundance, biomass and associations with bacteria and particulate fluxes. *Deep-Sea Research* 36, 1567–1577.

Timper, P. (2014) Conserving and enhancing biological control of nematodes. *Journal of Nematology* 46, 75–89.

Togashi, K. and Shigesada, N. (2006) Spread of the pinewood nematode vectored by the Japanese pine sawyer: modeling and analytical approaches. *Population Ecology* 48, 271–283.

Townshend, J.L. and Blackith, R.E. (1975) Fungal diet and the morphometric relationships in *Aphelenchus avenae*. *Nematologica* 11, 395–408.

Traunspurger, W. (2000) The ecology of lotic nematodes. *Freshwater Biology* 44, 29–45.

Traunspurger, W., Bergtold, M. and Goedkoop, W. (1997) The effects of nematodes on bacterial activity and abundance in a freshwater sediment. *Oecologia* 112, 118–122.

Umesh, K.C., Ferris, H. and Bayer, D. (1994) Competition between plant-parasitic nematodes *Pratylenchus neglectus* and *Meloidogyne chitwoodi*. *Journal of Nematology* 26, 286–295.

Van den Hoogen, J., Geisen, S., Routh, D., Ferris, H., Traunspurger, W. *et al.* (2019) Soil nematode abundance and functional group composition at a global scale. *Nature* 572, 194–198.

Van der Putten, W.H. (2003) Plant defense belowground and spatio-temporal processes in natural vegetation. *Ecology* 84, 2269–2280.

Van der Putten, W.H. and Van der Stoel, C.D. (1998) Plant parasitic nematodes on spatio-temporal variation in natural vegetation. *Applied Soil Ecology* 10, 253–262.

Van der Putten, W.H., Van Dijk, C. and Peters, B.A.M. (1993) Plant-specific soil-borne diseases contribute to succession in foredune vegetation. *Nature* 362, 53–56.

Van der Putten, W.H., Yeates, G.W., Duyts, H., Schreck Reis, C. and Karssen G. (2005) Invasive plants and their escape from root herbivory: a worldwide comparison of the root-feeding nematode communities of the dune grass *Ammophila arenaria* in natural and introduced ranges. *Biological Invasions* 7, 733–746.

Van der Putten, W.H., Cook, R., Costa, S., Davies, K.G., Fargette, M. *et al.* (2006) Nematode interactions in nature: models for sustainable control of nematode pests of crop plants? *Advances in Agronomy* 89, 227–260.

Van Gaever, S., Vanreusel, A., Hughes, J.A., Bett, B.J. and Kiriakoulakis, K. (2004) The macro- and micro-scale patchiness of meiobenthos associated with the Darwin Mounds (north-east Atlantic). *Journal of the Marine Biological Association of the United Kingdom* 84, 547–556.

Van Gaever, S., Moodley, L., De Beer, D. and Vanreusel, A. (2006) Meiobenthos at the Arctic Håkon Mosby Mud Volcano, with a parental caring nematode thriving in sulphide-rich sediments. *Marine Ecology Progress Series* 321, 143–155.

Van Oevelen, D., Moodley, L., Soetaert, K. and Middelburg, J.J. (2006) The trophic significance of bacterial carbon in a marine intertidal sediment: results of an *in situ* stable isotope labeling study. *Limnology and Oceanography* 51, 2349–2359.

Van Ruijven, J., De Deyn, G.B., Raaijmakers, C.E., Berendse, F. and Van der Putten, W.H. (2005) Interactions between spatially separated herbivores indirectly alter plant diversity. *Ecology Letters* 8, 30–37.

Vanfleteren, J.R. (1980) Nematodes as nutritional models. In: Zuckerman, B.M. (ed.) *Nematodes as Biological Models*, Vol. 2. Academic Press, New York, pp. 47–79.

Venette, R.C. and Ferris, H. (1998) Influence of bacterial type and density on population growth of bacterial-feeding nematodes. *Soil Biology and Biochemistry* 30, 949–960.

Verschoor, B.C. (2002) Carbon and nitrogen budgets of plant-feeding nematodes in grasslands of different productivity. *Applied Soil Ecology* 20, 15–25.

Verschoor, B.C., De Goede, R.G.M., De Vries, F.W. and Brussaard, L. (2001) Changes in the composition of the plant-feeding nematode community in grasslands after cessation of fertiliser application. *Applied Soil Ecology* 17, 1–17.

Viglierchio, D.R. (1991) *The World of Nematodes*. David R. Viglierchio, Davis, California.

Viketoft, M., Palmborg, C., Sohlenius, B., Huss-Danell, K. and Bengtsson, J. (2005) Plant species effects on soil nematode communities in experimental grasslands. *Applied Soil Ecology* 30, 90–103.

Wall, D.H. (ed.) (2004) *Sustaining Biodiversity and Ecosystem Services in Soils and Sediments*. Island Press, Washington, DC.

Waller, P.J., Schwan, O., Ljungström, B.-L., Rydzik, A. and Yeates, G.W. (2004) Evaluation of biological control of sheep parasites using *Duddingtonia flagrans* under commercial farming conditions on the island of Gotland, Sweden. *Veterinary Parasitology* 126, 199–315.

Wang, J., Johnson, C.S. and Eisenback, J.D. (1997) Enhanced hatching of *Globodera tabacum solanacearum* juveniles by root exudates of flue-cured tobacco. *Journal of Nematology* 29, 484–490.

Wardle, D.A., Yeates, G.W., Watson, R.N. and Nicholson, K.S. (1995) The detritus food-web and diversity of soil fauna as indicators of disturbance regimes in agro-ecosystems. *Plant and Soil* 170, 35–43.

Wardle, D.A., Bardgett, R.D., Klironomos, J.N., Setala, H., Van der Putten, W.H. *et al.* (2004) Ecological linkages between aboveground and belowground biota. *Science* 304, 1629–1633.

Wilschut, R.A. and Giesen, S. (2021) Nematodes as drivers of plant performance in natural systems. *Trends in Plant Science* 26, 237–247.

Wilschut, R.A., Silva, J.C.P., Garbeva, P.V. and Van der Putten, W.H. (2017) Belowground plant–herbivore interactions vary among climate-driven range-expanding plant species with different degrees of novel chemistry. *Frontiers in Plant Science* 8, 1861. DOI: 10.3389/fpls.2017.01861.

Wilson, M.J. and Grewal, P.S. (2005) Biology, production and formulation of slug-parasitic nematodes. In: Grewal, P.S., Ehlers, R.-U. and Shapiro-Ilan, D.I. (eds) *Nematodes as Biocontrol Agents*. CAB International, Wallingford, UK, pp. 421–429.

Wubben, M.J.E., Su, H., Rodermel, S.R. and Baum, T.J. (2001) Susceptibility to the sugar beet cyst nematode is modulated by ethylene signal transduction in *Arabidopsis thaliana*. *Molecular Plant-Microbe Interactions* 14, 1206–1212.

Wubs, E.R.J., Van der Putten, W.H., Mortimer, S.R., Korthals, G.W., Duyts, H. *et al.* (2019) Single introductions of soil biota and plants generate long-term legacies in soil and plant community assembly. *Ecology Letters* 22, 1145–1151.

Yeates, G.W. (1969a) Predation by *Mononchoides potohikus* (Nematoda: Diplogasteridae) in laboratory culture. *Nematologica* 15, 1–9.

Yeates, G.W. (1969b) A note on the function of the prerectum in *Diplenteron potohikus* (Yeates, 1969) (Nematoda: Diplogasteridae). *Nematologica* 15, 430–431.

Yeates, G.W. (1984) Variation in soil nematode diversity under pasture with soil and year. *Soil Biology and Biochemistry* 16, 95–102.

Yeates, G.W. (2003) Nematodes as soil indicators: functional and biodiversity aspects. *Biology and Fertility of Soils* 37, 199–210.

Yeates, G.W. and Boag, B. (2006) Female size shows similar trends in all clades of the phylum Nematoda. *Nematology* 8, 111–127.

Yeates, G.W., Bongers, T., de Goede, R.G.M., Freckman, D.W. and Georgieva, S.S. (1993) Feeding habits in soil nematode families – an outline for ecologists. *Journal of Nematology* 25, 315–331.

Yeates, G.W., Wardle, D.A. and Watson, R.N. (1999a) Responses of soil nematode populations, community structure and temporal variability to agricultural intensification over a seven-year period. *Soil Biology and Biochemistry* 31, 1721–733.

Yeates, G.W., Saggar, S., Hedley, C.B. and Mercer, C.F. (1999b) Increase in ^{14}C-carbon translocation to the soil microbial biomass when five plant-parasitic nematode infect roots of white clover. *Nematology* 1, 295–300.

Zasada, I.A., Peetz, A., Wade, N., Navarre, R.A. and Ingham, R.E. (2013) Host status of different potato (*Solanum tuberosum*) varieties and hatching in root diffusates of *Globodera ellingtonae*. *Journal of Nematology* 45, 195–201.

Zhang, X., Ferris, H., Mitchell, J. and Liang, W. (2017) Ecosystem services of the soil food web after long-term application of agricultural management practices. *Soil Biology and Biochemistry* 111, 36–43.

Zoon, F.C., Troelstra, S.R. and Maas, P.W.Th. (1993) Ecology of the plant-feeding nematode fauna associated with sea buckthorn (*Hippophaë rhamnoides* L. ssp. *rhamnoides*) in different stages of dune succession. *Fundamental and Applied Nematology* 16, 247–258.

Appendix 1: Outline Classification of the Phylum Nematoda

Molecular approaches have permitted an integrated view of the phylum. While results are generally congruent with traditional relationships, on one hand they so far have not resolved uncertainty among marine groups while on the other hand the order Rhabditida is a monophyletic group equivalent with all of 'Secernentea' (= Phasmidia) in many previous classifications (Blaxter *et al.*, 1998; De Ley and Blaxter, 2002). The classification below, taken from De Ley and Blaxter (2004), indicates the position of many families of interest to readers of this volume.

Class Enoplea

Subclass Enoplia

Order Enoplida: Ironidae, Alaimidae

Order Triplonchida: Diphtherophoridae, Trichodoridae, Tobrilidae, Prismatolaimidae, Tripylidae

Subclass Dorylaimia

Order Dorylaimida: Dorylaimidae, Aporcelaimidae, Longidoridae, Belondiridae, Leptonchidae, Tylencholaimidae, Nygolaimidae

Order Mononchida: Bathyodontidae, Anatonchidae, Mononchidae, Mylonchulidae

Order Isolaimida: Isolaimiidea

Order Mermithida: Mermithidae, Tetradonematidae

Order Trichinellida: Capillariidae, Trichinellidae, Trichuridae

Class Chromadorea

Subclass Chromadoria

Order Desmoscolecida: Desmoscolecidae

Order Chromadorida: Chromadoridae, Ethmolaimidae, Cyatholaimdae

Order Desmodorida: Desmodoridae, Microlaimidae

Order Monhysterida: Monhysteridae, Sphaerolaimidae

Order Araeolaimida: Axonolaimidae, Diplopeltidae

Order Plectida: Leptolaimidae, Bastianiidae, Rhabdolaimidae, Plectidae, Chronogasteridae, Metateratocephalidae, Haliplectidae, Aulolaimidae

Order Rhabditida:

Suborder Spirurina: Thelastomatidae, Oxyuridae, Rhigonematidae, Hethidae, Camallanidae, Hedruridae, Tetrameridae, Filariidae, Ascarididae, Heterakidae

Suborder Tylenchina: Panagrolaimidae, Steinernematidae, Cephalobidae, Aphelenchidae, Criconematidae, Anguinidae, Hoplolaimidae, Meloidogynidae, Tylenchidae, Pratylenchidae, Drilonematidae

Suborder Rhabditina: Bunonematidae, Diplogastridae, Mesorhabditidae, Peloderidae, Rhabditidae, Heterorhabditidae, Trichostrongylidae, Metastrongylidae

2 Nematode Diversity in Marine, Freshwater and Terrestrial Ecosystems

Nabil Majdi[1]*, Michaela Schratzberger[2], Dorota L. Porazinska[3], Walter Traunspurger[4], Jan Henrik Schmidt[5], Jeroen Ingels[6], Stefan Geisen[7] and Mike Hodda[8]

[1]Réserve Naturelle Nationale de la Forêt de la Massane, Laboratoire Arago, Banyuls-sur-Mer, France; [2]Centre for Environment, Fisheries and Aquaculture Science, Lowestoft, UK; [3]Department of Entomology and Nematology, University of Florida, Florida, USA; [4]Animal Ecology, Bielefeld University, Germany; [5]Institute for Epidemiology and Pathogen Diagnostics, Julius Kühn Institute, Braunschweig, Germany; [6]Earth Sciences New Zealand, Wellington, New Zealand; [7]Laboratory of Nematology, Wageningen University and Research, The Netherlands; [8]National Research Collections Australia, CSIRO, Canberra, Australia

Abstract

This chapter explores nematode diversity patterns across biomes and their dependency on prevailing physico-chemical and biotic conditions. Currently, approximately 30,000 nematode species have been described worldwide, but research efforts are not distributed homogeneously across continents, latitudes, habitats and lifestyles (e.g. free-living nematodes are less investigated than are parasitic species). We present the objectives and methodologies used in nematode diversity studies before providing an overview of nematode biodiversity through the lens of factors and processes affecting the distribution of nematodes in marine, freshwater and terrestrial ecosystems. Finally, we highlight similarities and differences between nematode biodiversity patterns across biomes, highlighting the critical importance of nematodes in globally important ecosystem processes. We conclude that more attention should therefore be placed on studying nematode diversity (of species and ecological roles) across biomes to better understand their biogeography as well as their contribution to the delivery of ecosystem services.

2.1 Introduction

Every species, including humans, depends on other species to survive. This vast interdependence means that the variety, quantity and distribution of life on Earth (the biodiversity) is critical for the maintenance of life as we know it. The United Nation's Convention on Biological Diversity, ratified in 1992, defines biological diversity as 'the variability among living organisms from all sources including, *inter alia*, terrestrial, marine, and other aquatic ecosystems and the ecological complexes of which they are part. This includes diversity within species, among species, and of ecosystems' (Josefsson, 2018).

Nematodes are often presented as the numerically most abundant group of multicellular animals on Earth. Nematodes are present globally

*Email: n.majdi@hotmail.fr

 Nematodes as Environmental Indicators: From Theory to Practice, 2nd Edition (T. Kakouli-Duarte *et al.* eds)
DOI: 10.1079/9781800624221.0002

in habitats including sea floors, inland waters, soils, endo- and epiphytic, endo- and epizooic spaces, and even the most inhospitable places including deep hypoxic environments (cf. Nathan Cobb's famous quote: 'In short, if all the matter in the universe except the nematodes were swept away, our world would still be dimly recognizable, and if, as disembodied spirits, we could then investigate it, we should find its mountains, hills, vales, rivers, lakes, and oceans represented by a film of nematodes'). Such a large occupancy of the biosphere begets a great diversity of morphologies (Hodda, 2022a) and a diversity of roles in ecosystems (see Chapter 1, this volume). Nematoda is considered the fifth most-diverse phylum of Animalia (Zhang, 2011) and estimates of global nematode diversity range from 50,000 to 10,000,000 species (Appeltans *et al.*, 2012; Hodda, 2022b) with a prediction of 1,000,000 species but with low data support (Anthony *et al.*, 2023). Among those, *c.*30,000 species are described (Hodda, 2022c), leaving the majority of species yet unknown.

The diversity of nematodes, including that of parasitic and free-living species, is of great scientific and societal interest in an ever more rapidly changing world (Coleman *et al.*, 1992; Wardle *et al.*, 1998; Vanreusel *et al.*, 2010; Gingold *et al.*, 2013; Lazarova *et al.*, 2021; Traunspurger and Majdi, 2021). While the link between nematode diversity, and functions and services relevant to human societies may not be immediately apparent, the critical importance of nematodes has been demonstrated in decomposition processes (e.g. De Mesel *et al.*, 2006; Majdi and Traunspurger, 2017; Taylor *et al.*, 2020), nutrient cycling (e.g. DuPont *et al.*, 2009; Ristau *et al.*, 2013a; Bonaglia *et al.*, 2014), the structure and function of food webs (e.g. Pausch *et al.*, 2016; Menzel *et al.*, 2018; Majdi *et al.*, 2021; Ptatscheck, 2021), and parasitism (e.g. Davis *et al.*, 2004; Wei *et al.*, 2007; Dillman *et al.*, 2012; Pulavarty *et al.*, 2021). Nematode diversity is critical for the maintenance of many ecosystem processes that ultimately affect ecosystem productivity and resilience, and thus human well-being (e.g. Lavelle *et al.*, 2006; Yeates *et al.*, 2009; Majdi *et al.*, 2017; Schratzberger and Ingels, 2018).

Nematodes are ubiquitous but, due to their microscopic size and diverse morphological features contained within relatively simple body plans, they can be laborious to identify. Indeed, pervasive, yet out of plain sight, nematodes are hard to relate to the macroscopic world we are a part of and observe daily. Moreover, the decline in taxonomic expertise suggests that nematology is becoming an increasingly challenging field of science; however, this trend can be reversed by simple measures like facilitative environments, increased student training and greater public exposure (e.g. Cortada *et al.*, 2019). It remains that despite their ubiquitous distribution, and their high taxonomic, functional and ecological diversity, comparing nematode diversity across ecosystems has been challenging. Studies differ regarding the taxa identified, the taxonomic resolution, and the spatial and temporal scales investigated. Furthermore, standardization in terms of the type of survey, extraction, identification methods, and the definitions of taxonomic and functional categories has been difficult to achieve and remains variable among studies (e.g. Hodda, 2022a).

In this chapter we bring together information from the literature on nematode diversity in marine, freshwater and terrestrial ecosystems. We include selected studies to illustrate general trends and hypotheses, rather than provide an exhaustive list of studies:

1. We summarize and discuss the key variables investigated in nematode diversity studies, including different measures of diversity considered, the methods used, and/or the spatial and temporal scales examined.
2. We compare nematode diversity patterns in terrestrial and aquatic ecosystems, as well as across environmental gradients.
3. We identify the main factors and processes affecting nematode diversity within marine, freshwater and terrestrial ecosystems.
4. We synthesize findings from steps 2 and 3 to emphasize the parallels of diversity patterns among ecosystems, and we discuss the importance of nematode diversity for the resilience of ecosystem functions.

2.2 Variables in Studies on Nematode Diversity

2.2.1 How is nematode diversity defined?

In its simplest form, diversity is the number of different taxa (e.g. species, genera, families or other taxonomic units), termed richness. The biological species – defined as the set of potentially interbreeding organisms (Mayr, 1942) – is the basic unit of biology on which evolution operates. However, very few nematode species have been experimentally verified by breeding studies (e.g. De Guiran and Bruguier, 1989; Derycke *et al.*, 2008a,b) and not all species interbreed. Although morphological species (i.e. 'morpho-species', which are defined taxonomically based on morphological features) have been used most frequently in nematode diversity studies, molecular operational taxonomic units (i.e. MOTUs, defined by molecularly distinct clades) are becoming more common in the literature (e.g. Holovachov *et al.*, 2017). The greatest limitation in using morpho-species is logistics: a great deal of expertise and resources is required to identify any substantial number of morpho-species from an environmental sample. Moreover, DNA-based approaches have increasingly questioned the concept of 'what is a species' (De Queiroz, 2007). For example, two specimens present in a sample may have undistinguishable morphologies thus being identified and counted as the same morpho-species. However, if barcoded and if showing significant nucleotide variations in non-variable DNA regions, they may well belong to genetically distinct lineages, thus representing a case of a 'species complex' comprising two or more 'cryptic species'. Cryptic species complexes have been well studied in plant-parasitic nematodes (Palomares-Rius *et al.*, 2014), and although similar studies in free-living nematodes are still scarce, there is mounting evidence of cryptic species complexes in marine (e.g. *Litoditis marina* species complex: Derycke *et al.*, 2016), freshwater (e.g. *Tobrilus gracilis* species complex: Ristau *et al.*, 2013b) and terrestrial habitats (e.g. *Plectus* spp. complex: Borgmeier *et al.*, 2022). Molecular delineation of nematode species holds promises to improve nematode biogeography and biomonitoring because cryptic species may show different ecological preferences and responses to stressors (e.g. Guden *et al.*, 2018, 2021; Monteiro *et al.*, 2018).

The genus level has been commonly used in nematode biogeography and biomonitoring studies (e.g. Wasilewska, 1997; Ridall and Ingels, 2021). Identifying nematodes to genus level requires far less time and taxonomic expertise than identifying them to species level and is therefore more practical for large-scale or more regular surveys. In addition, owing to anthropogenic impacts often being too recent for the evolution of suitably adapted species, higher-level identifications (above species level) can be sufficiently, or even more, informative in impact-monitoring studies (Warwick, 1988; Bongers, 1990; Höss *et al.*, 2017).

The taxonomic level of family is meaningful as functional grouping in nematode diversity studies, but probably less so to document more subtle differences in biogeographic distribution patterns. Nevertheless, the phylogenetic placement of free-living nematodes into families may represent one of the most promising fast-diagnostic tools for large-scale evaluations of ecosystem functionalities using molecular methods (e.g. Calderón-Sanou *et al.*, 2024).

Finally, a variety of traits (e.g. morphological, physiological, behavioural and trophic) has been used to describe nematode diversity in the context of their roles in ecosystems (e.g. Ristau *et al.*, 2015) and their responses to environmental processes and changes (Semprucci *et al.*, 2018; Hou *et al.*, 2023). This has resulted in the development of functional diversity indices (reviewed in Du Preez *et al.*, 2022). The strength of functional diversity is its link to ecological processes and ultimately to ecosystem functions and services (e.g. nutrient mineralization or top-down control of pests) or management (e.g. extent of pest pressure). Frequently used traits relate to population dynamics resulting from behavioural and physical characteristics of nematodes (see Chapter 4, this volume). For example, the colonizer–persister (*c-p*) spectrum categorizes nematodes from colonizers (opportunists, vagrants, r-strategists) to persisters (perpetuators, competitors, stayers, K-strategists). Based on relative abundance of *c-p* taxa, various diversity indices collectively known as 'maturity indices' can be calculated (see Bongers and Ferris, 1999; Du Preez *et al.*, 2022 and references therein). Such indices have found powerful applications in soil biomonitoring (e.g. Sieriebriennikov *et al.*, 2014). The abundance and prevalence of trophic traits (typically based on mouth and oesophagus morphology: Hodda, 2022a) can be used to match with ecosystem productivity in lakes and rivers (Traunspurger and Majdi, 2021) or alterations of food webs triggered by eutrophication processes in aquatic ecosystems (Schratzberger *et al.*, 2008; Ristau *et al.*, 2013a), or shifts in resources in terrestrial ecosystems (Yeates,

2003). The empirical evidence surrounding such indices is not well developed and there are numerous examples of studies where different conclusions were drawn based on different measures of diversity (e.g. Wright and Coleman, 2002; Viketoft *et al.*, 2005).

2.2.2 Area sampled

The area within which diversity of nematodes (or other organisms) should be assessed is as uncertain as the 'correct' measure of diversity. In practice, the scale at which nematode diversity has been studied varies from individual samples covering 1–10 cm^2 surface area (Gansfort *et al.*, 2018a), through 10–100 m^2 (e.g. Traunspurger *et al.*, 2020; Porazinska *et al.*, 2021), to 10–100 km^2 (e.g. Fonseca *et al.*, 2010; Ingels and Vanreusel, 2013; Dümmer *et al.*, 2016; Gansfort and Traunspurger, 2019). The choice of scale is often driven by study-specific aims and the scale at which ecological processes or environmental changes of interest occur. For example, a particular vegetation type may occupy entire regions, pollution may affect a moderate space near the source, and organisms (plant roots or biofilms) may affect a very small space.

Ecological theory posits that different processes control diversity at different spatial scales (α-, β-, and γ-diversity, representing local, regional and inter-regional scales, respectively) (Crist *et al.*, 2003). While these scales may be easily defined for macroscopic organisms like vertebrates or plants, they are ill-defined for microscopic organisms such as nematodes. It is currently unclear whether nematodes respond to any particular combination of spatial scales (Hodda, 1990; Ettema, 1998; Zullini, 2014). However, some endemism can be observed in lakes or extreme environments (Zullini, 2018). Emerging patterns were shown in a spatial hierarchical design using lake nematode communities by Flach *et al.* (2012). They observed low beta diversity at centimetre scales (between sediment cores, and sediment layers) and higher-than-expected beta diversity at the largest scale between the different lakes studied (10–35 km) and concluded that nematode species distribution could be generally similar to the distribution of macroscopic organisms. In fact, because their long-distance dispersal is so effective (Fig. 2.1), nematodes also show a high degree of cosmopolitanism, which is a feature they share with many other microscopic organisms (Azovsky *et al.*, 2020). In this context, as proposed by Zullini (2018), nematode distribution should be regarded as an intermediate distribution between that of microorganisms (tendency towards cosmopolitanism) and that of macroscopic organisms (endemism under certain circumstances). Rather than geographic distance, environmental filtering (e.g. resource availability, competition and temporal fluctuations) emerges as a potent driver of nematode distribution patterns (Gansfort and Traunspurger, 2019). It is therefore not surprising that, at least for free-living marine nematodes, strong drivers for community patterns are those parameters that define the immediate environment, such as food proxies, oxygen availability, sediment characteristics, microhabitat heterogeneity and bioturbation by larger organisms (Ingels and Vanreusel, 2013; Rosli *et al.*, 2018).

2.2.3 Number of samples and/or individuals

For morphological analyses, the number of specimens to be identified from each sample is an important parameter. Usually, the abundance and diversity of nematodes are so high that subsampling is necessary prior to more detailed studies of diversity. But in the case of subsampling, the number of individuals identified to species level needs to be sufficiently high to represent rare as well as abundant species. Generally, 50 to 100 specimens identified from each sample provides a suitable investment:accuracy ratio, but this number can vary according to the level of diversity that is being considered (species, genus, family, trait) and should be verified with rarefaction or species accumulation curves (see Chapter 3, this volume). Aside from the selection of a representative, but fixed, number of nematodes to identify, a percentage of the sampled community can be considered; for example, an assemblage consisting of 10% or 20% within the total sample is identified across all samples so that relative proportions per sample are maintained within the study. The number of samples (e.g. sediment cores) varies largely depending on the study objectives, and, more importantly, depending on

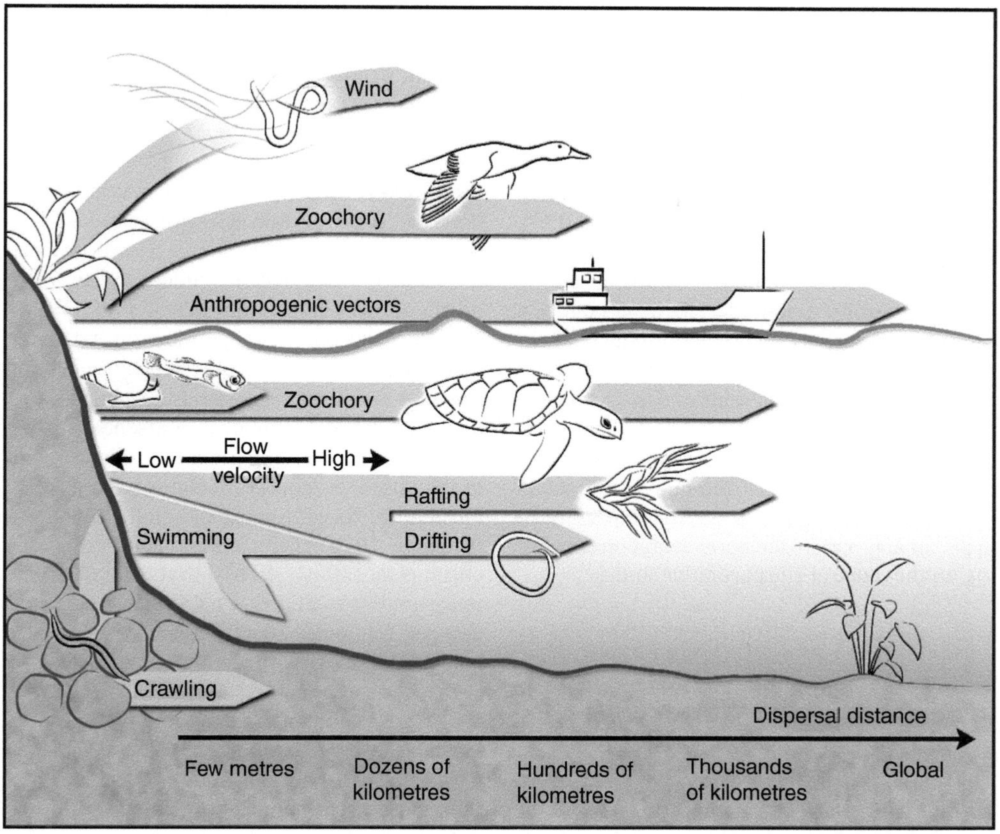

Fig. 2.1. Dispersal modes and potential geographic ranges of aquatic free-living nematodes. (After Ptatscheck and Traunspurger, 2020. CC BY 4.0.)

laboratory resources and staff expertise, but usually between ten and 1000 samples may be processed in a single study of a lake or river ecosystem for instance, meaning most 'morphological' studies may be based on the identification of between 1000 and 100,000 individuals (Traunspurger *et al.*, 2020; Traunspurger and Majdi, 2021). More recent DNA-based methods (e.g. high-throughput sequencing, HTS) remove this limitation, especially for large numbers of samples. For example, it was budgeted that HTS may allow for scaling up to identification of all specimens across 100 samples at about half the cost of morphological identification (Geisen *et al.*, 2018). However, this can potentially be at the expense of quantitative data such as abundance, and may risk generating genetic diversity data that is disconnected from taxonomic information, depending on the quality of the reference databases the molecular identification relies on.

2.2.4 Time span

An important component of the concept of diversity is the time span over which it is measured. Although many nematode species have short population turnover rates (days–weeks), the effect of time (this may be a single time point, across seasons or years) on measures of diversity, beyond the simple effect of adding more individuals, is a key variable to consider in many environments that experience seasonal fluctuations (e.g. in rivers: Majdi *et al.*, 2011; in lakes: Schroeder *et al.*, 2012). Moreover, some environments (e.g. profundal zone of lakes or sea bottoms) may appear stable over long periods of time if sampled irregularly or based on samples collected over longer time periods. However, interannual variability, such as the variation between seasonal productivity cycles, may go unnoticed if sampling frequency is poorly planned.

Irregular or infrequent sampling, or sampling asynchronously with the timelines of phenomena that can exert considerable influence on nematode communities, may not capture the desired temporal variability. In addition, stochastic or extreme events can occur locally and disrupt assemblages (e.g. tsunami impacts: Grzelak *et al.*, 2014; anoxia in pond bottoms: Majdi *et al.*, 2024), and different types of physical disturbance regimes may cause different temporal signals in nematode community dynamics. Likewise, pollution impacts act on short timescales initially and are detectable at different taxonomic levels, whereas recovery from pollution events and the associated nematode community dynamics, both at taxonomic and functional levels, are observable over longer (Stark *et al.*, 2020) and shorter timescales (e.g. Höss *et al.*, 2015). It is thus important to apply an appropriate temporal definition depending on the scope of the particular study.

2.2.5 Sampling and extraction techniques affect nematode recovery

The sampling depth for sediments and soils is important and must be adapted on account of knowledge of the local substrate and nematode communities that are targeted. For example, in lakes, most nematodes occur in the upper sediment layer (0–5 cm) (e.g. Traunspurger and Drews, 1996) but their vertical distribution can extend to hyporheic (<50 cm) and phreatic (>50 cm) zones under stream beds (e.g. Boulton and Stanley, 1995). In estuaries, redox potential can explain most of the nematode assemblage variability between 0 and 15 cm depth (Vieira and Fonseca, 2013), but redox chemistry is not the whole story, and seasonality, food availability and population densities exert strong controls on nematode vertical distribution (Steyaert *et al.*, 2001) as do rates and patterns of deposition and physical disturbances on the sea floor. Knowledge of biogeochemical gradients that determine the vertical distribution of nematode taxa can inform sampling strategies. Thus, without *a priori* knowledge, it is advisable to perform preliminary measures of vertical distributions. Similar ideas apply to soils as some nematodes occur near the soil surface only seasonally, and some remain at substantial depth depending on the extent of the rhizosphere or preference for soil moisture and temperature. There are fewer assumptions for sampling nematodes in biofilms growing on hard substrates where the habitat is collected as a whole by scraping the entire organic film (including nematodes) off the substrate surface.

Disturbance of samples (e.g. vigorous homogenization or sieving) prior to nematode extraction can reduce certain sensitive species such as large ones (Maas and Brinkman, 1980), or when soils are very dry and nematodes are in a dry, brittle anhydrobiotic state (Slack *et al.*, 1972). The use of different sieve mesh sizes during nematode extraction also has an important effect on the recovery of many species from samples and, hence, diversity estimates (Ptatscheck *et al.*, 2020). Mesh sizes of 63 μm and 32 μm are commonly used to collect nematodes and other meiobenthic organisms from aquatic sediments (Fenchel, 1978; Giere, 2009). However, Ptatscheck *et al.* (2020) showed that 23% of nematodes are lost when using a 44 μm sieve. Hence, they suggest using a 20 μm mesh more routinely to balance accuracy with resolving mesh-clogging issues.

Many nematode extraction methods are available, with Baermann funnels, Whitehead trays, silica suspension (Ludox) or sugar flotation, Oostenbrink elutriation, decantation and root incubation are also commonly used (see Chapter 18, this volume). Compared to the decantation method, Ludox flotation may achieve a more 'quantitative' assessment of nematode diversity in aquatic sediments (de Jonge and Bouwman, 1977). It is thus important that the chosen extraction method suits the objectives of the study.

2.2.6 Molecular methods

Molecular methods have been proposed to avoid some of the issues described above, particularly regarding availability of taxonomic/analytical expertise and comparability of the measures used to evaluate biodiversity (see Chapters 8, 14, 15 and 18, this volume). However, many of the issues remain: genetic markers can vary enormously in their efficiency to recover specific taxa (Powers *et al.*, 1997; Schenk and Traunspurger, 2021); different species may have different amounts of genetic variation (Zauner *et al.*, 2007); the resource requirements for molecular studies may be even greater than for morphological ones, especially for the screening of a low number of samples (Geisen *et al.*, 2018); DNA reference sequence databases are incomplete (Gattoni *et al.*, 2023; Gendron *et al.*, 2023). Consequently, molecular methods may generate a different view of nematode community diversity and composition than those based on morphology (e.g. Ahmed *et al.*, 2016; Treonis *et al.*, 2018; Schenk *et al.*, 2020; Kitagami *et al.*, 2022). Despite these issues, differences in the representation of biodiversity patterns have been generally negligible and as both approaches have specific advantages, they should be treated as complementary.

2.2.7 Global nematode diversity

Currently, a total of 28,537 nematode species belonging to 3030 genera, 276 families, 32 orders and three classes have been listed by Hodda (2022c). This list includes known families, and genera with their species, and provides an overview of their occurence across marine, freshwater and terrestrial (mostly soil) environments. In this list, more than half (52%) of currently described nematode species are terrestrial, while more than a quarter (8137, i.e. 28.5%) are from freshwater environments. Potential rationales for disparities include habitat accessibility contraints or economic interests.

Reviewing recent trends of nematode species descriptions during the period 2011–2019, Hodda (2022b) reported that 44% of new descriptions were free-living, 28% vertebrate parasites, 15% plant parasites and 12% invertebrate parasites. Moreover, description efforts varied largely across continents (and their adjacent seas). For example, 47% of those new species were described from Asia between 2011 and 2019, 14% from Australasia, 12% from Europe, 12% from South America, 9% from North America and 5% from Africa.

2.3 Responses to Environmental Factors in Marine, Freshwater and Terrestrial Ecosystems at Different Scales (Global to Micro)

A wide range of factors affects nematode diversity, including physical and chemical properties of the environment, spatial and temporal heterogeneity, competition, and distribution of prey, predators and pathogens (e.g. Stuart *et al.*, 2015; Gansfort and Traunspurger, 2019; Zhang and Hu, 2019; Lazarova *et al.*, 2021). Identifying and understanding the processes underlying the effects of those factors, and their interactions, is important for predicting nematode diversity in response to environmental change (e.g. climate warming, pollution or species invasions) (Porazinska *et al.*, 2022). Below, we summarize some of the main factors and processes that drive nematode diversity. For each factor, we distinguish effects reported on marine, freshwater and terrestrial nematode communities. Our understanding of nematode diversity is derived from mostly spatial (with a small proportion of temporal) studies that compare populations from one place (or time) to another, and very few experimental studies that identify and test the underlying mechanisms.

2.3.1 On the existence of latitudinal patterns of nematode diversity at the global scale

Marine nematodes

Latitude determines climate, which subsequently should affect nematode communities. However, in coastal areas, an experimental approach using colonization of artificial substrates failed to generate evidence of latitudinal gradients in nematode diversity (Gobin and Warwick, 2006), while *in situ* field studies also failed to reveal unequivocal evidence that latitudinal gradients per se influence the diversity of marine nematode assemblages (e.g. Vanhove *et al.*, 1999; Lambshead *et al.*, 2000, 2002; Mokievsky and Azovsky, 2002). For example, an apparent increase in species numbers with decreasing latitude in sandy beaches in Australia and Chile was confounded with a concomitant gradient of disturbance (Nicholas and Trueman, 2005) and sea surface temperature (Lee and Riveros, 2012), respectively.

Different trends have been reported for deep-sea nematodes, including a unimodal pattern of diversity peaking in temperate regions (Mokievsky *et al.*, 2007) and bimodal patterns with abundance and diversity peaks around the equator and at temperate latitudes (Lambshead *et al.*, 2000, 2002). Differences in the geographical coverage of the studies, treatment of the data prior to analysis and differences in analytical methods preclude generalized statements regarding latitudinal patterns, although less productive zones generally support fewer individuals and species than more productive ones (Lambshead *et al.*, 2000; Danovaro *et al.*, 2008; Fonseca *et al.*, 2010). Food availability is a recurrent driver of nematode abundance, biomass and diversity; thus a signature of productivity tends to dominate observed nematode communities' structural and functional characteristics (e.g. Guilini *et al.*, 2013; Ingels *et al.*, 2023).

Freshwater nematodes

There are too few studies concentrated within too narrow geographic space – mostly across central and northern Europe – to draw any conclusion about the effects of latitudinal gradients on limno-nematode diversity (Eyualem-Abebe *et al.*, 2008; Zullini, 2018; Traunspurger and Majdi, 2021). Moreover, freshwater habitats (streams, lakes and ponds) are essentially discontinuous over large distances and differ substantially in their trophic state, discharge regime and level of anthropization at local and regional scales, so that latitudinal gradients are unlikely to be easily separated from the influences of other variables and from the effect of geographical bias in taxonomic expertise and sampling effort.

Terrestrial nematodes

There are several studies exploring latitudinal trends of soil nematode biodiversity. In a recent study at a global scale, climate, soil organic carbon and soil texture were the most significant factors explaining latitudinal patterns of nematode abundances (Van Den Hoogen *et al.*, 2019), with higher latitudes and subarctic ecosystems hosting highest nematode abundances. However, greater nematode abundance does not necessarily mean greater number of species. Temperature and water availability have been previously identified as the most important drivers of nematode species richness in farmlands, parks and forests in urban and urban-surrounding ecosystems (Gong *et al.*, 2023), while rainfall drove nematode diversity in grassland systems (Cui *et al.*, 2022) and water-holding capacity in alpine habitats (Porazinska *et al.*, 2021). We may therefore expect a sine-curve-like function for nematode diversity along the latitudinal gradient (Fig. 2.2).

Starting in the polar region, nematode diversity will be low due to the stable frost periods and low rainfall. This is commensurate with a study in the Antarctic, revealing very low nematode densities and species richness (three to five species in total) in dry and seeping permafrost soils (Smith *et al.*, 2012). Also, a meta-analysis of 137 sites located at different latitudes reported lower species richness near the poles (–60 to –90°S, 70 to 90°N) (Boag and Yeates, 1998). Diversity

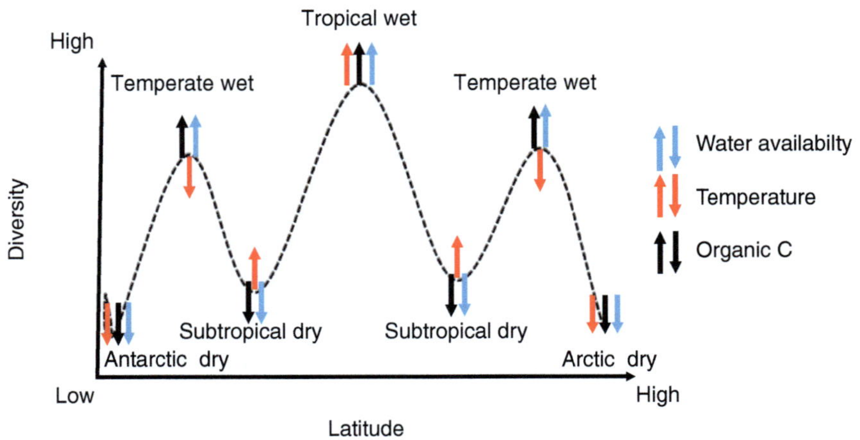

Fig. 2.2. Simplified and theoretical model illustrating the effect of latitude on terrestrial nematode diversity. In latitudes with sufficient (↑) water, temperature and organic carbon, nematode diversity peaks while diversity decreases if one or more factors are lacking (↓). (Author's own image.)

will increase towards temperate climates due to increasing temperatures and water availability. Towards the subtropical deserts, a strong decline of diversity is expected, entailed by high temperature fluctuations (day–night) and water limitation (Fig. 2.2) that foster soil degradation processes such as crust formation, erosion and salinization, among others (Batterbury and Warren, 2001). Subtropical hot deserts often have a specialized and sensitive soil food web that is strongly conditioned by few species (e.g., Santos *et al.*, 1981; Whitford *et al.*, 1982). In contrast, the tropical wet (rainforest-dominated) regions have constant high temperatures and rainfall, both being important drivers of competition-derived biodiversity. Hence, nematode diversity is likely the highest along the equator. Indeed, a nematode survey conducted within three vertical strata (soil, litter and canopy) of temperate (the Olympic National Forest, Washington, USA) and tropical (La Selva Biological Station, Costa Rica) rainforests showed that the total species richness was 300% higher in the tropical than temperate rainforest, albeit 80% of the species in the temperate rainforest resided in the soil, and only 20% in the tropics (Porazinska *et al.*, 2012). Similarly, higher beta and gamma diversity of nematodes was observed in tropical forests in Malaysia than in the arctic tundra in Svalbard (Kerfahi *et al.*, 2016) and greater alpha diversity in tropical forest soils than in temperate forests in China (Shao *et al.*, 2023), further supporting our hypothesis. Soil organic carbon played a major role for nematode diversity in the investigated forests. In our model, higher levels of soil organic carbon in temperate soils benefit nematode diversity while organic carbon limitations in (sub)tropical soils have adverse effects (Fig. 2.2). More comprehensive analyses of latitudinal effects on nematode diversity should standardize important factors like plant cover, microbial communities, land use, soil moisture, pH and texture that might obscure general biodiversity patterns (e.g. Smith *et al.*, 2012).

2.3.2 Effects of water depth or altitude on nematode diversity

Marine nematodes

Biodiversity data spanning the entire bathymetric range from coastal or shelf environments to the abyss and even hadal depths is limited, but most marine nematode studies along depth gradients across habitats have indicated a notable influence of water depth on nematode diversity, mediated by associated changes in temperature, salinity, dissolved oxygen and sediment variables (e.g. grain size distribution and organic content). A recent study in the Southwestern Gulf of Mexico from the upper continental slope to the abyssal plain (from 186 to 3774 m depth), for example, recorded declines in abundance and species richness with increasing water depth (Armenteros *et al.*, 2022). When rarefied to a standardized level of abundance (e.g. expected number of species), the relationship between water depth and species richness of nematode assemblages is not universal. Expected number of species has been shown to increase with depth (dos Santos *et al.*, 2020), decrease (Danovaro *et al.*, 2008; Trebukhova *et al.*, 2013) or remain unaltered (Lambshead *et al.*, 2000). Regional features of the seabed and strong selective pressures associated with depth-related gradients of food and oxygen availability appear decisive in yielding a specific diversity pattern. Of particular importance is the relationship between food availability and nematode abundance, biomass, and structural and functional diversity in the context of water depth. As the sea floor deepens, the quality and quantity of food resources generally decline, influencing nematode populations accordingly. However, the link between productivity and richness has been recognized as a unimodal curve, where intermediate productivity supports highest richness providing the full spectrum of productivities is considered (Ingels *et al.*, 2023).

Freshwater nematodes

Consistent patterns of nematode species diversity have been identified across rivers and lakes. In lakes, the distribution and abundance of nematodes within the sediment are influenced by depth (Traunspurger, 1996a,b; Eyualem-Abebe *et al.*, 2006; Traunspurger *et al.*, 2020). Species richness is significantly lower in profundal than in littoral zones (Traunspurger *et al.*, 2020, 2021), perhaps as a result of the reduction and homogenization of ecological niches in the profundal compared to the littoral. Moreover, nematode community structure changes along a depth gradient, presumably as environmental conditions in the littoral zones of lakes (e.g. wave disturbance, land-use pressure, seasonal fluctuations, light exposure, and presence of macrophytes and algal biofilms) differ substantially from the conditions in profundal zones (e.g. no light, stable temperature, muddy substrates, and trophic dependence on sedimentation of planktonic 'blooms'). This pattern shows some similarities with marine systems, where a combination of different factors changing along depth ultimately structures the assemblages of nematodes via adaptation to a particular set of environmental factors.

Altitude is known to shape river ecosystems following the 'river continuum concept' (Vannote *et al.*, 1980), although not specifically investigated in freshwater nematode studies. The latter assumes that headwater streams are small, fast-flowing, generally oligotrophic, have more connection with the riparian forests (e.g. importance of plant-derived organic matter), and have a more heterogeneous, coarser bed mostly composed of large rocks, cobbles and gravel. Conversely, at downstream reaches (lower altitudes), rivers become larger, meander, are richer in nutrients, contain more algal-derived organic matter, and the bed is mostly made of sandy to silty deposits. Gansfort and Transpurger (2019) emphasized that the watercourse distance (i.e. the river continuum), and thus presumably altitude, is an important driver of nematode community structure.

Terrestrial nematodes

The effects of elevation on nematode diversity are wide ranging. Studies reported decreased (e.g. Afzal *et al.*, 2021; Zhang *et al.*, 2021; Kashyap *et al.*, 2022), increased nematode diversity (Kouser *et al.*, 2021), unimodal effects (Dong *et al.*, 2017) and no changes (e.g. Tong *et al.*, 2010; Traunspurger *et al.*, 2017; Kitagami *et al.*, 2022). Results reported in the literature are difficult to interpret for several reasons. First, although nematode studies have received more attention recently, the number of studies to draw conclusions from is still limited. Second, published studies examined a range of elevations (most at low and medium, but few at high elevations), ecosystems (forests, alpine grasslands and meadows, agricultural fields), identification methods (morphology and metabarcoding) and taxonomic resolution (genus, family, operational taxonomic unit). Interestingly, although alpha diversity patterns have varied among studies, a pattern of distinct community compositions among elevations is common. Differences in community composition are generally explained by factors confounded with elevation, including a suite of abiotic (e.g. pH, soil moisture, soil organic matter, nutrients) (e.g. Traunspurger *et al.*, 2017; Afzal *et al.*, 2021; Kashyap *et al.*, 2022; Kitagami *et al.*, 2022), biotic (e.g. hyphal length and root biomass) (Traunspurger *et al.*, 2017; Kitagami *et al.*, 2022) and climatic factors (soil temperature) (Kouser *et al.*, 2021).

2.3.3 Local effects of habitat disturbance on nematode diversity

Marine nematodes

Effects of the most widespread habitat disturbances in the marine environment, including bottom fishing, pollution, the introduction of invasive species and climate change, on nematode diversity are complex and context-dependent (see Schratzberger *et al.*, 2023 and references therein). In general, the taxonomic and functional breadth of nematode assemblages is reduced by disturbance and is a consequence of the progressive decline of sensitive versus the proliferation of tolerant and opportunistic species. The widespread replacement of habitat-specific ecological specialists by more broadly adapted ecological generalists and opportunists often results in biotic and functional homogenization of once-diverse assemblages. A distinction in nematode community responses can also be made between disturbances that operate at different timescales. Pollution events that have an immediate impact can cause assemblage shifts on higher taxonomic levels (e.g. family level), whereas disturbances over longer periods of time (e.g. ocean acidification) may result in more subtle changes at the species level (Warwick, 1988). Hence, the (taxonomic) resolution of scientific study of marine nematode communities should correspond with the scale at which different types of disturbances are investigated.

Freshwater nematodes

In rivers, flow velocity shapes the assemblages near the water interface. For example, Majdi *et al.* (2012a) determined a water flow threshold of

30 cm s^{-1} in the Garonne River (France) above which algal biofilms (and their associated nematodes) were entirely resuspended in the water column and after which a new biofilm colonization cycle starts. Hence, floods may have a negative effect on biofilm-associated nematode abundances and diversity in the short term; but post-flood, the early biofilm recolonization process involves the highest number of species and over a long period of biofilm stability, diversity declines and the assemblage becomes dominated by a few algal-feeding specialists like *Chromadorina bioculata* (Majdi *et al.*, 2011). Droughts also have an intriguing effect on nematode diversity. For example, in a study investigating 22 Mediterranean streams, Majdi *et al.* (2020) found more diverse nematode assemblages in the bed of intermittent streams than in perennial streams. Intermittent streams provided an interesting combination of terrestrial and freshwater habitats and trophic channels, and as a result, they harboured more diverse assemblages of nematodes.

In lakes, the littoral zone is subjected to a range of disturbances including water-level fluctuations and wave-induced shear stress (either of natural origin or induced by boat circulation). Following a gradient of moss habitats from sublittoral to the terrestrial in Lake Erken (Sweden), Kreuzinger-Janik *et al.* (2021) found that the splash-water zone was the most diverse, combining a diversity of resources and a mixture of species from both the terrestrial and aquatic ecosystems. However, in a study along a habitat gradient that spanned lake sediments, shorelines and surrounding prairies in the Nebraska Sandhills, littoral diversity was intermediate and fell between the species-poor lake and species-rich prairie communities (Gattoni *et al.*, 2022), suggesting that, similar to marine studies, diversity patterns of freshwater nematodes are context-dependent.

Terrestrial nematodes

In general, substrates disturbed by human activity have fewer nematode species than undisturbed ones. For example, 30% more species were observed in a regenerating prairie than in adjacent agricultural land (Todd *et al.*, 2006). Similarly, a managed meadow supported nearly twice as many species than cultivated fields, with intermediate diversity levels in fields where cultivation ceased two years prior to the study (Háněl, 2003). More species were found in undisturbed tropical forest than after clearing, cultivation or during regeneration (Bloemers *et al.*, 1997). In deserts, too, diversity of nematodes was greater where disturbance was low and soil crusts were well developed (Darby *et al.*, 2007). Similar results were obtained in soybean fields (Okada and Harada, 2007) and a range of other terrestrial habitats (Pate *et al.*, 2000; Cadet *et al.*, 2003; Háněl, 2003). Soil disturbance appears to reduce the number of genera as well, a result reported in a considerable number of studies and contexts (e.g. Háněl, 2003; Kardol *et al.*, 2005; Korenko and Schmidt, 2006; Tsiafouli *et al.*, 2006). However, decreases in taxon richness might depend on latitude and be influenced by soil physicochemical properties. Gong *et al.* (2021) reported that agricultural practices reduced taxon richness primarily in the most species-rich latitudes (30°N). In contrast, under unfavourable climatic conditions, agricultural practices were associated with an increase of nematode species richness (Li *et al.*, 2020, 2022). Together, anthropogenic factors are not per se having adverse effects on soil nematode communities, but responses depend on the biodiversity facet studied. Consolidation of nematode biodiversity and abundance, as suggested by Ferris and Tuomisto (2015) and discussed in Chapter 4 of this volume, may support a more holistic approach.

2.3.4 Effects of substrate texture and food availability

Marine nematodes

The physical dynamics of the seabed, its spatial heterogeneity and the resident biotic components together determine the number and type of available niches and hence influence the diversity and composition of nematode assemblages. The interaction of local hydrodynamics and sediment transport is a key driver of the processes involved. Their influence occurs in a variety of different ways, and, importantly, in ways that are not predictable from using summary descriptors derived from sediment grain size alone. Fonseca *et al.* (2014) demonstrated that hypotheses postulating a linear (positive or negative) relationship between nematode species richness and sediment grain size have not been corroborated empirically. While some nematode species may show a preferred mean grain size (i.e. fundamental niche), associated environmental factors (e.g. organic carbon content, oxygen availability, macrofauna abundance and composition, microbial biomass and activity) can interact with granulometry and affect nematode fitness, thereby limiting their local population growth (i.e. realized niche). However, grain size structure (and its associated biogeochemical and trophic conditions) may remain the most important determinative factor in driving nematode communities when other stressors are superimposed (Ingels *et al.*, 2018). Meaningful and predictive explanations for nematode distributions are likely to emerge if these are also evaluated relative to the suite of hydrodynamic and sediment transport processes that are responsible for sediment distributions. Also, for offshore habitats it seems important to investigate the physical form of the sediments (e.g. bedforms) at a range of spatio-temporal scales when aiming to provide meaningful explanations for the associations of nematodes with their sedimentary environment (Schratzberger and Larcombe, 2014).

Freshwater nematodes

Based on a previous review of 75 lakes in central and northern Europe (Traunspurger *et al.*, 2020), there were significantly more nematode species dwelling in soft substrates (i.e. coarse, fine sediment, and silts) than on hard substrates (i.e. biofilms growing on stones, cobbles, wood or leaf litter) (Fig. 2.3). Soft substrates possibly offer a larger habitable volume than the (usually millimetric) layer of biofilm growing on hard substrates. Moreover, soft substrates may offer more ecological niches than hard substrates. Traunspurger *et al.* (2020) also found that nematode communities were significantly different among soft versus hard susbstrates, suggesting specialized lifestyles (Fig. 2.3). For example, *Chromadorina* and *Punctodora* tended to dominate epilithic algal biofilms in lakes (Peters and Traunspurger, 2005; Schroeder *et al.*, 2012), and in rivers (Gaudes *et al.*, 2006; Majdi *et al.*, 2011), possibly due to their ability to stick to stony substrates via spinneret secretions (Croll and Zullini, 1972) or to use the local microalgal resource (Majdi *et al.*, 2012b; Kazemi-Dinan *et al.*, 2014). Many studies in freshwater lakes have investigated the relationship between lake

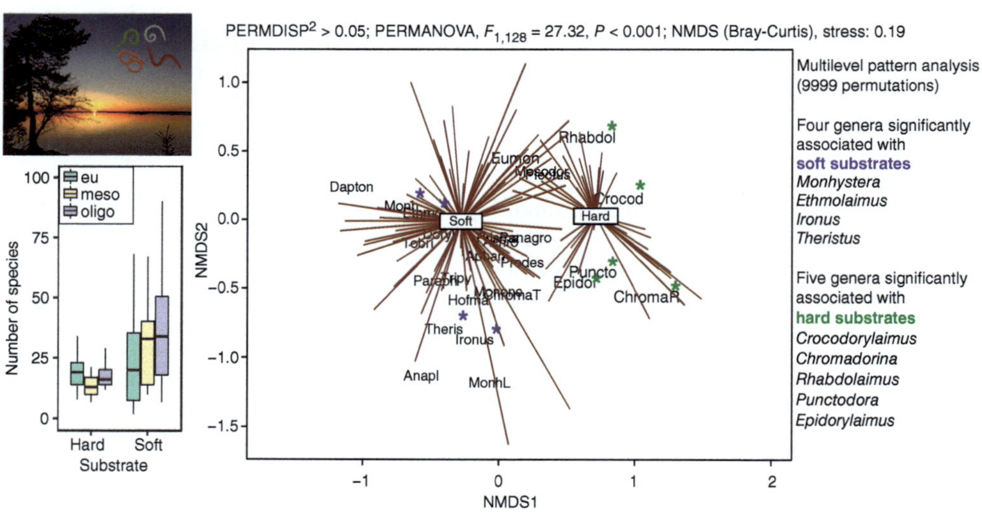

Fig. 2.3. Diversity and community structure of benthic nematodes dwelling on soft and hard substrates in 75 European lakes of different trophic states (eu = eutrophic, meso = mesotrophic, oligo = oligotrophic). Box plots show median and interquartile range. NMDS = non-metric multidimensional analysis of the Bray–Curtis similarities of genera structure across lakes; PERMANOVA = PERMutational ANalysis Of Variance; PERMDISP² = PERmutational Multivariate analysis of DISPersion, which is a multivariate counterpart of the Levene test for homogeneity of multivariate variances. For methodology and full results, see Traunspurger *et al.* (2020).

productivity and the diversity of benthic nematodes. Fitting with the idea that nutrients increase the dominance of a few species, the oligotrophic lakes tended to be more diverse than eutrophic lakes, and mesotrophic lakes showed intermediate diversity (Fig. 2.3). But this pattern was only significant for soft substrates. On hard substrates, diversity was lower overall, and no such pattern was observed (Fig. 2.3). In rivers, a similar trend of higher diversity in oligotrophic systems was found (see in Traunspurger and Majdi, 2021).

The generally low availability of resource might prevent competitive exclusion by a smaller subset of specialized species (Michiels and Traunspurger, 2005). This was confirmed in a long-term (16 months) microcosm experiment, designed to assess the effects of five levels of nutrients on nematode diversity (Ristau *et al.*, 2012). Over the course of the experiment, 46 nematode species were recorded. Within the first five months, nematode species richness remained fairly constant, at all nutrient concentrations, and then began to decline at month 8. At months 12 and 16, species richness decreased consistently with increasing nutrient concentrations.

Terrestrial nematodes

Both soil texture and food availability determine nematode communities. For example, soil pore size and aggregate structure act as a size filter for nematodes (i.e. the number of larger species may be reduced in finer texture soils) (e.g. de Goede and Bongers, 1994; Yeates *et al.*, 1997; Quist *et al.*, 2019; Martin and Sprunger, 2023). However, the complexity of the soil matrix has been suggested to promote nematode diversity (Ettema, 1998). In contrast, soil compaction and alteration of biogeochemistry can have varying effects on nematode diversity ranging from small and inconsistent (Yeates *et al.*, 1997; Alphei, 1998; Althoff and Thien, 2005; Rantalainen *et al.*, 2005; Gibson *et al.*, 2023) to strong, especially on trophic group abundance and food web simplifications (Choudhary *et al.*, 2023; Ma *et al.*, 2024). For example, agricultural management can profoundly alter nematode diversity through changes in soil texture (Puissant *et al.*, 2021) as well as bacterial and fungal prey (J.H. Schmidt, Braunschweig, 2024, personal communication).

Often, soil nematode diversity is directly related to the diversity of above-ground vegetation. Such relationships have been observed in natural grasslands, alpine tundra, and in experimental mesocosms (e.g. De Deyn *et al.*, 2007; Porazinska *et al.*, 2021). While in some studies nematode diversity measured by richness was related to plant diversity, in other studies it was evenness that matched with plant diversity (Cadet *et al.*, 2007; Viketoft *et al.*, 2005). Soil nematode diversity tends to be higher in areas of natural vegetation with many plant species compared to managed areas with fewer taxa, especially feeding on roots (Van Gundy and Freckman, 1977; Wallace, 1987). The hypothesized mechanism is the lack of competition resulting from division of roots into very precisely defined niches under natural vegetation (Van Gundy and Freckman, 1977), but evidence is limited and other confounding factors, such as the level of disturbance, could play a role as well (see above).

Even at the tree scale, studies have revealed diverse nematode assemblages inhabiting tree-related microhabitats (TreM) above ground. Remarkably, very distinct assemblages were found to coexist at short distances, on the same tree, their composition seemingly influenced by TreM type and position (e.g. epi- versus endophytic), the stage of wood decay, the availability of microbial resources, and the occupation by other insects or birds (Majdi *et al.*, 2025).

2.3.5 Are biological interactions driving nematode diversity at the micro scale?

Marine nematodes

Facilitative and competitive interactions among marine nematodes and benthic organisms create an intricate web of relationships that, collectively, are hypothesized to affect a range of ecosystem processes (Fig. 2.4) (Schratzberger and Ingels, 2018 and references therein).

For example, feeding on deposits controls access to food sources (especially particulate organic matter) for competitors and may reduce the availability of particulate organic substrates for oxygen-consuming microbes. Macrofaunal predation potentially reduces nematode prey populations and alters community structure although nematode responses to species-specific macrofauna changes may be difficult to observe, and responses may be more general as a result of

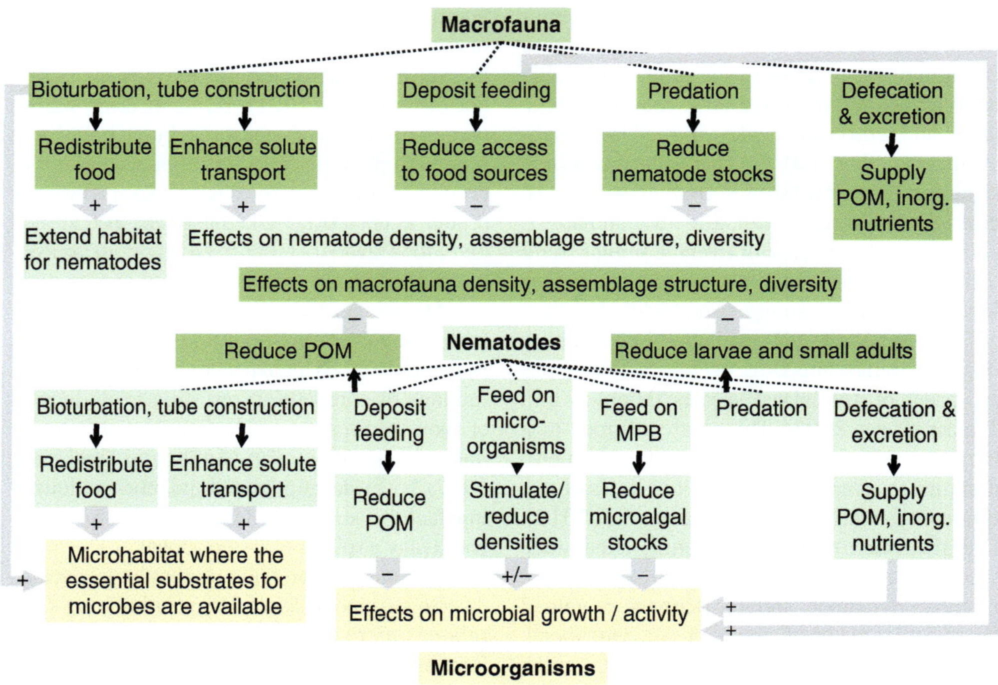

Fig. 2.4. Conceptual diagram, illustrating some of the multifaceted top-down and bottom-up interactions between microbiota, meiofauna (including nematodes) and macrofauna. POM = particulate organic matter; inorg. = inorganic; MPB = microphytobenthos. Effects of microbiota on meio- and macrofauna are not shown. Arrows are not indicative of effect size and competition for space is not shown. +, facilitative interaction; –, competitive interaction. (After Schratzberger and Ingels, 2018. Used with permission from Elsevier.)

ecosystem engineering, predation and competition processes (Ingels *et al.*, 2014). Conversely, nematode (and turbellarian) predation on small macrofauna and larvae can potentially reduce macrofaunal prey populations of less mobile species and affect their settlement success (Watzin, 1983, 1986). Although experimental evidence of this 'meiofauna bottleneck' process remains equivocal (Zobrist and Coull, 1992), this can affect nematode diversity.

Nematode grazing on microbiota can either stimulate or inhibit the growth and activity of oxygen-consuming microbes with resultant effects on a range of marine ecosystem processes (Schratzberger and Ingels, 2018). Nematode grazing on microphytobenthos may reduce or stimulate microalgal stocks, thereby affecting oxygen production in the upper layers of the sediment and sediment stability (Pinckney *et al.*, 2003; D'Hondt *et al.*, 2018). Nematodes defecate organic matter, which microbes use in oxygen-consuming reactions and excrete nitrogen which microphytes use in oxygen-producing reactions. The meio- and macrofauna-mediated efflux of ammonium from the sediment to the water column subsidizes overlying waters with a readily utilizable form of inorganic nitrogen.

These intra- and interspecific interactions, comprising both negative (competitive) and positive (facilitative) components, affect the spatial distribution of nematodes in terms of density, diversity and species composition. Marine nematodes often simultaneously compete with and facilitate each other, which makes the net effects of species interactions on the diversity of their assemblages difficult to measure and predict.

Freshwater nematodes

Freshwater nematodes interact closely with bacteria and micro-eukaryotes (ciliates, diatoms, chlorophytes) within biofilm matrices in interstitial spaces and on a variety of hard substrates (Bergtold *et al.*, 2005; Weitere *et al.*, 2018; Majdi *et al.*, 2021). In the chemotrophic environment of the Movile cave (Romania), for example, and further confirmed in the laboratory (Weber and Traunspurger, 2013), two bacterivores *Panagrolaimus* cf. *thienemanni* and *Poikilolaimus* cf. *regenfussi* can coexist based on cyclic fluctuations of bacterial density (Muschiol *et al.*, 2015): high bacteria density is preferred by *Panagrolaimus*, low densities by *Poikilolaimus*. Interestingly, further laboratory experiments showed that population growth was maximized for both species when they coexisted equally, suggesting the existence of a positive feedback loop in terms of habitat colonization and maintenance of optimal bacterial density (Gansfort *et al.*, 2018b).

Freshwater nematodes are prey for a variety of other metazoans (including other nematodes, but also other meiofauna, macrofauna and even fish) (Ptatscheck, 2021). Generally, the ecological role of nematodes in rivers, lakes and ponds might be comparable with their roles in marine ecosystems as conceptualized in Fig. 2.4. Some evidence suggests that predation by other nematodes may increase, or at least stabilize, nematode diversity. For example, across 13 alpine lakes, more nematode species were found where more predatory nematodes were present (Fig. 2.5) (Michiels and Traunspurger, 2003). However, whether predation favours diversity, or diversity favours predators, should be tested in laboratory-controlled conditions. How the set of biotic interactions in an ecosystem can be shifted by pollution is also an interesting question. For example, Heininger *et al.* (2007) speculated that the dominance of predatory nematodes in heavily polluted river sites could be the result of pollution-induced decrease of macrobenthic predators competing with predatory nematodes.

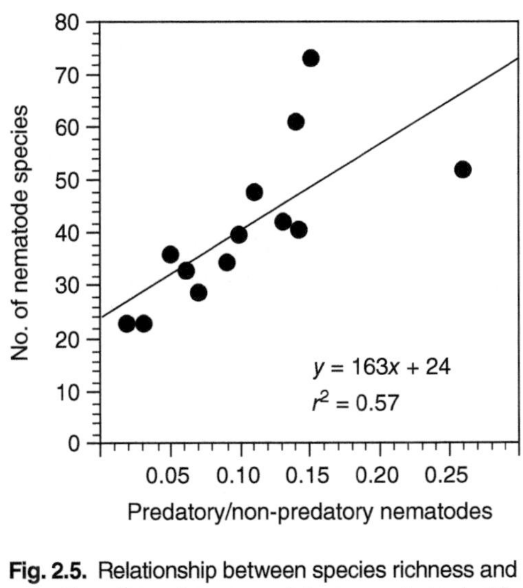

Fig. 2.5. Relationship between species richness and ratio of predatory to non-predatory nematodes in 13 alpine lakes. (From Michiels and Traunspurger, 2003.)

Terrestrial nematodes

There are direct interactions between nematodes that affect nematode diversity across terrestrial habitats. For example, the diversity of the omnivorous or predatory Mononchidae closely follows that of the Cephalobidae (Yeates and Bongers, 1999). Interspecific competition may limit diversity in entomophilic and root-feeding nematodes in soil (e.g. Neumann and Shields, 2006). Furthermore, entomophilic nematodes may at least temporarily reduce the diversity of other groups of nematodes (omnivorous Dorylaimida) (Chevalier and Webster, 2006). More ecologically relevant are interactions of soil nematodes and predatory mites. The latter can be reared on bacterivorous nematodes in the laboratory and there is much evidence that this can be transferred to natural ecosystems. By increasing the total abundance of nematodes in soil, predatory mites in the next trophic level of the soil food web will benefit and may support more ecosystem services, such as pest control (Rueda-Ramírez *et al.*, 2023).

As microbial-feeding nematodes consume, grow or reproduce differently according to the particular microorganisms available as food in laboratory conditions (e.g. Ruess and Dighton, 1996; Newsham *et al.*, 2004; Tahseen *et al.*, 2004), one might expect that microbial diversity would exert a strong influence on the diversity of microbial-feeding nematodes. In very simple communities in the Antarctic, however, no link existed between bacterial and nematode diversity (Barrett *et al.*, 2006). However, in a recent study along two environmental gradients in the western Nebraska Sandhills, biotic factors (i.e. diversity of bacteria, fungi and microbial eukaryotes) played a significant role in the assembly of nematode communities in all habitat types, although the extent of explained variation was context-dependent (Gattoni *et al.*, 2022). Moreover, by monitoring these communities over three years, nematode–nematode interactions were more important than interactions with microbes, especially in soils (Gattoni *et al.*, 2024). Certainly, more studies are needed before more general conclusions can be drawn.

2.4 Are There Comparable Diversity Patterns Across Marine, Freshwater and Terrestrial Communities?

In this last section of this chapter, we tentatively seek to draw parallels between marine, freshwater and terrestrial nematode diversity based on previous sections on global nematode diversity and the effects of some key factors on nematode diversity in the different biomes (Table 2.1). We then begin to discuss similarities in the way nematode diversity affects the provision of ecosystem services and informs on the resilience of ecosystems and their capacity to deliver desired ecosystem services (Table 2.1).

Despite the disparate methods used and environments investigated, parallels exist between marine and freshwater ecosystems (especially lakes) regarding the suite of factors (temperature, light, pollution, sedimentation regime, amount and quality of resources reaching the bottom) that primarily affect the taxonomic and functional structure of nematode communities. In soils, temperature and precipitation regime, soil texture, the amount of organic carbon and vegetation cover (Fig. 2.2) are key variables driving below-ground nematode diversity. A recent large-scale compilation of data has enabled the modelling of nematode abundances and biomass at the global scale (Van Den Hoogen *et al.*, 2019) and has allowed researchers to hypothesize latitudinal diversity patterns based on responses to vegetation cover and soil organic carbon content. Similar compilations and analyses of data in aquatic habitats would be helpful to evidence similarities/differences between nematode diversity patterns across habitats.

Bottom-up effects of resources (both in terms of quantity and quality) drive diversity patterns of nematodes across realms. Overall, ecosystem productivity does not necessarily increase diversity, and as observed in lakes, rivers and soils, oligotrophic habitats tend to harbour more diverse nematode assemblages than heavily subsidized ones. However, for bacterivorous nematodes cultured under laboratory conditions, there is mounting evidence that cyclic changes in food density can enable coexistence of more species based on different food density preferences (e.g. Schroeder *et al.*, 2010; Weber and Traunspurger, 2013; Derycke *et al.*, 2016; Gansfort *et al.*, 2018b; Guden *et al.*, 2021). In the field, agricultural practices consisting of high nutrient loads tend to reduce general nematode richness, but at the same time, stimulate especially bacterial trophic channels and associated bacterivorous nematodes through possibly increased evolutionary rates of fast-growing bacterivore species (Gong *et al.*, 2021).

Regardless of their habitat, nematodes form intricate food webs with other organisms (bacteria, fungi, microalgae, protists, predatory mites, etc.). Local nematode diversity will presumably show congruent responses to various examples of maintenance of diversity mechanisms such as predator density and diversity (Weber and Traunspurger, 2015), predation strength and flexibility (Eitzinger *et al.*, 2018), ecological engineering effects (Majdi *et al.*, 2014; Wulfes *et al.*, 2023), and competition strength and dispersal triggers (Kreuzinger-Janik *et al.*, 2022). As such, studies should integrate different components of diversity including richness, abundance and biomass to fully understand nematode biodiversity and the role it plays in ecosystems.

The relationships between nematode diversity and ecosystem services can be exemplified easily by the relationship between bacterivores and bacterial productivity, which has been observed frequently and in a variety of substrates (Freckman, 1988; Ferris *et al.*, 1997; Fu *et al.*, 2005; Schmidt *et al.*, 2020). There is some experimental evidence that nematodes can increase bacterial productivity locally (Traunspurger *et al.*, 1997), presumably through preventing bacterial overgrowth and providing bacteria with nitrogen-rich excretion products and mucus secretions (Riemann and Helmke, 2002; Moens *et al.*, 2005). Similarly, nematodes have been found to increase primary production of microphytobenthos in marine and freshwater biofilms (Mathieu *et al.*, 2007; D'Hondt *et al.*, 2018), presumably through increasing the porosity of biofilm matrices to light and nutrients. This stimulation of microbial activity and alteration of the extracellular biofilm matrix by

Table 2.1. Summary of some important drivers of nematode diversity at different scales and in different types of habitats, and parallels betwen habitats and perspectives for future research. *NS = not enough studies, or studies too patchy or discontinuous to highlight meaningful patterns. (Author's own table.)

		on Nematode diversity in			
Scale	Effect of	Marine	Freshwater	Terrestrial	Parallels and perspectives
Global	Latitude	Coastal – not significant Deep sea – unimodal (peaking at temperate latitudes) or bimodal (peaking at equator and temperate latitudes) patterns	*NS	Abundances peak at higher latitudes, but diversity may show bimodal pattern (peaking at equator and temperate latitudes, see Fig. 2.2)	Temperature and light regime, climate stability, access to water, diversity of microbial resources and magnitude of primary production by plants (or algae) are among the key factors explaining similar nematode diversity patterns in marine and terrestrial ecosystems
Regional	Altitude/ depth	Nematode abundances and biomass decrease with depth (mediated by temperature, food diversity, salinity, oxygen availability and sedimentation regime), but effects on diversity are unclear, mostly invoking oxygen and food availability as key factors. Diversity may be highest at intermediate levels of productivity	Abundance and species richness are lower in profundal than in littoral zones of lakes which may be due to habitat homogenization in the profundal. Altitude influences the river continuum and thus may affect river nematode communities, but *NS	Positive, negative and no effect of altitude on soil nematode diversity have been reported. *NS Only few studies include high altitudes and a sufficient altitudinal range. Many other factors mask the effect of altitude (pH, soil moisture, soil organic matter, plant cover, root mass, temperature)	The limited number of studies across meaningful gradients and the many biotic and abiotic factors confounding the effects of elevation or depth make it difficult to generalize or identify any robust patterns. Nevertheless, nematode communities are generally less abundant and more specialized under extreme conditions
Local	Habitat disturbance	Effects of physicochemical disturbance of the habitat on nematode diversity are largely context-dependent (type and spatio-temporal scale of the disturbance). Ephemeral disturbances may have immediate measurable effects even at the family level. Disturbances over longer periods of time (e.g. ocean acidification) can trigger more subtle changes at the species level	Effects are context-dependent. The hydrological regime (floods, droughts) imposes disturbance to riverine nematofaunas. Intermediate flooding and drought seem to favour the largest number of coexisting nematode species. In lakes, splash-water zones are the most species-rich	Soil disturbance (especially compaction) can reduce nematode diversity. However, this is also context-dependent, as the type of the agricultural practice, the intensity of the soil disturbance and the physicochemical background of the soil do matter. In some cases soil disturbance can increase nematode diversity	Physicochemical disturbance of the habitat has clear effects on nematode community composition, but the effects on diversity are inconsistent and largely context-dependent. More studies should specifically address effects of short- versus long-term disturbances

Continued

Table 2.1. Continued.

		on Nematode diversity in			
Scale	Effect of	Marine	Freshwater	Terrestrial	Parallels and perspectives
Local/ micro	Substrate texture and food availability	Grain size is important but probably not the sole factor affecting nematode diversity. Effects are non-linear, intermediate granulometries having the largest potential to support more nematode species	In lakes and rivers, interstitial habitats support more nematode species than biofilms growing on a substratum's surface. Highest productivities are associated with the poorest nematode communities (see Fig. 2.3). Oligotrophic lake bottoms tend to support the largest number of species	Clear effects of soil texture on nematode community composition and functionalities. Soil nematode diversity is often related to plant diversity (especially for plant-parasitic species). Above-ground nematode diversity depends on tree-related microhabitat position and use by other biota	Habitat space (i.e. pore size in the case of interstitial habitats) is an important driver of nematode communities. It may well be that intermediate granulometries or heterogeneous substrates have the potential to support the maximum number of nematode species, but *NS
Micro	Biological interactions	Marine nematodes are largely affected by top-down predation, competition, but also facilitation processes with other meiofauna and the macrofauna. Interactions with microbes (bottom-up processes) are also complex and affect nematode community composition (see Fig. 2.4)	As for marine environments, nematodes play a central role in the benthic food web and, as such, they are largely affected by top-down and bottom-up processes. Inter-nematode interactions are also obvious and have the potential to modify or stabilize diversity at the micro scale	Nematodes play an important role in the soil food web. Bottom-up and top-down processes affect community composition. Some studies suggest that microbial diversity is linked to nematode diversity at the micro scale	Both top-down and bottom-up effects on nematodes have been studied extensively across habitats, evidencing important effects on nematode development, population dynamics, species coexistence and community assembly. As an integral part of below-ground food webs, nematode diversity is tightly linked with that of other biota. Future studies integrating nematodes within a food web framework have the potential to better link diversity patterns to ecosystem processes

nematode (and other meiofauna) activities (aka 'meioturbation') seems to have larger implications for benthic functions (nitrogen and carbon cycles) as shown experimentally in sediment cores (Nascimento *et al.*, 2012; Bonaglia *et al.*, 2014).

Understanding the effects of habitat disturbance on nematodes means understanding how anthropogenic changes at local, regional and global scales affect individuals and impose selective pressures upon their populations (Schratzberger *et al.*, 2023). The challenge for future nematode studies is to bridge across very different spatial and temporal scales, from one cubic centimetre of sediment or soil to ecosystem or indeed global scales (see Chapter 12, this volume), while considering disturbance regimes and potential regime shifts due to global change. In marine, freshwater and terrestrial habitats, physical disturbance often leads to a common pattern of immediate reduction of nematode diversity through a reduction of habitat size or an alteration of resources that tends to foster opportunistic species. But, by breaking parity, disturbances may also restart ecological successions, thus opening new opportunities for a diversity of colonizers to settle and form more diverse assemblages (e.g. in peat soils after tillage) until habitat specialists take over once again in the longer term. However, questions that explore short-term positive impacts of disturbance on nematode diversity through colonization and community assembly must be investigated in relation to disturbance magnitude, frequency, and the speed and dynamics of the ecological succession that ensues.

Building a stronger body and weight of evidence on nematode diversity across realms is essential to help address (and eventually answer) fundamental questions like, 'Why are nematodes so successful on Earth across different biomes?' and 'What are the mechanisms that lead to nematodes as a life form having such great potential for adaptation?' In a more applied perspective, better understanding of the drivers of nematode diversity and commmunity composition at global scales will improve their use as potent biological or ecological indicators across biomes.

References

Afzal, S., Nesar, H., Imran, Z. and Ahmad, W. (2021) Altitudinal gradient affect abundance, diversity and metabolic footprint of soil nematodes in Banihal-Pass of Pir-Panjal mountain range. *Scientific Reports* 11(1), 16214. DOI: 10.1038/s41598-021-95651-x.

Ahmed, M., Sapp, M., Prior, T., Karssen, G. and Back, M.A. (2016) Technological advancements and their importance for nematode identification. *Soil* 2(2), 257–270. DOI: 10.5194/soil-2-257-2016.

Alphei, J. (1998) Differences in soil nematode community structure of beech forests: comparison between a mull and a moder soil. *Applied Soil Ecology* 9(1), 9–15. DOI: 10.1016/S0929-1393(98)00047-X.

Althoff, P.S. and Thien, S.J. (2005) Impact of M1A1 main battle tank disturbance on soil quality, invertebrates, and vegetation characteristics. *Journal of Terramechanics* 42(3), 159–176. DOI: 10.1016/j.jterra.2004.10.014.

Anthony, M.A., Bender, F.S. and van der Heijden, M.G.A. (2023) Enumerating soil biodiversity. *Proceedings of the National Academy of Sciences USA* 120, e2304663120.

Appeltans, W., Ahyong, S.T., Anderson, G., Angel, M.V., Artois, T. *et al.* (2012) The magnitude of global marine species diversity. *Current Biology* 22(23), 2189–2202.

Armenteros, M., Quintanar-Retama, O. and Gracia, A. (2022) Depth-related patterns and regional diversity of free-living nematodes in the deep-sea Southwestern Gulf of Mexico. *Frontiers in Marine Science* 9, 1023996. DOI: 10.3389/fmars.2022.1023996.

Azovsky, A.I., Chertoprud, E.S., Garlitska, L.A., Mazei, Y.A. and Tikhonenkov, D.V. (2020) Does size really matter in biogeography? Patterns and drivers of global distribution of marine micro- and meiofauna. *Journal of Biogeography* 47, 1180–1192.

Barrett, J.E., Virginia, R.A., Wall, D.H., Cary, S.C., Adams, B.J. *et al.* (2006) Co-variation in soil biodiversity and biogeochemistry in northern and southern Victoria Land, Antarctica. *Antarctic Science* 18(4), 535–548. DOI: 10.1017/S0954102006000587.

Batterbury, S. and Warren, A. (2001) Desertification. In: Smelser, N and Baltes, P. (eds) *International Encyclopedia of the Social and Behavioral Sciences*. Elsevier, Amsterdam, pp. 3526–3529.

Bergtold, M., Gunther, V. and Traunspurger, W. (2005) Is there competition among ciliates and nematodes? *Freshwater Biology* 50, 1351–1359. DOI: DOI 10.1111/j.1365-2427.2005.01402.x.

Bloemers, G.F., Hodda, M., Lambshead, P.J.D., Lawton, J.H. and Wanless, F.R. (1997) The effects of forest disturbance on diversity of tropical soil nematodes. *Oecologia* 111(4), 575–582. DOI: 10.1007/s004420050274.

Boag, B. and Yeates, G.W. (1998) Soil nematode biodiversity in terrestrial ecosystems. *Biodiversity and Conservation* 7(5), 617–630. DOI: 10.1023/A:1008852301349.

Bonaglia, S., Nascimento, F.J.A. and Bartoli, M. (2014) Meiofauna increases bacterial denitrification in marine sediments. *Nature Communications* 5, 5133.

Bongers, T. (1990) The maturity index: an ecological measure of environmental disturbance based on nematode species composition. *Oecologia* 83, 14–19.

Bongers, T. and Ferris, H. (1999) Nematode community structure as a bioindicator in environmental monitoring. *Tree* 14, 224–228.

Borgmeier, A., Gattoni, K., Harris, T., Higgins, R., Mullin, P. *et al.* (2022) *Plectus* of the Prairie: a case study of taxonomic resolution from a nematode biodiversity survey. *Journal of Nematology* 54(1), 29–71.

Boulton, A.J. and Stanley, E.H. (1995) Hyporheic processes during flooding and drying in a Sonoran Desert stream. 2. Faunal dynamics. *Archiv für Hydrobiologie* 134(1), 27–52.

Cadet, P., Pate, E. and N'Diaye-Faye, N. (2003) Nematode community changes and survival rates under natural fallow in the sudano-sahelian area of Senegal. *Pedobiologia* 47(2), 149–160. DOI: 10.1078/0031-4056-00178.

Cadet, P., Berry, S.D., Leslie, G.W. and Spaull, V.W. (2007) Management of nematodes and a stalk borer by increasing within-field sugarcane cultivar diversity. *Plant Pathology* 56(3), 526–535. DOI: 10.1111/j.1365-3059.2006.01553.x.

Calderón-Sanou, I., Ohlmann, M., Münkemüller, T., Zinger, L., Hedde, M. *et al.* (2024) Mountain soil food webs shaped by the interplay between habitat and pedoclimatic conditions. *Soil Biology and Biochemistry* 190, 109282. DOI: 10.1016/j.soilbio.2023.109282

Chevalier, R. and Webster, J.M. (2006) Temporary changes in populations of soil organisms after field application of entomopathogenic nematodes. *Journal of Nematology* 38(4), 466–473.

Choudhary, F., Bhardwaj, A., Sayeed, I., Rather, S.A., Khan, M.A.H. and Shah, A.A. (2023) Elevational patterns of soil nematode diversity, community structure and metabolic footprint in the Trikuta mountains of North-western Himalaya. *Frontiers in Forests and Global Change* 6, 1135219. DOI: 10.3389/ffgc.2023.1135219.

Coleman, D.C., Odum, E.P. and Crossley, D.A. (1992) Soil biology, soil ecology, and global change. *Biology and Fertility of Soils* 14, 104–111.

Cortada, L., Dehennin, I., Bert, W. and Coyne, D. (2019) Integration of nematology as a training and research discipline in sub-Saharan Africa: progress and prospects. *Nematology* 22, 1–21.

Crist, T.O., Veech, J.A., Gering, J.C. and Summerville, K.S. (2003) Partitioning species diversity across landscapes and regions: a hierarchical analysis of α, β, and γ diversity. *The American Naturalist* 162(6), 734–743. DOI: 10.1086/378901.

Croll, N.A. and Zullini, A. (1972) Observations on the bionomics of the freshwater nematode *Chromadorina bioculata*. *Journal of Nematology* 4, 256–260.

Cui, S., Han, X., Xiao, Y., Wu, P., Zhang, S. *et al.* (2022) Increase in rainfall intensity promotes soil nematode diversity but offset by nitrogen addition in a temperate grassland. *Science of The Total Environment* 825, 154039. DOI: 10.1016/j.scitotenv.2022.154039.

D'Hondt, A.-S., Stock, W., Blommaert, L., Moens, T. and Sabbe, K. (2018) Nematodes stimulate biomass accumulation in a multispecies diatom biofilm. *Marine Environmental Research* 140, 78–89.

Danovaro, R., Gambi, C., Lampadariou, N. and Tselepides, A. (2008) Deep-sea nematode biodiversity in the Mediterranean basin: testing for longitudinal, bathymetric and energetic gradients. *Ecography* 31(2), 231–244. DOI: 10.1111/j.0906-7590.2008.5484.x.

Darby, B.J., Neher, D.A. and Belnap, J. (2007) Soil nematode communities are ecologically more mature beneath late- than early-successional stage biological soil crusts. *Applied Soil Ecology* 35(1), 203–212. DOI: 10.1016/j.apsoil.2006.04.006.

Davis, E.L., Hussey, R.S. and Baum, T.J. (2004) Getting to the roots of parasitism by nematodes. *Trends in Parasitology* 20(3), 134–141.

De Deyn, G.B., Van Ruijven, J., Raaijmakers, C.E., De Ruiter, P.C. and Van der Putten, W.H. (2007) Above- and belowground insect herbivores differentially affect soil nematode communities in species-rich plant communities. *Oikos* 116, 923–930.

de Goede, R.G.M. and Bongers, T. (1994) Nematode community structure in relation to soil and vegetation characteristics. *Applied Soil Ecology* 1(1), 29–44. DOI: 10.1016/0929-1393(94)90021-3.

De Guiran, G. and Bruguier, N. (1989) Hybridization and phylogeny of the pine wood nematode (*Bursaphelenchus* spp.). *Nematologica* 35(3), 321–330.

de Jonge, V.N. and Bouwman, L.A. (1977) A simple density separation technique for quantitative isolation of meiobenthos using the colloidal silica Ludox-TM. *Marine Biology* 42, 143–148.

De Mesel, I., Derycke, S., Swings, J., Vincx, M. and Moens, T. (2006) Role of nematodes in decomposition processes: does within-trophic group diversity matter? *Marine Ecology Progress Series* 321, 157–166.

De Queiroz, K. (2007) Species concepts and species delimitation. *Systematic Biology* 56(6), 879–886.

Derycke, S., Fonseca, G., Vierstraete, A., Vanfleteren, J., Vincx, M. and Moens, T. (2008a) Disentangling taxonomy within the *Rhabditis* (*Pellioditis*) *marina* (Nematoda, Rhabditidae) species complex using molecular and morhological tools. *Zoological Journal of the Linnean Society* 152(1), 1–15.

Derycke, S., Remerie, T., Backeljau, T., Vierstraete, A., Vanfleteren, J. *et al.* (2008b) Phylogeography of the *Rhabditis* (*Pellioditis*) *marina* species complex: evidence for long-distance dispersal, and for range expansions and restricted gene flow in the northeast Atlantic. *Molecular Ecology* 17(14), 3306–3322.

Derycke, S., De Meester, N., Rigaux, A., Creer, S., Bik, H. *et al.* (2016) Coexisting cryptic species of the *Litoditis marina* complex (Nematoda) show differential resource use and have distinct microbiomes with high intraspecific variability. *Molecular Ecology* 25(9), 2093–2110.

Dillman, A.R., Guillermin, M.L., Lee, J.H., Kim, B., Sternberg, P.W. and Hallem, E.A. (2012) Olfaction shapes host–parasite interactions in parasitic nematodes. *Proceedings of the National Academy of Sciences USA* 109(35), E2324–E2333.

Dong, K., Moroenyane, I., Tripathi, B., Kerfahi, D., Takahashi, K. *et al.* (2017) Soil nematodes show a mid-elevation diversity maximum and elevational zonation on Mt. Norikura, Japan. *Scientific Reports* 7(1), 3028. DOI: 10.1038/s41598-017-03655-3.

dos Santos, G.A.P., Silva, A.C., Esteves, A.M., Ribeiro-Ferreira, V.P., Neres, P.F. *et al.* (2020) Testing bathymetric and regional patterns in the Southwest Atlantic deep sea using infaunal diversity, structure, and function. *Diversity* 12(12), 485. DOI: 10.3390/d12120485.

Dümmer, B., Ristau, K. and Traunspurger, W. (2016) Varying patterns on varying scales: a metacommunity analysis of nematodes in European Lakes. *PLoS One* 11(3), e0151866. DOI: 10.1371/journal.pone.0151866.

DuPont, S.T., Ferris, H. and Van Horn, M. (2009) Effects of cover crop quality and quantity on nematode-based soil food webs and nutrient cycling. *Applied Soil Ecology* 41(2), 157–167.

Du Preez, G., Daneel, M., De Goede, R., Du Toit, M.J., Ferris, H. *et al.* (2022) Nematode-based indices in soil ecology: application, utility, and future directions, *Soil Biology and Biochemistry* 169, 108640. DOI: 10.1016/j.soilbio.2022.108640.

Eitzinger, B., Rall, B.C., Traugott, M. and Scheu, S. (2018) Testing the validity of functional response models using molecular gut content analysis for prey choice in soil predators. *Oikos* 127(7), 915–926. DOI: 10.1111/oik.04885.

Ettema, C.H. (1998) Soil nematode diversity: species coexistence and ecosystem function. *Journal of Nematology* 30, 159–169.

Eyualem-Abebe, Traunspurger, W. and Michiels, I.C. (2006) Dynamics of freshwater nematodes: abundance, biomass and diversity, In: Abebe, E., Andrássy, I. and Traunspurger, W. (eds) *Freshwater Nematodes: Ecology and Taxonomy*. CAB International, Wallingford, UK, pp. 77–93.

Eyualem-Abebe, Decraemer, W. and De Ley, P. (2008) Global diversity of nematodes (Nematoda) in freshwater. *Hydrobiologia* 595, 67–78.

Fenchel, T.M. (1978) The ecology of micro- and meiobenthos. *Annual Review of Ecology and Systematics* 9, 99–121.

Ferris, H. and Tuomisto, H. (2015) Unearthing the role of biological diversity in soil health. *Soil Biology and Biochemistry* 85, 101–109. DOI: 10.1016/j.soilbio.2015.02.037.

Ferris, H., Venette, R.C. and Lau, S.S. (1997) Population energetics of bacterial-feeding nematodes: carbon and nitrogen budgets. *Soil Biology and Biochemistry* 29(8), 1183–1194. DOI: 10.1016/S0038-0717(97)00035-7.

Flach, P.Z.S., Ozorio, C.P. and Melo, A.S. (2012) Alpha and beta components of diversity of freshwater nematodes at different spatial scales in subtropical coastal lakes. *Fundamental and Applied Limnology* 180, 249–258.

Fonseca, G., Soltwedel, T., Vanreusel, A. and Lindegarth, M. (2010) Variation in nematode assemblages over multiple spatial scales and environmental conditions in Arctic deep seas. *Progress in Oceanography* 84(3), 174–184. DOI: 10.1016/j.pocean.2009.11.001.

Fonseca, G., Maria, T.F., Kandratavicius, N. and Venekey, V. (2014) Testing for nematode–granulometry relationships. *Marine Biodiversity* 44(3), 435–443. DOI: 10.1007/s12526-014-0241-4.

Freckman, D.W. (1988) Bacterivorous nematodes and organic-matter decomposition. *Agriculture, Ecosystems and Environment* 24, 195–217.

Fu, S., Ferris, H., Brown, D. and Plant, R. (2005) Does the positive feedback effect of nematodes on the biomass and activity of their bacteria prey vary with nematode species and population size? *Soil Biology and Biochemistry* 37(11), 1979–1987. DOI: 10.1016/j.soilbio.2005.01.018.

Gansfort, B. and Traunspurger, W. (2019) Environmental factors and river network position allow prediction of benthic community assemblies: a model of nematode metacommunities. *Scientific Reports* 9, 14716.

Gansfort, B., Traunspurger, W., Threis, I. and Majdi, N. (2018a) Wide variation in a tiny space: the microdistribution of meiobenthos in an artificial pond. *Freshwater Biology* 63(5), 420–431.

Gansfort, B., Uthoff, J. and Traunspurger, W. (2018b) Interactions among competing nematode species affect population growth rates. *Oecologia* 187(1), 75–84.

Gattoni, K., Gendron, E.M.S., Borgmeier, A., McQueen, J.P., Mullin, P.G. *et al.* (2022) Context-dependent role of abiotic and biotic factors structuring nematode communities along two environmental gradients. *Molecular Ecology* 31(14), 3903–3916. DOI: 10.1111/mec.16541.

Gattoni, K., Gendron, E.M.S., Sandoval-Ruiz, R., Borgemeier, A., McQueen, J.P. *et al.* (2023) 18S-NemaBase: curated 18S rRNA database of nematode sequences. *Journal of Nematology* 55, e2023-1. DOI: 10.2478/jofnem-2023-0006.

Gattoni, K., Gendron, E.M.S., Powers, K., Powers, T.O., Harner, M.J. and Porazinska, D.L. (2024) Effects of drought-induced stress on nematode communities in aquatic and terrestrial habitats of the Nebraska Sandhills. *Frontiers in Ecology and Evolution* 12, 1305930. DOI: 10.3389/fevo.2024.1305930.

Gaudes, A., Sabater, S., Vilalta, E. and Muñoz, I. (2006) The nematode community in cyanobacterial biofilms in the river Llobregat, Spain. *Nematology* 8, 909–919. DOI: 10.1163/156854106779799169.

Geisen, S., Snoek, L.B., ten Hooven, F.C., Duyts, H., Kostenko, O. *et al.* (2018) Integrating quantitative morphological and qualitative molecular methods to analyse soil nematode community responses to plant range expansion. *Methods in Ecology and Evolution* 9(6), 1366–1378. DOI: 10.1111/2041-210X.12999.

Gendron, E.M., Sevigny, J.L., Byiringiro, I., Thomas, W.K., Powers, T.O. and Porazinska, D.L. (2023) Nematode mitochondrial metagenomics: a new tool for biodiversity analysis. *Molecular Ecology Resources* 23(5), 975–989. DOI: 10.1111/1755-0998.13761.

Gibson, K.S., Neher, D.A., Johnson, N.C., Parmenter, R.R. and Antoninka, A.J. (2023) Heavy logging machinery impacts soil physical properties more than nematode communities. *Forests* 14(6), 1205. DOI: 10.3390/f14061205.

Giere, O. (2009) *Meiobenthology. The Microscopic Motile Fauna of Aquatic Sediments*. Springer, Berlin.

Gingold, R., Moens, T. and Rocha-Olivares, A. (2013) Assessing the response of nematode communities to climate change-driven warming: a microcosm experiment. *PLoS One* 8(6), e66653.

Gobin, J.F. and Warwick, R.M. (2006) Geographical variation in species diversity: a comparison of marine polychaetes and nematodes. *Journal of Experimental Marine Biology and Ecology* 330(1), 234–244. DOI: 10.1016/j.jembe.2005.12.030.

Gong, X., Chen, X., Geisen, S., Zhang, J., Zhu, H. *et al.* (2021) Agricultural habitats are dominated by rapidly evolving nematodes revealed through phylogenetic comparative methods. *Soil Biology and Biochemistry* 155, 108183. DOI: 10.1016/j.soilbio.2021.108183.

Gong, X., Sun, X., Thakur, M.P., Qiao, Z., Yao, H. *et al.* (2023) Climate and edaphic factors drive soil nematode diversity and community composition in urban ecosystems. *Soil Biology and Biochemistry* 180, 109010. DOI: 10.1016/j.soilbio.2023.109010.

Grzelak, K., Szczuciński, W., Kotwicki, L. and Sugawara, D. (2014) Ecological status of sandy beaches after tsunami events: insights from meiofauna investigations after the 2011 Tohoku-oki tsunami, Sendai Bay, Japan. In: Kontar, Y.A., Santiago-Fandiño, V. and Takahashi, T. (eds) *Tsunami Events and Lessons Learned: Environmental and Societal Significance*. Springer, Dordrecht, The Netherlands, pp. 177–191. DOI: 10.1007/978-94-007-7269-4_9.

Guden, R.M., Vafeiadou, A.-M., De Meester, N., Derycke, S. and Moens, T. (2018) Living apart-together: microhabitat differentiation of cryptic nematode species in a saltmarsh habitat. *PLoS One* 13(9), e0204750.

Guden, R.M., Derycke, S. and Moens, T. (2021) A multi-faceted approach to understand how resource diversity can mediate the coexistence of cryptic marine nematode species. *Frontiers in Marine Science* 8, 777425.

Guilini, K., Veit-Köhler, G., De Troch, M., Van Gansbeke, D. and Vanreusel, A. (2013) Latitudinal and temporal variability in the community structure and fatty acid composition of deep-sea nematodes in the Southern Ocean. *Progress in Oceanography* 110, 80–92. DOI: 10.1016/j.pocean.2013.01.002.

Háněl, L. (2003) Recovery of soil nematode populations from cropping stress by natural secondary succession to meadow land. *Applied Soil Ecology* 22(3), 255–270. DOI: 10.1016/S0929-1393(02)00152-X.

Heininger, P., Höss, S., Claus, E., Pelzer, J. and Traunspurger, W. (2007) Nematode communities in contaminated river sediments. *Environmental Pollution* 146, 64–76.

Hodda, M. (1990) Variation in estuarine littoral nematode populations over three spatial scales. *Estuarine, Coastal and Shelf Science* 30(4), 325–340. DOI: 10.1016/0272-7714(90)90001-8.

Hodda, M. (2022a) Phylum Nematoda: feeding habits for all valid genera using a new, universal scheme encompassing the entire phylum, with descriptions of morphological characteristics of the stoma, a key, and discussion of the evidence for trophic relationships. *Zootaxa* 5114(1), 318–451.

Hodda, M. (2022b) Phylum Nematoda: trends in species descriptions, the documentation of diversity, systematics, and the species concept. *Zootaxa* 5114(1), 290–317.

Hodda, M. (2022c) Phylum Nematoda: a classification, catalogue and index of valid genera, with a census of valid species. *Zootaxa* 5114(1), 1–289.

Holovachov, O., Haenel, Q., Bourlat, S.J. and Jondelius, U. (2017) Taxonomy assignment approach determines the efficiency of identification of OTUs in marine nematodes. *Royal Society Open Science* 4(8), 170315.

Höss, S., Frank-Fahle, B., Lueders, T. and Traunspurger, W. (2015) Response of bacteria and meiofauna to iron oxide colloids in sediments of freshwater microcosms. *Environmental Toxicology and Chemistry* 34, 2660–2669. DOI: 10.1002/etc.3091.

Höss, S., Heininger, P., Claus, E., Möhlenkamp, C., Brinke, M. and Traunspurger, W. (2017) Validating the NemaSPEAR[%]-index for assessing sediment quality regarding chemical-induced effects on benthic communities in rivers. *Ecological Indicators* 73, 52–60.

Hou, W., Kuzyakov, Y., Qi, Y., Liu, X., Zhang, H. and Zhou, S. (2023) Functional traits of soil nematodes define their response to nitrogen fertilization. *Functional Ecology* 37, 1197–1210.

Ingels, J. and Vanreusel, A. (2013) The importance of different spatial scales in determining structural and functional characteristics of deep-sea infauna communities. *Biogeosciences* 10(7), 4547–4563. DOI: 10.5194/bg-10-4547-2013.

Ingels, J., Dashfield, S.L., Somerfield, P.J., Widdicombe, S. and Austen, M.C. (2014) Interactions between multiple large macrofauna species and nematode communities – mechanisms for indirect impacts of trawling disturbance. *Journal of Experimental Marine Biology and Ecology* 456, 41–49. DOI: 10.1016/j.jembe.2014.03.009.

Ingels, J., dos Santos, G., Hicks, N., Vazquez, Y.V., Neres, P.F. *et al.* (2018) Short-term CO_2 exposure and temperature rise effects on metazoan meiofauna and free-living nematodes in sandy and muddy sediments: results from a flume experiment. *Journal of Experimental Marine Biology and Ecology* 502, 211–226. DOI: 10.1016/j.jembe.2017.07.012.

Ingels, J., Leduc, D., Zeppilli, D. and Vanreusel, A. (2023) Deep-sea meiofauna – a world on its own or deeply connected? In: Giere, O. and Schratzberger, M. (eds) *New Horizons in Meiobenthos Research: Profiles, Patterns and Potentials*. Springer, Cham, Switzerland, pp. 257–283. DOI: 10.1007/978-3-031-21622-0_8.

Josefsson, H. (2018) The simplification of biological diversity in international and EU law. *Sustainability* 10(10), 3424.

Kardol, P., Bezemer, T.M., van der Wal, A. and Van der Putten, W.H. (2005) Successional trajectories of soil nematode and plant communities in a chronosequence of ex-arable lands. *Biological Conservation* 126(3), 317–327. DOI: 10.1016/j.biocon.2005.06.005.

Kashyap, P., Afzal, S., Rizvi, A.N., Ahmad, W., Uniyal, V.P. and Banerjee, D. (2022) Nematode community structure along elevation gradient in high altitude vegetation cover of Gangotri National Park (Uttarakhand), India. *Scientific Reports* 12(1), 1428. DOI: 10.1038/s41598-022-05472-9.

Kazemi-Dinan, A., Schroeder, F., Peters, L., Majdi, N. and Traunspurger, W. (2014) The effect of trophic state and depth on periphytic nematode communities in lakes. *Limnologica* 44, 49–57.

Kerfahi, D., Tripathi, B.M., Porazinska, D.L., Park, J., Go, R. and Adams, J.M. (2016) Do tropical rain forest soils have greater nematode diversity than High Arctic tundra? A metagenetic comparison of Malaysia and Svalbard. *Global Ecology and Biogeography* 25(6), 716–728. DOI: 10.1111/geb.12448.

Kitagami, Y., Obase, K. and Matsuda, Y. (2022) High-throughput sequencing and conventional morphotyping show different soil nematode assemblages but similar community responses to altitudinal gradients on Mt. Ibuki, Japan. *Pedobiologia* 90, 150788. DOI: 10.1016/j.pedobi.2021.150788.

Korenko, V. and Schmidt, C. (2006) Effects of agricultural practices in the rice crop system on nematode communities in Uruguay. *Nematologia Mediterranea* 34, 151–159.

Kouser, Y., Shah, A.A. and Rasmann, S. (2021) The functional role and diversity of soil nematodes are stronger at high elevation in the lesser Himalayan Mountain ranges. *Ecology and Evolution* 11(20), 13793–13804. DOI: 10.1002/ece3.8061.

Kreuzinger-Janik, B., Majdi, N. and Traunspurger, W. (2021) Distribution and diversity of meiofauna along an aquatic–terrestrial moss ecotone. *Nematology* 23(6), 695–714.

Kreuzinger-Janik, B., Gansfort, B. and Ptatscheck, C. (2022) Population density, bottom-up and top-down control as an interactive triplet to trigger dispersal. *Scientific Reports* 12, 5578. DOI: 10.1038/s41598-022-09631-w.

Lambshead, P.J.D., Tietjen, J., Ferrero, T. and Jensen, P. (2000) Latitudinal diversity gradients in the deep sea with special reference to North Atlantic nematodes. *Marine Ecology Progress Series* 194, 159–167.

Lambshead, P.J.D., Brown, C.J., Ferrero, T.J., Mitchell, N.J., Smith, C.R. *et al.* (2002) Latitudinal diversity patterns of deep-sea marine nematodes and organic fluxes: a test from the central equatorial Pacific. *Marine Ecology Progress Series* 236, 129–135.

Lavelle, P., Decaëns, T., Aubert, M., Barot, S., Blouin, M. *et al.* (2006) Soil invertebrates and ecosystem services. *European Journal of Soil Biology* 42, S3–S15.

Lazarova, S.. Coyne, D., Rodríguez, M.G., Peteira, B. and Ciancio, A. (2021) Functional diversity of soil nematodes in relation to the impact of agriculture – a review. *Diversity* 13(2), 64.

Lee, M.R. and Riveros, M. (2012) Latitudinal trends in the species richness of free-living marine nematode assemblages from exposed sandy beaches along the coast of Chile (18–42°S). *Marine Ecology* 33(3), 317–325. DOI: 10.1111/j.1439-0485.2011.00497.x.

Li, X., Zhu, H., Geisen, S., Bellard, C., Hu, F. *et al.* (2020) Agriculture erases climate constraints on soil nematode communities across large spatial scales. *Global Change Biology* 26(2), 919–930. DOI: 10.1111/gcb.14821.

Li, X., Liu, T., Li, H., Geisen, S., Hu, F. and Liu, M. (2022) Management effects on soil nematode abundance differ among functional groups and land-use types at a global scale. *Journal of Animal Ecology* 91(9), 1770–1780. DOI: 10.1111/1365-2656.13744.

Ma, Q., Zhu, Y., Wang, Y., Liu, T., Qing, X. *et al.* (2024) Livestock grazing modifies soil nematode body size structure in mosaic grassland habitats. *Journal of Environmental Management* 351, 119600. DOI: 10.1016/j.jenvman.2023.119600.

Maas, P.W.T. and Brinkman, H. (1980) Sampling of soil for nematode vectors of plant viruses in the Netherlands. *Mededelingen van de Faculteit Landbouwwetenschappen Rijksuniversiteit Gent* 45, 769–773.

Majdi, N. and Traunspurger, W. (2017) Leaf fall affects the isotopic niches of meiofauna and macrofauna in a stream food web. *Food Webs* 10, 5–14. DOI: 10.1016/j.fooweb.2017.01.002.

Majdi, N., Traunspurger, W., Boyer, S., Mialet, B., Tackx, M. *et al.* (2011) Response of biofilm-dwelling nematodes to habitat changes in the Garonne River, France: influence of hydrodynamics and microalgal availability. *Hydrobiologia* 673, 229–244.

Majdi, N., Mialet, B., Boyer, S., Tackx, M., Leflaive, J. *et al.* (2012a) The relationship between epilithic biofilm stability and its associated meiofauna under two patterns of flood disturbance. *Freshwater Science* 31(1), 38–50. DOI: 10.1899/11-073.1.

Majdi, N., Tackx, M., Traunspurger, W. and Buffan-Dubau, E. (2012b) Feeding of biofilm-dwelling nematodes examined using HPLC-analysis of gut pigment contents. *Hydrobiologia* 680, 219–232. DOI: 10.1007/s10750-011-0920-0.

Majdi, N., Boiché, A., Traunspurger, W. and Lecerf, A. (2014) Predator effects on a detritus-based food web are primarily mediated by non-trophic interactions. *Journal of Animal Ecology* 83(4), 953–962. DOI: 10.1111/1365-2656.12189.

Majdi, N., Threis, I. and Traunspurger, W. (2017) It's the little things that count: meiofaunal density and production in the sediment of two headwater streams. *Limnology and Oceanography* 62(1), 151–163. DOI: 10.1002/lno.10382.

Majdi, N., Colls, M., Weiss, L., Acuña, V., Sabater, S. and Traunspurger, W. (2020) Duration and frequency of non-flow periods affect the abundance and diversity of stream meiofauna. *Freshwater Biology* 65(11), 1906-1922.

Majdi, N., Moens, T. and Traunspurger, W. (2021) Feeding ecology of free-living nematodes. In: Traunspurger, W. (ed.) *Ecology of Freshwater Nematodes*. CAB International, Wallingford, UK, pp. 185–215.

Majdi, N., Gansfort, B., Gehner, S. and Traunspurger, W. (2024) Nematode production in artificial ponds: a long-term perspective. *Nematology* 26(3), 319–339.

Majdi, N., Traunspurger, W., Garrigue, J. and Larrieu, L. (2025) Tree-related microhabitats host specific micro-invertebrate communities involved in complex food webs. *Oecologia* 207, 148. DOI: 10.1007/s00442-025-05774-5.

Martin, T. and Sprunger, C.D. (2023) Nematodes require space: the relationship between nematode community assemblage and soil carbon across varying aggregate fractions. *Geoderma* 436, 116536. DOI: 10.1016/j.geoderma.2023.116536.

Mathieu, M., Leflaive, J., Ten-Hage, L., de Wit, R. and Buffan-Dubau, E. (2007) Free-living nematodes affect oxygen turnover of artificial diatom biofilms. *Aquatic Microbial Ecology* 49, 281–291. DOI: 10.3354/ame01150.

Mayr, E. (1942) *Systematics and the Origin of Species*. Columbia University Press, New York.

Menzel, R., Geweiler, D., Sass, A., Simsek, D. and Ruess, L. (2018) Nematodes as important source for omega-3 long-chain fatty acids in the soil food web and the impact in nutrition for higher trophic levels. *Frontiers in Ecology and Evolution* 6, 96.

Michiels, I.C. and Traunspurger, W. (2003) Maintenance of biodiversity through predation in freshwater nematodes? *Nematology Monographs and Perspectives* 2, 1–15.

Michiels, I.C. and Traunspurger, W. (2005) Impact of resource availability on species composition and diversity in freshwater nematodes. *Oecologia* 142, 98–103.

Moens, T., dos Santos, G.A.P., Thompson, F., Swings, J., Fonsêca-Genevois, V. *et al.* (2005) Do nematode mucus secretions affect bacterial growth? *Aquatic Microbial Ecology* 40, 77–83.

Mokievsky, V. and Azovsky, A. (2002) Re-evaluation of species diversity patterns of free-living marine nematodes. *Marine Ecology Progress Series* 238, 101–108.

Mokievsky, V.O., Udalov, A.A. and Azovsky, A.I. (2007) Quantitative distribution of meiobenthos in deep-water zones of the World Ocean. *Oceanology* 47(6), 797–813. DOI: 10.1134/S0001437007060057.

Monteiro, L., Van Butsel, J., De Meester, N., Traunspurger, W., Derycke, S. and Moens, T. (2018) Differential heavy-metal sensitivity in two cryptic species of the marine nematode *Litoditis marina* as revealed by developmental and behavioural assays. *Journal of Experimental Marine Biology and Ecology* 502, 203–210.

Muschiol, D., Giere, O. and Traunspurger, W. (2015) Population dynamics of a cavernicolous nematode community in a chemoautotrophic groundwater system. *Limnology and Oceanography* 60(1), 127–135.

Nascimento, F.J.A., Näslund, J. and Elmgren, R. (2012) Meiofauna enhances organic matter mineralization in soft sediment ecosystems. *Limnology and Oceanography* 57, 338–346.

Neumann, G. and Shields, E.J. (2006) Interspecific interactions among three entomopathogenic nematodes, *Steinernema carpocapsae* Weiser, *S. feltiae* Filipjev, and *Heterorhabditis bacteriophora* Poinar, with different foraging strategies for hosts in multipiece sand columns. *Environmental Entomology* 35(6), 1578–1583. DOI: 10.1093/ee/35.6.1578.

Newsham, K.K., Rolf, J., Pearce, D.A. and Strachan, R.J. (2004) Differing preferences of Antarctic soil nematodes for microbial prey. *European Journal of Soil Biology* 40(1), 1–8. DOI: 10.1016/j.ejsobi.2004.01.004.

Nicholas, W.L. and Trueman, J.W.H. (2005) Biodiversity of marine nematodes in Australian sandy beaches from tropical and temperate regions. *Biodiversity and Conservation* 14(4), 823–839. DOI: 10.1007/s10531-004-0656-5.

Okada, H. and Harada, H. (2007) Effects of tillage and fertilizer on nematode communities in a Japanese soybean field. *Applied Soil Ecology* 35(3), 582–598. DOI: 10.1016/j.apsoil.2006.09.008.

Palomares-Rius, J.E., Cantalapiedra-Navarrete, C. and Castillo, P. (2014) Cryptic species in plant-parasitic nematodes. *Nematology* 16(10), 1105–1118.

Pate, E., Ndiaye-Faye, N., Thioulouse, J., Villenave, C., Bongers, T. *et al.* (2000) Successional trends in the characteristics of soil nematode communities in cropped and fallow lands in Senegal (Sonkorong). *Applied Soil Ecology* 14(1), 5–15. DOI: 10.1016/S0929-1393(99)00051-7.

Pausch, J., Hofmann, S., Scharroba, A., Kuzyakov, Y. and Ruess, L. (2016) Fluxes of root-derived carbon into the nematode micro-food web of an arable soil. *Food Webs* 9, 32–38.

Peters, L. and Traunspurger, W. (2005) Species distribution of free-living nematodes and other meiofauna in littoral periphyton communities of lakes. *Nematology* 7, 267–280. DOI: 10.1163/1568541054879520.

Pinckney, J.L., Carman, K.R., Lumsden, S.E. and Hymel, S.N. (2003) Microalgal–meiofaunal trophic relationships in muddy intertidal estuarine sediments. *Aquatic Microbial Ecology* 31, 99–108. DOI: 10.3354/ame031099.

Porazinska, D.L., Giblin-Davis, R.M., Powers, T.O. and Thomas, W.K. (2012) Nematode spatial and ecological patterns from tropical and temperate rainforests. *PLoS One* 7(9), e44641. DOI: 10.1371/journal.pone.0044641.

Porazinska, D.L., Bueno de Mesquita, C.P., Farrer, E.C., Spasojevic, M.J., Suding, K.N. and Schmidt, S.K. (2021) Nematode community diversity and function across an alpine landscape undergoing plant colonization of previously unvegetated soils. *Soil Biology and Biochemistry* 161, 108380. DOI: 10.1016/j.soilbio.2021.108380.

Porazinska, D.L., Seastedt, T.R., Gendron, E.M.S. and Schmidt, S.K. (2022) Invasive annual cheatgrass enhances the abundance of native microbial and microinvertebrate eukaryotes but reduces invasive earthworms. *Plant and Soil* 473(1), 591–604. DOI: 10.1007/s11104-022-05312-9.

Powers, T.O., Todd, T.C., Burnell, A.M., Murray, P.C., Fleming, C.C. *et al.* (1997) The rDNA internal transcribed spacer region as a taxonomic marker for nematodes. *Journal of Nematology* 29(4), 441–450.

Ptatscheck, C. (2021) Role of nematodes in the food web: nematodes as predator and prey. In: Traunspurger, W. (ed.) *Ecology of Freshwater Nematodes*. CAB International, Wallingford, UK, pp. 216–246.

Ptatscheck, C. and Traunspurger, W. (2020) The ability to get everywhere: dispersal modes of free-living, aquatic nematodes. *Hydrobiologia* 847, 3519–3547. DOI: 10.1007/s10750-020-04373-0.

Ptatscheck, C., Gehner, S. and Traunspurger, W. (2020) Should we redefine meiofaunal organisms? The impact of mesh size on collection of meiofauna with special regard to nematodes. *Aquatic Ecology* 54, 1135–1143. DOI: 10.1007/s10452-020-09798-2.

Puissant, J., Villenave, C., Chauvin, C., Plassard, C., Blanchart, E. and Trap, J. (2021) Quantification of the global impact of agricultural practices on soil nematodes: a meta-analysis. *Soil Biology and Biochemistry* 161, 108383. DOI: 10.1016/j.soilbio.2021.108383.

Pulavarty, A., Egan, A., Karpinska, A., Horgan, K. and Kakouli-Duarte, T. (2021) Plant parasitic nematodes: a review on their behaviour, host interaction, management approaches and their occurrence in two sites in the Republic of Ireland. *Plants* 10(11), 2352.

Quist, C.W., Gort, G., Mooijman, P., Brus, D.J., van den Elsen, S. *et al.* (2019) Spatial distribution of soil nematodes relates to soil organic matter and life strategy. *Soil Biology and Biochemistry* 136, 107542. DOI: 10.1016/j.soilbio.2019.107542.

Rantalainen, M.-L., Fritze, H., Haimi, J., Pennanen, T. and Setälä, H. (2005) Species richness and food web structure of soil decomposer community as affected by the size of habitat fragment and habitat corridors. *Global Change Biology* 1614–1627.

Ridall, A. and Ingels, J. (2021) Suitability of free-living marine nematodes as bioindicators: status and future considerations. *Frontiers in Marine Science* 8, 685327.

Riemann, F. and Helmke, E. (2002) Symbiotic relations of sediment-agglutinating nematodes and bacteria in detrital habitats: the enzyme-sharing concept. *Marine Ecology* 23, 93–113. DOI: 10.1046/j.1439-0485.2002.02765.x.

Ristau, K., Faupel, M. and Traunspurger, W. (2012) The effects of nutrient enrichment on a freshwater meiofaunal assemblage. *Freshwater Biology* 57(4), 824–834.

Ristau, K., Faupel, M. and Traunspurger, W. (2013a) Effects of nutrient enrichment on the trophic structure and species composition of freshwater nematodes – a microcosm study. *Freshwater Science* 32, 155–168.

Ristau, K., Steinfartz, S. and Traunspurger, W. (2013b) First evidence of cryptic species diversity and significant population structure in a widespread freshwater nematode morphospecies (*Tobrilus gracilis*). *Molecular Ecology* 22, 4562–4575.

Ristau, K., Spann, N. and Traunspurger, W. (2015) Species and trait compositions of freshwater nematodes as indicative descriptors of lake eutrophication. *Ecological Indicators* 53, 196–205.

Rosli, N., Leduc, D., Rowden, A.A. and Probert, P.K. (2018) Review of recent trends in ecological studies of deep-sea meiofauna, with focus on patterns and processes at small to regional spatial scales. *Marine Biodiversity* 48, 13–34.

Rueda-Ramírez, D., Palevsky, E. and Ruess, L. (2023) Soil nematodes as a means of conservation of soil predatory mites for biocontrol. *Agronomy* 13(1), 32. DOI: 10.3390/agronomy13010032.

Ruess, L. and Dighton, J. (1996) Cultural studies on soil nematodes and their fungal hosts,. *Nematologica* 42(3), 330–346. DOI: 10.1163/004425996X00065.

Santos, P.F., Phillips, J. and Whitford, W.G. (1981) The role of mites and nematodes in early stages of buried litter decomposition in a desert. *Ecology* 62(3), 664–669. DOI: 10.2307/1937734.

Schenk, J. and Traunspurger, W. (2021) Sampling and processing of freshwater nematodes with emphasis on molecular methods. In: Traunspurger, W. (ed.) *Ecology of Freshwater Nematodes*. CAB International, Wallingford, UK, pp. 31–57.

Schenk, J., Kleinbölting, N. and Traunspurger, W. (2020) Comparison of morphological, DNA barcoding, and metabarcoding characterizations of freshwater nematode communities. *Ecology and Evolution* 10(6), 2885–2899.

Schmidt, J.H., Hallmann, J. and Finckh, M.R. (2020) Bacterivorous nematodes correlate with soil fertility and improved crop production in an organic minimum tillage system. *Sustainability* 12(17), 6730. DOI: 10.3390/su12176730.

Schratzberger, M. and Ingels, J. (2018) Meiofauna matters: the roles of meiofauna in benthic ecosystems. *Journal of Experimental Marine Biology and Ecology* 502, 12–25.

Schratzberger, M. and Larcombe, P. (2014) The role of the sedimentary regime in shaping the distribution of subtidal sandbank environments and the associated meiofaunal nematode communities: an example from the southern North Sea. *PLoS One* 9(10), e109445.

Schratzberger, M., Forster, R.M., Goodsir, F. and Jennings, S. (2008) Nematode community dynamics over an annual production cycle in the central North Sea. *Marine Environmental Research* 66(5), 508–519. DOI: 10.1016/j.marenvres.2008.08.004.

Schratzberger, M., Danovaro, R., Ingels, J., Montagna, P.A., Lupher, M.R. *et al.* (2023) Hidden players – meiofauna mediate ecosystem effects of anthropogenic disturbances in the ocean. In: Giere, O. and Schratzberger, M. (eds) *New Horizons in Meiobenthos Research: Profiles, Patterns and Potentials*. Springer, Cham, Switzerland, pp. 175–255. DOI: 10.1007/978-3-031-21622-0_7.

Schroeder, F., Muschiol, D. and Traunspurger, W. (2010) Fluctuating food availability may permit coexistence in bacterivorous nematodes. *Fundamental and Applied Limnology* 178, 59–66.

Schroeder, F., Peters, L. and Traunspurger, W. (2012) Temporal variations in epilithic nematode assemblages in lakes of different productivities. *Fundamental and Applied Limnology* 181, 143–157. DOI: 10.1127/1863-9135/2012/0332.

Semprucci, F., Cesaroni, L., Guidi, L. and Balsamo, M. (2018) Do the morphological and functional traits of free-living marine nematodes mirror taxonomical diversity? *Marine Environmental Research* 135, 114–122.

Shao, Y., Wang, Z., Liu, T., Kardol, P., Ma, C. *et al.* (2023) Drivers of nematode diversity in forest soils across climatic zones. *Proceedings of the Royal Society B: Biological Sciences* 290(1994), 20230107. DOI: 10.1098/rspb.2023.0107.

Sieriebriennikov, B., Ferris, H. and de Goede, R.G. (2014) NINJA: an automated calculation system for nematode-based biological monitoring. *European Journal of Soil Biology* 61, 90–93.

Slack, D.A., Riggs, R.D. and Hamblen, M.L. (1972) The effect of temperature and moisture on the survival of heterodera glycines in the absence of a host. *Journal of Nematology* 4(4), 263–266.

Smith, T.E., Wall, D.H., Hogg, I.D., Adams, B.J., Nielsen, U.N. and Virginia, R.A. (2012) Thawing permafrost alters nematode populations and soil habitat characteristics in an Antarctic polar desert ecosystem. *Pedobiologia* 55(2), 75–81. DOI: 10.1016/j.pedobi.2011.11.001.

Stark, J.S., Mohammad, M., McMinn, A. and Ingels, J. (2020) Diversity, abundance, spatial variation, and human impacts in marine meiobenthic nematode and copepod communities at Casey Station, East Antarctica. *Frontiers in Marine Science* 7, 480. DOI: 10.3389/fmars.2020.00480.

Steyaert, M., Herman, P.M.J., Moens, T., Widdows, J. and Vincx, M. (2001) Tidal migration of nematodes on an estuarine tidal flat (the Molenplaat, Schelde Estuary, SW Netherlands). *Marine Ecology Progress Series* 224, 299–304.

Stuart, R.J., Barbercheck, M.E. and Grewal, P.S. (2015) Entomopathogenic nematodes in the soil environment: distributions, interactions and the influence of biotic and abiotic factors. In: Campos-Herrera, R. (ed.) *Nematode Pathogenesis of Insects and Other Pests: Ecology and Applied Technologies for Sustainable Plant and Crop Protection*. Springer, Cham, Switzerland, pp. 97–137. DOI: 10.1007/978-3-319-18266-7_4.

Tahseen, Q., Sheridan, J. and Perry, R. (2004) Electropharyngeograms of pharyngeal pumping activity in six species of free-living nematodes. *Nematology* 6(1), 49–54. DOI: 10.1163/156854104323072919.

Taylor, L.S., Phillips, G., Bernard, E.C. and DeBruyn, J.M. (2020) Soil nematode functional diversity, successional patterns, and indicator taxa associated with vertebrate decomposition hotspots. *PLoS One* 15(11), e0241777.

Todd, T.C., Powers, T.O. and Mullin, P.G. (2006) Sentinel nematodes of land-use change and restoration in tallgrass prairie. *Journal of Nematology* 38(1), 20–27.

Tong, F., Xiao, Y. and Wang, Q. (2010) Soil nematode community structure on the northern slope of Changbai Mountain, Northeast China. *Journal of Forestry Research* 21(1), 93–98. DOI: 10.1007/s11676-010-0016-0.

Traunspurger, W. (1996a) Distribution of benthic nematodes in the littoral of an oligotrophic lake (Konigssee, National Park Berchtesgaden, FRG). *Archiv für Hydrobiologie* 135, 393–412.

Traunspurger, W. (1996b) Distribution of benthic nematodes in the littoriprofundal and profundal of an oligotrophic lake (Konigssee, National Park Berchtesgaden, FRG). *Archiv für Hydrobiologie* 135, 557–575.

Traunspurger, W. and Drews, C. (1996) Vertical distribution of benthic nematodes in an oligotrophic lake: seasonality, species and age segregation. *Hydrobiologia* 331, 33–42.

Traunspurger, W. and Majdi, N. (2021) Species composition and distribution of free-living nematodes in lakes and streams. In: Traunspurger, W. (ed.) *Ecology of Freshwater Nematodes*. CAB International, Wallingford, UK, pp. 58–108.

Traunspurger, W., Bergtold, M. and Goedkoop, W. (1997) The effects of nematodes on bacterial activity and abundance in a freshwater sediment. *Oecologia* 112, 118–122. DOI: 10.1007/s004420050291.

Traunspurger, W., Reiff, N., Krashevska, V., Majdi, N. and Scheu, S. (2017) Diversity and distribution of soil micro-invertebrates across an altitudinal gradient in a tropical montane rainforest of Ecuador, with focus on free-living nematodes. *Pedobiologia* 62, 28–35. DOI: 10.1016/j.pedobi.2017.04.003.

Traunspurger, W., Wilden, B. and Majdi, N. (2020) An overview of meiofaunal and nematode distribution patterns in lake ecosystems differing in their trophic state. *Hydrobiologia* 847, 2665–2679. DOI: 10.1007/s10750-019-04092-1.

Traunspurger, W., Witthöft-Mühlmann, A. and Höss, S. (2021) Free-living nematode communities in a large and deep oligotrophic lake in Europe: comparison of different depth zones of Lake Constance (Germany). *Nematology* 23, 69–87. DOI 10.1163/15685411-bja10030.

Trebukhova, Y.A., Miljutin, D.M., Pavlyuk, O.N., Mar'yash, A.A. and Brenke, N. (2013) Changes in deep-sea metazoan meiobenthic communities and nematode assemblages along a depth gradient (North-western Sea of Japan, Pacific). *Deep Sea Research Part II: Topical Studies in Oceanography* 86–87, 56–65. DOI: 10.1016/j.dsr2.2012.08.015.

Treonis, A.M., Unangst, S.K., Kepler, R.M., Buyer, J.S., Cavigelli, M.A. *et al.* (2018) Characterization of soil nematode communities in three cropping systems through morphological and DNA metabarcoding approaches. *Scientific Reports* 8(1), 2004. DOI: 10.1038/s41598-018-20366-5.

Tsiafouli, M.A., Argyropoulou, M.D., Stamou, G.P. and Sgardelis, S.P. (2006) Soil nematode biodiversity in organic and conventional agroecosystems of Northern Greece. *Russian Journal of Nematology* 14, 159–169.

Van Den Hoogen, J., Geisen, S., Routh, D., Ferris, H., Traunspurger, W. *et al.* (2019) Soil nematode abundance and functional group composition at a global scale. *Nature* 572(7768), 194–198.

Van Gundy, S.D. and Freckman, D.W. (1977) Phytoparasitic nematodes in belowground agroecosystems. *Ecological Bulletins* 25, 320–329.

Vanhove, S., Arntz, W. and Vincx, M. (1999) Comparative study of the nematode communities on the south-eastern Weddell Sea shelf and slope (Antarctica). *Marine Ecology Progress Series* 181, 237–256.

Vannote, R.L., Minshall, G.W., Cummins, K.W., Sedell, J.R. and Cushing, C.E. (1980) The river continuum concept. *Canadian Journal of Fisheries and Aquatic Sciences* 37, 130–137.

Vanreusel, A., Fonseca, G., Danovaro, R., Da Silva, M.C., Esteves, A.M. *et al.* (2010) The contribution of deep-sea macrohabitat heterogeneity to global nematode diversity. *Marine Ecology* 31(1), 6–20.

Vieira, D.C. and Fonseca, G. (2013) The importance of vertical and horizontal dimensions of the sediment matrix in structuring nematodes across spatial scales. *PLoS One* 8(10), e77704. DOI: 10.1371/journal.pone.0077704.

Viketoft, M., Palmborg, C., Sohlenius, B., Huss-Danell, K. and Bengtsson, J. (2005) Plant species effects on soil nematode communities in experimental grasslands. *Applied Soil Ecology* 30(2), 90–103. DOI: 10.1016/j.apsoil.2005.02.007.

Wallace, H.R. (1987) A perception of tolerance. *Nematologica* 33(4), 419–432. DOI: 10.1163/187529287X00083.

Wardle, D.A., Verhoef, H.A. and Clarholm, M. (1998) Trophic relationships in the soil microfood-web: predicting the responses to a changing global environment. *Global Change Biology* 4(7), 713–727.

Warwick, R.M. (1988) The level of taxonomic discrimination required to detect pollution effects on marine benthic communities. *Marine Pollution Bulletin* 19, 259–268.

Wasilewska, L. (1997) Soil invertebrates as bioindicators, with special reference to soil-inhabiting nematodes. *Russian Journal of Nematology* 5, 113–126.

Watzin, M.C. (1983) The effects of meiofauna on settling macrofauna: meiofauna may structure macrofaunal communities. *Oecologia* 59(2), 163–166. DOI: 10.1007/BF00378833.

Watzin, M.C. (1986) Larval settlement into marine soft-sediment systems: interactions with the meiofauna. *Journal of Experimental Marine Biology and Ecology* 98(1), 65–113. DOI: 10.1016/0022-0981(86)90076-6.

Weber, S. and Traunspurger, W. (2013) Food choice of two bacteria-feeding nematode species dependent on food source, food density and interspecific competition. *Nematology* 15, 291–301.

Weber, S. and Traunspurger, W. (2015) The effects of predation by juvenile fish on the meiobenthic community structure in a natural pond. *Freshwater Biology* 60(11), 2392–2409.

Wei, J., Wang, L., Zhu, J., Zhang, S., Nandi, O.I. and Kang, L. (2007) Plants attract parasitic wasps to defend themselves against insect pests by releasing hexenol. *PLoS One* 2(9), e852.

Weitere, M., Erken, M., Majdi, N., Arndt, H., Norf, H. *et al.* (2018) The food web perspective on aquatic biofilms. *Ecological Monographs* 88(4), 543–559.

Whitford, W.G., Freckman, D.W., Santos, P.F., Elkins, N.Z. and Parker, L.W. (1982) The role of nematodes in decomposition in desert ecosystems. In: Freckman, D.W. (ed.) *Nematodes in Soil Ecosystems*. University of Texas Press, New York, pp. 98–116. DOI: 10.7560/755260-009.

Wright, C.J. and Coleman, D.C. (2002) Responses of soil microbial biomass, nematode trophic groups, N-mineralization, and litter decomposition to disturbance events in the southern Appalachians. *Soil Biology and Biochemistry* 34(1), 13–25. DOI: 10.1016/S0038-0717(01)00128-6.

Wulfes, R., Traunspurger, W. and Ptatscheck, C. (2023) The armored catfish *Corydoras aeneus* shapes the meiofaunal community due to predation and disturbance-induced water column dispersal, *Hydrobiologia* 851, 1565–1576. DOI: 10.1007/s10750-023-05408-y.

Yeates, G.W. (2003) Nematodes as soil indicators: functional and biodiversity aspects. *Biology and Fertility of Soils* 37(4), 199–210. DOI: 10.1007/s00374-003-0586-5.

Yeates, G.W. and Bongers, T. (1999) Nematode diversity in agroecosystems. In: Paoletti, M.G. (ed.) *Invertebrate Biodiversity as Bioindicators of Sustainable Landscapes*. Elsevier, Amsterdam, pp. 113–135. DOI: 10.1016/B978-0-444-50019-9.50010-8.

Yeates, G.W., Bardgett, R.D., Cook, R., Hobbs, P.J., Bowling, P.J. and Potter, J.F. *et al.* (1997) Faunal and microbial diversity in three Welsh grassland soils under conventional and organic management regimes. *Journal of Applied Ecology* 34(2), 453–470. DOI: 10.2307/2404890.

Yeates, G.W., Ferris, H., Moens, T. and Van der Putten, W.H. (2009) The role of nematodes in ecosystems. In: Wilson, M.J. and Kakouli-Duarte, T. (eds) *Nematodes as Environmental Indicators*. CAB International, Wallingford, UK, pp. 1–44.

Zauner, H., Mayer, W.E., Herrmann, M., Weller, A., Erwig, M. and Sommer, R.J. (2007) Distinct patterns of genetic variation in *Pristionchus pacificus* and *Caenorhabditis elegans*, two partially selfing nematodes with cosmopolitan distribution. *Molecular Ecology* 16(6), 1267–1280. DOI: 10.1111/j.1365-294X.2006.03222.x.

Zhang, Q. and Hu, G. (2019) Structural analysis of the essential environmental factors associated with the marine meiofauna. *Water Science & Technology* 79(3), 566–572. DOI: 10.2166/wcc.2018.158.

Zhang, Y., Ji, L. and Yang, L. (2021) Abundance and diversity of soil nematode community at different altitudes in cold-temperate montane forests in northeast China. *Global Ecology and Conservation* 29, e01717. DOI: 10.1016/j.gecco.2021.e01717.

Zhang, Z.-Q. (2011) Animal biodiversity: an outline of higher-level classification and survey of taxonomic richness. *Zootaxa* 3148, 7–12.

Zobrist, E.C. and Coull, B.C. (1992) Meiobenthic interactions with macrobenthic larvae and juveniles: an experimental assessment of the meiofaunal bottleneck. *Marine Ecology Progress Series* 88(1), 1–8.

Zullini, A. (2014) Is a biogeography of freshwater nematodes possible? *Nematology* 16(1), 1–8.

Zullini, A. (2018) Cosmopolitism and endemism in free-living nematodes. *Biogeographia – The Journal of Integrative Biogeography* 33, 33–39.

3 The Evolution of Nematode Diversity

Sara Sánchez-Moreno[1]*, Dorota L. Porazinska[2] and Lee Robertson[3]
[1]*Department of Biodiversity and Evolutionary Biology, National Museum of Natural Sciences (MNCN-CSIC), Madrid, Spain;* [2]*Department of Entomology and Nematology, University of Florida, Gainesvile, Florida, USA;* [3]*Forestal Pathology Group, Center of Forest Sciences, National Institute for Agricultural and Food Research and Technology (INIA-CSIC), Madrid, Spain*

Abstract
This chapter briefly explores the remarkable evolutionary success of nematodes. Although nematodes are extremely abundant, their true diversity and the levels of endemism among terrestrial and aquatic species remain largely unknown. Parasitism is widespread within the phylum, but other ecological interactions, such as associations with insects and bacteria, are also common. Nematodes and bacteria have co-evolved, leading to the development of varying relationships among nematodes and bacterial species and communities. Nematodes exhibit the ability to colonize new habitats, and some nematode species may act as invasive organisms and quarantine species whose spread must be controlled. Despite their ecological importance, our current knowledge of nematode diversity, distribution and conservation status remains limited. The ongoing global biodiversity crisis also affects nematode diversity, yet nematodes are often overlooked as victims of this crisis. This chapter highlights the urgent need to enhance global efforts to protect nematodes and the largely unexplored biodiversity of soils.

3.1 Evolution of Nematode Diversity and Speciation

Nematodes might look like simple organisms for those unfamiliar with the phylum. However, nematodes are extremely diverse organisms that hide a large complexity in terms of ecological, physiological and genetic adaptation to the numerous habitats they inhabit and life forms they present. Like most commonly encountered metazoan life forms that exist on Earth today, nematodes most likely appeared during the Cambrian explosion (600–550 million years ago) as marine organisms. Their microscopic and soft bodies hinder fossilization, so their fossil records are scarce. However, nematode fossils have been found in amber, rocks (especially in fine-grained deposits) and coprolites (Poinar, 2015). The oldest fossil (Early Devonian, 396 million years ago) belongs to the plant-parasitic nematode *Palaeonema phyticum* (Poinar *et al.*, 2008) found in the stomatal chambers of the plant *Aglaophyton major*. Nematode animal parasites have been recovered from 240-million-year-old cynodont coprolites in Brazil (Hugot *et al.*, 2014), while insect parasites in their insect hosts have been preserved in Baltic, Dominican and Mexican amber older than 100 million years (Poinar, 2015).

Nematodes are an extraordinarily successful group of animals occupying most ecosystems on Earth, including extreme polar ice caps, hot

*Email: sarasm@inia.csic.es

DOI: 10.1079/9781800624221.0003

springs and desserts, and deep-sea vents. Although plant- and animal-parasitic nematodes have received much scientific attention due to their impact on human, animal and plant health, most nematodes are beneficial organisms with relevant roles in ecosystem functions. Nematodes are ubiquitous components of aquatic, marine and soil food webs. They are vastly abundant in soils (up to 3000 individuals per 100 g of soil; van den Hoogen *et al.*, 2019) and compose up to 90% of all marine meiofauna, with abundances of up to 10,000 individuals per 10 cm^2 in marine sediments (Ridall and Ingels, 2021). By being the most abundant animals on Earth, they represent 80% of all metazoan abundance. Their global estimated abundance reaches 4.4×10^{20} individuals and total biomass 0.3 gigatonnes (van den Hoogen *et al.*, 2019). In contrast to typical latitudinal patterns of increasing abundance from polar to tropical regions observed for most animals and plants, nematodes are most abundant in arctic/subarctic soils that are rich in organic matter (van den Hoogen *et al.*, 2019). So far, approximately 30,000 nematode species have been described. While this is only a fraction of the actual number of species, estimates of the expected number of existing species vary greatly. Larsen *et al.* (2017) estimated the number of nematode species parasitizing insects, vertebrates and plants at 40.8 million nematode species, but more modest estimates of total nematode species hover around 1 million (Scheffers *et al.*, 2012). There are many factors contributing to this lack of an accurate estimate of nematode diversity, including the presence of cryptic and endemic species. With only 30,000 nematode species officially described (Kiontke and Fitch, 2013), the extent of cryptic species for most of the nematode diversity has not been characterized. However, the existence of numerous cryptic species among marine nematodes is well known. Species complexes such as *Litoditis marina sensu lato* presents up to four different species with varying dispersal behaviour, resource use and environmental requirements (De Meester *et al.*, 2015, 2016, 2018; Derycke *et al.*, 2016). It is possible that the diversity of terrestrial, freshwater and animal parasitic nematodes has been also heavily underestimated due to the presence of cryptic species (Nadler and Pérez-Ponce De León, 2011; Ristau *et al.*, 2013; Palomares-Rius *et al.*, 2014).

Like crypticity, the extent of nematode endemicity (and cosmopolitism) has rarely been studied. The limited information available mostly relates to the dispersal of marine nematode species. In the case of cosmopolitanism, we know that it can be found, for example, in deep-sea nematode genera such as *Manganonema*, which are able to colonize sediments at different water depths. Indeed, species within the *Manganonema* genus show an impressive range of distributions from endemic species to cosmopolitan species, with still today half of the species encountered being new to science (Derycke *et al.*, 2013), indicating the potential for hidden biodiversity, even in well-studied deep-sea ecosystems. Turbidity currents and benthic storms that can transport organisms, including nematodes, over long distances are one of the reasons for the high dispersal of deep-sea nematodes despite their limited adult dispersal capabilities. Further, sediment resuspension is also thought to contribute to the dispersal of marine nematodes. This explains why marine nematodes are commonly considered highly cosmopolitan, although some endemism might occur (Bik *et al.*, 2010). Unfortunately, the cosmopolitanism and/or endemicity of terrestrial nematode species has not been studied in great detail, probably due to the difficulty of compiling information on the presence and distribution of soil nematodes to species level and the high number of undescribed species. However, we know that endemicity is more common than what was expected, and probably common in isolated territories such as islands, extreme environments, and pre-Quaternary lakes (Zullini, 2018).

Nematode evolutionary success has been attributed to their diversity of forms and feeding traits, underlined by a complex evolutionary history. De Ley (2006) described nematodes as 'the haiku among multicellular animals, combining endless variation with a deceptively simple underlying anatomical pattern'. Nematodes are extremely abundant at most trophic levels, by feeding on bacteria, fungi, algae, plants and other soil animals. Parasites of plants, vertebrates, molluscs and insects initially evolved from ancient free-living nematodes (Enoplia), but eventually all types of parasitism evolved multiple times within most of the major nematode clades (Blaxter and Koutsovoulos, 2015). Plant-parasitic nematodes probably evolved from fungivorous ancestors (Holterman *et al.*, 2006), likely more than once (Blaxter and Koutsovoulos, 2015). Nematode clades show great convergent evolution, and it is difficult to find biological characteristics that did not arise more than once during evolution (van Megen *et al.*, 2009).

3.1.1 Evolution of nematode plant parasitism

One of the most striking events within the evolution of nematodes is the degree of parasitism within the phylum. Among the 40 animal phyla, Nematoda is the phylum with the highest number of parasitic species, and parasitism seems to have been an extremely successful strategy in nematode evolutionary history. Parasitism can be defined as a relationship between two living species in which one benefits at the expense of the other. The evolutionary history of parasitism in nematode parasites and their hosts is complex and multifaceted; nematodes can parasitize a wide range of hosts including plants, vertebrates and invertebrates (Fig. 3.1), each having a unique life cycle making direct comparisons very difficult. Understanding the evolutionary origins of animal and plant parasitism and the mechanisms by which these parasites localize and invade their hosts, evade host detection and immunity, and acquire food are important goals not only of basic science but also of medical, veterinary and agricultural sciences.

The phylum Nematoda can be divided into two primary classes, Enoplea and Chromadorea (De Ley and Blaxter, 2002). Each class consists of parasitic and free-living nematodes that live in a wide variety of habitats. Originally, nematodes were classified based on anatomical characteristics, life cycle and habitats, but the advent of molecular techniques based on 18S and 28S ribosomal gene sequences (Holterman *et al.*, 2006, 2009; van Megen *et al.*, 2009) enabled a better understanding of nematode relationships and as a result nematodes are nowadays divided into five different clades. Interestingly, phylogenetic analysis demonstrates that: (i) parasitic nematodes are generally found in the same clades irrespective of the host they parasitize, whether it be plants, vertebrates or invertebrates; (ii) parasitism is a polyphyletic trait that evolved multiple times in the evolutionary history of nematodes; and (iii) plant parasitism is linked to the evolution of several morphological specializations. All plant-parasitic nematodes present a stylet-type structure, an odontostyle,

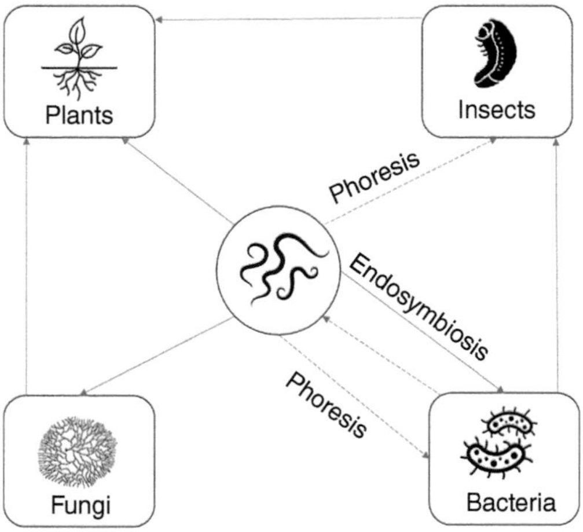

Fig. 3.1. Biotic interactions between nematodes and other biota as described in this chapter. Solid lines represent feeding relationships; dashed lines represent non-feeding relationships such as phoretic and symbiotic relations. Arrow heads point to the prey species or to the dependent species. (Author's own image.)

stomastostyle or onchiostyle, that allows cell-wall piercing, and an oesophagus with accessorial glands (Baldwin *et al.*, 2004).

Parasitic evolution occurred, however, in intimate relationship with the host. Root-knot and cyst nematodes present the most complex host–parasite interactions, in which nematodes regulate host plant genes in their favour, creating a complex feeding site that is the source of nutrition for the nematode (Abad and Williamson, 2010). Plant-parasitic nematodes commonly present some peculiarities, for example their genomes are generally small compared to free-living counterparts (often smaller than 100 Mb) and present an AT (adenine–thymine) bias. The plant-parasitic nematode *Pratylenchus coffeae* has the smallest metazoan genome sequenced to date (19.7 Mbp), with less than half the genes present in, for example, cyst nematodes (Burke *et al.*, 2015). Also, many plant-parasitic nematodes are polyploid and present a wide range of reproduction modes (Kikuchi *et al.*, 2017). The transition from sexual forms with a low number of chromosomes to polyploid apomictic species, which represent some of the most damaging pests worldwide, also occurred in the evolution of parasitism. Although exceptions are numerous, mitotic species commonly present wider host ranges than meiotic ones (Castagnone-Sereno *et al.*, 2013). Horizontal gene transfer from bacteria and fungi to nematodes was also a driving force in the evolution of parasitism in nematodes, and allowed nematodes to acquire genes that, for example, synthesize cell-wall-degrading enzymes, especially cellulases (Jones *et al.*, 2005).

Nematode associations with insects are diverse and spread across numerous nematode orders, mainly in the subclass Chromadoria. Tri-trophic interactions in which nematodes associate with both insects and plants only occur in the families Aphelenchidae and Tylenchidae. Insect-associated plant-parasitism probably evolved from fungal-feeding nematodes that presented preadaptations with phoretic transport, including developmental synchrony with the insect host, the presence of dispersal-stage juveniles (dauers) and the ability to survive environmental stress (Giblin-Davis *et al.*, 2003). The best-known nematodes that present this association with plants and insects are in the genera *Bursaphelenchus* (the pine wood nematode (PWN) *Bursaphelenchus xylophilus* and the red ring nematode *Bursaphelenchus cocophilus*) and the fig-associated *Schistonchus* nematodes. In *Bursaphelenchus*, dauer juveniles are phoretically associated with insects such as sawyer beetles and weevils (*Monochamus* spp. in *B. xylophilus*, *Rhynchophorus palmarum* in *B. cocophilus*), which transport the nematode to new plant hosts necessary for the completion of their life cycle. *Bursaphelenchus xylophilus* has recently converted into an invasive species that presents a global threat to pine forests (see Section 3.2 in this chapter). Similarly, nematodes in the genus *Schistonchus* are transported by Agaonidae wasps that pollinize figs, in which the nematodes feed and complete their life cycle.

An extreme case of tri-trophic co-evolution occurred in nematode species of the genus *Fergusobia*. Such nematodes present a dicyclic life cycle with an obligate stage as a plant-parasitic nematode in a number of Myrtaceae plant hosts, mainly *Eucalyptus* species in Australia. Species of the *Fergusobia* genus enter in species-specific mutualistic relationships with *Fergusonina* flies (Fergusoninidae, Diptera), together being responsible for gall formation and maintenance in the plant. The *Fergusonina* flies lay their eggs, together with juvenile nematodes, in undifferentiated meristematic plant tissue, in which nematodes feed, inducing gall formation, and develop into parthenogenetic females. When insect eggs hatch, the larvae feed on gall tissue. Female parthenogenetic nematodes lay eggs that develop into amphimictic male and female nematodes, and inseminated females enter third-instar fly larvae. The parasitic female nematodes lay eggs into the haemolymph of the fly, and emerged nematodes move to the oviducts to be deposited with its eggs into a new plant (Giblin-Davis *et al.*, 2003).

3.1.2 Nematode–bacteria co-evolution

Co-evolution did not only occur between parasitic nematodes and their hosts. Within terrestrial soil and aquatic sediments, nematodes routinely associate with microbes in at least two main ways: (i) as predators of fungi and bacteria (Yeates *et al.*, 1993); and (ii) as symbionts (Ogier *et al.*, 2020) (Fig. 3.1). Our current knowledge about the symbiotic association between nematodes and bacteria is mainly limited to studies of *Caenorhabditis elegans* and a handful of other nematode species, leaving large gaps in the current knowledge for the vast majority of phylogenetically and functionally diverse nematode species. Do all nematodes associate with microbes? Are these associations predictable? Do they affect nematode biology and their ecological relationships? Do they influence nematode evolution, diversification, and their contribution to ecosystem functioning? Are the symbiotic relationships subject to shifts driven by global change factors? Most of these questions do not have well-defined answers yet.

Generally, nematode–microbe associations can span a wide spectrum of interactions ranging from mutualistic to commensal to pathogenic. These interactions can be simple by involving just a single microbial species or be much more complex and extend to diverse microbial communities. Nevertheless, nematode–microbe associations are increasingly recognized as non-random ecological units (holobionts) (Zilber-Rosenberg and Rosenberg, 2008) with microbes playing important roles in nematode fitness and evolution by protecting nematodes against pathogens, regulating their development, reproduction and longevity, assisting with digestion and provision of vitamins, nutrients and energy, and modulating responses to biotic and abiotic stresses (Cabreiro and Gems, 2013).

Among the best-documented associations to date are those involving single symbionts of filarial, plant-parasitic and entomopathogenic nematodes. Thus far, these single associates have been connected to only four phylogenetically divergent bacterial phyla, namely Bacteroidota, Firmicutes, Proteobacteria and Verrucomicrobia (Brown, 2018). Although the nature of their relationships with nematode hosts has been defined as generally species-specific, it can range from strictly obligatory to facultative and from beneficial to antagonistic. For example, the human parasite *Brugia malayi* requires rickettsial *Wolbachia* (e.g. *Wolbachia bancrofti*) endosymbionts for development, fertility and survival (Taylor *et al.*, 2005). Similarly, verrucomicrobial *Xiphinematobacter* (e.g. *Xiphinematobacter americani*, *Xiphinematobacter brevicolli* and *Xiphinematobacter rivesi*) associated with plant-parasitic dagger nematodes (*Xiphinema*) likely supplement their hosts with nutrients (Brown *et al.*, 2015). Although such types of intracellular endosymbionts have been thought to be rare in other nematodes, *Wolbachia* and *Cardinium* have been recently recovered from plant parasites such as cyst (*Globodera* and *Heterodera*), lesion (*Pratylenchus*) and burrowing (*Radopholus*) plant-pathogenic nematodes as potentially facultative rather than obligatory mutualists (Brown, 2018). New discoveries of endosymbionts will likely continue as we examine more nematodes in more detail on bacteria (e.g. *Ricketsiella* in omnivorous nematodes) (Ladygina *et al.*, 2009).

The best-studied bacterial symbionts belong to *Xenorhabdus* and *Photorhabdus*, both known to exclusively associate with entomopathogenic nematodes (*Steinernema* and *Heterorhabditis*). While the exact nature of the relationship is still not fully known, bacterial symbionts are thought to benefit the nematodes by suppressing the immunity of insect hosts, killing them and themselves serving as a food source for the nematodes (Goodrich-Blair and Clarke, 2007). The nematode benefits the bacteria by serving as a host during dispersal and by invading an insect victim. As exemplified by the exclusive interaction between *Steinernema carpocapsae* and *Xenorhabdus nematophila*, relationships between entomopathogenic nematode hosts and bacterial symbionts appear to be highly species-specific.

Over the last decade, there has been increasing evidence that nematodes not only associate with single bacterial species but also with rather complex bacterial communities (microbiomes). Extracellular microbiomes (i.e. intestinal lumen and cuticular surface) are of particular interest. The best-described microbiomes thus far belong to the model nematode *C. elegans*. We now know that the microbiome of *C. elegans* is highly diverse, dominated by four major bacterial phyla (i.e. Proteobacteria, Bacteroidota, Firmicutes and Actinobacteria) (Samuel *et al.*, 2016), and distinct from both the microbiome of its natural environment (rotting fruit) and the microbiome of its close relative *Caenorhabditis remanei* (Dirksen *et al.*, 2016; Schulenburg and Félix, 2017). Although microbiomes can vary considerably among individual specimens (Taylor and Vega, 2021), natural populations of *C. elegans* from distant geographic locations with different environmental microbial communities showed remarkably similar microbiomes. In addition, replacement of the natural gut microorganisms with *Escherichia coli* not only reduced the fitness of *C. elegans*, but also its resistance to pathogens (Dirksen *et al.*, 2016). Likewise, while the microbiomes of *C. elegans* and other species such as *Caenorhabditis briggsae* protect each species against pathogenic infections, the microbiome of one species failed to protect the other species (Berg *et al.*, 2016). All this evidence suggests that nematodes likely assemble a non-random microbiome shaped by deterministic factors, including the environment (source community), host-driven selection (e.g. genetics, feeding behaviour and metabolism) and biotic interactions within the microbiome (Berg *et al.*, 2016; Ortiz *et al.*, 2021).

Individual components of *C. elegans* gut microbiomes can have neutral, beneficial and detrimental effects on the host (Radeke and Herman, 2021). First, the fitness of *C. elegans* appears to depend on live intact bacteria as lysed or killed bacteria were found to arrest nematode development (Cabreiro and Gems, 2013; Shapira, 2017). Second, the nature of individual bacterial isolates can vary, with up to 80% being largely beneficial while 20% detrimental to the growth of *C. elegans* (Samuel *et al.*, 2016; Xue *et al.*, 2023). Moreover, *C. elegans* raised on microbial monocultures produce different life-history traits than when raised on mixed cultures (Urquiza-Zurich *et al.*, 2023). Finally, the nature of bacterial effects appears to be dynamic through the life cycle of *C. elegans* and can shift from primarily beneficial in the early life stages to dysbiotic in later life stages (Radeke and Herman, 2021).

More recently, nematode microbiome studies have expanded beyond *C. elegans* to include plant-parasitic nematodes (e.g. *Meloidogyne incognita* and *Pratylenchus penetrans*) (Elhady *et al.*, 2017), entomopathogenic nematodes (e.g. *S. carpocapsae*) (Ogier *et al.*, 2020), free-living nematodes (e.g. *Acrobeloides maximus*, *Plectus murrayi* and *Eudorylaimus antarcticus*) (Baquiran *et al.*, 2013; McQueen *et al.*, 2013, 2022), insect-associated nematodes (e.g. *Pristionchus harborus*) (Meyer *et al.*, 2017) and marine nematodes (e.g. *L. marina*) (Derycke *et al.*, 2016; Xue *et al.*, 2023). Generally, the patterns observed are largely consistent with the findings in studies of *C. elegans*. First, all nematode species studied, including entomopathogenic nematodes, have been confirmed to have a microbiome. In addition to nematodes with only one expected symbiont (*X. nematophila*), *S. carpocapsae* was found to support at least a dozen Proteobacteria species (e.g. *Pseudomonas* spp., *Alcaligenes* spp. and *Ochrobactrum* spp.), albeit with unknown roles (Ogier *et al.*, 2020). Second, nematode microbiomes have been clearly distinct from and generally less diverse than the microbiomes of their environments (Meyer *et al.*, 2017; Elhady *et al.*, 2017; Schuelke *et al.*, 2018; McQueen *et al.*, 2022), confirming some level of environmental filtering playing a role in the assembly of nematode microbiomes. And third, nematode microbiomes appear to be nematode-species-specific (Baquiran *et al.*, 2013; Elhady *et al.*, 2017; Meyer *et al.*, 2017; McQueen *et al.*, 2022; Xue *et al.*, 2023), indicating that host-specific factors such as genotype, phylogeny, feeding preferences or other life traits also likely shape the microbiomes. Interestingly, the lack of distinct roles of host-specific factors in the microbiome of marine and estuarine nematodes (Schuelke *et al.*, 2018; Vafeiadou *et al.*, 2022) suggests that nematode–microbe relationships can vary significantly among species, feeding groups, life strategies and habitats, and that more studies are needed before generalizable patterns can be described. Nevertheless, microbiomes appear to be a normal part of nematode body, biology and ecology, and are hence potentially key to nematode evolution, diversification and function in ecosystems. Direct effects of global change factors (e.g. climate warming and pollution) on nematodes (Porazinska *et al.*, 2021; Shepherd *et al.*, 2021) may be amplified by indirect effects via microbe symbionts, potentially deepening the biodiversity crisis.

3.2 Nematodes as Invasive Species

Recent ecological changes due to global change, together with current human mobility around the globe, have produced the expansion of some nematode species to new areas. This has occasionally resulted in biological invasions of exotic diseases that pose a major threat to modern agriculture and food security. The success in the establishment of a parasitic nematode in a new area depends on several factors, including the quantity and quality of propagules, the host range, the duration of the life cycle, the presence of specific adaptations for survival in new niches, the presence of competitors and abiotic factors. When these factors permit the establishment of a new plant-parasitic nematode in a territory, it might become an invasive species.

There are no global lists of potentially invasive nematode species as invasiveness of a species will depend on the suitability of the new area, which cannot be easily deduced from the conditions of the area of origin, namely where the species is native. However, the European and Mediterranean Plant Protection Organization (EPPO) lists five nematode genera recommended for regulation as quarantine pests, including *Ditylenchus destructor*, *Ditylenchus dipsaci*, *Bursaphelenchus xylophilus*, *Xiphinema americanum sensu lato*, *Aphelenchoides besseyi*, *Aphelenchoides fragariae*, *Aphelenchoides ritzemabosi*, and the species within the genus *Anguina*. To our current knowledge, such nematodes do not occur on European territory, and their introduction must be prevented at all costs. Another 11 nematode species are locally present in the EPPO area but should also be regulated as quarantine pests and are considered invasive. In contrast, the US Department of Agriculture (USDA) only considers two *Globodera* species as invasive to the USA. Independently of the length of such lists, it is mandatory to avoid the dispersal of nematode species that may harm humans, crops and ecosystems. International standards to avoid the dispersal of such and other pests have been brought into force with the aim of reducing the introduction of invasive species, thereby protecting sustainable agriculture and global food security, the environment, forests and biodiversity, and safeguarding trade and economic development.

In recent years, the most-notorious invasive nematode has been the wood-feeding pine wood nematode, *Bursaphelenchus xylophilus*. The species is native to North America, where most conifer species are resistant or tolerant to the species. Outside North America, it affects forest stands of multiple conifer species and has produced a tremendous decline of conifer species in Eurasia, especially in Asian countries such as Japan, China, Korea and Taiwan. In recent years, the nematode has established in Southern Europe with a great damage potential. Its aggressiveness has forced the European Union (EU) to include *B. xylophilus* among other pests whose economic, environmental and social impact on EU territory is the most severe. The nematode, which attacks live standing trees, does not generally thrive once the tree is felled. However, it can be found present in a latent state in dead wood (e.g. stumps, cut wood in the field or sawmills, wood chips, pallets), allowing it to spread and affect new areas with economic, ecological and social impacts (Lawson and Sathyapala, 2008). Prediction models of the economic impact for the period 2008–2030 at a European level indicate that, if no control measures for *B. xylophilus* are employed, the European forest industry might accumulate losses of €3 billion in the initial years, which would progressively decrease to €300–800 million annually in the following ones, estimating a loss of €218 million in 2030 due to a shortage of timber, with all the consequences for related industries and society (Soliman *et al.*, 2012). However, this model does not take into account the effects of climate change. The increase in global temperature is predicted to be 1.1–6.4°C at the end of the 21st century (IPCC, 2013). In such scenario, average summer temperatures are expected to increase, which will probably result in the expansion of the northern border of the nematode's current distribution, and in an increase in the breeding cycle of the insect vector, which might thus lead to an increase in both the distribution area and the transmission of the nematode (de la Fuente *et al.*, 2018). In recent years, the expansion of *B. xylophilus* showed how nematode invaders might harm local economies and global ecosystems. To mitigate the impacts of nematode invasive species, prevention of dispersion and sanitation are the most important strategies to control such nematodes, as it is extremely difficult to control nematode populations once they become established in a given location.

3.3 Nematode Diversity under Current Global Change Conditions

The distribution of soil nematodes and their diversity at global scales depend on a number of large-scale environmental factors including climate, vegetation, land use and edaphic properties (see Chapter 12, this volume), although local factors (e.g. local climate) might affect local nematode diversity as well. Among these factors, climate might be of particular importance. For example, nematode community composition across 12 sites on five continents was in large part explained by mean annual rainfall and temperature (65% and 58%, respectively) (Nielsen *et al.*, 2014). Similarly, at a local scale climate-driven shifts in plant communities explained shifts in abundance and diversity patterns of high-alpine nematode communities (Porazinska *et al.*, 2021). Thus, shifts in both local and global climate will likely result in parallel shifts in global nematode distribution patterns. As mentioned in the previous section, the alarming spread of the pine wilt disease caused by *B. xylophilus* is mainly due to the current warmer and drier climate that enhances the activation of latent nematodes (Hirata *et al.*, 2017) although global trading of wood-related products plays a role as well (Woodward *et al.*, 2022). Both local and global changes in climatic conditions are expected to alter the distribution of nematodes worldwide (see Chapter 12, this volume). Other global change factors, such as nitrogen deposition, also reduce nematode diversity (Zhou *et al.*, 2022), while sulfur deposition also alters the composition of nematode assemblages (Zhang *et al.*, 2021). Although the specific effects of the numerous global change factors that are transforming the anthroposphere might be unknown, the fact that global change is declining soil diversity worldwide is undeniable.

The notion of declining biological diversity is not new. Already in his 1985 paper, 'The biological diversity crisis', E.O. Wilson (1985) warned of an alarmingly accelerated disappearance of species (400 times faster than expected) even before they had a chance to be discovered and described. Indeed, although only around 1.5 million species have been described, some estimates suggest there are up to 300 million species of protists, fungi and animals, and 1 billion species of bacteria, with a total biological diversity of 1–6 billion species (Larsen *et al.*, 2017).

Forty years after Wilson's notice, it is widely acknowledged that the actual rate of species loss is one of the most important threats to the welfare of humankind. Insect decline, for example, might have catastrophic consequences on bats, reptiles, amphibians, small mammals and fish, while 75% of all crop species depend on insect pollinators (Goulson, 2019). Biodiversity loss might also increase the risk of human exposure to new and existing zoonoses (Keesing and Ostfeld, 2021). Unfortunately, the loss of biodiversity has largely continued since the publication of Wilson's paper, and the global consequences for the functioning of our planet have become even greater (Mace *et al.*, 2014). Moreover, the rate of biodiversity loss of cryptobiotic species, such as most soil invertebrates including soil nematodes (a quarter of total terrestrial biodiversity), is completely unknown and unrecognized. To prevent the 'quiet extinction of invertebrates', Eisenhauer *et al.* (2019) proposed three main global actions: (i) the mobilization and integration of invertebrate biodiversity data; (ii) the development of biodiversity surveys of previously studied sites to analyse current temporal trends of invertebrate biodiversity loss; and (iii) the establishment of harmonized national monitoring programmes. Such actions would help protect soil biota as we realize our dependence on global biodiversity. Urgent action is needed to protect soil biodiversity worldwide.

References

Abad, P. and Williamson, V.M. (2010) Plant nematode interaction: a sophisticated dialogue. In: Kader, J.C. and Delseny, M. (eds) *Advances in Botanical Research*, Vol. 53. Elsevier, Amsterdam, pp. 147–192. DOI: 10.1016/S0065-2296(10)53005-2.

Baquiran, J.P., Thater, B., Sedky, S., De Ley, P., Crowley, D. and Orwin, P.M. (2013) Culture-independent investigation of the microbiome associated with the nematode *Acrobeloides maximus*. *PLoS One* 8, e67425.

Baldwin, J.G., Nadler, S.A. and Adams, B.J. (2004) Evolution of plant parasitism among nematodes. *Annual Review of Phytopathology* 42, 83–105. DOI: 10.1146/annurev.phyto.42.012204.130804.

Berg, M., Stenuit, B., Ho, J., Wang, A., Parke, C. *et al.* (2016) Assembly of the *Caenorhabditis elegans* gut microbiota from diverse soil microbial environments. *The ISME Journal* 10, 1998–2009.

Bik, H.M., Thomas, W., Lunt, D.H. and Lambshead, P.J.D. (2010) Low endemism, continued deep–shallow interchanges, and evidence for cosmopolitan distributions in free-living marine nematodes (order Enoplida). *BMC Evolutionary Biology* 10, 389. DOI: 10.1186/1471-2148-10-389.

Blaxter, M. and Koutsovoulos, G. (2015) The evolution of parasitism in Nematoda. *Parasitology* 142, S26–S39. DOI: 10.1017/S0031182014000791.

Brown, A.M. (2018) Endosymbionts of plant-parasitic nematodes. *Annual Review of Phytopathology* 56, 225–242.

Brown, A.M., Howe, D.K., Wasala, S.K., Peetz, A.B., Zasada, I.A. and Denver, D.R. (2015) Comparative genomics of a plant-parasitic nematode endosymbiont suggest a role in nutritional symbiosis. *Genome Biology and Evolution* 7, 2727–2746.

Burke, M., Scholl, E.H., Bird, D.M., Schaff, J.E., Colman, S.D. *et al.* (2015) The plant parasite *Pratylenchus coffeae* carries a minimal nematode genome. *Nematology* 17, 621–637. DOI: 10.1163/15685411-00002901.

Cabreiro, F. and Gems, D. (2013) Worms need microbes too: microbiota, health and aging in *Caenorhabditis elegans*. *EMBO Molecular Medicine* 5, 1300–1310.

Castagnone-Sereno, P., Danchin, E.G.J., Perfus-Barbeoch, L. and Abad, P. (2013) Diversity and evolution of root-knot nematodes, genus *Meloidogyne*: new insights from the genomic era. *Annual Review of Phytopathology* 51, 203–220. DOI: 10.1146/annurev-phyto-082712-102300.

De la Fuente, B., Saura, S. and Beck, P.S. (2018) Predicting the spread of an invasive tree pest: the pine wood nematode in Southern Europe. *Journal of Applied Ecology* 55, 2374–2385.

De Ley, P. (2006) A quick tour of nematode diversity and the backbone of nematode phylogeny (January 25, 2006). In: WormBook, (ed.) The *C. elegans* Research Community, WormBook, DOI: 10.1895/Wormbook.1.41.1, http://www.wormbook.org.

De Ley, P. and Blaxter, M. (2002) Systematic position and phylogeny. In: Lee, D.L. (ed.) *The Biology of Nematodes*. Taylor & Francis, London, pp. 1–30.

De Meester, N., Derycke, S., Rigaux, A. and Moens, T. (2015) Temperature and salinity induce differential responses in life histories of cryptic nematode species. *Journal of Experimental Marine Biology and Ecology* 472, 54–62. DOI: 10.1016/j.jembe.2015.07.002.

De Meester, N., Gingold, R., Rigaux, A., Derycke, S. and Moens, T. (2016) Cryptic diversity and ecosystem functioning: a complex tale of differential effects on decomposition. *Oecologia* 182, 559–571. DOI: 10.1007/s00442-016-3677-3.

De Meester, N., Van Daele, T., Van Malderen, J., Monteiro, L., Van Colen, C. *et al.* (2018) Active and species-specific dispersal behaviour in a marine nematode cryptic species complex. *Marine Ecology Progress Series* 600, 71–83. DOI: 10.3354/meps12658.

Derycke, S., Backeljau, T. and Moens, T. (2013) Dispersal and gene flow in free-living marine nematodes. *Frontiers in Zoology* 10, 1. DOI: 10.1186/1742-9994-10-1.

Derycke, S., De Meester, N., Rigaux, A., Creer, S., Bik, H. *et al.* (2016) Coexisting cryptic species of the *Litoditis marina* complex (Nematoda) show differential resource use and have distinct microbiomes with high intraspecific variability. *Molecular Ecology* 25, 2093–2110.

Dirksen, P., Marsh, S.A., Braker, I., Heitland, N., Wagner, S. *et al.* (2016) The native microbiome of the nematode *Caenorhabditis elegans*: gateway to a new host–microbiome model. *BMC Biology* 14, 38.

Eisenhauer, N., Bonn, A. and Guerra, C.A. (2019) Recognizing the quiet extinction of invertebrates. *Nature Communications* 10, 50. DOI: 10.1038/s41467-018-07916-1.

Elhady, A., Giné, A., Topalovic, O., Jacquiod, S., Sørensen, S.J. *et al.* (2017) Microbiomes associated with infective stages of root-knot and lesion nematodes in soil. *PLoS One* 12, e0177145.

Giblin-Davis, R.M., Davies, K.A., Morris, K. and Thomas, W.K. (2003) Evolution of parasitism in insect-transmitted plant nematodes. *Journal of Nematology* 35, 133–141.

Goodrich-Blair, H. and Clarke, D.J. (2007) Mutualism and pathogenesis in *Xenorhabdus* and *Photorhabdus*: two roads to the same destination. *Molecular Microbiology* 64, 260–268.

Goulson, D. (2019) The insect apocalypse, and why it matters. *Current Biology* 29, R967–R971. DOI: 10.1016/j.cub.2019.06.069.

Hirata, A., Nakamura, K., Nakao, K., Kominami, Y., Tanaka, N. *et al.* (2017) Potential distribution of pine wilt disease under future climate change scenarios. *PLoS One* 12, e0182837. DOI: 10.1371/journal.pone.0182837.

Holterman, M., van der Wurff, A., van den Elsen, S., van Megen, H., Bongers, T. *et al.* (2006) Phylum-wide analysis of SSU rDNA reveals deep phylogenetic relationships among nematodes and accelerated evolution toward crown clades. *Molecular Biology and Evolution* 23, 1792–1800. DOI: 10.1093/molbev/msl044.

Holterman, M., Karssen, G., van den Elsen, S., van Megen, H., Bakker, J. and Helder, J. (2009) Small subunit rDNA-based phylogeny of the Tylenchida sheds light on relationships among some high-impact plant-parasitic nematodes and the evolution of plant feeding. *Phytopathology* 99, 227–235. DOI: 10.1094/PHYTO-99-3-0227.

Hugot, J.-P., Gardner, S.L., Borba, V., Araujo, P., Leles, D. *et al.* (2014) Discovery of a 240 million year old nematode parasite egg in a cynodont coprolite sheds light on the early origin of pinworms in vertebrates. *Parasites and Vectors* 7, 486. DOI: 10.1186/s13071-014-0486-6.

IPCC (International Panel on Climate Change) (2013) *Climate Change 2013: The Physical Science Basis. Contribution of Working Group I to the Fifth Assessment Report of the Intergovernmental Panel on Climate Change* [Stocker, T.F., Qin, D., Plattner, G.-K., Tignor, M., Allen, S.K. *et al.* (eds)]. Cambridge University Press, Cambridge and New York, p. 1535.

Jones, J.T., Furlanetto, C. and Kikuchi, T. (2005) Horizontal gene transfer from bacteria and fungi as a driving force in the evolution of plant parasitism in nematodes. *Nematology* 7, 641–646. DOI: 10.1163/156854105775142919.

Keesing, F. and Ostfeld, R.S. (2021) Impacts of biodiversity and biodiversity loss on zoonotic diseases. *Proceedings of the National Academy of Sciences USA* 118, e2023540118. DOI: 10.1073/pnas.2023540118.

Kikuchi, T., Eves-van den Akker, S. and Jones, J.T. (2017) Genome evolution of plant-parasitic nematodes. *Annual Review of Phytopathology* 55, 333–354. DOI: 10.1146/annurev-phyto-080516-035434.

Kiontke, K. and Fitch, D.H.A. (2013) Nematodes. *Current Biology* 23, R862–R864. DOI: 10.1016/j.cub.2013.08.009.

Ladygina, N., Johansson, T., Canbäck, B., Tunlid, A. and Hedlund, K. (2009) Diversity of bacteria associated with grassland soil nematodes of different feeding groups. *FEMS Microbiology Ecology* 69, 53–61.

Larsen, B.B., Miller, E.C., Rhodes, M.K. and Wiens, J.J. (2017) Inordinate fondness multiplied and redistributed: the number of species on Earth and the new pie of life. *Quarterly Review of Biology* 92, 229–265. DOI: 10.1086/693564.

Lawson, S.A. and Sathyapala, S. (2008) The risk of pine wilt disease to Australia and New Zealand. In: Mota, M.M. and Vieira, P. (eds) *Pine Wilt Disease: A Worldwide Threat to Forest Ecosystems*. Springer, Dordrecht, The Netherlands, pp. 41–58. DOI: 10.1007/978-1-4020-8455-3_3.

Mace, G.M., Reyers, B., Alkemade, R., Biggs, R., Chapin III, F.S. *et al.* (2014) Approaches to defining a planetary boundary for biodiversity. *Global Environmental Change* 28, 289–297. DOI: 10.1016/j.gloenvcha.2014.07.009.

McQueen, J.P., Gattoni, K., Gendron, E.M., Schmidt, S.K., Sommers, P. and Porazinska, D.L. (2013) External and internal microbiomes of Antarctic nematodes are distinct, but more similar to each other than the surrounding environment. *Journal of Nematology* 55, 1–28.

McQueen, J.P., Gattoni, K., Gendron, E.M., Schmidt, S.K., Sommers, P. and Porazinska, D.L. (2022) Host identity is the dominant factor in the assembly of nematode and tardigrade gut microbiomes in Antarctic Dry Valley streams. *Scientific Reports* 12, 20118.

Meyer, J.M., Baskaran, P., Quast, C., Susoy, V., Rödelsperger, C. *et al.* (2017) Succession and dynamics of *Pristionchus* nematodes and their microbiome during decomposition of *Oryctes borbonicus* on La Réunion Island. *Environmental Microbiology* 19, 1476–1489.

Nadler, S.A. and Pérez-Ponce De Leon, G. (2011) Integrating molecular and morphological approaches for characterizing parasite cryptic species: implications for parasitology. *Parasitology* 138, 1688–1709. DOI: 10.1017/S003118201000168X.

Nielsen, U.N., Ayres, E., Wall, D.H., Li, G., Bardgett, R.D. *et al.* (2014) Global-scale patterns of assemblage structure of soil nematodes in relation to climate and ecosystem properties. *Global Ecology and Biogeography* 23, 968–978. DOI: 10.1111/geb.12177.

Ogier, J.C., Pagès, S., Frayssinet, M. and Gaudriault, S. (2020) Entomopathogenic nematode-associated microbiota: from monoxenic paradigm to pathobiome. *Microbiome* 8, 25.

Ortiz, A., Vega, N.M., Ratzke, C. and Gore, J. (2021) Interspecies bacterial competition regulates community assembly in the *C. elegans* intestine. *The ISME Journal* 15(7), 2131–2145.

Palomares-Rius, J.E., Cantalapiedra-Navarrete, C. and Castillo, P. (2014) Cryptic species in plant-parasitic nematodes. *Nematology* 16, 1105–1118. DOI: 10.1163/15685411-00002831.

Poinar, G.O. Jr (2015) The geological record of parasitic nematode evolution. *Advances in Parasitology* 90, 53–92. DOI: 10.1016/bs.apar.2015.03.002.

Poinar, G. Jr, Kerp, H. and Hass, H. (2008) *Palaeonema phyticum* gen. n., sp n. (Nematoda: Palaeonematidae fam. n.), a Devonian nematode associated with early land plants. *Nematology* 10, 9–14. DOI: 10.1163/156854108783360159.

Porazinska, D.L., de Mesquita, C.P., Farrer, E.C., Spasojevic, M.J., Suding, K.N. and Schmidt S.K. (2021) Nematode community diversity and function across an alpine landscape undergoing plant colonization of previously unvegetated soils. *Soil Biology and Biochemistry* 1, 108380.

Radeke, L.J. and Herman, M.A. (2021) Take a walk to the wild side of *Caenorhabditis elegans*–pathogen interactions. *Microbiology and Molecular Biology Reviews* 85, 10–128.

Ridall, A. and Ingels, J. (2021) Suitability of free-living marine nematodes as bioindicators: status and future considerations. *Frontiers in Marine Science* 8, 685327. DOI: 10.3389/fmars.2021.685327.

Ristau, K., Steinfartz, S. and Traunspurger, W. (2013) First evidence of cryptic species diversity and significant population structure in a widespread freshwater nematode morphospecies (*Tobrilus gracilis*). *Molecular Ecology* 22, 4562–4575. DOI: 10.1111/mec.12414.

Samuel, B.S., Rowedder, H., Braendle, C., Félix, M.A. and Ruvkun, G. (2016) *Caenorhabditis elegans* responses to bacteria from its natural habitats. *Proceedings of the National Academy of Sciences USA* 113, E3941–E3949.

Scheffers, B.R., Joppa, L.N., Pimm, S.L. and Laurance, W.F. (2012) What we know and don't know about Earth's missing biodiversity. *Trends in Ecology and Evolution* 27, 501–510. DOI: 10.1016/j.tree.2012.05.008.

Schuelke, T., Pereira, T.J., Hardy, S.M. and Bik, H.M. (2018) Nematode-associated microbial taxa do not correlate with host phylogeny, geographic region or feeding morphology in marine sediment habitats. *Molecular Ecology* 27, 1930–1951.

Schulenburg, H. and Félix, M.A. (2017) The natural biotic environment of *Caenorhabditis elegans*. *Genetics* 206, 55–86.

Shapira, M. (2017) Host–microbiota interactions in *Caenorhabditis elegans* and their significance. *Current Opinion in Microbiology* 38, 142–147.

Shepherd, R., Brigham, L.M., Bueno de Mesquita, C.P., Gattoni, K.M., Gendron, E.M. *et al.* (2021) Long-term nitrogen deposition in alpine tundra alters nematode communities. *Journal of Nematology* 53, 33.

Soliman, T., Mourits, M.C.M., van der Werf, W., Hengeveld, G.M., Robinet, C. and Lansink, A.G.J.M.O. (2012) Framework for modelling economic impacts of invasive species, applied to pine wood nematode in Europe. *PLoS One* 7, e45505.

Taylor, M. and Vega, N.M. (2021) Host immunity alters community ecology and stability of the microbiome in a *Caenorhabditis elegans* model. *mSystems* 6, e00608-20.

Taylor, M.J., Bandi, C. and Hoerauf, A. (2005) *Wolbachia*. Bacterial endosymbionts of filarial nematodes. *Advances in Parasitology* 60, 245–284.

Urquiza-Zurich, S., Garcia-Angulo, V.A., Burdisso, P., Palominos, M.F., Fernandez-Hubeid, L. *et al.* (2023) The assembly of bacteria living in natural environments shapes neuronal integrity and behavioral outputs in *Caenorhabditis elegans*. *mBio* 14, e03402-22.

Vafeiadou, A.M., Derycke, S., Rigaux, A., De Meester, N., Guden, R.M. and Moens, T. (2022) Microbiome differentiation among coexisting nematode species in estuarine microhabitats: a metagenetic analysis. *Frontiers in Marine Science* 30, 881566.

van den Hoogen, J.S., Geisen, D., Routh, H., Ferris, W., Traunspurger, D.A. *et al.* (2019) Soil nematode abundance and functional group composition at a global scale. *Nature* 572, 194–198.

van Megen, H., van den Elsen, S., Holterman, M., Karssen, G., Mooyman, P. *et al.* (2009) A phylogenetic tree of nematodes based on about 1200 full-length small subunit ribosomal DNA sequences. *Nematology* 11, 927–950. DOI: 10.1163/156854109X456862.

Wilson, E.O. (1985) The biological diversity crisis: despite unprecedented extinction rates, the extent of biological diversity remains unmeasured. *BioScience* 35, 700–706. DOI: 10.2307/1310051.

Woodward, S., Beram, R.C. and Dogmus, H.T. (2022) Drivers of forest pathogen invasions: the roles of global trade and climate change. *SEEFOR – South-East European Forestry* 13, 1–18. DOI: 10.15177/seefor.22-03.

Xue, Y., Xie, Y., Cao, X. and Zhang, L. (2023) The marine natural microbiome mediates physiological outcomes in host nematodes [preprint]. *BioRxiv*, posted 11 May 2023. DOI: 10.1101/2023.05.10.540299.

Yeates, G.W., Bongers, T., De Goede, R.G., Freckman, D.W. and Georgieva, S. (1993) Feeding habits in soil nematode families and genera – an outline for soil ecologists. *Journal of Nematology* 25, 315–331.

Zhang, A., Olatunji, O.A., Tariq, A., Li, T., Wang, R. and Jiang, Y. (2021) Sulfur deposition changed the community structure of soil nematodes by affecting omnivores-predators. *Science of The Total Environment* 771, 144912. DOI: 10.1016/j.scitotenv.2020.144912.

Zhou, J., Wu, J., Huang, J., Sheng, X., Dou, X. and Lu, M. (2022) A synthesis of soil nematode responses to global change factors. *Soil Biology and Biochemistry* 165, 108538. DOI: 10.1016/j.soilbio.2021.108538.

Zilber-Rosenberg, I. and Rosenberg, E. (2008) Role of microorganisms in the evolution of animals and plants: the hologenome theory of evolution. *FEMS Microbiology Reviews* 32, 723–735.

Zullini, A. (2018) Cosmopolitanism and endemism in free-living nematodes. *Biogeographia* 33, 33–39.

4 General Community Indices for the Analysis of Nematode Assemblages

Deborah A. Neher[1]* and Brian J. Darby[2†]
[1]*Department of Agriculture, Landscape and Environment, University of Vermont, Vermont, USA;* [2]*Formerly Department of Biology, University of North Dakota, North Dakota, USA*

Abstract
This chapter reviews classical community indices that condense community data into one or a few meaningful variables to simplify analysis and interpretation. Given that no indices can conclusively reveal all ecological processes, the recommendation is to complement univariate approaches that disregard taxon identity with multivariate approaches that preserve taxon identity to improve one's understanding of both the autecology of individual community members and synecology of the community. Common univariate indices include index families such as diversity or maturity. Multivariate approaches include clustering and ordination. Recommendations for computer software and R scripts are included.

4.1 Introduction

To be successful as an indicator, a single index must perform one of two functions: either reflect a past ecological process or predict a future ecological process. The success of community indices to reflect ecological processes or predict patterns depends on the relative completeness of ecological knowledge. Limitations of community indices are that they are density-independent and rely on *pattern* to reflect process, and often several processes can result in similar patterns. Productivity, resilience and stability are some of the ecological characteristics relevant to ecosystem management, and some early successful attempts to link diversity with function include Rosenberg (1976) and Schafer (1973) and continue to be investigated (e.g. Lazarova *et al.*, 2021). However, the link between ecosystem processes and diversity is not always clear even for well-studied communities, so it is not surprising that linkages between ecosystem processes and nematode diversity are also unclear (Ettema, 1998; Brussaard *et al.*, 2004). Appropriate sampling and statistical techniques are critical to valid interpretation of diversity indices. Generally, stratified- or simple-stage cluster sampling are touted as generating less bias in diversity estimates than simple random sampling (Gimaret-Carpentier *et al.*, 1998). Systematic sampling with equal sampling effort (volume, area) is necessary for comparison among samples (Neher and Campbell, 1996). Although nematode communities vary by season (Neher *et al.*, 2005), a general recommendation to sample at the end of a growing season prevails. Furthermore,

*Email: Deborah.Neher@uvm.edu

DOI: 10.1079/9781800624221.0004

one must consider the time lag that exists between nematode community composition and evidence of ecosystem function (Neher *et al.*, 2012).

Like other microscopic groups, nematodes can be tedious and laborious to extract from their environment, count, preserve and identify. Most data sets that characterize nematode communities are either: (i) quantitative, with several replicates from several sites, but performed at coarse taxonomic resolution (e.g. family or genus); or (ii) have a fine level of taxonomic resolution (e.g. species) but either lack replication or are not quantitative. Admittedly, although molecular tools are not to be used exclusively but rather as additional characters to identify nematodes, the advent of high-throughput amplicon sequencing creates the potential to characterize nematode communities, as well as associated bacteria, fungi and protozoans, for many sites with a level of taxonomic precision that was not feasible previously. Likely, as high-throughput amplicon sequencing becomes more popular with nematode community characterizations, data sets with many sites and identifications with species-level precision will become more popular and may incite the use of a variety of tests and indices that are otherwise commonplace in general ecological literature but scarcely used in the context of nematode ecology.

4.2 Univariate Identity-Independent Indices

Depending on the context, the term 'diversity' is sometimes used to simply describe the number of taxa. In the context of quantitative community ecology, the term 'diversity' is more commonly, and appropriately, used to describe an integration of both numbers of taxa (*species richness*) and equitability among taxa (*species evenness*) (Hurlbert, 1971). Nematode diversity can be computed at the species, genus or trophic level. Most nematode communities are enumerated at coarser resolutions because species identification based on morphology is difficult. Besides, functional groups are a practical necessity because the effect of individual species on ecosystem processes has yet to be determined (Chapin *et al.*, 1992). At the trophic level, diversity is a measure of food chain length and food web complexity (Neher and Campbell, 1994). A trophic diversity index assumes that greater diversity of trophic groups in soil food webs (i.e. complexity) and longer food chains reflect improved ecosystem function (Moore, 1993; Neher *et al.*, 2019). Furthermore, the appropriate spatial resolution of diversity should be consistent with the objectives of the study. *Alpha* diversity reflects the species diversity of individual localities, *beta* diversity reflects the difference of communities across landscapes, and *gamma* diversity reflects the differences between landscapes across regions (e.g. Kerfahi *et al.*, 2016). In the context of soils, the concept of a landscape could have a variety of interpretations but would generally be operationally defined and would likely involve a treatment or comparison of research interest, potentially even as close as two experimental plots.

4.2.1 Identity-independent indices and their calculation

A variety of identity-independent indices is available to serve different purposes in different circumstances (Table 4.1). Each diversity index weights richness and evenness uniquely, but all diversity indices generally function so that an increase in either richness or evenness will always increase diversity. In some reports, the term diversity continues to refer simply to the total number of species; it is preferable, however, to restrict the use of 'diversity' to incorporate both the number of species and evenness. Formulae for calculating several common indices are summarized (Table 4.1) and accompanied by a customized R-studio script written to compute all indices (Table 4.2).

Determining numbers of species (richness) requires standardization and clear reporting for each experiment or sampling regime to prevent artefacts of sampling effort when comparing richness and diversity indices. Nematode density generally varies widely from sample to sample, so the number of nematodes enumerated is a representative subset of the total number extracted, i.e. an unknown number at the time of sampling. Therefore, species *richness* is the appropriate term to refer to the total number of species found when enumerating a uniform *number* of extracted individuals (e.g. 200 from each sample) from samples of a uniform initial mass or volume. Species *density* differs by referring to the total number of species expressed as a

Table 4.1. Selected richness, diversity and evenness indices that can be calculated for nematode communities. (Author's own table.)

Name	Equation[a]	Application	Reference
Margalef's richness	$D_{Marg} = \frac{(S-1)}{\ln(N)}$	Its use should be restricted to comparing species richness among large communities	Margalef (1958)
Shannon's diversity	$H' = -\Sigma(p_i \ln p_i)$	Sensitive to rare taxa. This widely used and versatile index can be applied for both large and small sample sizes. The Shannon index is generally more influenced by rare species than the Simpson index	Shannon (1948)
Hill's N1	$N1 = \exp[-\Sigma(p_i \ln p_i)] = \exp(H')$	The value of this index can be interpretated as the number of abundant taxa	Hill (1973)
Simpson's dominance (infinite community)	$D = \Sigma p_i^2$	Weights common taxa. Probability that two randomly chosen individuals of an infinite community belong to the same class, thus inversely related to diversity	Simpson (1949)
Simpson's dominance (finite community)	$\lambda = \frac{\Sigma n_i(n_i - 1)}{N(N-1)}$	Like Simpson's D but corrected for finite communities. Mathematically, it is usually more appropriate in ecological studies than Simpson's D but is used less often	Simpson (1949)
Hill's N2	$N2 = (\Sigma p_i^2)^{-1} = 1/D$	The value of this index can be interpreted as the number of very abundant taxa	Hill (1973)
Brillouin's diversity	$H = \frac{1}{N}\log\frac{N!}{\prod N_i!}$	Use only on fully censused communities because it is a true statistic and, thus, free from statistical error	Brillouin (1962); Pielou (1975)
Brillouin's maximum diversity	$H_{max} = \frac{1}{N}\ln\frac{N!}{(X!)^{S-r}(Y!)^r}$	Represents maximum possible evenness of a sample of N individuals and S species	Brillouin (1962); Pielou (1975)
Brillouin's minimum diversity	$H_{min} = \frac{1}{N}\ln\frac{N!}{(N-S+1)!}$	Represents minimum possible evenness of a sample of N individuals and S species	Brillouin (1962); Pielou (1975)
Brillouin's evenness	$J = \frac{H}{H_{max}}$ or $J' = \frac{H'}{\ln S}$	Evenness represents equality of abundances in a community. Use J for samples (and J′ for collections) to determine the evenness portion of diversity; J or J′ represent observed and maximum diversity, respectively	Brillouin (1962); Pielou (1975)

Continued

Table 4.1. Continued.

Name	Equation[a]	Application	Reference
Brillouin's relative evenness	$V = \frac{H - H_{min}}{Hmin_{max}}$	Unlike J and J′, V is not influenced by species richness (S)	Hurlbert (1971); Pielou (1975)
Hill's evenness	$E_{2,1} = \frac{(N_2)}{(N_1)}$	Ratio of very abundant taxa to rare taxa. Approaches value of 1 as a single species become more dominant in a community	Hill (1973)
Heip's evenness	$E_{Heip} = \frac{(e^{H'} - 1)}{(S - 1)}$	More sensitive to variations in rare species richness and/or abundance	Heip (1974)

[a]p_i represents the proportion of the i-th taxa in a sample, or n_i the number, with N individuals and S total species. X (in Brillouin's maximum diversity) is the integer portion of (N/S), $Y = X + 1$, and r = the remainder of X.

Table 4.2. R code to compute the various diversity indices from Table 4.1. (Author's own table.)

```
library(vegan)
library(ggplot2)

## Diversity Indices

# Read in data and separate labels from read counts
speciesraw = read.csv('sequences.csv')

spdata = speciesraw[,-c(1:5)]
labels = speciesraw[,c(1:5)]

# Prepare commonly used calculations

S = ncol(spdata)
N = rowSums(spdata)
pind = spdata/N

# Calculate indices

DMarg = (S - 1) / log(N)
ShannonH = -(rowSums(pind*log(pind), na.rm = TRUE))
N1 = exp(ShannonH)
SimpsonD = rowSums(pind**2)
N2 = 1 / SimpsonD
BrillouinJ = ShannonH / log(S)
HillE = N2 / N1
HeipE = (N1 - 1) / (S - 1)

## Using vegan package

veganShannonH = diversity(spdata, index="shannon")
veganN2 = diversity(spdata, index="invsimpson")

## Graph selected indices

df = cbind(labels, ShannonH)
ggplot(df, aes(x=SEASON, y=ShannonH, fill=SEASON))+
  geom_boxplot()
```

uniform *portion* of all extracted individuals (e.g. 20% of the individuals from each sample). This distinction is important because species richness and density are not necessarily linear in relationship. For example, 20 species found among 200 individuals does not necessarily mean that one will find 40 species from 400 individuals. This type of extrapolation requires rarefaction of original data to estimate the number of species collected from a hypothetical number of individuals or samples. Ecologists use this approach to generate *species area curves* which plot the number of unique species accumulated for each sample as they are sampled (Gotelli and Colwell, 2001). A rarefaction curve used to define uniform portions of extracted individuals acts as an accumulation curve, and is performed retrospectively, after samples have been collected (Fig. 4.1A). Curves that do not fully approach a hypothetical asymptote suggest that additional

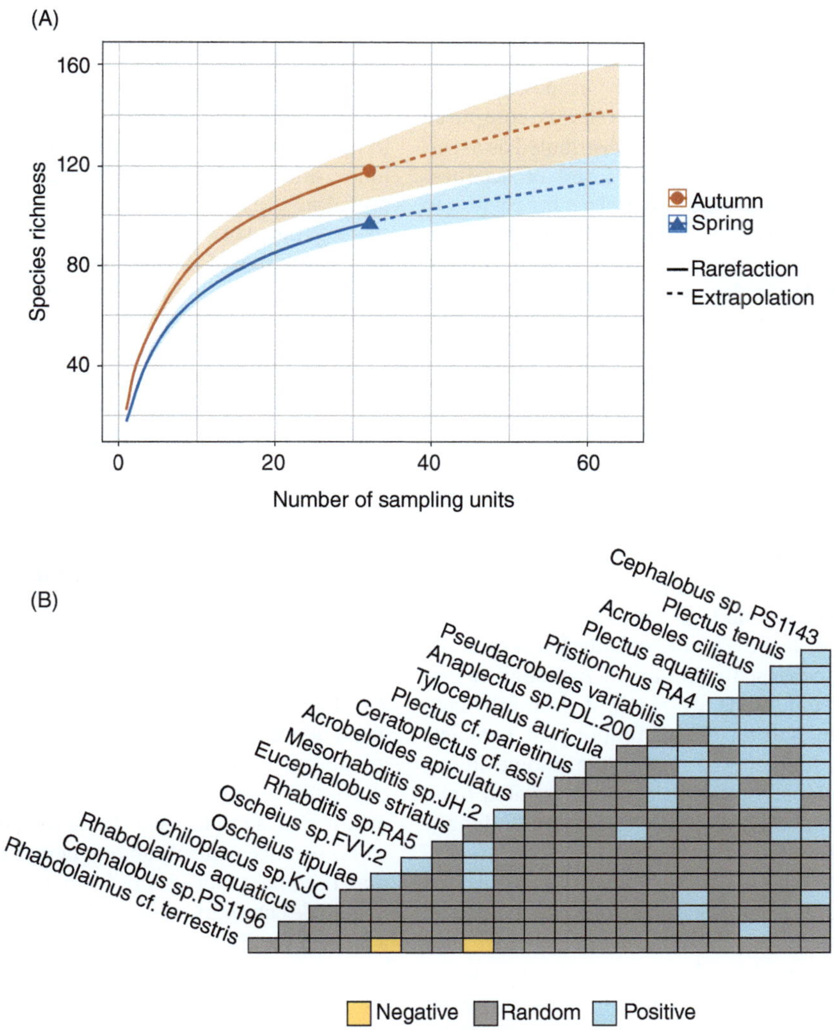

Fig. 4.1. Species rarefaction curves and co-occurrence analysis. (A) Species rarefaction curves for nematode species sampled in the spring and autumn. Solid lines indicate the rarefaction curves, dotted lines indicate extrapolated predictions, and shaded fill indicates bootstrapped confidence intervals. (B) Visualization of the results of a species-based co-occurrence analysis using the R-language package *cooccur* (Griffith *et al.*, 2016). Pairwise comparisons with a blue box indicate a positive co-occurrence pattern, while yellow boxes indicate negative co-occurrence patterns. (Data for both (A) and (B) come from nematodes sampled at the Konza tallgrass prairie; Darby *et al.*, 2013.)

sampling is likely to yield additional species. The Margalef index (Table 4.1) is useful to adjust the number of species (*S*) for the number of individuals enumerated (*N*).

Evenness represents the relative uniformity in abundance of each taxon within a sample. Heip proposed an evenness index (Table 4.1) to standardize the Shannon's diversity index (H′) by total number of species (*S*). Alternatively, Brillouin developed a series of statistics for censused communities that are computationally complex (Table 4.1). For example, Brillouin's maximum theoretical diversity (= H_{max}) is computed with the assumption that all individuals are distributed as uniformly as possible, and minimum theoretical diversity is computed assuming all individuals are distributed as asymmetrically as possible. Two forms of evenness can be computed, the first as diversity relative to maximum diversity (= J) and the second ('relative evenness') as diversity relative to maximum diversity but scaled to minimum diversity (= V). The first type (not scaled to minimum diversity) can be based on either of two estimates of diversity depending on whether the user wishes to assume a finite or infinite community enumeration. We recommend using Brillouin's sample diversity relative to Brillouin's maximum diversity (= J) when assuming a finite community enumeration, or Shannon's population diversity relative to the natural logarithm of richness (= J′) when assuming infinite community enumeration. The second type (= V, scaled to minimum diversity) uses Brillouin's calculation of diversity from a censused community. Although nematode communities are rarely, if ever, fully censused in nature, the assumption of complete enumeration may be appropriate in some unique applications; for example, small, isolated habitats or virtual individuals in a computationally simulated model community.

There are a variety of diversity indices that incorporate both richness and evenness, and they differ mostly as to the degree to which they are influenced by dominant and rare species. Shannon's diversity (Table 4.1) is a popular diversity index. The exponent of Shannon's index (Hill's N1) can be interpreted as the number of uniformly distributed species that would produce an identical Shannon's index as the non-uniformly distributed community. For example, consider a community with 20 non-uniformly distributed species and a Shannon's index of 2.3. The exponent of 2.3 (Hill's N1) equals 9.97, so, intuitively, approximately 10 uniformly distributed species would be needed to produce a Shannon's index like the community of 20 non-uniformly distributed species. Furthermore, Heip's evenness index = [(9.97 – 1)/(20 – 1)] = 0.47, indicating that about half of the observed species would be necessary to produce a similar Shannon's index if they were distributed uniformly. Simpson's D index (Table 4.1) is considered a dominance index because it increases as species are distributed more unevenly (increasing dominance) and can be interpreted intuitively as the probability that two randomly selected individuals from an infinite community will be the same. The reciprocal of Simpson's index (Hill's N2) is often reported as a diversity index, and like Hill's N1, Hill's N2 can be interpreted as the number of uniformly distributed species that would produce a Simpson's index identical to that of the non-uniform community. Notice that the minimum Simpson's D possible (i.e. least dominance by any taxa) is $1/S$ and the maximum Hill's N2 possible (greatest equitability) is *S*, so we could compute an evenness index similar to Heip's approach as $N2/S$. See Neher and Darby (2006) for a deeper explanation of interpretation of various diversity indices.

4.3 Community Assemblage Models

4.3.1 Species co-occurrence patterns

As high-throughput molecular sequencing gains popularity in nematode ecology, allowing greater taxonomic resolution and facilitating characterization of nematode species for a greater number of samples, we think that analysing co-occurrence patterns among nematode species may become a more fruitful way to understand nematode community assemblages. For example, a co-occurrence matrix from data at the species level using presence–absence data obtained from amplicon sequencing is helpful to define species associates (Fig. 4.1B). Species co-occurrence patterns have an important historical place in the field of ecology and represent an attempt to understand how communities form and are assembled. Species co-occurrence patterns can be linked back to the earliest concepts of community formation, including the competing Gleason

(1926) and Clements (1936) concepts of community assemblage. Co-occurrence patterns were again made the subject of ecologists' attention by Diamond (1975) who suggested that communities are structured with patterns of positive or negative co-occurrences akin to a 'checkerboard' pattern, reflecting cooperative or competitive relationships, respectively. This theory, and its associated models, have been debated extensively (Simberloff, 1978) and generated several studies on the proper use of null models against which to test empirical presence–absence data (Gotelli and Graves, 1996; Weiher and Keddy, 1999). Veech (2013) developed a probabilistic model for analysing co-occurrence patterns and provided software for the R environment (Griffith *et al.*, 2016). This package computes rates of all occurrences between each pairwise species combination from a site by species presence–absence matrix, plus their probability. In this case, pairwise probabilities are obtained from a hypergeometric distribution, much like performing a Fisher's exact test. Other software programs designed for more speciose communities, like bacteria or fungi, compute co-occurrence as correlation. However, read counts from high-throughput amplicon sequencing are poorly correlated with individual abundance, making correlations between nematode species only possible with specimen counts. Species differ in their number of genomic rRNA copy numbers, which is the locus typically used for identification by amplicon sequencing. Therefore, it would take species-specific 'copy-number corrections' to convert high-throughput sequencing reads to specimen counts (Darby *et al.*, 2013). Unfortunately, the accuracy of such copy-number correction factors will vary by season if the species itself varies seasonally in its number of somatic cells per individual (i.e. differences in the number of juveniles) or in the number of reproductive cells (i.e. season reproductive patterns). Ultimately, the most precise community identifications will likely come from a combination of specimen counts by morphology but using sequencing data to assist the identifications.

4.3.2 Ecological succession

Ecological succession refers to a relatively predicable or directional sequence of spatio-temporal patterns of ecological interactions within a community. As species composition changes, it alters the abiotic environment, which in turn selects against the existing community favouring a community composition that performs better under the newly created abiotic environment. The concept originated in plant ecology (Whittaker, 1975) but also applies to invertebrate communities in soil and sediment. Succession usually progresses directionally unless set back by an environmental disturbance such as cultivation, pollution or nutrient enrichment (Neher, 1999). Therefore, quantitative measures of ecological succession can serve as indicators of disturbance. With improved knowledge of synecology of nematode communities, one could identify the type and intensity of disturbance based on an index of succession. Bongers (1990) proposed an index of ecological succession for application to nematodes and Ruf (1998) applied a similar approach to mesostigmatid mites. Maturity indices are used as a measure of the ecological successional status of a soil community. They are based on the principle that different taxa have contrasting sensitivities to stress or disruption of the successional sequence because of their life-history characteristics. Successional indices are described in greater detail in Chapter 5 of this volume.

4.3.4 Beta diversity

Beta diversity is another way to examine patterns of ecological succession in nematodes. Whereas alpha diversity reflects the communities at individual localities, beta diversity describes the comparison between communities across a landscape. For example, Podani and Schmera (2011) developed indices that separated the various components of beta diversity into species similarity, richness differences and species replacements (Fig. 4.2). These three indices describe different aspects of beta diversity, or the potential differences between communities at different localities. For a community ecologist, calculating these indices across different communities may help to identify mechanisms of community assemblage or ecological succession. For example, as a community naturally shifts in time, these indices reflect whether taxa maintain similar richness and progress by

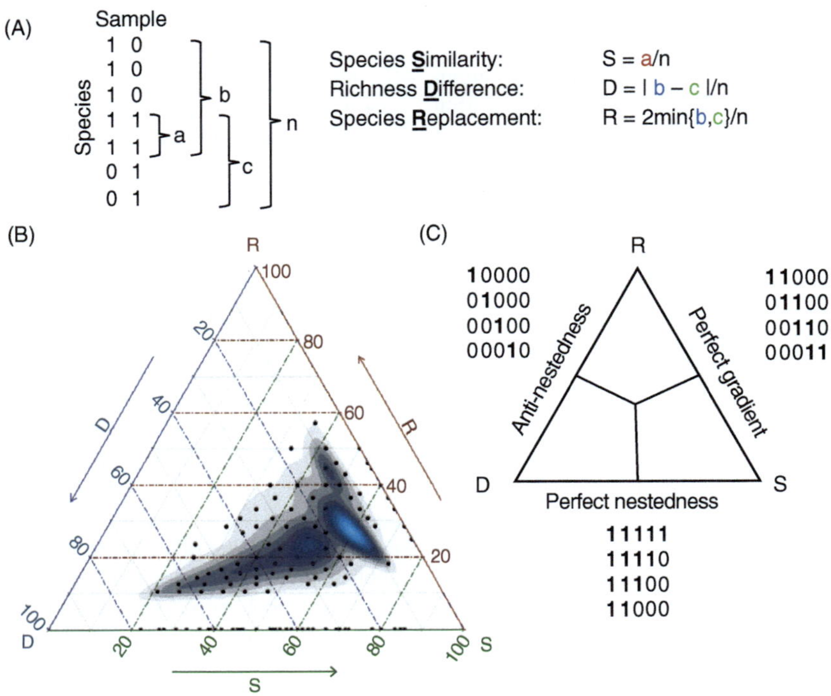

Fig. 4.2. Illustrating the use of species similarity, richness difference and species replacement indices of beta diversity. (A) Indices are calculated for each pairwise sample in a site-by-species presence–absence matrix. (B) Index values for all pairwise comparisons can be plotted on a ternary diagram. (C) The aggregate distribution of indices can reflect beta diversity patterns of anti-nestedness, perfect gradient or perfect nestedness between communities. (The *ggtern* package of R software was used to generate the graphs from unpublished data of co-author Brian Darby.)

species replacement, or whether they change by the addition or subtraction of individual species. One helpful feature of these indices is that the three values sum to 1.0, which allows one to plot the values on a ternary plot to illustrate the dominant component of beta diversity between each pairwise comparison (Fig. 4.2).

4.4 Multivariate Techniques

Multivariate analysis offers both descriptive and inferential procedures to analyse multiple variables simultaneously to reveal the collective interactions of all variables and the effect each variable has on the others. *Descriptive* procedures help to illustrate the overall structure of a data set while *inferential* procedures help to test hypotheses of interactions. Therefore, multivariate analysis has two complementary applications, *exploratory hypothesis-generating* and *inferential hypothesis-testing*, that can be combined into a two-phase approach that might begin with an exploratory phase that seeks patterns in nature by asking 'To what can I ascribe the variation in my data?' The second phase, then, tests the hypotheses that were generated by asking 'Can I reject the null hypothesis that species are unrelated to each other or postulated environmental factor(s)?' In this way, multivariate analysis is useful in evaluating nematode community structure as a biological indicator by keeping the identity of individual taxa explicit throughout the analysis. Below, we discuss two types of multivariate analysis commonly applied to nematode communities: cluster analysis and ordination.

4.4.1 Cluster analysis

Cluster analysis treats each multivariate observation (sample) as a vector and attempts to

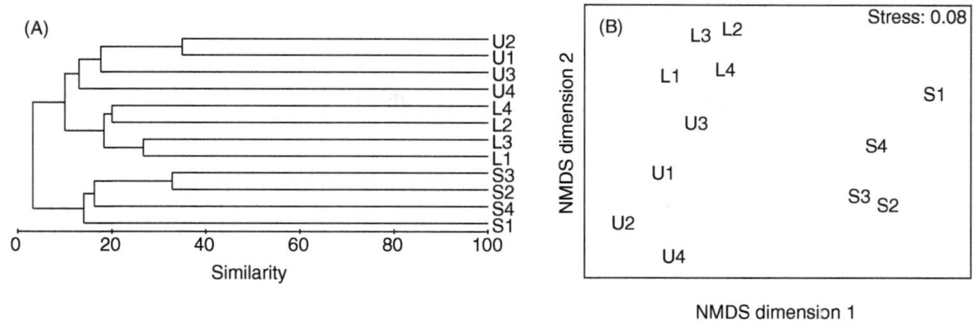

Fig. 4.3. Graphic representations among samples of nematode communities illustrated either as (A) similarities in a dendrogram or (B) dissimilarities on a multidimensional scaling (MDS) biplot. Bray–Curtis similarity was computed on non-transformed abundance data for each pairwise combination of samples. A dendrogram and biplot show the same data in two formats of duplicative information. (Illustrations were created using ANOSIM and CLUSTER modules of Primer-E Version 5.2.9 software; Clarke and Gorley, 2001.)

group vectors that are like each other into clusters (Fig. 4.3). Cluster analysis begins with a (dis)similarity matrix, often computed as the Euclidean distance or Bray–Curtis similarity among all pairs of vectors. *Hierarchical clustering* algorithms are either agglomerative or divisive. *Agglomerative clustering* begins with each vector representing a unique cluster and sequentially combining the two nearest clusters into one until an optimal number of clusters is attained. *Divisive clustering* begins with one cluster containing all vectors and sequentially divides the cluster into two until an optimal number of clusters has been obtained. Agglomerative clustering is most common and there are several methods of determining the distance of vector clusters from each other. The *single linkage (or nearest neighbour) method* determines the distance between two clusters as the minimum distance (e.g. Euclidean) between the two most similar vectors of each cluster, while the *complete linkage (e.g. furthest neighbour) method* determines the distance between two clusters as the maximum distance (e.g. Euclidean) between the two most dissimilar vectors of each cluster. The *average linkage method* defines the distance between two clusters as the average distance of all elements from each cluster, while the *centroid method* defines the distance between two clusters as the distance between the two mean (or median) vectors of a cluster, called the centroids. Finally, *Ward's method* joins clusters to minimize the increase in sum of squares within and between clusters. The result of hierarchical cluster analysis is a dendrogram (i.e. tree diagram) that shows each step of the clustering procedure and the distance at which the clusters merge (e.g. Fig. 4.3A).

Discriminant analysis is a related approach based on an a priori expectation of group members whereas cluster analysis has no preconceived expectation of group members and therefore conducts a posteriori aggregation. With discriminant analysis, one hypothesizes that there are two or more distinct groups and then determines whether the observations divide significantly among those two predicted groups (Afifi *et al.*, 2020).

4.4.2 Ordination

Ordination techniques are popular in community analysis due to their ability to visualize multidimensional data in two-dimensional space (Afifi *et al.*, 2020). There are two main classes of ordination techniques: direct and indirect gradient analysis. *Indirect gradient analysis*, also called *unconstrained*, seeks to interpret patterns from within a data set. *Direct gradient analysis* seeks to extract patterns from known gradients and is therefore *constrained* by the environmental variables supplied. Indirect gradient analysis is divided into distance-based and eigenanalysis-based methods, whereas all direct gradient analyses are eigenanalysis-based methods. Examples of distance-based indirect gradient ordination include polar ordination (PO), principal coordinates analysis (PCoA) and non-metric multidimensional scaling (NMDS). In PO, two samples most different from each other based on their species composition serve as end points and all other samples are plotted relative to them. In this way, new samples can be added to polar ordination without changing the structure of the ordination diagram. PCoA simply maximizes linear distance measures of the ordination in metric space (using a Euclidean or Bray–Curtis distance matrix), while NMDS is analogous to a non-parametric variant of PCoA by maximizing *rank* distance measures of the ordination in non-metric space (e.g. Fig. 4.3B).

Redundancy analysis (RDA) and canonical correspondence analysis (CCA) are two types of direct gradient analysis that constrains the distribution of taxa by environmental variables (Fig. 4.4). They vary by whether there is a linear (RDA) or unimodal (CCA) link between suites of taxon data with suites of environmental variables. Environmental variables can include treatment classes (coded as nominal 0 or 1 variables) or chemical or physical properties (such as pollutants or temperature) as continuous variables. All these procedures can be performed in R, either through Base R software, or through packages such as *labdsv* or *vegan* (Oksanen *et al.*, 2022). Alternatively, Canoco (ter Braak and Šmilauer, 2012) and Primer-E (Clarke and Gorley, 2001) software packages are simple tools to perform these procedures. In Canoco, abundances are transformed as $\log(x + 1)$ prior to analysis, which is a historically used transformation in nematology. Transformations are unnecessary in Primer-E because the scaling is non-metric multidimensional. CCA results are displayed graphically with biplots. In CCA biplots, each vector for an environmental variable defines an axis, and site or taxa scores can be projected on to that axis. An indication of relative importance of a vector is its length; the angle indicates correlation with other vectors and CCA axes. Eigenvalues for CCA axes indicate the importance of the axes in explaining relationships in the genera–environment data matrices. The first axis represents the greatest explained variation, and subsequent axes represent progressively less

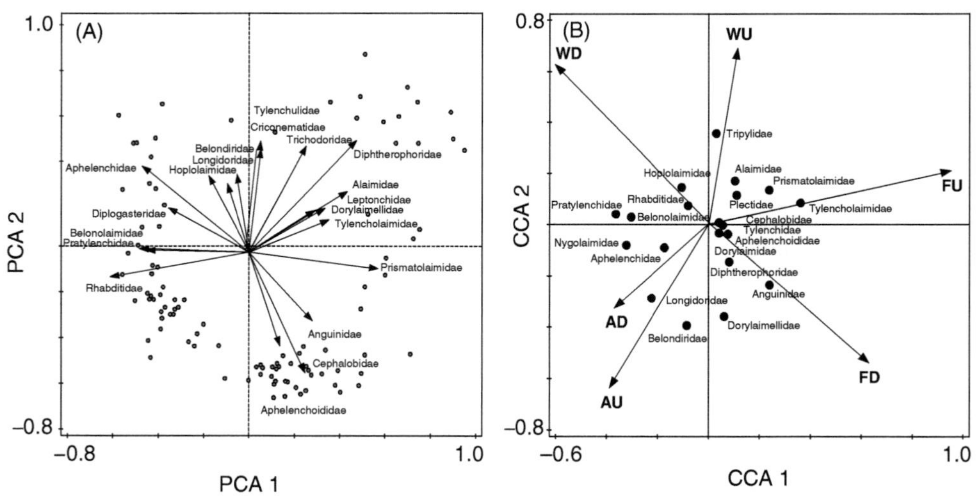

Fig. 4.4. Nematodes in 20 families (out of 39) are illustrated as either (A) principal components analysis (PCA) or (B) canonical correspondence analysis (CCA). PCA is unconstrained, illustrating taxa (arrows) versus samples (grey dots). CCA is constrained by environmental variables representing factorial combinations of three ecosystems (A = agriculture, F = forest, W = wetland) and two disturbance levels (D = disturbed, U = undisturbed) sampled 12 times over two years in North Carolina. The longer vectors explain more variation than shorter vectors. Vectors with acute angles are correlated positively and those in opposite quadrants are correlated negatively. Right angles are orthogonal or independent of each other. (Biplots were generated using Canoco Version 5 software; ter Braak and Šmilauer, 2012. Data from Neher *et al.*, 2005.)

variance. Unfortunately, CCA analyses are restricted to illustrating one instance in time. Therefore, repeated measures of communities through time are consolidated into a single biplot that loses information about temporal patterns.

Principal response curves (PRC) are a multivariate method for the analysis of repeated-measurement designs (van den Brink and ter Braak, 1998, 1999). PRC is based on RDA; each experimental unit and sampling times and unit-by-time interactions are treated as dummy explanatory variables. The result is a diagram showing the sampling periods on the x-axis and the first principal component of the variance explained by treatment on the y-axis (Fig. 4.5). For illustrative purposes, undisturbed condition was treated as a 'control', representing a zero baseline, and 'disturbed' of the same experimental unit as the 'treatment' to focus on the *differences* between the two states of condition through time. Monte Carlo permutation tests permuting whole time series are applied to compute statistical significance.

The type of analysis chosen depends on the research question but is critical in terms of output and interpretation. Nematode community

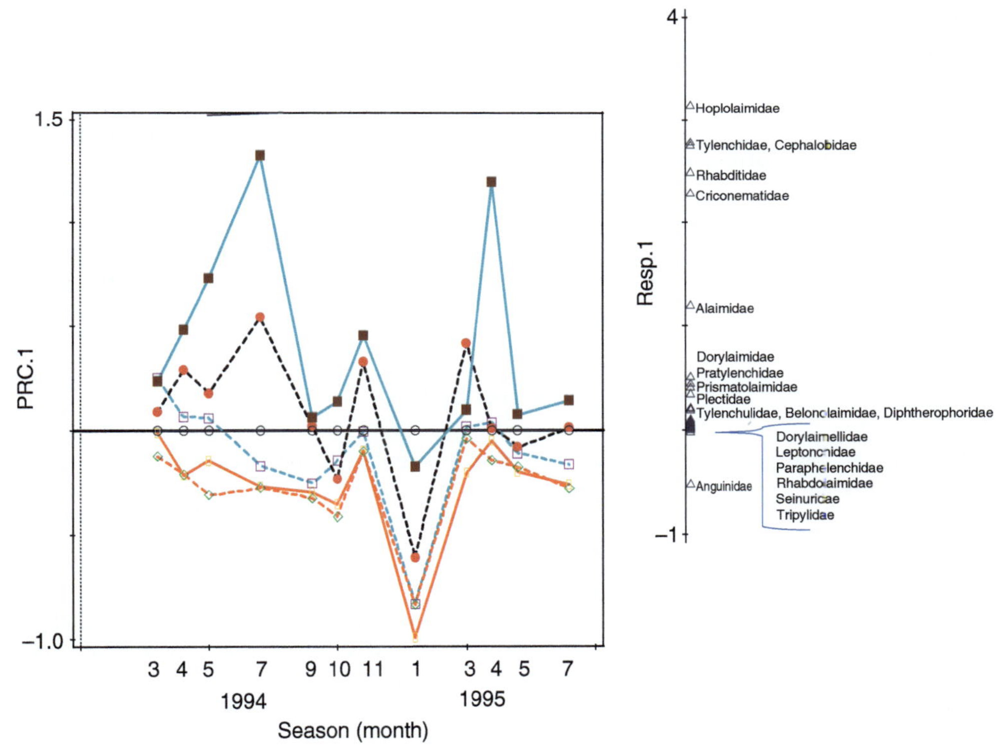

Fig. 4.5. Nematodes in 20 families (out of 39) are illustrated that were sampled repeatedly for 12 times over two years in factorial combinations of three ecosystems (A = agriculture, F = forest, W = wetland) and two disturbance levels (D = disturbed, U = undisturbed). The x-axis (black, solid line) represents undisturbed agriculture as a reference baseline for comparison of relative differences for disturbed agriculture (black, dashed), disturbed (dashed) and undisturbed (solid) forests (orange), and disturbed (dashed) and undisturbed (solid) wetlands (blue). On the right are the species scores (Resp. 1) for taxa that explain the fluctuations and contrasts of the ecosystem communities through time. For example, the cluster of taxa at the top (Hoplolaimidae, Tylenchidae, Cephalobidae, Rhabditidae and Criconematidae) distinguish UW from others; the Alaimidae through Diphtherophoridae are associated with the baseline reference; and Dorylamellidae through Tripylidae and Anguinidae follow dips in AD, WD and forests. (Principal response curve (PRC) plot was generated using Canoco Version 5 software; ter Braak and Šmilauer, 2012. Data from Neher *et al.*, 2005.)

data collected 12 times over a period of two years was analysed three different ways (Figs 4.4 and 4.5). For example, there is only 50% of taxa in common when unconstrained (Fig. 4.4A) or constrained (Fig. 4.4B). By constraining, one gains information about which taxa are associated with environmental variables. For example, Anguinidae, Longidoridae and Hoplolaimidae are relatively abundant in disturbed forest, agriculture and wetland sites, respectively (Fig. 4.4). PRC plots separate times sampled to further refine which taxa explain differences among treatments and how those varied through time (Fig. 4.5). This information is lost in a CCA (Fig. 4.4B) and there is a 20% difference in taxa explaining variation. From the PCA, we learn that Hoplolaimidae and Anguinidae are associated with fluctuating communities in wetlands and forests, respectively.

4.5 Conclusion

Classical community composition can be analysed using metrics that either disregard or preserve the identity of taxon within the community. Identity-independent methods such as diversity and evenness indices are relatively simple to compute and analyse statistically. However, the user must exercise caution by selecting the form of index most appropriate to the goals of the study and resisting the temptation to singularly extrapolate to a greater ecological meaning without substantial supplementary evidence. Alternatively, indices that incorporate and/or maintain taxon identity can more convincingly be linked to ecological process and function. Measures of ecological succession and species assemblage are univariate forms that can be analysed using traditional statistical tools such as regression and analysis of variance. A variety of multivariate methods are accessible through commercial software packages. Many multivariate approaches capture a one-time snapshot of community composition. However, repeated-measures approaches are becoming available to evaluate changes in community composition through time. Practitioners should be aware of the many limitations, assumptions and caveats of community assemblage and multivariate techniques by consulting with expert statisticians.

References

Afifi, A., May, S., Donatello, R.A. and Clark, V.A. (2020) *Practical Multivariate Analysis*. CRC Press, Boca Raton, Florida.

Bongers, T. (1990) The maturity index: an ecological measure of environmental disturbance based on nematode species composition. *Oecologia* 83, 14–19.

Brillouin, L. (1962) *Science and Information Theory*, 2nd edn. Academic Press, New York.

Brussaard, L., Kuyper, T.W., Didden, W.A.M., de Goede, R.G.M. and Bloem, J. (2004) Biological soil quality from biomass to biodiversity – importance and resilience to management stress and disturbance. In: Schjoenning, P., Elmholt, S. and Christensen, B.T. (eds) *Managing Soil Quality: Challenges in Modern Agriculture*. CAB International, Wallingford, UK, pp. 139–171.

Chapin, F.S., Schultz, E.D. and Mooney, H.A. (1992) Biodiversity and ecosystem processes. *Trends in Ecology and Evolution* 7, 107–108.

Clarke, K.R. and Gorley, R.N. (2001) *Plymouth Routines in Multivariate Ecological Research (Primer) v5: User Manual/Tutorial*. Primer-E Ltd, Plymouth, UK.

Clements, F.E. (1936) Nature and structure of the climax. *Journal of Ecology* 24, 252–284. DOI: 10.2307/2256278.

Darby, B.J., Todd, T.C. and Herman, M.A. (2013) High-throughput amplicon sequencing of rRNA genes requires a copy number correction to accurately reflect the effects of management practices on soil nematode community structure. *Molecular Ecology* 22, 5456–5471. DOI: 10.1111/mec.12480.

Diamond, J.M. (1975) *Ecology and Evolution of Communities*. Harvard University Press, Cambridge, Massachusetts.

Ettema, C.H. (1998) Soil nematode diversity: species coexistence and ecosystem function. *Journal of Nematology* 30, 159–169.

Gimaret-Carpentier, C., Pélissier, R., Pascal, J.-P. and Houllier, F. (1998) Sampling strategies for the assessment of tree species diversity. *Journal of Vegetation Science* 9, 161–172. DOI: 10.2307/3237115.

Gleason, H.A. (1926) The individualistic concept of the plant association. *Bulletin of the Torrey Botanical Club* 53, 7–26. DOI: 10.2307/2479933.
Gotelli, N.J. and Colwell, R.K. (2001) Quantifying biodiversity: procedures and pitfalls in the measurement and comparison of species richness. *Ecology Letters* 4, 379–391. DOI: 10.1046/j.1461-0248.2001.00230.x.
Gotelli, N.J. and Graves, G.R. (1996) *Null Models in Ecology*. Smithsonian Institution Press, Washington, DC.
Griffith, D.M., Veech, J.A. and Marsh, C.J. (2016) Cooccur: probabilistic species co-occurrence analysis in R. *Journal of Statistical Software* 69, 1–17. DOI: 10.18637/jss.v069.c02.
Heip, C. (1974) A new index measuring evenness. *Journal of the Marine Biological Association* 54, 555–557. DOI: 10.1017/S0025315400022736.
Hill, M.O. (1973) Diversity and evenness: a unifying notation and its consequences. *Ecology* 54, 427–432. DOI: 10.2307/1934352.
Hurlbert, S.H. (1971) The nonconcept of species diversity: a critique and alternative parameters. *Ecology* 52, 577–586. DOI: 10.2307/1934145.
Kerfahi, D., Tripathi, B.M., Porazinska, D.L., Park, J., Go, R. and Adams, J.M. (2016) Do tropical rain forest soils have greater nematode diversity than High Arctic tundra? A metagenetic comparison of Malaysia and Svalbard. *Global Ecology and Biogeography* 25, 716–728. DOI: 10.1111/geb.12448.
Lazarova, S., Coyne, D., Rodriguez, M.G., Peteira, B. and Ciancio, A. (2021) Functional diversity of soil nematodes in relation to the impact of agriculture – a review. *Diversity* 13, 64. DOI: 10.3390/d13020064.
Margalef, R. (1958) Information theory in ecology. *General Systems* 3, 36–71.
Moore, J.C. (1993) Impact of agricultural practices on soil food web structure: theory and application. *Agriculture, Ecosystems and Environment* 51, 239–247. DOI: 10.1016/0167-8809(94)90047-7.
Neher, D.A. (1999) Soil community composition and ecosystem processes: comparing agricultural ecosystems with natural ecosystems. *Agroforestry Systems* 45, 159–185. DOI: 10.1023/A:1006299100678.
Neher, D.A. and Campbell, C.L. (1994) Nematode communities and microbial biomass in soils with annual and perennial crops. *Applied Soil Ecology* 1, 17–28. DOI: 10.1016/0929-1393(94)90020-5.
Neher, D.A. and Campbell, C.L. (1996) Sampling for regional monitoring of nematode communities in agricultural soils. *Journal of Nematology* 28, 196–208.
Neher, D.A. and Darby, B.J. (2006) Computation and application of nematode community indices: general guidelines. In: Eyualem, A., Traunspurger, W. and Andrássy, I. (eds) *Freshwater Nematodes: Ecology and Taxonomy.* CAB International, Wallingford, UK, pp. 211–222.
Neher, D.A., Wu, J., Barbercheck, M.E. and Anas, O. (2005) Ecosystem type affects interpretation of soil nematode community measures. *Applied Soil Ecology* 30, 47–64. DOI: 10.1016/j.apsoil.2005.01.002.
Neher, D.A., Weicht, T.R. and Barbercheck, M.E. (2012) Linking invertebrate communities to decomposition rate and nitrogen availability in pine forest soils. *Applied Soil Ecology* 54, 14–23.
Neher, D.A., Nishanthan, T., Grabau, Z.J. and Chen, S.Y. (2019) Crop rotation and tillage affect nematode communities more than biocides in monoculture soybean. *Applied Soil Ecology* 140, 89–97. DOI: 10.1016/j.apsoil.2019.03.016.
Oksanen, J., Simpson, G.L., Blanchet, F.G., Kindt, R., Legendre, P. *et al.* (2022) Package 'Vegan'. Available at: https://cran.r-project.org/web/packages/vegan/vegan.pdf (accessed 24 October 2023).
Pielou, E.C. (1975) *Ecological Diversity*. Wiley, New York.
Podani, J. and Schmera, D. (2011) A new conceptual and methodological framework for exploring and explaining pattern in presence–absence data. *Oikos* 120, 1625–1638. DOI: 10.1111/j.1600-0706.2011.19451.x.
Rosenberg, R. (1976) Benthic faunal dynamics during succession following pollution abatement in a Swedish estuary. *Oikos* 27, 414–427. DOI: 10.2307/3543460.
Ruf, A. (1998) A maturity index for predatory soil mites (Mesostigmata: Gamasina) as an indicator of environmental impacts of pollution on forest soils. *Applied Soil Ecology* 9, 447–452. DOI: 10.1016/S0929-1393(98)00103-6.
Schafer, C.T. (1973) Distribution of foraminifera near pollution sources in Chaleur Bay. *Water, Air, and Soil Pollution* 2, 219–233. DOI: 10.1007/BF00655698.
Shannon, C.E. (1948) A mathematical theory of communication. *The Bell System Technical Journal* 27, 379–423.
Simberloff, D. (1978) Using island biogeographic distributions to determine if colonization is stochastic. *The American Naturalist* 112, 713–726.
Simpson, E.H. (1949) Measurement of diversity. *Nature* 163, 688. DOI: 10.1038/163688a0.
ter Braak, C.J.F. and Šmilauer, P. (2012) *Canoco Reference Manual and User's Guide: Software for Ordination (version 5.0)*. Biometris, Wageningen, The Netherlands.

van den Brink, P.J. and ter Braak, C.J.F. (1998) Multivariate analysis of stress in experimental ecosystems by principal response curves and similarity analysis. *Aquatic Ecology* 32, 163–178. DOI: 10.1023/A:1009944004756.

van den Brink, P.J. and ter Braak, C.J.F. (1999) Principal response curves: analysis of time-dependent multivariate responses of biological community to stress. *Environmental Toxicology and Chemistry* 18, 138–148. DOI: 10.1002/etc.5620180207.

Veech, J.A. (2013) A probabilistic model for analysing species co-occurrence. *Global Ecology and Biogeography* 22, 252–260. DOI: 10.1111/j.1466-8238.2012.00789.x.

Weiher, E. and Keddy, P. (1999) Assembly rules as general constraints on community composition. In: Weiher, E. and Keddy, E.C.P. (eds) *Ecological Assembly Rules: Perspectives, Advances, Retreats*. Cambridge University Press, Cambridge, pp. 251–271.

Whittaker, R.H. (1975) *Communities and Ecosystems*, 2nd edn. Macmillan, New York.

5 Indices Developed Specifically for Analysing Nematode Assemblages

Jan Henrik Schmidt[1]*, Gerhard du Preez[2], Sara Sánchez-Moreno[3] and Howard Ferris[4]

[1]*Institute for Epidemiology and Pathogen Diagnostics, Julius Kühn Institute, Braunschweig, Germany;* [2]*Unit for Environmental Sciences and Management, North-West University, Potchefstroom, South Africa;* [3]*Department of Biodiversity and Evolutionary Biology, National Museum of Natural Sciences (MNCN-CSIC), Madrid, Spain;* [4]*Department of Nematology, University of California Davis, California, USA*

Abstract

Nematodes are well established as bioindicators of soil ecosystem status, exemplified by the development of nematode-based indices (NBIs). This chapter aims to: (i) detail the several NBIs derived from nematode assemblage data and based on nematode functional traits, ecophysiology and metabolic activity; (ii) highlight the advantages of NBIs over general community indices; and (iii) emphasize the power and versatility of newly described NBIs (second generation, since 2010) in inferring ecosystem status. A case study exemplifies the potential application of nematode metabolic footprints for indicating connections between soil food webs and ecosystem functioning. The discovery and validation of links between NBIs and agronomic, ecological and economic parameters will be an ongoing mission in soil ecology and may aid decision support tools for soil quality assessments and soil management practices.

5.1 Foundational Achievements that Established Nematodes as Bioindicators

The development of the 'De Man formulae' during the period between Bastian (1866) and De Man (1884–1893) simplified nematode taxonomy and introduced a period of discovery of terrestrial, freshwater and marine nematodes. Detail-rich taxonomic descriptions and advances made by nematologists in that period and thereafter (e.g. de Man, 1884) provided the scientific basis for the development of nematodes as bioindicators during the second half of the 20th century.

At the beginning of the 'green revolution' in the 1960s, studies on nematodes focused on plant-parasitic taxa in terrestrial ecosystems. However, research on nematode ecology exploded from 1970 onwards. The dominance of plant parasites (feeders) over other nematode functional groups was the first indication that management practices had the capacity to reduce functional diversity in the soil food web. Studies of anthropogenic impacts on soil nematode assemblages appeared around 1960, including the effects that liming and fertilization had on forest soils (Bassus, 1960, 1967). Other research demonstrated the restorative effects that woodlands

*Email: jan-henrik.schmidt@julius-kuehn.de

DOI: 10.1079/9781800624221.0005

and grasslands had on nematode assemblages after conversion from intensive agriculture (Pick-Hoong, 1966). Lastly, the identification of nematode assemblages in sand dunes, forests and grasslands (Yeates, 1968, 1972) demonstrated differences in community assemblages among contrasting ecosystems. For instance, Wasilewska (1979) and Johnson *et al.* (1972) studied nematode assemblages of forests and concluded that omnivorous dorylaimids were most sensitive to disturbances resulting from grazing (cattle, sheep) or cutting (human interference).

A few years later, the first studies appeared revealing that the incidence and prevalence of nematode species reflect the nature and quality of rivers (Zullini, 1976) and lakes (Prejs, 1977). As in terrestrial ecosystems, nematodes are omnipresent and more consistent than insects, molluscs and annelids in muddy, stony or sandy sediments of freshwater systems. Although some species were endemic to aquatic systems, most taxa found in Italian rivers were also terrestrial soil inhabitants (Zullini, 1976). Several authors made use of the work of Zullini to study the relationships between water quality and nematode assemblages (Cantelmo and Rao, 1978; Lambshead, 1986; Zullini and Peretti, 1986).

Based on these and many other fundamental studies (e.g. Freckman, 1982) in terrestrial and aquatic environments, Bongers *et al.* (1989) arranged nematode taxa into five categories along a colonizer–persister (*c-p*) scale that ultimately evolved into the Maturity Index (MI) for terrestrial and marine nematodes (Bongers, 1990; Bongers *et al.*, 1991). The index was developed as a measure of ecological succession. The MI and related indices (PPI, ΣMI) were the first set of nematode-based indices (NBIs) that served as indicators of disturbance intensity. Over the subsequent three decades, substantial advancements have been made in the development of NBIs for terrestrial nematodes (Du Preez *et al.*, 2022). With the development of nematode metabolic footprints (NMFs), Ferris (2010b) and Ferris and Tuomisto (2015) provided the first indices that evaluate the diversity-weighted magnitude of ecosystem functions fulfilled by soil biota.

NBIs represent a powerful ecological toolset that facilitate the study of ecological states, processes and functions. However, careful consideration should be given towards the correct time of investigation and selecting an appropriate subset of NBIs, which is dependent on the research question (Du Preez *et al.*, 2022). While this chapter focuses on terrestrial systems, the use and formalization of nematodes as bioindicators in aquatic systems are addressed in Chapters 6, 10 and 16 of this volume.

5.2 Indices of Ecosystem Status

NBIs are calculated using the *c-p* classification system based on the ecophysiological (life-history trait) characteristics of nematode communities (Du Preez *et al.*, 2022). The concept for the development and application of NBIs originated from the need to assess and monitor soil pollution in the 1980s. For this, nematodes proved to be sensitive indicators. Different taxa have varying sensitivities to biotic and abiotic stressors, leading to changes in index values in disturbed environments (Wasilewska, 1979; Bongers, 1990; Du Preez *et al.*, 2022). The advantage of using these indices over simple diversity indices (e.g. species richness, Simpson's diversity index) was statistically supported at scales that ranged from fields to regions (Freckman and Ettema, 1993; Neher *et al.*, 1995). Indices developed later expressed the potential of fungal- and bacterial-feeding nematodes to indicate prevalence of fungal or bacterial decomposition channels and nutrient turnover speed in soil. Omnivore and predatory nematodes, in turn, were seen as indicators of trophic connectance within the soil food web. Over time, a growing toolset of NBIs of diverse indicator functions crystallized and include the MI family (i.e. MI, Maturity Index 2-5 (MI2-5), Plant-Parasitic Index (PPI) and Sigma Maturity Index (ΣMI)), the food web diagnostic indices (i.e. Enrichment Index (EI), Structure Index (SI), Basal Index (BI) and Channel Index (CI)) and the NMFs. For a detailed description of the development (and evolution) of NBIs, please see Du Preez *et al.* (2022).

5.2.1 Basics of the *c-p* series and evolving taxon classifications

For the calculation of NBIs, soil nematodes are categorized into a *c-p* series (Fig. 5.1) ranging from r to K life-history strategists (Bongers *et al.*,

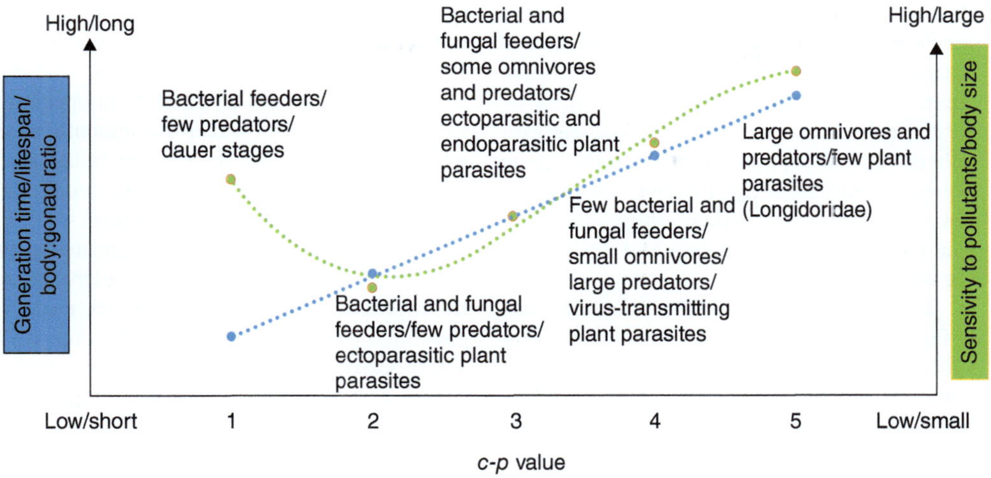

Fig. 5.1. Hypothetical concept of the colonizer–persister (c-p) values and their relation to nematode body characteristics, ecology and trophic group classification. (Author's own image.)

1989). 'Colonizer' (or early colonizer) nematodes are weighted 1 or 2 and indicate greater disturbance and/or resource availability; 'persister' (late successional) nematodes are weighted (3–5 and indicate increasing system stability, food web complexity and connectance among different trophic-level organisms therein.

As recognized early in the development of the *c-p* series (Bongers, 1990; Bongers *et al.*, 1991), categorization at the genus or species level would be more informative than at the family level. However, early attempts to assign *c-p* values at the genus level (Bongers *et al.*, 1989) proved difficult due to lack of information on the biology and sensitivity of the individual genera. Consequently, family-level assignments to *c-p* classes were used in the formal introduction of the MI (Bongers, 1990). The relevance of the family-level assignments has been justified on the basis that nematodes with similar morphology and feeding habits, and with similar life-history traits, have a high probability of similar sensitivity and responsiveness to environmental change (Bongers and Ferris, 1999). Empirical testing supported the *c-p* value assignments at the family level (e.g. Ettema and Bongers, 1993) with a few exceptions that became apparent as knowledge increased on the biology and sensitivity of individual genera and species (e.g. Fiscus and Neher, 2002). For example, bacterial feeders can be distinguished as enrichment opportunistic (*c-p*1) or general opportunists (*c-p*2-4) with the former having a non-feeding dauer stage. Because Monhysteridae species do not have a dauer stage and are tolerant of unfavourable conditions, the family was moved to the *c-p*2, a generalist bacterial-feeding category (Bongers *et al.*, 1995). Another example is that of the genus *Filenchus*, which was originally classified as an epidermal-cell and root-hair feeder but later shown capable of reproducing when feeding on fungi (Okada *et al.*, 2005). Some researchers also recategorized individual species. For example, Beier and Traunspurger (2001) changed *Tobrilus diversipapillatus* from *c-p*3 to *c-p*2 when the species was recorded at higher densities in polluted river sites. Over time, by the experimental generation of data on feeding and ecological preferences, morphologically inferred *c-p* values will be either confirmed or adjusted.

In the publication year of the first edition of this book, Andrássy (2009) reassigned several families and subfamilies within the order Dorylaimida. This led to several changes of *c-p* classes among dorylaimid families. For example, the predatory Discolaimidae were reclassified as subfamilies Discolaiminae and Carcharolaiminae in the family Qudsianematidae and, consequently, were downgraded from *c-p*5 to *c-p*4. Similarly, *Ecumenicus*, formerly classified within the family Thornenematidae, was moved to the new subfamily Ecumenicinae of the Qudsianematidae. Furthermore, several genera of Thornenematidae (*Prodorylaimus*, *Prodorylaimium* and

Mesodorylaimus) were moved into the family Dorylaimidae. More recently, Peña-Santiago and Álvarez-Ortega (2014) transferred multiple genera from Qudsianematidae to the family of Dorylaimidae based on morphological features, embryonic development and 28 rDNA gene sequences. Although thit has not affected current *c-p* assignments, further changes based on gene sequences and developmental studies can be expected.

Therefore, as the original *c-p* classifications by Bongers and Bongers (1998) and Bongers *et al.* (1991) evolve, referral to the updated nematode ecophysiology data sets provided by the Nemaplex website of University of California, Davis (http://nemaplex.ucdavis.edu, accessed 9 August 2025) is recommended. For this reason, the preservation, evolution and accessibility of this web page should be a primary consideration for nematode ecologists.

5.2.2 Calculation and use of the Maturity Index family

All the indices are based on the weighted proportion of nematodes in the fauna that meet the index criteria. The generic formula for calculation of indices in the MI family is:

$$XI = \sum_{i=1,f} v_i n_i / \sum_{i=1,f} n_i$$

where XI is the index of interest, v_i is the *c-p* value assigned to taxon i and n_i is the number of nematodes in each of the f taxa *that meet the criteria of the index* (Table 5.1).

Maturity Index (MI)

The MI is based on nematode taxa that are not parasitic on plants, insects or animals, and is considered a measure of environmental disturbance. Low MI values indicate a disturbed and/or enriched environment, while high MI values indicate a stable environment (Bongers, 1990). In essence, the MI is an ecological indicator of the state of succession of a system whereby disturbance and its consequent enrichment effects result in a setback of succession to an earlier state. In the case of the nematode assemblage, the successional setback is reflected in a lower MI effected by an increase of r-strategists, characterized by shorter lifespans and smaller body sizes (Odum, 1985; Bongers *et al.*, 1997) (Fig. 5.1).

The dauer stages of enrichment opportunists (animal parasites and entomopathogenic nematodes are rare in soil samples) (Jaffuel *et al.*, 2018) may be included or excluded from the calculation of the MI (Bongers and Bongers, 1998). An abundance of dauer stages indicates a system that was enriched (increased microbial activity) previously, which shifted to a less enriched phase at the time of sampling. Such a phase may also be seasonal and caused by either temperature flux or drought and the corresponding change in microbial activity. The ratio of dauer stages to active stages of rhabditids, as an indicator of resource availability, was introduced and tested by Sohlenius (1969). Temporal assessments of this ratio may provide insights into the resource dynamics of a system. However, a difficulty with this approach is identifying dauer stages of different nematode taxa. For example, dauer stages of entomopathogenic nematodes may occur in soil in the absence of insect hosts but are not indicators of food web enrichment.

Plant-Parasitic Index (PPI)

The PPI is comparable to the MI but is computed using only plant-parasitic (feeding) nematodes. The rationale behind this is that their abundance is determined by the vigour of their host plants, which, in turn, is influenced by system enrichment and other management factors. Consequently, under nutrient-poor conditions of natural ecosystems, the PPI is often low (compared to fertilized or organically amended agricultural systems) and associated with a high proportion of Tylenchidae (*c-p*2) in the nematode assemblage. Furthermore, this index shows an inverse response compared to the MI when it comes to enrichment (Bongers, 1990; Bongers *et al.*, 1997). However, it should be noted that the PPI depends on the host suitability of plants and the type of (agro)ecosystem in which they occur. This may result in variance between studies. Any inference of the PPI should consider specific information on the taxa present in the plant-parasitic assemblage. Also, the reports that *Filenchus* spp. feed on fungi and were recovered from leaf litter (Brzeski, 1998; Okada *et al.*, 2005) underscore the need for further studies on the

Table 5.1. Calculation of the Maturity Index (MI), Maturity Index 2-5 (MI2-5), Sigma Maturity Index (ΣMI), Sigma Maturity Index 2-5 (ΣMI2-5), Plant-Parasitic Index (PPI), and general community indices (see Chapter 4, this volume), for two nematode assemblages (samples A and B). (Author's own table.)

	Abundance				MI		MI2-5		ΣMI		ΣMI2-5		PPI	
					A	B	A	B	A	B	A	B	A	B
Nematode taxon	Sample A	Sample B	*c-p*	Feeding group	*c-p* wtd propn	*c-p* wtd propn	*c-p* wtd propn	*c-p* wtd propn	*c-p* wtd propn	*c-p* wtd propn	*c-p* wtd propn	*c-p* wtd propn	*c-p* wtd propn	*c-p* wtd propn
Hoplolaimidae	**5**	**15**	3	H					0.1	0.4	0.1	0.8	1.5	1.5
Pratylenchidae	**5**	**15**	3	H					0.1	0.4	0.1	0.8	1.5	1.5
Aphelenchidae	**15**	**5**	2	F	0.3	0.1	0.3	0.3	0.3	0.1	0.3	0.2		
Cephalobidae	**15**	**2**	2	B	0.3	0.1	0.3	0.1	0.3	0.0	0.3	0.1		
Plectidae	**2**	**15**	2	B	0.0	0.4	0.0	1.0	0.0	0.3	0.0	0.5		
Rhabditidae	**2**	**50**	1	B	0.0	0.6			0.0	0.5				
Dorylaimidae	**50**	**5**	4	O	2.0	0.3	2.1	0.7	1.8	0.2	1.9	0.3		
Aporcelaimidae	**15**	**2**	5	P	0.8	0.1	0.8	0.3	0.7	0.1	0.7	0.2		
Relevant total	**109**	**109**			**99**	**79**	**97**	**29**	**109**	**109**	**107**	**59**	**10**	**30**
MI	**3.4**	**1.6**												
MI2-5	**3.5**	**2.6**												
ΣMI	**3.4**	**2.0**												
ΣMI2-5	**3.4**	**2.8**												
PPI	**3.0**	**3.0**												
Simpson	**0.27**	**0.27**												
Shannon	**1.61**	**1.61**												
Hill N1	**1.31**	**1.31**												
Pielou J´	**0.77**	**0.77**												

Feeding group: H = plant parasites (feeders); F = fungal feeders; B = bacterial feeders; O = omnivores; P = predators. In the columns in which each index is calculated, the total number of nematodes meeting the criteria of the index is first determined (relevant total). Next, the proportion (propn) of that total represented by taxa that meet the criteria is calculated and weighted (wtd) by the *c-p* values of those taxa. The index value is the sum of these weighted proportions.

feeding habits of the many genera and species in the ubiquitous Tylenchidae group (Qing and Bert, 2019). Such nematodes, able to feed on both root hairs and hyphal fungi, can be considered as root associates. There is an ongoing debate about whether to include (Bongers, 1990) or exclude (Neher and Campbell, 1996) Tylenchidae in the PPI. The controversy stems from the classification of Tylenchidae members as fungal, root, algal or lichen feeders. To date, this issue has been resolved for only a minority of the 44 genera in this highly diverse family (Qing and Bert, 2019).

Maturity Index 2-5 (MI2-5)

This index is identical to the MI but excludes the *c-p*1 enrichment opportunists. The index was first proposed by Popovici (1992) and was empirically tested and confirmed during studies of the relationship between MI and copper concentration under agricultural conditions (Korthals *et al.*, 1996a,b). In those studies, it was apparent that there was a strong relationship between decreased abundance in higher *c-p* value nematodes and pollution-induced stress while the *c-p*1 nematodes responded to the presence of decomposing organic material. In some cases, the pollutant may become a resource for a component of the microbial community which, in turn, acts as a resource for the *c-p*1 nematodes. The use of this index (instead of or supplementary to MI) should thus be considered in studies evaluating the effect of heavy metals, and potentially other pollutants, or when the effects of ephemeral nutrient resources on similarly ephemeral *c-p*1 nematodes can be ignored.

Sigma Maturity Index (ΣMI)

This index was proposed by Yeates (1994) and is equivalent to the Total MI of Wasilewska (1994). It represents the MI for all nematodes in the system, including plant parasites, based on the assertion that the complete assemblage provides integral information. This includes the assumption of a congruent relationship between MI and PPI, offering insight into disturbance and environmental conditions. However, sampling of more than 150 agricultural sites in North Carolina and Nebraska (Neher and Campbell, 1996) revealed that the use of either the MI or ΣMI may be plant-host-dependent. When crops that are good hosts for *c-p*3 plant parasites are grown, the relationship between MI and PPI may be incongruent. In contrast, in (agro)ecosystems with high plant diversity (e.g. extensive crop rotations, natural ecosystems), an inverse relationship of MI and PPI does not occur (Neher and Campbell, 1996).

Sigma Maturity Index 2-5 (ΣMI2-5)

This index computes the MI for all nematodes within the *c-p*2-5 range (Neher and Campbell, 1996). It acknowledges that higher *c-p* values of plant-parasitic species also provide valuable information on environmental stress. However, the index may be influenced by the presence of large numbers of *c-p*3 plant parasites, which can affect its overall interpretation, like the ΣMI.

For ease of calculating NBIs, and to ensure accuracy, the use of the Nematode Indicator Joint Analysis (NINJA) online tool is recommended (Sieriebriennikov *et al.*, 2014) (https://shiny.wur.nl/ninja/, accessed 9 August 2025). This tool processes, calculates and displays the NBIs by making use of the large, updated repository of nematode ecophysiological data included in Nemaplex (http://nemaplex.ucdavis.edu, accessed 9 August 2025). A thorough interpretation support for each NBI is provided by Du Preez *et al.* (2022).

5.2.3 Indicators of soil food web condition: Enrichment, Structure, Basal and Channel indices

The evolution of concepts, research and model validation associated with the development of the MI family led to a functional guild classification of nematodes, i.e. 'groups of nematodes that have similar effects on ecosystem processes', as a basis for studying and comparing ecosystem services (Bongers and Bongers, 1998; Bongers and Ferris, 1999). This functional redundancy creates a high probability that the absence of a guild is a reliable indicator of disturbance and that the presence of a guild is a reliable indicator of lack of perturbation or of recovery from perturbation. Hence, the nematode fauna indicates two major characteristics of the soil environment and its resident communities. One characteristic is the flow of resources into the food web system as indicated by enrichment opportunists; the other is the trophic connectance of the system as indicated by the prevalence and abundance of higher-trophic-level organisms with multichannel (omnivory) feeding characteristics (Wolkovich, 2016). Considering soil nematode taxa as representatives of functional guilds generates an indicator profile that is not constrained by population distribution patterns and microenvironment effects (Ferris and Bongers, 2006).

The Enrichment, Structure and Basal indices rely on the relative abundance of functional guilds of nematodes and are descriptors of food web condition (Table 5.2, Fig. 5.2). Nematodes of all feeding habits classified as *c-p*2 are considered basal (b) to both enrichment and structure trajectories, i.e. they exist in most soil food webs, are almost always present and survive the most adverse conditions. Ferris *et al.* (2001) hypothesized that bacterial-feeding *c-p*1 and fungal feeders in *c-p*2 are indicators of enrichment (e), while nematodes of all feeding habits in *c-p*3-5 are indicators of structure (s). Functional guild indicators are weighted according to growth and metabolic rates (resource utilization) on the enrichment axis. On the structure axis, they are weighted by the number of nematodes in higher *c-p* classes, reflecting the degree of connectance in food webs of increasing complexity (Fig. 5.2). Greater detail on derivation of the structure and enrichment weightings (W_i) is provided in Ferris *et al.* (2001).

The nematode fauna is comprised of basal, enrichment and structural components (b, e, s), calculated as follows:

$$b = (Ba_2 + Fu_2) \times W_2, \text{ where } W_2 = 0.8$$

$$e = (Ba_1 \times W_1) + (Fu_2 \times W_2), \text{ where } W_1 = 3.2 \text{ and } W_2 = 0.8$$

$$s = (Ba_n \times W_n + Ca_n \times W_n + Fu_n \times W_n + Om_n \times W_n), \text{ where } n = 3-5,\ W_3 = 1.8,\ W_4 = 3.2 \text{ and } W_5 = 5.0$$

The Enrichment (EI), Structure (SI), Basal (BI) and Channel (CI) indices are calculated from the weighted faunal components (Ferris *et al.*, 2001; Berkelmans *et al.*, 2003):

$$EI = 100 \times [e/(e + b)]$$

$$SI = 100 \times [s/(s + b)]$$

$$BI = 100 \times [b/(e + s + b)]$$

$$CI = 100 \times [(Fu_2 \times W_2)/(Ba_1 \times W_1 + Fu_2 \times W_2)]$$

We used the nematode assemblages of Table 5.1 to calculate the above-described indices (Table 5.3). Accordingly, sample A represents an environment with an abundance of omnivore and predator nematodes, suggesting greater connectance in the soil food web and the probable top-down regulation of opportunistic species (Quadrat C in Fig. 5.2). Sample B represents a disturbed and enriched condition in which the disturbance has had detrimental effects on higher trophic levels but fostered quickly developing *c-p*1 bacterial feeders (Quadrat B in Fig. 5.2). Faunal analyses based on these indices provide insights into food web enrichment and structure and allow derivation of testable hypotheses (e.g. Table 5.2) based on the relative enrichment and structure of the system.

The CI represents a partitioning of resource flow through fungal and bacterial decomposition channels. Indices of fungal and bacterial activity based on the relative abundance of fungal- and bacterial-feeding nematodes have been proposed several times following the calculation of their relative proportions in grasslands, woodlands and cultivated fields by Twinn (1974). These approaches included a ratio of F/B (Sohlenius and Bostrom, 1984), which has been defined most recently as the Nematode Channel Ratio (NCR) = B/(B + F), where B and F represent the abundance of bacterial- and fungal-feeding nematodes, respectively (Yeates, 2003). The CI differs

Table 5.2. Inferred condition of the soil food web and its environment based on weighted nematode faunal analysis. Quadrats refer to faunal ordination in the faunal profile (Fig. 5.2). (Modified from Ferris *et al.*, 2001.)

General diagnosis	Quadrat A	Quadrat B	Quadrat C	Quadrat D
Disturbance	High	Low to mod.	Undisturbed	Stressed
Enrichment	N-enriched	N-enriched	Moderate	Depleted
Decomposition channels	Bacterial	Balanced	Fungal	Fungal
C:N ratio	Low	Low	Mod. to high	High
Food web condition	Disturbed	Maturing	Structured	Degraded

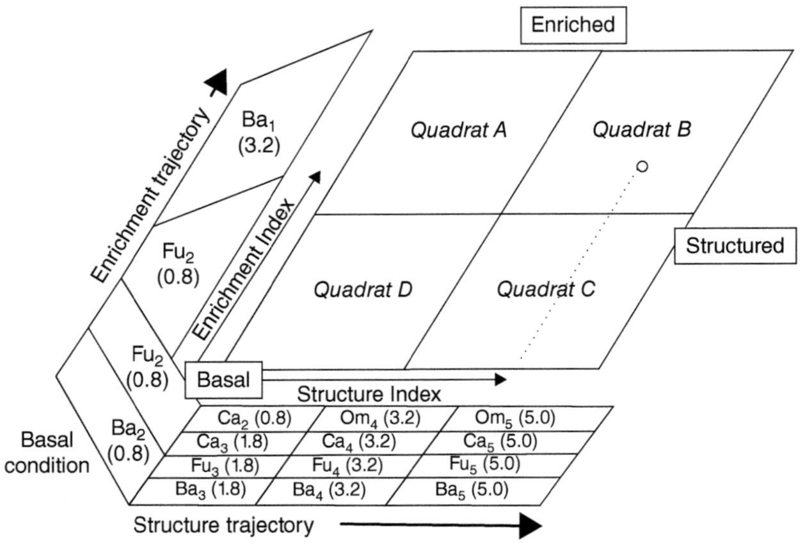

Fig. 5.2. A graphical representation of the nematode faunal profile indicates whether the soil community is enriched but unstructured (Quadrat A), enriched and structured (Quadrat B), resource-limited and structured (Quadrat C), or resource-depleted with minimal structure (Quadrat D): functional guilds of soil nematodes are characterized by diet (trophic group) and life-history characteristics, after Bongers and Bongers (1998). Indicator guilds of soil food web condition (basal, structured, enriched) are designated and weightings of the guilds along the structure and enrichment trajectories are provided for determination of the Enrichment Index and Structure Index of the food web. (Modified from Ferris *et al.*, 2001.)

in including weighting parameters for the size and *c-p* values (for detailed information, see Ferris *et al.*, 2001) of the nematode functional guilds. As such, values above 50 indicate predominantly fungal-mediated (sample A) and values below 50 indicate predominantly bacterial-mediated (sample B) decomposition pathways (Table 5.2).

When resources become available to soil organisms through external input, disturbance, organism mortality, turnover or environment changes, there is an enrichment pulse of early successional taxa (opportunists). The pulse is followed by heterotrophic succession whereby the predominance of organisms changes through time depending on trophic roles, life-course dynamics and prevailing environmental conditions (Ferris, 2010a). Substrates with high C:N ratios may favour the fungal rather than the bacterial decomposition channel (Ruess and Ferris, 2004).

Further understanding of enrichment is determined by the carbon flow through active fungal and bacterial channels using their biomass and respiration rates. These concepts led to the development of metabolic footprints which measure the magnitude of ecosystem functions fulfilled by soil biota (Ferris, 2010b). It is also the first NBI that increases proportionally to the abundance (and biomass) of nematode taxa.

5.2.4 Abundance, biomass and functional diversity – the metabolic footprint concept

To study ecosystem functions of nematodes, and their roles in contributing to ecosystem services, it is important to consider their biomass and biomass-dependent activity. Since previous NBIs are based on relative abundances or proportions, communities with different abundances may present the same index values. Consequently, correlating the relative abundances of nematode feeding guilds and/or indices to microbial activity and biomass, and magnitudes of their functions (e.g. nutrient mineralization), is problematic. Even within bacterial-feeding nematode families, size and biomass may vary greatly between genera. For example, the biomass for

Table 5.3. Calculation of the Basal Index (BI), Enrichment Index (EI), Structure Index (SI) and Channel Index (CI) for two nematode assemblages (samples A and B) of the same number of taxa and the same number of individuals (cf. Table 5.1). (Author's own table.)

								A			B		
Nematode taxon	Sample A	Sample B	*c-p*	Feeding group	b wtd	e wtd	s wtd	b	e	s	b	e	s
Hoplolaimidae	**5**	**15**	3	H				0.0	0.0	0.0	0.0	0.0	0.0
Pratylenchidae	**5**	**15**	3	H				0.0	0.0	0.0	0.0	0.0	0.0
Aphelenchidae	**15**	**5**	2	F	0.8	0.8		12.0	12.0	0.0	4.0	4.0	0.0
Cephalobidae	**15**	**2**	2	B	0.8			12.0	0.0	0.0	1.6	0.0	0.0
Plectidae	**2**	**15**	2	B	0.8			1.6	0.0	0.0	12.0	0.0	0.0
Rhabditidae	**2**	**50**	1	B		3.2		0.0	6.4	0.0	0.0	160.0	0.0
Dorylaimidae	**50**	**5**	4	O			3.2	0.0	0.0	160.0	0.0	0.0	16.0
Aporcelaimidae	**15**	**2**	5	P			5.0	0.0	0.0	75.0	0.0	0.0	10.0
Total	**109**	**109**											
								25.6	18.4	235.0	17.6	164.0	26.0
Index components													
BI	**9.2**	**8.5**											
EI	**41.8**	**90.3**											
SI	**90.2**	**59.6**											
CI	**65.2**	**2.4**											

Feeding group: H = plant parasites (feeders); F = fungal feeders; B = bacterial feeders; O = omnivores; P = predators. In the columns in which each index is calculated, the total number of nematodes in each taxon meeting the criteria of the index is weighted (wtd) in terms of the basal (b), enrichment (e) and structure (s) characteristics of that taxon. The index values are calculated from the index components, i.e. the sum of the b, e and s values (see text).

Mesorhabditis (= *Bursilla*) *labiata* and *Cruznema tripartitum* adult females, both belonging to the family Rhabditidae, is 0.48 µg and 7.7 µg, respectively (Ferris *et al.*, 1995). Consequently, the magnitude of certain functions (e.g. regulation of microbial populations) might vary with the absolute abundance or biomass of a given species population in a given soil sample (16 *M. labiata* = 1 *C. tripartitum*).

To overcome this limitation, Ferris (2010b) developed NMFs as an assessment of the metabolic activity of nematode functional guilds. These footprints rely on the assessment of the carbon utilization of nematodes on growth and egg production (production component) and respiration (respiration component). The production component is based on the calculation of nematode carbon biomass according to the individual taxon body volume calculated with the Andrássy formula (Andrássy, 1956) and weighted by the taxon lifespan inferred from its *c-p* value (Fig. 5.1). The respiration component is estimated through the allometric power dependence of basal metabolism and body weight. Therefore, NMFs are calculated as the sum of both components and as an index of carbon utilization of each taxon. Applied to nematode functional guilds, the Enrichment Footprint is the estimation of carbon utilization by enrichment-opportunistic nematodes, and the Structure Footprint the estimation of carbon utilization by higher trophic links which are indicators of regulatory (and/or disease-suppressive) functions. Such footprints can be graphically depicted as a rhomboid shape around the (EI, SI) point in a biplot, and the Functional NMF calculated as the area of the rhombus (Ferris, 2010b).

Another option is to link NMFs directly with their expected function (Sánchez-Moreno and Ferris, 2018), exemplified here with radar charts with data from Tables 5.1 and 5.3 (Fig. 5.3) as well as from Schmidt *et al.* (2020, 2022) (Fig. 5.4). These graphs provide additional information on individual trophic-group footprints: the Herbivore, Bacterivore, Fungivore, Omnivore and Predator NMFs refer to the estimation of carbon and energy flow through the soil food web by different functional groups. The Composite Footprint represents the metabolic footprint of the entire nematode community.

This resource inflow into and within the soil food web can be visualized using a radar chart. The underlying data for this representation may include metabolic footprints (Figs 5.3 and 5.4). An alternative is the use of single components of the metabolic footprint, namely nematode body mass, respired carbon and incorporated carbon (all obtainable from http://nemaplex.ucdavis.edu/,

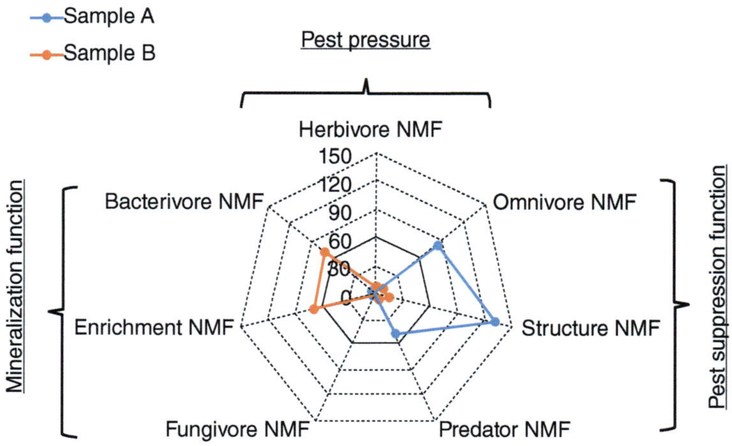

Fig. 5.3. Nematode metabolic footprint (NMF) family comprising Herbivore, Omnivore, Structure, Predator, Fungivore, Enrichment and Bacterivore NMFs. NMFs (µg carbon per functional group respired and incorporated in biomass) were calculated for the nematode assemblages of samples A and B (Tables 5.1 and 5.3). Individual nematode family characteristics were extracted from http://nemaplex.ucdavis.edu/ (accessed 9 August 2025) (note that the Predator and Omnivore NMFs are not separated on this web page and need manual reassignments according to the feeding codes of Yeates *et al.*, 1993). Ecosystem functions (underlined) are from Sánchez-Moreno and Ferris (2018).

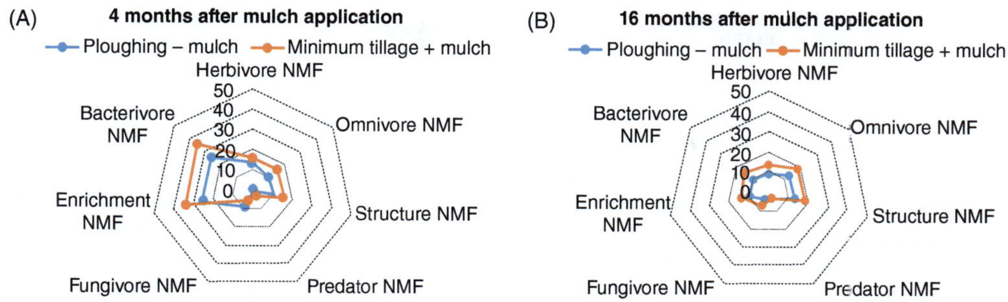

Fig. 5.4. Long-term field experiments conducted by the University of Kassel compared a plough system, where potatoes were frequently hilled in rotation, with a minimum tillage system that applied cover crop mulch (10–15 t ha^{-1}, C:N = 20–25) on top of the potato hills before tiller emergence. Square-root-transformed nematode metabolic footprints (NMFs) are presented for better visualization (μg C 100 ml $soil^{-1}$) at 4 months (A) and 16 months (B) after the last mulch application. (Data from Schmidt *et al*., 2020, 2022.)

accessed 9 August 2025). The arrangement of the radar chart allows the inference of two specific ecosystem functions assessed by the nematode assemblage: the left side (Bacterivore, Enrichment and Fungivore NMFs) represents the nutrient turnover function that is more pronounced in sample B than in sample A (Fig. 5.3). In contrast, the right side of the radar chart reflects the degree of disease suppressiveness (regulatory function) indicated by the Omnivore, Structure and Predator NMFs. According to Fig. 5.3, the disease-suppressive ability is greater in sample A than in sample B. In contrast, the Herbivore NMF is interpreted as the disservice that indicates whether plant-parasitic nematodes interfere with plant growth and/or yield production.

Furthermore, we used nematode community data from two long-term German field experiments to demonstrate how NMF analyses can be applied to natural (agro)ecosystems (Fig. 5.4). Four months after application, the mulch in the potato cropping system with minimum tillage increased the NMFs of bacterial feeders substantially compared to the plough system without mulch (Fig. 5.4A). The Bacterivore NMF collapsed 16 months after mulch application (Fig. 5.4B). However, the slight increase of the Omnivore and Fungivore NMFs and the maintained Herbivore NMF indicate a growing food web connectance in the minimum tillage system. This may explain the greater resilience to fungal-mediated pea root rots in this particular system (Schmidt *et al*., 2022). The low Predator NMF relative to the food resource (prey) indicates the effect of environmental perturbations on nematodes of higher *c-p* classes (here, intensive soil tillage).

In an attempt to improve the correlations of NMFs, biomass and abundance with ecosystem functions, Ferris and Tuomisto (2015) proposed weighting these indices by nematode diversity:

$$\theta = \left({}^{q}\mathrm{D}\right)^{b} \times \sum_{i=1}^{R} \left(A_i\right),$$

where θ is the diversity-weighted abundance (or biomass or NMF), $\left({}^{q}\mathrm{D}\right)^{b}$ is the true diversity and A_i is the value for biomass, abundance or NMF of species *i* of the *R* (R = species richness) species in that sample that fulfil a specific functional trait. In the simplest case, $\left({}^{q}\mathrm{D}\right)^{b}$ reflects the Shannon diversity index of all species that are, for example, responsible for mineralization of nutrients. According to Fig. 5.3, A_i may be then either Fungivore, Bacterivore and Enrichment NMFs, or their corresponding biomass or abundance. The rationale behind this is based on the concept of functional redundancy, as discussed earlier. For a given function, higher nematode diversity allows for the occupation of more niches (e.g. soil pores of different sizes). In our example, as nematode diversity increases, the potential for available prey (bacteria and fungi) also increases, which in turn enhances the nutrient mineralization potential.

Although diversity-weighted indicators have been used in several studies (Landi *et al*., 2018; Mejía-Madrid and Sánchez-Moreno, 2022), more evidence is necessary to ascertain their relationships to soil functioning.

5.3 Conclusions

The evolution of functional indices based on nematode faunal analysis provides insights into the functioning and services of ecosystems. It has been greatly advanced by inference and observation of nematode feeding habits in relation to stomal architecture and by knowledge of the life-history traits of nematode functional guilds. Undoubtedly, calibration of indices by quantitative levels of disturbance, ecosystem type and major land-resource regions guided by new information on feeding habits, life-history traits and the assignment of taxa to functional guilds is necessary for improving the interpretative value of NBIs. There are other examples of the use of the community structure of various organism groups for environmental monitoring. The advantage of those based on nematode functional guilds derives from the abundance and ubiquity of nematodes, the relationships between form and function, the differences among taxa in sensitivity to environmental disturbance, and the ease with which nematodes can be separated from substrate and categorized into taxonomic groups or functional guilds. The NMFs and diversity-weighted abundances represent potentially useful tools for ecologists, agronomists and (soil) biologists in that they contribute to further understanding of nematode biodiversity and links to ecosystem functions in soils shaped by natural events and anthropogenic activities.

Acknowledgements

The authors express their gratitude to Dr Tom Bongers for the fruitful discussions that contributed to this book chapter, as well as to the two independent reviewers for their valuable comments and insights.

References

Andrássy, I. (1956) Die Rauminhalts- und Gewichtsbestimmung der Fadenwürmer (Nematoden). *Acta Zoologica* 2, 1–13.

Andrássy, I. (2009) *Free-Living Nematodes of Hungary, III (Nematoda errantia)*. Hungarian Natural History Museum, Budapest.

Bassus, W. (1960) Die Nematodenfauna des Fichtenrohhumus unter dem Einfluss der Kalkdüngung. *Nematologica* 5, 86–91. DOI: 10.1163/187529260X00442.

Bassus, W. (1967) Der Einfluss von Meliorations- und Düngungsmaßnahmen auf die Nematodenfauna verschiedener Waldböden. *Pedobiologia* 7, 280–295.

Bastian, H.C. (1866) *Monograph on the Anguillulidae, or Free Nematoids, Marine, Land, and Freshwater with Descriptions of 100 New Species*. Linnean Society of London, London. DOI: 10.5962/bhl.title.14153.

Beier, S. and Traunspurger, W. (2001) The meiofauna community of two small German streams as indicator of pollution. *Journal of Aquatic Ecosystem Stress and Recovery* 8, 387–405. DOI: 10.1023/A:1012965424527.

Berkelmans, R., Ferris, H., Tenuta, M. and van Bruggen, A.H.C. (2003) Effects of long-term crop management on nematode trophic levels other than plant feeders disappear after 1 year of disruptive soil management. *Applied Soil Ecology* 23, 223–235. DOI: 10.1016/S0929-1393(03)00047-7.

Bongers, T. (1990) The maturity index: an ecological measure of environmental disturbance based on nematode species composition. *Oecologia* 83, 14–19. DOI: 10.1007/BF00324627.

Bongers, T. and Bongers, M. (1998) Functional diversity of nematodes. *Applied Soil Ecology* 10, 239–251. DOI: 10.1016/S0929-1393(98)00123-1.

Bongers, T. and Ferris, H. (1999) Nematode community structure as a bioindicator in environmental monitoring. *Trends in Ecology & Evolution* 14, 224–228. DOI: 10.1016/S0169-5347(98)01583-3.

Bongers, A.M.T., de Goede, R.G.M., Kappers, F.I. and Manger, R. (1989) *Ecologische typologie van de Nederlandse bodem op basis van de vrijlevende nematodenfauna*. Rapportnummer 718602002. RIVM, Bilthoven, The Netherlands.

Bongers, T., Alkemade, R. and Yeates, G.W. (1991) Interpretation of disturbance-induced maturity decrease in marine nematode assemblages by means of the Maturity Index. *Marine Ecology Progress Series* 76, 135–142. DOI: 10.3354/meps076135.

Bongers, T., de Goede, R.G.M., Korthals, G.W. and Yeates, G.W. (1995) Proposed changes of c-p classification for nematodes. *Russian Journal of Nematology* 3, 61–62.

Bongers, T., van der Meulen, H. and Korthals, G. (1997) Inverse relationship between the nematode maturity index and plant parasite index under enriched nutrient conditions. *Applied Soil Ecology* 6, 195–199. DOI: 10.1016/S0929-1393(96)00136-9.

Brzeski, M.W. (1998) *Nematodes of Tylenchina in Poland and Temperate Europe*. Muzeum i Instytut Zoologii Polska Akademia Nauk, Warsaw, Poland.

Cantelmo, F.R. and Rao, K.R. (1978) Effect of pentachlorophenol (PCP) on meiobenthic communities established in an experimental system. *Marine Biology* 46, 17–22. DOI: 10.1007/BF00393815.

Du Preez, G., Daneel, M., de Goede, R., Du Toit, M.J., Ferris, H. *et al.* (2022) Nematode-based indices in soil ecology: application, utility, and future directions. *Soil Biology and Biochemistry* 169, 108640. DOI: 10.1016/j.soilbio.2022.108640.

Ettema, C.H. and Bongers, T. (1993) Characterization of nematode colonization and succession in disturbed soil using the Maturity Index. *Biology and Fertility of Soils* 16, 79–85. DOI: 10.1007/BF00369407.

Ferris, H. (2010a) Contribution of nematodes to the structure and function of the soil food web. *Journal of Nematology* 42, 63–67.

Ferris, H. (2010b) Form and function: metabolic footprints of nematodes in the soil food web. *European Journal of Soil Biology* 46, 97–104. DOI: 10.1016/j.ejsobi.2010.01.003.

Ferris, H. and Bongers, T. (2006) Nematode indicators of organic enrichment. *Journal of Nematology* 38, 3–12.

Ferris, H. and Tuomisto, H. (2015) Unearthing the role of biological diversity in soil health. *Soil Biology and Biochemistry* 85, 101–109. DOI: 10.1016/j.soilbio.2015.02.037.

Ferris, H., Lau, S. and Venette, R. (1995) Population energetics of bacterial-feeding nematodes: respiration and metabolic rates based on CO_2 production. *Soil Biology and Biochemistry* 27, 319–330. DOI: 10.1016/0038-0717(94)00186-5.

Ferris, H., Bongers, T. and de Goede, R.G.M. (2001) A framework for soil food web diagnostics: extension of the nematode faunal analysis concept. *Applied Soil Ecology* 18, 13–29. DOI: 10.1016/S0929-1393(01)00152-4.

Fiscus, D.A. and Neher, D.A. (2002) Distinguishing sensitivity of free-living soil nematode genera to physical and chemical disturbances. *Ecological Applications* 12, 565–575. DOI: 10.1890/1051-0761(2002)012[0565:DSOFLS]2.0.CO;2.

Freckman, D.W. (ed.) (1982) *Nematodes in Soil Ecosystems*. University of Texas Press, Austin, Texas.

Freckman, D.W. and Ettema, C.H. (1993) Assessing nematode communities in agroecosystems of varying human intervention. *Agriculture, Ecosystems and Environment* 45, 239–261. DOI: 10.1016/0167-8809(93)90074-Y.

de Man, J.G. (1884) *Die frei in der reinen Erde und im süssen Wasser lebenden Nematoden der niederländischen Fauna*. Brill, Leiden, Germany.

Jaffuel, G., Blanco-Pérez, R., Hug, A.-S., Chiriboga, X., Meuli, R.G. *et al.* (2018) The evaluation of entomopathogenic nematode soil food web assemblages across Switzerland reveals major differences among agricultural, grassland and forest ecosystems. *Agriculture, Ecosystems and Environment* 262, 48–57. DOI: 10.1016/j.agee.2018.04.008.

Johnson, S.R., Ferris, V.R. and Ferris, J.M. (1972) Nematode community structure of forest woodlots. I. Relationships based on similarity coefficients of nematode species. *Journal of Nematology* 4, 175–183.

Korthals, G., Bongers, T., Kammenga, J.E., Alexiev, A.D. and Lexmond, T.M. (1996a) Long-term effects of copper and pH on the nematode community in an agroecosystem. *Environmental Toxicology and Chemistry* 15, 979–985. DOI: 10.1002/etc.5620150621.

Korthals, G., van de Ende, A., van Megen, H., Lexmond, T.M., Kammenga, J.E. and Bongers, T. (1996b) Short-term effects of cadmium, copper, nickel and zinc on soil nematodes from different feeding and life-history strategy groups. *Applied Soil Ecology* 4, 107–117. DOI: 10.1016/0929-1393(96)00113-8.

Lambshead, P.J. (1986) Sub-catastrophic sewage and industrial waste contamination as revealed by marine nematode faunal analysis. *Marine Ecology Progress Series* 29, 247–259. DOI: 10.3354/meps029247.

Landi, S., Papini, R., d'Errico, G., Barzanti, G.P. and Roversi, P.F. (2018) Nematode indicators as integrative measures of soil condition during conversion from conventional to organic rice production in Italy: a case study. *Biological Agriculture & Horticulture* 34, 141–153. DOI: 10.1080/01448765.2017.1399166.

Mejía-Madrid, H.H. and Sánchez-Moreno, S. (2022) Natural ecosystem heterogeneity diversity and functioning of nematode communities in a semi-desert ecosystem in Mexico. *Applied Soil Ecology* 176, 104477. DOI: 10.1016/j.apsoil.2022.104477.

Neher, D.A. and Campbell, C.L. (1996) Sampling for regional monitoring of nematode communities in agricultural soils. *Journal of Nematology* 28, 196–208.

Neher, D.A., Peck, S.L., Rawlings, J.O. and Campbell, C.L. (1995) Measures of nematode community structure and sources of variability among and within agricultural fields. *Plant and Soil* 170, 167–181.

Odum, E.P. (1985) Trends expected in stressed ecosystems. *BioScience* 35, 419–422. DOI: 10.2307/1310021.

Okada, H., Harada, H. and Kadota, I. (2005) Fungal-feeding habits of six nematode isolates in the genus *Filenchus*. *Soil Biology and Biochemistry* 37, 1113–1120. DOI: 10.1016/j.soilbio.2004.11.010.

Peña-Santiago, R. and Álvarez-Ortega, S. (2014) An integrative approach to assess the phylogeny and the systematics of rounded-tailed genera of the subfamily Qudsianematinae (Nematoda, Dorylaimida). *Zoologica Scripta* 43, 418–428. DOI: 10.1111/zsc.12058.

Pick-Hoong, Y. (1966) The nematode fauna of the regenerated woodland and grassland of Broadbalk Wilderness. *Nematologica* 12, 195–214. DOI: 10.1163/187529266X00617.

Popovici, I. (1992) Nematodes as indicators of ecosystem disturbance due to pollution. *Studia Universitatis Babes-Bolyai Biologia* 37, 15–27.

Prejs, K. (1977) The littoral and profundal benthic nematodes of lakes with different trophy. *Ekologia Polska* 25, 21–30.

Qing, X. and Bert, W. (2019) Family Tylenchidae (Nematoda): an overview and perspectives. *Organisms Diversity & Evolution* 19, 391–408. DOI: 10.1007/s13127-019-00404-4.

Ruess, L. and Ferris, H. (2004) Decomposition pathways and successional changes. In: Cook, R.C. and Hunt, D.J. (eds) *Proceedings of the Fourth International Congress of Nematology*. Brill, Tenerife, Spain, pp. 547–556.

Sánchez-Moreno, S. and Ferris, H. (2018) Nematode ecology and soil health. In: Sikora, R.A., Coyne, D., Hallmann, J. and Timper, P. (eds) *Plant Parasitic Nematodes in Subtropical and Tropical Agriculture*, 3rd edn. CAB International, Wallingford, UK, pp. 62–86. DOI: 10.1079/9781786391247.0062.

Schmidt, J.H., Hallmann, J. and Finckh, M.R. (2020) Bacterivorous nematodes correlate with soil fertility and improved crop production in an organic minimum tillage system. *Sustainability* 12, 6730. DOI: 10.3390/su12176730.

Schmidt, J.H., Theisgen, L.V., Finckh, M.R. and Šišić, A. (2022) Increased resilience of peas toward root rot pathogens can be predicted by the nematode metabolic footprint. *Frontiers in Sustainable Food Systems* 6, 881520. DOI: 10.3389/fsufs.2022.881520.

Sieriebriennikov, B., Ferris, H. and de Goede, R.G.M. (2014) NINJA: an automated calculation system for nematode-based biological monitoring. *European Journal of Soil Biology* 61, 90–93. DOI: 10.1016/j.ejsobi.2014.02.004.

Sohlenius, B. (1969) The monoxenic cultivation of some rhabditid nematodes. *Oikos* 20, 287–293. DOI: 10.2307/3543195.

Sohlenius, B. and Bostrom, S. (1984) Colonization, population development and metabolic activity of nematodes in buried barley straw. *Pedobiologia* 27, 67–78.

Twinn, D.C. (1974) Nematodes. In: Dickinson, C.H. and Pugh, G.J.F. (eds) *Biology of Plant Litter Decomposition*, 2nd edn. Academic Press, London, pp. 421–465.

Wasilewska, L. (1979) The structure and function of soil nematode communities in natural ecosystems and agrocenoses. *Polish Ecological Studies* 5, 97–146.

Wasilewska, L. (1994) The effect of age of meadows on succession and diversity in soil nematode communities. *Pedobiologia* 38, 1–11.

Wolkovich, E.M. (2016) Reticulated channels in soil food webs. *Soil Biology and Biochemistry* 102, 18–21. DOI: 10.1016/j.soilbio.2016.06.021.

Yeates, G.W. (1968) An analysis of annual variation of the nematode fauna in dune sand, at Himantangi beach. *Pedobiologia* 8, 173–207.

Yeates, G.W. (1972) Nematoda of a Danish beech forest. I. Methods and general analysis. *Oikos* 23, 178–189. DOI: 10.2307/3543403.

Yeates, G.W. (1994) Modification and qualification of the nematode maturity index. *Pedobiologia* 38, 97–101.

Yeates, G.W. (2003) Nematodes as soil indicators: functional and biodiversity aspects. *Biology and Fertility of Soils* 37, 199–210. DOI: 10.1007/s00374-003-0586-5.

Yeates, G.W., Bongers, T., de Goede, R.G.M., Freckman, D.W. and Georgieva, S.S. (1993) Feeding habits in soil nematode families and genera – an outline for soil ecologists. *Journal of Nematology* 25, 315–331.

Zullini, A. (1976) Nematodes as indicators of river pollution. *Nematologia Mediterranea* 4, 13–22.

Zullini, A. and Peretti, E. (1986) Lead pollution and moss-inhabiting nematodes of an industrial area. *Water, Air, and Soil Pollution* 27, 403–410. DOI: 10.1007/BF00649421.

6 Nematode-Based Pollution Indices for Aquatic Ecosystems

Sebastian Höss[1]*, Walter Traunspurger[2], Federica Semprucci[3] and Michaela Schratzberger[4]

[1]Ecossa, Starnberg, Germany; [2]Animal Ecology, Bielefeld University, Germany; [3]Department of Biomolecular Sciences (DiSB), University of Urbino, Italy; [4]Centre for Environment, Fisheries and Aquaculture Science, Lowestoft, UK

Abstract

Bioindication is an important tool for assessing the risks of anthropogenic impact on freshwater and marine ecosystems and for triggering appropriate water management actions to improve the ecological quality. However, in the European Union, only approximately 40% of all surface waterbodies have achieved good ecological and chemical status. Current bioindication systems are mainly based on macroinvertebrates, neglecting the prevailing microscopic meiofauna in fine, cohesive sediments, which are often hotspots of chemical contamination. This chapter highlights the urgent need for including meiofaunal organisms, such as nematodes, in routine biomonitoring programmes to achieve: (i) a more accurate classification of ecological status; and (ii) a better cause–effect-based identification of anthropogenic impacts on rivers, lakes, seas, and the ocean. Considering also the bio-indicative power of nematodes in terrestrial systems, nematodes are deemed promising, ubiquitous and habitat-overarching bioindicators that can be applied worldwide to monitor the ecological status of biofilms, sediments and soils.

6.1 Pollution of Aquatic Ecosystems

Important functions of aquatic ecosystems, such as nutrient cycling and biomass production, depend on water and sediment quality, which can deteriorate as a result of anthropogenically induced stress, such as climate change and pollution. Ultimately, human health can be affected by pollutants that accumulate in aquatic food webs. Important regulations for evaluating the risk of chemicals in the environment, both proactively and retrospectively, have been adopted within the European Union (EU) (e.g. European Chemical Agency, European Food Safety Authority, EU Water Framework Directive (WFD) and EU Marine Strategy Framework Directive (MSFD)), by the USA (e.g. US Clean Water Act) and at international level (Organisation for Economic Co-operation and Development (OECD)). Combining WFD and MSFD, the EU has moved towards integrative catchment-to-coast management (Borja *et al.*, 2010). However, to date, only approximately 40% of all European surface waterbodies have achieved good ecological and chemical status, with rivers generally achieving a lower status than lakes and coastal waters (European Environment Agency, 2018). The limited success in improving the ecological status is partly due to the lack of appropriate bioindicators that provide robust information on cause–effect

*Email: hoess@ecossa.de

DOI: 10.1079/9781800624221.0006

relationships between chemical and ecological status, which would allow to apply the appropriate water management actions to improve the latter. Aquatic environments are among the most complex to monitor, requiring an array of tools to assess their status and to define conservation targets. A single biological metric is often not enough to describe the complexity and dynamics of these ecosystems (Semprucci *et al.*, 2015).

Sediments, which can be hotspots of pollution due to their ability to bind chemicals, are not appropriately evaluated for identifying chemical and ecological status. Polluted sediments often impede the achievement of good ecological status in the whole waterbody (De Zwart *et al.*, 2009), because: (i) chemicals are constantly remobilized (e.g. by flood events) from the sediments, which serve as a chemical reservoir; and (ii) an impoverished benthic community is not able to maintain critical ecosystem functions, such as nutrient cycling. Ignoring the ability of sediments to store, but also release, contaminants can lead to erroneous conclusions in status assessments (Förstner, 2002; Heininger *et al.*, 2007).

6.2 *In Situ* Bioindication

In situ bioindication is an essential component of environmental quality assessments, including those aimed at determining the ecological status of aquatic ecosystems (e.g. EU WFD, EU MSFD; Borja *et al.*, 2010). Assessments of *in situ* benthic communities or assemblages integrate interactions with their biotic and abiotic environment, including the presence of pollutants, thus providing an ecologically more relevant approach than laboratory bioassays involving single species.

There is a long tradition in using benthic invertebrate fauna as a biological quality element in determinations of the ecological status of marine and freshwater ecosystems (Rosenberg and Resh, 1993; De Boer *et al.*, 2001). Many macrofauna-based indices, such as the saprobic index, are used routinely to detect organic stressors (Hering *et al.*, 2004; Sandin and Hering, 2004; Sanchis *et al.*, 2021). For example, the macroinvertebrate-based *Perlodes* system with its modular design is currently used to evaluate the impact of different stressors (organic pollution, acidification and general degradation) on the ecological quality of flowing waters (Meier *et al.*, 2006), so as to meet demands of the EU WFD. There are other, more stressor-specific indicators, such as the SPEAR (SPEcies At Risk)-index utilizing specific traits of macroinvertebrate species to identify the stress on freshwater systems posed by anthropogenic organic pollution (Von der Ohe *et al.*, 2007), or specifically by pesticides ($SPEAR_{pesticides}$; Liess and Von der Ohe, 2005; Knillmann *et al.*, 2018). Several indicator systems have also been developed to evaluate the ecological status in marine ecosystems (e.g. Borja *et al.*, 2000, 2008; Simboura and Zenetos, 2002).

However, current bioindication systems are mainly based on macroinvertebrates, neglecting the prevailing meiofauna in fine, cohesive sediments (López-Doval *et al.*, 2010; Wolfram *et al.*, 2010), which are often hotspots of chemical contamination (Salomons and Förstner, 1984). Therefore, this chapter highlights the urgent need for including meiofaunal organisms, such as nematodes, in routine biomonitoring programmes to achieve: (i) a more accurate classification of ecological status; and (ii) a better cause–effect-based identification of anthropogenic stress in freshwater and marine ecosystems.

Nematodes are key players in both marine and freshwater sediments and, as endobenthic organisms, are exposed throughout their whole life to chemicals that are either bound to sediment particles or dissolved in the interstitial water. Moreover, marine and freshwater nematodes offer many features that are beneficial for pollution studies (Höss and Traunspurger, 2021; Schratzberger *et al.*, 2000, 2023 and references therein; Table 6.1).

Physical, chemical and biological factors (e.g. sediment type, concentration of pollutants, competition) vary among aquatic environments. Over time, resident nematode species have evolved strategies to maximize growth and reproduction within a specific range of environmental factors. Outside these environmental tolerance ranges, a species' physiology and behaviour may be altered, leading to changes in the performance of individuals, either in the short term through detrimental effects on growth and development, or in the long term through genetic selection. Species-specific sensitivities to environmental conditions in general, and pollution in

Table 6.1. Advantages and disadvantages associated with the use of nematodes in pollution studies. (Adapted from Höss and Traunspurger, 2021; Schratzberger *et al.*, 2000, 2023 and references therein.)

Advantages	Disadvantages
Small size	
Can be maintained in relatively small volumes of sediment; intensive repeated sampling is possible with minor disruption to the sampling site because the sample size required is small; follow-up studies in the laboratory are possible under controlled and repeatable conditions	Taxonomic problems increase with smaller body size, whereas ecological knowledge decreases; preparation for identification can be time-consuming; a high-power microscope is required for species identification. Small-scale spatial variability may be high
Ubiquitous distribution	
Occur in many environments that provide a source of organic carbon, under all climatic conditions, and in habitats that vary from pristine to degraded. These features of nematodes can be exploited to investigate changes that occur across a broad spectrum of stress conditions in different ecosystems	Individual, population and community responses to environmental change are not well documented
High abundance and diversity	
A generally large number of individuals and species gives high intrinsic information value to each sample and ensures statistical validity of the data. Diversity covers a broad range of physiological and feeding types and thus allows a balanced assessment of the effects of prevailing conditions on food webs and community processes	High abundance and diversity together with lack of taxonomic expertise make the analysis of nematode community structure difficult, time-consuming and labour-intensive
Short generation times	
Most species have short life cycles (from a few days to three months) so that changes in community structure can be observed in short-term studies	Population density is affected by a variety of abiotic and biotic factors so that densities may fluctuate over small spatial scales
Direct benthic development, limited mobility	
Holobenthic lifestyle, and direct contact of many species with the interstitial water, makes them sensitive to changes in local conditions. Although nematodes can move relatively rapidly within the sediment, their community structure is generally more directly related to the physicochemical conditions of the habitat sampled compared with more mobile macrofaunal communities	Separating nematodes from the sediment matrix requires a carefully controlled laboratory protocol and the life cycle for most species is unknown

particular, propagate through population dynamics to community-level responses, mediated by shifts in interspecific interactions (Schratzberger *et al.*, 2023; see Fig. 6.1). Bioindicator species effectively indicate the condition of the environment because of their moderate tolerance to environmental conditions (including pollution). However, the use of aquatic nematodes as bioindicators is not restricted to single species and their populations. Entire nematode assemblages, encompassing diverse species with a broad range of environmental tolerances, can serve as bioindicators in aquatic environments, and represent multiple sources of data to assess environmental condition in biotic indices (see Fig. 6.1).

In this chapter, we focus on the use of nematodes to indicate the condition of aquatic ecosystems. We describe the main characteristics of such biological indicators, paying particular attention to indicators that are most applied in the management of aquatic ecosystems. We identify their strengths and weaknesses (also in comparison with established bioindicator systems), explain their background, development and applications, and describe their interpretations within specific management contexts.

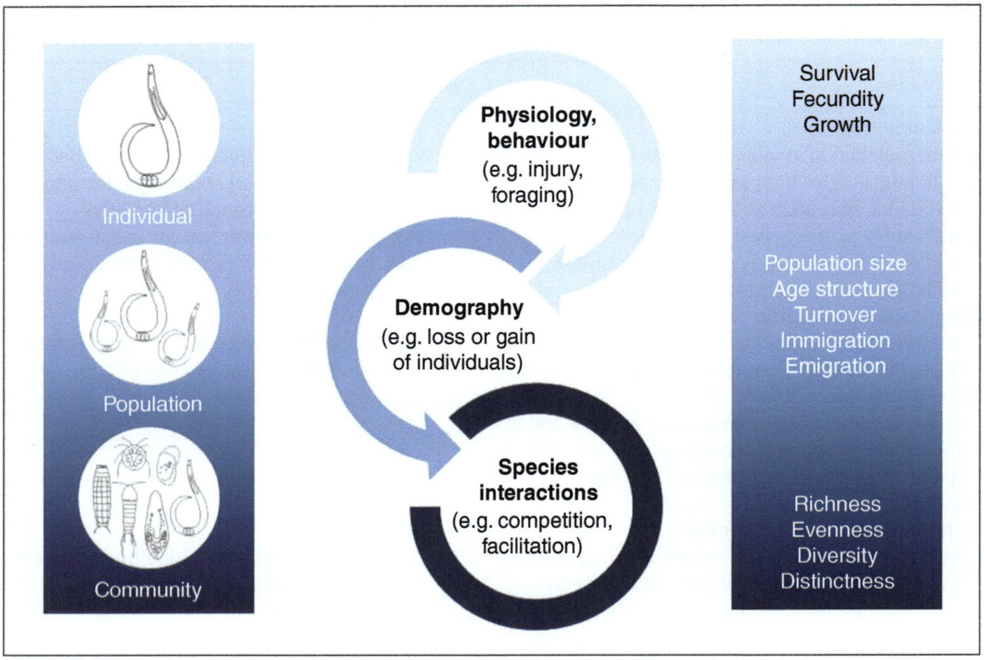

Fig. 6.1. Responses of nematodes to disturbance are the result of sequential mechanisms that operate at progressively higher levels of biological organization. See text for details. (From Schratzberger *et al.*, 2023 with permission from Springer Nature.)

6.3 Nematode-Based Indices for Classification of the Quality of Waterbodies

The close link between changes in nematode assemblage structure and altered functioning of aquatic ecosystems has been encouraging nematologists to explore this taxon as an additional descriptor of the ecological quality status (EQS) of waterbodies as required by European Directives (WFD, 2000/60/EC and MSFD, 2008/56/EC).

6.3.1 The NemaSPEAR[%]-index – an index for assessing chemical-induced effects on freshwater sediments

Although nematode-based indices do exist for assessing the quality of freshwater sediments, the motivation for developing the NemaSPEAR[%]-index as a pollution-sensitive index for freshwater systems was the ambiguity of the existing indicator systems in terms of indicating chemical stress (Höss *et al.*, 2011). For example, Zullini (1976) suggested a simple index to assess polluted habitats using the percentage of Secernentea (Zullini, 1976), a taxon of nematodes referred to in more recent phylogenetics as Rhabditida (De Ley and Blaxter, 2002). However, this index strongly correlated with the organic matter content in water and might therefore not be sufficiently sensitive to chemical pollution. Also, while soil-based indices such as the Maturity indices (MI, MI2-5; Bongers and Ferris, 1999) or the framework for soil food web diagnostics (Ferris *et al.*, 2001) are useful for assessing anthropogenic soil disturbance (see also Chapter 5 of this volume), they produced conflicting results for freshwater sediments (Den Besten *et al.*, 2000; Beier and Traunspurger, 2001). Heininger *et al.* (2007) showed that, for large German rivers, nematodes assigned to high colonizer–persister (*c-p*) values (see Bongers, 1990), such as Tobrilidae and Mononchidae, were more abundant at highly contaminated sites, a contradiction of the theoretical basis of the Maturity Index (MI).

Principles of the NemaSPEAR[%]-index

The NemaSPEAR[%]-index was therefore developed as a co-occurrence approach, based on a large data set comprising nematode species data, geophysicochemical data and chemical data for priority pollutants for >200 sediments sampled from >100 locations in German rivers and lakes (Höss *et al.*, 2011). Nematode species were classified according to their occurrence along a pollution gradient as species at risk (NemaSPEAR) and species not at risk (NemaSPEnotAR). The species at risk were predominantly found in sediments at reference sites with low levels of chemical contamination but only rarely occurred in polluted sediments (Fig. 6.2A). Thus, the NemaSPEAR[%]-index, in contrast to analogous indices for macroinvertebrates (SPEAR-indices; Liess and Von der Ohe, 2005; Schäfer and Liess, 2013), is based solely on empirical data for nematode communities rather than on trait-based data for single species. The index is calculated as follows:

$$NemaSPEAR[\%] = 100 \times \frac{\sum \log \%NemaSPEAR}{\sum \log \%All\ species}$$

Validation of the NemaSPEAR[%]-index

This NemaSPEAR[%]-index was validated with 85 sediment samples from 73 different locations, evidencing a clear dose-dependent decrease with increasing contamination of sediments (Höss *et al.*, 2017; see also Fig. 6.2A). Microcosm studies, where concentrations of single chemicals could be controlled, provided further experimental proof of the specific response of the NemaSPEAR[%]-index to chemical contamination. Zinc and the nematicide ivermectin, spiked to sediments including the native nematode fauna, induced a dose-dependent decrease of the NemaSPEAR[%]-index, with increasing effects at longer exposure times (Fig. 6.2B; Brinke *et al.*, 2010; Haegerbaeumer *et al.*, 2016; Höss *et al.*, 2017).

The index can also be calculated based on genus data, which increases the acceptance of the NemaSPEAR[%]-index for end users (Fig. 6.2A). Based on the variation of the

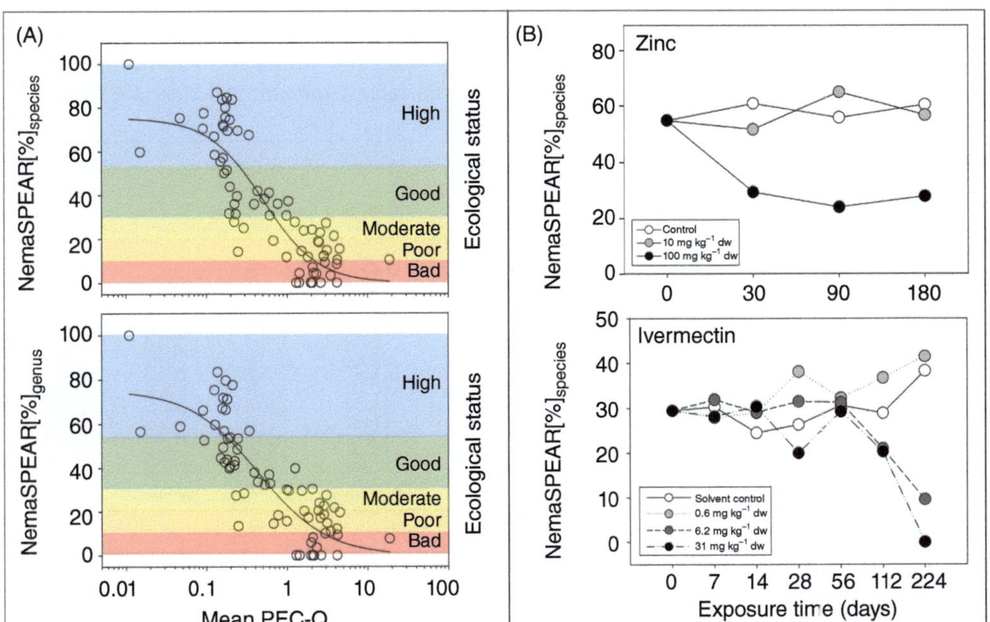

Fig. 6.2. (A) Correlation of the NemaSPEAR[%]-index on species and genus level to the toxic potential of sediments (mean quotient of measured sediment concentrations divided by sediment quality criteria (PEC = probable effect concentration) defined by De Deckere *et al.*, 2011), with thresholds for classes of ecological status: high, NemaSPEAR[%] ≥ 54; good, 30 to < 54; moderate, 20 to < 30; poor, 10 to < 20; bad, < 10. (Modified from Höss *et al.*, 2017.) (B) Response of the NemaSPEAR[%]-index in microcosm studies to zinc- (nominal concentrations; dw = dry weight; data from Haegerbaeumer *et al.*, 2016) and ivermectin-spiked sediments including the native nematode fauna (measured concentrations; data from Brinke *et al.*, 2010). (Modified from Höss *et al.*, 2017.)

NemaSPEAR[%]-index within reference sites with low-level pollution, thresholds could be defined for classifying sediments or sites according to their ecological status (Höss *et al.*, 2017; Fig. 6.2A). So, in the absence of reference sites to compare with, putatively polluted sites can be classified according to their ecological status.

In a recent study, Brüchner-Hüttemann *et al.* (2021) showed that the NemaSPEAR[%] was quite robust to seasonal variations within nematode communities. Although nematode species composition in river sediments collected monthly at a lowly polluted reference site (Furlbach, Germany) showed a statistically significant seasonal pattern (Fig. 6.3A), the NemaSPEAR[%]-index remained quite stable, ranging from 50 to 65, which indicates good to high ecological status throughout the year (Fig. 6.3B).

Comparison of the NemaSPEAR[%]-index with macrozoobenthos-based indices

In a study where nematodes were analysed along with the benthic macroinvertebrate fauna (Brüchner-Hüttemann *et al.*, 2021), the NemaSPEAR[%]-index provided additional, complementary information for the classification of sites with varying degrees of contamination. Two macroinvertebrate-based bioindication systems (SPEAR[%]-index; *Perlodes*) failed to distinguish highly from lowly contaminated sites, revealing contrasting results (SPEAR[%]: good and high ecological status; *Perlodes*: moderate to bad ecological status). In contrast, the NemaSPEAR[%] was clearly dose-dependently related to sediment contamination. Because of their close contact with the sediment matrix (including the sediment-bound contaminants), the exclusively endobenthic nematodes are exposed to potentially toxic chemicals for their entire life cycle. This might explain the higher sensitivity of the NemaSPEAR[%]-index to sediment contamination than the macroinvertebrate-based indices, which use key indicator species that are more associated with the water phase.

NemaSPEAR[%] based on molecular taxonomy

A major drawback for the use of nematodes as a biomonitoring tool is the challenges in nematode morphological identification, due to the minute body size and uniform structure, requiring considerable taxonomic expertise (see Table 6.1).

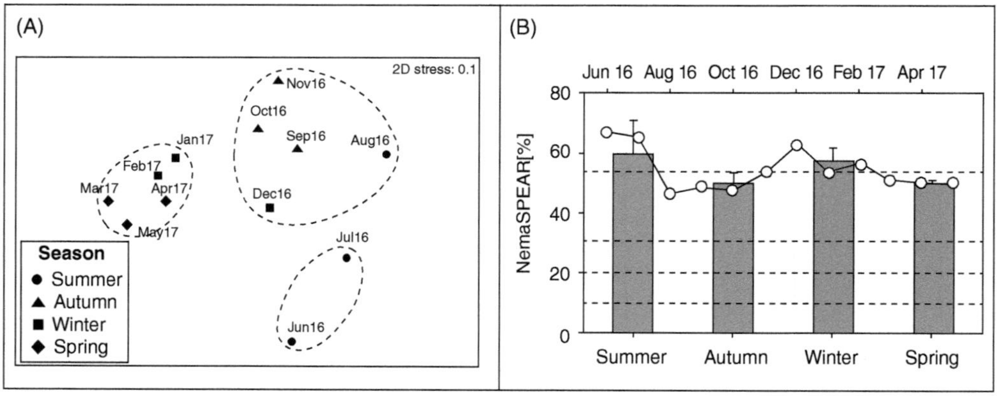

Fig. 6.3. (A) Non-metric multidimensional scaling (NMDS) plot of nematode species composition based on the Bray–Curtis similarity and using the non-transformed relative species abundance data from the Furlbach for 12 months (June 2016–May 2017). Superimposed dotted circles separate significantly different clusters (SIMPROF; $P < 0.05$) on a 50% similarity level as revealed by hierarchical cluster analysis. (B) Calculated NemaSPEAR[%]-index over the course of one year (June 2016–May 2017, —○—) and NemaSPEAR[%]-index (mean values, with their standard deviation represented by error bars, $n = 3$) calculated for the four seasons (summer = June–August 2016; autumn = September–November 2016; winter = December 2016–February 2017; spring = March 2017–May 2017) based on the nematode species of the Furlbach. The dotted lines represent the borders of the ecological quality classes (high, NemaSPEAR[%] ≥ 54; good, 30–54; moderate, 20–30; poor, 10–20, bad, < 10). (From Brüchner-Hüttemann *et al.*, 2021 with permission from Elsevier.)

However, along with the overall efforts to establish DNA-based methods for biomonitoring within the EU WFD (Leese *et al.*, 2016; Pawlowski *et al.*, 2022), the NemaSPEAR[%]-index could be applied successfully using metabarcoding data of freshwater nematode communities (Schenk *et al.*, 2020, 2022). For 39 samples collected in Germany from 38 sites within 25 different rivers, a good correlation of the NemaSPEAR[%] derived from morphological and molecular data has already been achieved (Fig. 6.4A; Schenk *et al.*, 2022). This resulted in a good correspondence of ecological quality classes based on molecular data when combined with morphological data, and especially when compared to biomass-based data (Fig. 6.4B; Schenk *et al.*, 2022).

6.3.2 Bioindication with marine nematodes

Principles of bioindication in marine systems

In marine systems, Moreno *et al.* (2011) applied a meta-analytical approach to evaluate the utility of nematodes as biological indicators of EQS in accordance with the EU WFD. Based on threshold values recommended by Long and co-workers (Long *et al.*, 1995; Long, 2006) and Marin *et al.* (2008), EQS was evaluated for 12 coastal sediments of the western Mediterranean Sea, ranging from pristine to highly polluted. Trends in biochemical (total organic matter) and chemical parameters (heavy metals and polycyclic aromatic hydrocarbons) were compared to those of nematode community structure and taxonomic and functional indices. Taxonomic approaches, including the occurrence of sentinel nematode genera, indicated a close link between ecological status and biological response. Indices based on the composition of nematode life-history traits (e.g. *c-p*, MI; see Chapter 5, this volume) and nematode taxa (e.g. Shannon diversity index H′; see Chapter 4, this volume) provided sensitive summary descriptors to evaluate the EQS. The resulting nematode descriptors and thresholds proposed by Moreno *et al.* (2011) were standardized by applying the same approach to different sediment types across habitats and geographical regions, ranging from the Mediterranean basin (Semprucci *et al.*, 2014a, 2018; Jouili *et al.*, 2017; Bianchelli *et al.*, 2018) to the Black Sea (Ürkmez *et al.*, 2014) and the Indian and Pacific Oceans (i.e. Maldives: Semprucci *et al.*, 2014b; South China Sea: Chen *et al.*, 2018; Yellow Sea: Hong *et al.*, 2020). Findings further supported the potential role of marine nematodes as promising

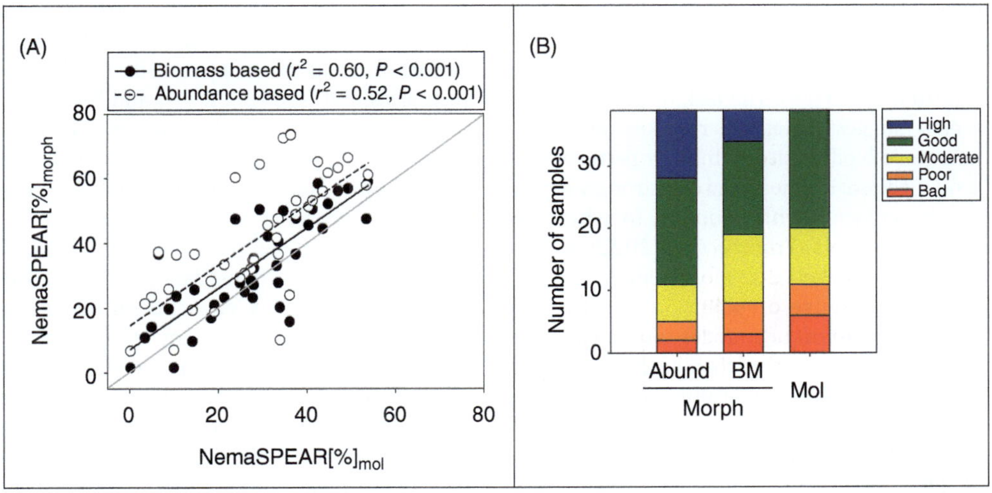

Fig. 6.4. (A) Linear regression between the NemaSPEAR[%] based on molecular and morphological data (●, morphological data based on biomass; ○, morphological data based on abundance); (B) the number of samples categorized in five classes of ecological quality. NemaSPEAR[%] thresholds for classes are those defined by Höss *et al.* (2017): high (> 54), good (30–54), moderate (20–30), poor (10–20) or bad (<10). morph = morphological approach; abund = abundance-based; BM = biomass-based; mol = molecular approach. (From Schenk *et al.*, 2022 with permission from Elsevier.)

biological quality elements and suggested that they can help to identify a priority list of vulnerable areas and inform management actions to achieve good ecological status.

Using nematodes as bioindicators across marine habitats

Studies using such marine nematode descriptors across a wide range of coastal sediments to assess ecological status between 2010 and 2022 (see Franzo *et al.*, 2022; also Semprucci *et al.*, 2015 for review) suggested that the application of diversity indices in trend-based pollution studies across habitat and sediment types is limited because: (i) statistically significant changes in biodiversity indices over time often require strong variations in nematode assemblage structure; and (ii) the often close relationship between nematode diversity and characteristics of the substrate in which they live masks the pollution response.

Furthermore, in a survey at three commercial ports in the Mediterranean basin, Franzo *et al.* (2022) found a significant positive relationship between nematode biodiversity and the concentrations of total organic carbon (TOC) and polycyclic aromatic hydrocarbons (ΣPAH). This is contrary to many previous studies where less diverse nematode communities are typically found in the most contaminated sediments (Losi *et al.*, 2013; Boufahja and Semprucci, 2015; Franzo *et al.*, 2018). The lack of empirical consensus suggests that the response of marine nematode biodiversity to human perturbations is not necessarily linear and can strongly depend on the environmental context in which the pollution occurs (Franzo *et al.*, 2022; see also Schratzberger *et al.*, 2023 for review).

The presence of pollution-sensitive and pollution-resistant nematodes can provide more accurate information on their ecology and adaptation to prevailing environmental conditions. Based on available literature, Moreno *et al.* (2011) proposed some nematode genera as indicators of good and poor ecological quality status (Table 6.2). The relative abundance of these proposed genera served as a more informative indicator of sediment quality than other indices based on taxonomic diversity. However, the limited number of marine nematode genera proposed initially constrained the wider application of this genus-based approach across habitat types or geographical regions (Table. 6.2). The initial set of potential indicator sentinels has therefore been progressively revised and updated (see Fig. 6.5). Recently, a co-occurrence analysis approach was utilized, for example, to integrate knowledge of nematode indicator genera in commercial ports (Franzo *et al.*, 2022). Only the most relevant outputs of the co-occurrence analysis were considered (relative abundance > 30%). As illustrated in Fig. 6.5, genera were subdivided into pollution-resistant (including genera primarily found in polluted sediments), opportunistic (including genera equally found in both polluted and unpolluted sediments) and relatively opportunistic (including genera primarily occurring in unpolluted sediments) behaviour. This approach is similar to the NemaSPEAR[%]-index, which also categorizes pollution-sensitive or -tolerant taxa according to their *in situ* occurrence in clean or polluted sediments, respectively. If using larger co-occurrence data sets of marine nematodes and sediment contamination, a marine NemaSPEAR[%]-index could be developed analogous to the existing freshwater index.

Trait-based indices, such as the MI, support consistent conclusions about ecological status in open coastal systems only. Generally, MI decreases with increasing levels of perturbation (Bongers *et al.*, 1991; see also Table 6.2). However, results are inconsistent and inconclusive in chronically disturbed areas such as transitional habitats (estuaries and lagoons), ports, and highly contaminated bays that lack water recycling (e.g. Hong *et al.*, 2020; Franzo *et al.*, 2022). In a study in the Varano lagoon (Adriatic basin, Mediterranean), Semprucci *et al.* (2014a) calculated the Pollution Load Index (PLI) considering As, Cd, Co, Cr, Cu, Ni, Pb, Zn and Mn in order to assess the influence of trace elements on nematodes. The MI, in particular, did not always permit a clear discrimination between the effects of anthropogenic and natural environmental factors on nematode communities in transitional habitats. A possible explanation is the 'estuarine quality paradox' (Elliott and Quintino, 2007): transitional communities are primarily adapted to fluctuations in physicochemical parameters and, thus, more tolerant to additional, often anthropogenically induced stress. Therefore, the combined use of MI and other metrics such as H′ is recommended in specific environmental

Table. 6.2. Thresholds proposed to evaluate the Ecological Quality Status (EQS) in marine coastal systems. ERL = effects range low; ERM = effects range median; PAHs = polycyclic aromatic hydrocarbons; MI = Maturity Index; *c-p* = colonizer–persister; H′ index = Shannon index of diversity (based on $\log_2$); ITD = Index of Trophic Diversity. (From Moreno *et al.*, 2011.)

Quality descriptors	High	Good	Moderate	Poor	Bad
Organic matter	<5%		5–10%	>10%	
Metals/PAHs	< ERL		ERL < conc. < ERM	> ERM	
MI	> 2.8	2.8 ≤ MI < 2.6	2.6 ≤ MI < 2.4	2.4 ≤ MI < 2.2	≤ 2.2
Colonizer–persister	*c-p*2 ≤ 50%; *c-p*4 > 10%	*c-p*2 ≥ 50%; *c-p*4 >10%	*c-p*2 ≥ 50%; 3 < *c-p*4 < 10%	*c-p*2 > 60%; *c-p*4 < 3%	*c-p*2 > 80%
H′ index	>4.5	3.5 < H′ < 4.5	2.5 < H′ < 3.5	1 < H′ ≤ 2.5	0 < H′ ≤ 1
ITD	0.25	0.25 < ITD ≤ 0.4	0.4 < ITD ≤ 0.6	0.6 < ITD ≤ 0.8	1
Sensitive/tolerant indicator taxa (>10%)	Desmoscolecidae, *Microlaimus*, *Richtersia*, *Oncholaimus*, *Pomponema*, *Epacanthion*	*Halalaimus*, *Setosabatieria*, *Ptycholaimellus*	*Anticoma*, *Desmodora*, *Spirinia*, *Marylynia*, *Prochromadorella*	*Daptonema/Theristus*, *Paralongicyatholaimus*, *Parodontophora*, *Odontophora*	*Paracomesoma*, *Terschellingia*, *Sabatieria*

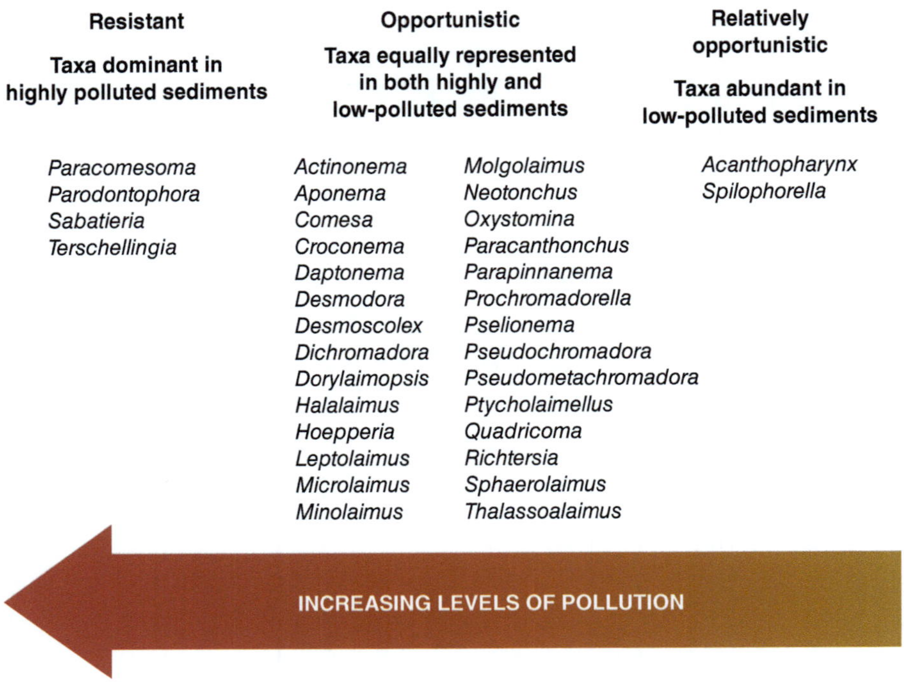

Fig. 6.5. Nematode genera assigned to three categories of resistance/opportunism in port sediments. (From Losi *et al.*, 2021; Franzo *et al.*, 2022.)

contexts (i.e. transitional environments and ports), providing a more comprehensive and consistent indication of ecological status. Furthermore, a revision or at least an update of the original assignment of the *c-p* classes for marine nematodes is timely (Bongers *et al.*, 1991; Losi *et al.*, 2021).

6.4 Conclusions and Perspectives

There are various promising nematode-based approaches for bioindication in marine and freshwater habitats, linking chemical and ecological status of ecosystems. While, to date, relevant research has been resulting in the development of habitat-specific indices (freshwater, marine, soil; see also Chapter 5, this volume), all indices are underpinned by the presence of taxon-specific traits related to the vulnerability to certain types of stress in their environment. In terrestrial systems, indices describing the ecosystem status, such as the Maturity, Structure, Enrichment and Channel indices (MI, SI, EI, CI), are categorizing terrestrial nematode taxa (i.e. genera/families) based on auto-ecological information (feeding behaviour and life-history traits). In turn, aquatic nematode indices (NemaSPEAR[%]-index; marine indicator species) are derived empirically using co-occurring data on nematode taxonomic composition and sediment contamination. Both approaches have benefits and drawbacks, which makes them only partly applicable across ecosystems. For example, specific traits of nematode taxa occurring in soils and aquatic habitats are related to sensitivity to pollution in soil, but not in freshwater systems. Heininger *et al.* (2007) explained this conflicting phenomenon by competitive interaction with pollution-sensitive predatory macroinvertebrates which are present in freshwater sediments, but not in soils. Instead, for estuarine nematode communities, these inconsistencies might result from the adaptation of transitional communities to fluctuating environmental conditions (i.e. stress; 'estuarine quality paradox'). These habitat-specific circumstances can be better factored in by using empirical, *in situ* approaches, which were developed for freshwater and marine systems, as here all co-occurring biotic and abiotic interactions are included. A reliable categorization of nematodes in species at risk (i.e. species present mainly at clean sites) and species not at risk (i.e. ubiquitous and tolerant species) would provide a robust basis for bioindication (Höss *et al.*, 2017; Brüchner-Hüttemann *et al.*, 2021). This would require large data sets that would have to be specifically derived once for marine and soil nematodes, as it was done for freshwater nematodes (Höss *et al.*, 2011). Thus, the future aim is to combine the advantages of trait-based and co-occurrence approaches for nematode-based bioindication and, thus, to provide a habitat-overarching indication system that, due to the ubiquity of nematodes (van den Hoogen *et al.*, 2019), could be applied worldwide to monitor the ecological status of biofilms, sediments and soils.

References

Beier, S. and Traunspurger, W. (2001) The meiofauna community of two small German streams as indicators of pollution. *Journal of Aquatic Ecosystem Stress and Recovery* 8(3–4), 387–405. DOI: 10.1023/A:1012965424527.

Bianchelli, S., Buschi, E., Danovaro, R. and Pusceddu, A. (2018) Nematode biodiversity and benthic trophic state are simple tools for the assessment of the environmental quality in coastal marine ecosystems. *Ecological Indicators* 95, 270–287. DOI: 10.1016/j.ecolind.2018.07.032.

Bongers, T. (1990) The maturity index: an ecological measure of environmental disturbance based on nematode species composition. *Oecologia* 83, 14–19. DOI: 10.1007/BF00324627.

Bongers, T. and Ferris, H. (1999) Nematode community structure as a bioindicator in environmental monitoring. *Trends in Ecology and Evolution* 14(6), 224–228. DOI: 10.1016/S0169-5347(98)01583-3.

Bongers, T., Alkemade, R. and Yeates, G.W. (1991) Interpretation of disturbance-induced maturity decrease in marine nematode assemblages by means of the Maturity Index. *Marine Ecology Progress Series* 76, 135–142. DOI: 10.3354/meps076135.

Borja, A., Franco, J. and Pérez, V. (2000) A marine Biotic Index to establish the ecological quality of soft-bottom benthos within European estuarine and coastal environments. *Marine Pollution Bulletin* 40(12), 1100–1114. DOI: 10.1016/S0025-326X(00)00061-8.

Borja, A., Mader, J., Muxika, I., Rodríguez, J.G. and Bald, J. (2008) Using M-AMBI in assessing benthic quality within the Water Framework Directive: some remarks and recommendations. *Marine Pollution Bulletin* 56(9), 1668–1677. DOI: 10.1016/j.marpolbul.2007.12.003.

Borja, Á., Elliott, M., Carstensen, J., Heiskanen, A.S. and van de Bund, W. (2010) Marine management – towards an integrated implementation of the European Marine Strategy Framework and the Water Framework Directives. *Marine Pollution Bulletin* 60(12), 2175–2186. DOI: 10.1016/j.marpolbul.2010.09.026.

Boufahja, F. and Semprucci, F. (2015) Stress-induced selection of a single species from an entire meiobenthic nematode assemblage: is this possible using iron enrichment and does pre-exposure affect the ease of the process? *Environmental Science and Pollution Research* 22(3), 1978–1998. DOI: 10.1007/s11356-014-3479-2.

Brinke, M., Höss, S., Fink, G., Ternes, T.A., Heininger, P. and Traunspurger, W. (2010) Assessing effects of the pharmaceutical ivermectin on meiobenthic communities using freshwater microcosms. *Aquatic Toxicology* 99, 126–137. DOI: 10.1016/j.aquatox.2010.04.008.

Brüchner-Hüttemann, H., Höss, S., Ptatscheck, C., Brinke, M., Schenk, J. and Traunspurger, W. (2021) Added value of the NemaSPEAR[%]-index to routinely used macrofauna-based indices for assessing the quality of freshwater sediments. *Ecological Indicators* 121, 107015. DOI: 10.1016/j.ecolind.2020.107015.

Chen, C.-A., Soo, C.-L., Balsamo, M. and Semprucci, F. (2018) An approach based on nematode descriptors for the classification of ecological quality (EcoQ) of the Malaysian coasts. *Marine Biodiversity* 48(1), 117–126. DOI: 10.1007/s12526-017-0813-1.

De Boer, W.F., Daniels, P. and Essink, K. (2001) *Towards Ecological Quality Objectives for North Sea Benthic Communities*. National Institute for Coastal and Marine Management, Haren, The Netherlands.

De Deckere, E., Cooman, W., Leloup, V., Meire, P., Schmitt, C. and Von der Ohe, P.C. (2011) Development of sediment quality guidelines for freshwater ecosystems. *Journal of Soils and Sediments* 11(3), 504–517. DOI: 10.1007/s11368-010-0328-x.

De Ley, P. and Blaxter, M. L. (2002) Systematic position and phylogeny. In Lee, D.L. (ed.) *The Biology of Nematodes*. Taylor & Francis, London, pp. 1–30.

De Zwart, D., Posthuma, L., Gevery, M., Von der Ohe, P.C. and De Deckere, E. (2009) Diagnosis of ecosystem impairment in a multiple-stress context – how to formulate effective river basin management plans. *Integrated Environmental Assessment and Management* 5(1), 38–49.

Den Besten, P.J., Postma, J.F., Wegener, J.W.M., Keidel, H., Klink, A. *et al.* (2000) Biological and chemical monitoring after pilot remediations in the delta of the rivers Rhine and Meuse. *Aquatic Ecosystem Health and Management* 3, 317–334. DOI: 10.1016/S1463-4988(00)00035-X.

Elliott, M. and Quintino, V. (2007) The estuarine quality paradox, environmental homeostasis and the difficulty of detecting anthropogenic stress in naturally stressed areas. *Marine Pollution Bulletin* 54(6), 640–645. DOI: 10.1016/j.marpolbul.2007.02.003.

European Environment Agency (2018) *European Waters: Assessment of Status and Pressures*. European Environment Agency, Copenhagen.

Falk, D.A., Watts, A.C. and Thode, A.E. (2019) Scaling ecological resilience. *Frontiers in Ecology and Evolution* 7, 275. DOI: 10.3389/fevo.2019.00275.

Ferris, H., Bongers, T. and De Goede, R.G.M. (2001) A framework for soil food web diagnostics: extension of the nematode faunal analysis concept. *Applied Soil Ecology* 18, 13–29. DOI: 10.1016/S0929-1393(01)00152-4.

Förstner, U. (2002) Sediments and the European Water Framework Directive. *Journal of Soils and Sediments* 2, 54. DOI: 10.1007/BF02987869.

Franzo, A., Guilini, K., Cibic, T. and Del Negro, P. (2018) Structure and function of nematode assemblages in contaminated sediments: what can we learn from the Mar Piccolo of Taranto (Ionian Sea)? *Journal of the Marine Biological Association of the United Kingdom* 98(8), 1845–1857. DOI: 10.1017/S0025315418000553.

Franzo, A., Baldrighi, E., Grassi, E., Grego, M., Balsamo, M. *et al.* (2022) Free-living nematodes of Mediterranean ports: a mandatory contribution for their use in ecological quality assessment. *Marine Pollution Bulletin* 180, 113814. DOI: 10.1016/j.marpolbul.2022.113814.

Haegerbaeumer, A., Höss, S., Ristau, K., Claus, E., Möhlenkamp, C. *et al.* (2016) A comparative approach using ecotoxicological methods from single-species bioassays to model ecosystems. *Environmental Toxicology and Chemistry* 35(12), 2987–2997. DOI: 10.1002/etc.3482.

Heininger, P., Höss, S., Claus, E., Pelzer, J. and Traunspurger, W. (2007) Nematode communities in contaminated river sediments. *Environmental Pollution* 146(1), 64–76. DOI: 10.1016/j.envpol.2006.06.023.

Hering, D., Moog, O., Sandelin, L. and Verdonschot, P.F.M. (2004) Overview and application of the AQEM assessment system. *Hydrobiologia* 516, 1–20. DOI: 10.1023/B:HYDR.0000025255.70009.a5.

Hong, J.-H., Semprucci, F., Jeong, R., Kim, K., Lee, S. *et al.* (2020) Meiobenthic nematodes in the assessment of the relative impact of human activities on coastal marine ecosystem. *Environmental Monitoring and Assessment* 192(2), 81. DOI: 10.1007/s10661-019-8055-2.

Höss, S. and Traunspurger, W. (2021) Freshwater nematodes as bioindicators in field studies – the NemaSPEAR[%]-index. In: Traunspurger, W. (ed.) *Ecology of Freshwater Nematodes*. CAB International, Wallingford, UK, pp. 323–340. DOI: 10.1079/9781789243635.0011.

Höss, S., Claus, E., Von der Ohe, P.C., Brinke, M., Güde, H. *et al.* (2011) Nematode species at risk: a metric to assess pollution in soft sediments of freshwaters. *Environment International* 37, 940–949. DOI: 10.1016/j.envint.2011.03.013.

Höss, S., Heininger, P., Claus, E., Möhlenkamp, C., Brinke, M. and Traunspurger, W. (2017) Validating the NemaSPEAR[%]-index for assessing sediment quality regarding chemical-induced effects on benthic communities in rivers. *Ecological Indicators* 73, 52–60. DOI: 10.1016/j.ecolind.2016.09.022.

Jouili, S., Essid, N., Semprucci, F., Boufahja, F., Nasri, A. *et al.* (2017) Environmental quality assessment of El Bibane lagoon (Tunisia) using taxonomic and functional diversity of meiofauna and nematodes. *Journal of the Marine Biological Association of the United Kingdom* 97(8), 1593–1603. DOI: 10.1017/S0025315416000990.

Knillmann, S., Orlinskiy, P., Kaske, O., Foit, K. and Liess, M. (2018) Indication of pesticide effects and recolonization in streams. *Science of The Total Environment* 630, 1619–1627. DOI: 10.1016/j.scitotenv.2018.02.056.

Leese, F., Altermatt, F., Bouchez, A., Ekrem, T., Hering, D. *et al.* (2016) DNAqua-Net: developing new genetic tools for bioassessment and monitoring of aquatic ecosystems in Europe. *Research Ideas and Outcomes* 2, e11321. DOI: 10.3897/rio.2.e11321.

Liess, M. and Von der Ohe, P.C. (2005) Analyzing effects of pesticides on invertebrates in streams. *Environmental Toxicology and Chemistry* 24(4), 954–965. DOI: 10.1897/03-652.1.

Long, E.R. (2006) Calculation and uses of mean Sediment Quality Guideline quotients: a critical review. *Environmental Science and Technology* 40(6), 1726–1736. DOI: 10.1021/es058012d.

Long, E.R., MacDonald, D.D., Smith, S.L. and Calder, F.D. (1995) Incidence of adverse biological effects within ranges of chemical concentrations in marine and estuarine sediments. *Environmental Management* 19(1), 81–97. DOI: 10.1007/BF02472006.

López-Doval, J., Großschartner, M., Höss, S., Orendt, C., Traunspurger, W. *et al.* (2010) Invertebrate communities in soft sediments along a pollution gradient in a Mediterranean river (Llobregat, NE Spain). *Limnetica* 29(2), 311–322. DOI: 10.23818/limn.29.26.

Losi, V., Ferrero, T.J., Moreno, M., Gaozza, L., Rovere, A. *et al.* (2013) The use of nematodes in assessing ecological conditions in shallow waters surrounding a Mediterranean harbour facility. *Estuarine and Coastal Shelf Science* 130, 209–221. DOI: 10.1016/j.ecss.2013.02.017.

Losi, V., Grassi, E., Balsamo, M., Rocchi, M., Gaozza, L. and Semprucci, F. (2021) Changes in taxonomic structure and functional traits of nematodes as tools in the assessment of port impact. *Estuarine, Coastal and Shelf Science* 260, 107524. DOI: 10.1016/j.ecss.2021.107524.

Marin, V., Moreno, M., Vassallo, P., Vezzulli, L. and Fabiano, M. (2008) Development of a multistep indicator-based approach (MIBA) for the assessment of environmental quality of harbours. *ICES Journal of Marine Science* 65(8), 1436–1441. DOI: 10.1093/icesjms/fsn117.

Meier, C., Böhmer, J., Biss, R., Feld, C., Haase, P. *et al.* (2006) *Weiterentwicklung und Anpassung des nationalen Bewertungssystems für Makrozoobenthos an neue internationale Vorgaben*. German Federal Environmental Protection Agency (Umweltbundesamt), Berlin.

Moreno, M., Semprucci, F., Vezzulli, L., Balsamo, M., Fabian, M. and Albertelli, G. (2011) The use of nematodes in assessing ecological quality status in the Mediterranean coastal ecosystems. *Ecological Indicators* 11, 328–336. DOI: 10.1016/j.ecolind.2010.05.011.

Pawlowski, J., Bruce, K., Panksep, K., Aguirre, F.I., Amalfitano, S. *et al.* (2022) Environmental DNA metabarcoding for benthic monitoring: a review of sediment sampling and DNA extraction methods. *Science of The Total Environment* 818, 151783. DOI: 10.1016/j.scitotenv.2021.151783.

Rosenberg, D.M. and Resh, V.H. (1993) *Freshwater Biomonitoring and Benthic Macroinvertebrates*. Chapman and Hall, New York.

Salomons, W. and Förstner, U. (1984) *Metals in the Hydrocycle*. Springer, New York.

Sanchis, C., Soto, E.H. and Quiroga, E. (2021) The importance of a functional approach on benthic communities for aquaculture environmental assessment: trophic groups – a polychaete view. *Marine Pollution Bulletin* 167, 112309. DOI: 10.1016/j.marpolbul.2021.112309.

Sandin, L. and Hering, D. (2004) Comparing macroinvertebrate indices to detect organic pollution across Europe: a contribution to the EC Water Framework Directive intercalibration. *Hydrobiologia* 516(1–3), 55–68. DOI: 10.1023/B:HYDR.0000025258.63416.11.

Schäfer, R.B. and Liess, M. (2013) Species at Risk (SPEAR) biomonitoring indicators. In: Ferard, J.-F. and Blaise, C. (eds) *Encyclopedia of Aquatic Ecotoxicology*. Springer, Dordrecht, The Netherlands, pp. 1063–1071. DOI: 10.1007/978-94-007-5704-2.

Schenk, J., Höss, S., Brinke, M., Kleinbölting, N., Brüchner-Hüttemann, H. and Traunspurger, W. (2020) Nematode-based bioindication of polluted sediments using metabarcoding and microscopic taxonomy. *Environment International* 143, 105922. DOI: 10.1016/j.envint.2020.105922.

Schenk, J., Höss, S., Kleinbölting, N. and Traunspurger, W. (2022) Suitability of molecular taxonomy for assessing polluted sediments using the NemaSPEAR [%] index. *Ecological Indicators* 137, 108761. DOI: 10.1016/j.ecolind.2022.108761.

Schratzberger, M., Gee, J.M., Rees, H.L., Boyd, S.E. and Wall, C.M. (2000) The structure and taxonomic composition of sublittoral meiofauna assemblages as an indicator of the status of marine environments. *Journal of the Marine Biological Association of the United Kingdom* 80(6), 969–980. DOI: 10.1017/S0025315400003039.

Schratzberger, M., Danovaro, R., Ingels, J., Montagna, P.A., Rohal Lupher, M. *et al.* (2023) Hidden players – meiofauna mediate ecosystem effects of anthropogenic disturbances in the ocean. In: Giere, O. and Schratzberger, M. (eds) *New Horizons in Meiobenthos Research: Profiles, Patterns and Potentials*. Springer, Cham, Switzerland, pp. 175–255. DOI: 10.1007/978-3-031-21622-0_7.

Semprucci, F., Balsamo, M. and Frontalini, F. (2014a) The nematode assemblage of a coastal lagoon (Lake Varano, southern Italy): ecology and biodiversity patterns. *Scientia Marina* 78(4), 579–588. DOI: 10.3989/scimar.04018.02A.

Semprucci, F., Colantoni, P., Sbrocca, C., Baldelli, G. and Balsamo, M. (2014b) Spatial patterns of distribution of meiofaunal and nematode assemblages in the Huvadhoo lagoon (Maldives, Indian Ocean). *Journal of the Marine Biological Association of the United Kingdom* 94(7), 1377–1385. DOI: 10.1017/S002531541400068X.

Semprucci, F., Losi, V. and Moreno, M. (2015) A review of Italian research on free-living marine nematodes and the future perspectives on their use as Ecological Indicators (EcoInds). *Mediterranean Marine Science* 16(2), 452–465. DOI: 10.12681/mms.1072.

Semprucci, F., Balsamo, M., Appolloni, L. and Sandulli, R. (2018) Assessment of ecological quality status along the Apulian coasts (eastern Mediterranean Sea) based on meiobenthic and nematode assemblages. *Marine Biodiversity* 48(1), 105–115. DOI: 10.1007/s12526-017-0745-9.

Simboura, N. and Zenetos, A. (2002) Benthic indicators to use in ecological quality classification of Mediterranean soft bottom marine ecosystems, including a new Biotic Index. *Mediterranean Marine Science* 3(2), 77–111. DOI: 10.12681/mms.249.

Ürkmez, D., Sezgin, M. and Bat, L. (2014) Use of nematode maturity index for the determination of ecological quality status: a case study from the Black Sea. *Journal of the Black Sea/Mediterranean Environment* 20(2), 96–107.

Van den Hoogen, J., Geisen, S., Routh, D., Ferris, H., Traunspurger, W. *et al.* (2019) Soil nematode abundance and functional group composition at a global scale. *Nature* 572(7768), 194–198. DOI: 10.1038/s41586-019-1418-6.

Von der Ohe, P.C., Prüß, A., Schäfer, R.B., Liess, M., De Deckere, E. and Brack, W. (2007) Water quality indices across Europe – a comparison of the good ecological status of five river basins. *Journal of Environmental Monitoring* 9(9), 970–978. DOI: 10.1039/B704699P.

Wolfram, G., Orendt, C., Höss, S., Großschartner, M., Adamek, Z. *et al.* (2010) The macroinvertebrate and nematode community from soft sediments in impounded sections of the river Elbe near Pardubice, Czech Republic. *Lauterbornia* 69, 87–105.

Zullini, A. (1976) Nematodes as indicators of river pollution. *Nematologia Mediterranea* 4, 13–22.

7 Nematode Contributions to Ecosystem Functioning and Bioindication Under Different Land Uses

Gerhard du Preez[1]*, Raquel Campos-Herrera[2], Mieke Daneel[1,3], Gerard Korthals[4], Jan Henrik Schmidt[5] and Christine D. Sprunger[6]

[1]*Unit for Environmental Sciences and Management, North-West University, Potchefstroom, South Africa;* [2]*Instituto de las Ciencias de la Vid y del Vino (Government of La Rioja, CSIC, University of La Rioja), Logroño, Spain;* [3]*Agricultural Research Council–Tropical and Subtropical Crops, Nelspruit, South Africa;* [4]*Wageningen Plant Research, Wageningen University and Research, The Netherlands;* [5]*Institute for Epidemiology and Pathogen Diagnostics, Julius Kühn Institute, Braunschweig, Germany;* [6]*Department of Plant, Soil, and Microbial Sciences, Michigan State University, Michigan, USA and W.K. Kellogg Biological Station, Michigan State University, Michigan, USA*

Abstract

Nematodes, representative of multiple soil trophic levels and performing important ecological processes, are ideally positioned as bioindicators for assessing soil ecosystem health and functioning. This chapter delves into the instrumental role that nematode trophic groups play in major soil ecosystem functions, specifically: (i) carbon and climate regulation; (ii) water regulation and purification; (iii) nutrient cycling; and (iv) disease and pest regulation. Emphasis is placed on understanding how different land-use types impact these functions while evaluating the efficacy of nematodes as bioindicators across various ecosystem contexts. The chapter concludes by suggesting future research areas to fill existing knowledge gaps relating to nematodes as bioindicators of soil ecosystem functioning.

7.1 Introduction

Biodiversity serves as the backbone of soil ecosystems (Wall *et al.*, 2015). Without functioning ecosystems, the services provided by nature (e.g. clean environments and the production of food and fibre) will collapse. Therefore, promoting environmental health and sustainable food and goods production hinges on our ability to understand and support complex and interconnected ecosystem functions (Creamer *et al.*, 2022). Equally important is the identification of suitable (bio)indicators to observe and measure changes in soil ecosystems, whether they stem from natural or anthropogenic origins. In this context, acknowledging the influence of environmental and anthropogenic factors on ecosystem status (health and functioning) becomes increasingly important (Bünemann *et al.*, 2018).

*Email: Gerhard.duPreez@nwu.ac.za

DOI: 10.1079/9781800624221.0007

Nematodes, as key players in multiple trophic levels and contributors to essential ecosystem processes, are valuable indicators of soil ecosystem status. Their widespread distribution and responsiveness to environmental change promote their versatility as indicators under different land-use practices. One noteworthy advancement since the 1990s is the development of nematode-based indices, a powerful ecological toolset that allows for a comprehensive assessment of soil food web status, as well as ecosystem health and functioning (Du Preez *et al.*, 2022).

This chapter explores the unique role that nematodes play in supporting soil ecosystem functioning and how their presence, community structure and ecological contributions are influenced under different land-use practices. We begin by outlining the processes that different nematode trophic groups fulfil in support of ecosystem functioning. Thereafter, we explore the current knowledge on nematode interactions with the four main soil ecosystem functions, namely: (i) carbon and climate regulation; (ii) water regulation and purification; (iii) nutrient cycling; and (iv) disease and pest regulation. This is considered in the context of different land-use practices, while also discussing the efficacy of nematodes as bioindicators per ecosystem function. Lastly, we propose avenues for future research, highlighting gaps in our current understanding and the potential research areas where our knowledge about these important organisms should be further explored.

7.2 Functional Role of Nematodes in Soil Ecosystems

Scientific and public interest in soil ecosystem functioning is linked to the growing focus on soil health, which is defined as 'the continued capacity of soil to function as a vital living ecosystem that sustains plants, animals, and humans' (NRCS, 2024). Historically, there has been substantial ambiguity relating to the term 'soil function'; however, Bünemann *et al.* (2018) redefined soil functions as 'bundles of soil processes that underpin the delivery of soil services'. These processes are mostly fulfilled by soil biota, while soil services, in turn, are the benefits that humans receive from soils.

Creamer *et al.* (2022) created an integrated framework linking actors or soil biota (e.g. nematodes) to processes (e.g. microbial grazing) to sub-functions (e.g. decomposition) to functions (e.g. carbon and climate regulation). The main soil functions in agricultural systems are: (i) carbon and climate regulation; (ii) water regulation and purification; (iii) nutrient cycling; and (iv) disease and pest regulation. Although bacteria, fungi and archaea were recorded as the major contributors to soil processes, nematodes among other actors (e.g. protozoa, enchytraeids, earthworms and plant roots) were also listed as key role players. According to Creamer *et al.* (2022), nematodes contribute to several processes, including food web assimilation, microbial grazing, predation, parasitism, and resistance and defence. Furthermore, an additional process is proposed in this chapter, namely microbial dispersion (Topalović *et al.*, 2020; Topalović and Geisen, 2023). These processes can be linked to specific nematode trophic groups, namely bacterivores, fungivores, omnivores, predators, entomopathogenic nematodes (EPNs) and plant-parasitic nematodes (PPNs) (Fig. 7.1).

All the nematode trophic groups contribute to *food web assimilation*, which refers to the uptake of carbon (C) and nutrients through feeding. This is a primary driver of nutrient assimilation, a sub-function of nutrient cycling that clusters the processes responsible for incorporating nutrients from the soil into the biomass of biota (Creamer *et al.*, 2023). Food web assimilation is also linked to resource reallocation, which influences the availability and replacement of mainly C and nitrogen (N). A major mechanism of resource reallocation by nematodes is the release of assimilated C and N through excretion, rendering these nutrients available for utilization by other soil biota (including plants) (Gebremikael *et al.*, 2016).

Bacterivores, fungivores and omnivores are important regulators of microbial populations through the process of *microbial grazing*. Bacteria and fungi are primarily responsible for the breakdown of labile and recalcitrant organic material, respectively, and therefore key role players in the ecosystem's ability to regulate C (Paterson *et al.*, 2008). By feeding on microbes, nematodes are thus responsible for regulating the structure and size of microbial communities, thereby influencing their ability to decompose

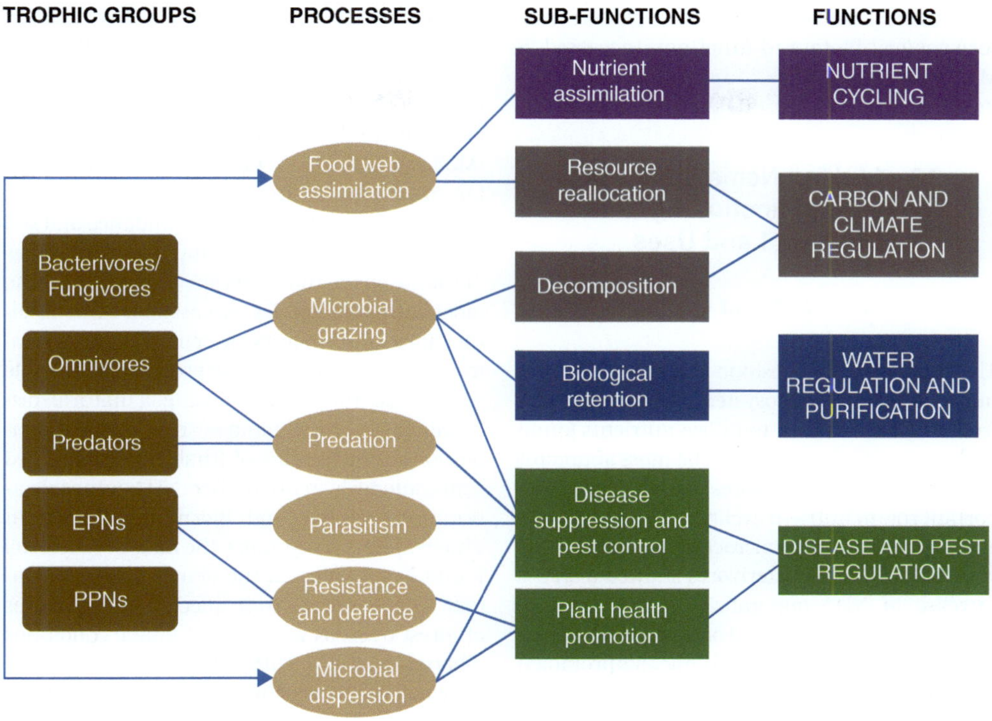

Fig. 7.1. Diagram illustrating the connections between nematode trophic groups and ecosystem processes, sub-functions and functions. EPNs = entomopathogenic nematodes; PPNs = plant-parasitic nematodes. (Author's own image.)

organic materials. Furthermore, microbial grazing by nematodes is linked to the sub-function biological retention, which occurs when soil biota remove toxic substances from soil water (Creamer *et al.*, 2023).

Through microbial grazing (bacterivores, fungivores and omnivores), *predation* of disease-causing organisms (omnivores and predators), and *parasitism* of insect pests (EPNs), nematodes also play a role in regulating disease suppression and pest control. Predation occurs as top-down control of lower-trophic-group organisms (including other nematodes) (Li *et al.*, 2024), while EPNs, specifically those from the *Steinernema* and *Heterorhabditis* species, establish symbiotic relationships with bacteria from the *Xenorhabdus* and *Photorhabdus* genera (Topalović and Geisen, 2023). Together, they effectively kill the insect host after entering its digestive system. Furthermore, EPNs and PPNs also contribute towards activating the *resistance and defence* mechanisms of plants. This includes the stimulation of induced systemic resistance (Ferguson, 2016; Kamali *et al.*, 2022), which is discussed in more depth later in this chapter.

Finally, we propose that nematodes also contribute to plant health promotion through *microbial dispersion* (feeding and excretion, as well as phoresy). For example, Topalović *et al.* (2020) reported that microbes attached to the cuticle of *Meloidogyne hapla* juveniles stage 2 triggered plant defence reactions during penetration, resulting in the suppression of nematode damage in roots. However, it should be noted that nematodes can also carry secondary plant pathogens. For example, the nematode families Trichodoridae and Longidoridae transmit nepoviruses and tobraviruses during feeding (Topalović and Geisen, 2023). While these examples highlight the potential dual role of nematodes in both promoting and suppressing plant health, the ability of especially free-living nematodes (FLNs) to mediate microbial dispersal and contribute to disease management is an emerging field that requires further exploration. For in-depth descriptions on each of the listed processes, please visit https://biosisplatform.eu/services/soil-biota-and-functions (accessed 5 May 2025).

7.3 Linking Nematodes with Ecosystem Functioning Under Different Land Uses

7.3.1 Nutrient cycling

The functional composition of nematode communities in natural ecosystems stems from each trophic group's ability to utilize nutrients found in their specific food source. The most abundant trophic group is bacterivores, which play an important role in nutrient cycling (Van den Hoogen *et al.*, 2019). In sterile microcosms, soils enriched with bacteria and bacterivores showed a 270% increase in NH_4^+ mineralization compared to soils with bacteria, but void of bacterivore nematodes (Ingham *et al.*, 1985). While this promoted plant growth, the authors further confirmed that the impact of bacterivore and fungivore nematodes on nutrient release can result in a short-term response that promotes plant N uptake.

In managed grassland systems the abundance of the entire nematode community may positively correlate with primary production (Yeates and Coleman, 1982). Positive correlations between nematodes and available phosphorus (P) in P-depleted tussock grasslands (Yeates and Coleman, 1982) suggests that both the nematode community and soil nutrient content may influence the relevance of nutrient turnover processes in primary (plant) production. This may further explain why correlations of nematodes and soil nutrients are study dependent, often connected to soil type and land use. For example, bacterivore footprints correlated with N ($R^2 > 0.53$), potassium (K) ($R^2 > 0.38$), magnesium (Mg) ($R^2 > 0.19$) and boron (B) ($R^2 > 0.29$), but not with any other macro- and micronutrients, in two organic fields in Germany (data from Schmidt *et al.*, 2020). Furthermore, Gebremikael *et al.* (2016) showed that the plant uptake of soil N and P is greater in the presence of nematodes. PPNs may also contribute to these cycling processes. For example, PPNs feeding on white clover roots supported the clover-derived ^{15}N-flux transfer to a companion crop (ryegrass) that accordingly reacted with a 21% increase in biomass production (Bardgett *et al.*, 1999).

Forest ecosystems, in turn, have greater proportions of fungivores than grass- and shrublands due to their abundance in decaying wood (Moll *et al.*, 2021). In natural Asian rainforests, Liu *et al.* (2022) found positive correlations between the total nematode abundance and soil N content. Considering that the fungivores *Tylencholaimus* and *Filenchus* were among the most abundant species, we assume these as either important players in or important indicators of the N mineralization process under stable conditions. However, following a natural disturbance (i.e. storm damage) in a Danish deciduous forest, the number of Rhabditidae increased simultaneously with higher NH_4^+ mineralization rates (Ritter and Bjørnlund, 2005). In Chinese forest ecosystems, the links between soil P and N content and the abundance of nematode trophic groups were affected by the species of forest trees, as well as the climatic conditions (Shao *et al.*, 2023). This highlights the complexity of nematode–nutrient (mineralization) interactions in forests, which impedes a generalization in this context.

Nematodes also play a nutrient cycling role in more extreme ecosystems, such as deserts, which can present stable populations of nematodes. Bacterivores can be found in areas where organic C and moisture are accumulated (e.g. roots and root deposits below shrubs, animal faeces, small water films) (Whitford *et al.*, 1982). Food webs in deserts are therefore highly specific and nutrient cycling occurs predominantly in microhabitats that are susceptible to food web disturbance. For example, the exclusion of predatory mites from buried desert shrubs in litter bags caused a rapid increase in bacterivore nematodes that in turn reduced bacterial populations affecting decomposition rates (Santos *et al.*, 1981). Therefore, resource limitations and food web disturbance may be linked to mineralization decline and the collapse of N cycles in microhabitats. In strong contrast are the low-productive Antarctic deserts with low nematode diversity and weak links to the resident nutrients (Freckman and Virginia, 1997).

Finally, in industrial areas such as mining sites the environmental conditions can be extreme with high concentrations of heavy metals. As a result, nematode abundances in such environments are often low and, as such, also their potential functions. Some frequently occurring heavy metals, such as copper, zinc and lead, are highly toxic to nematodes and therefore often negatively correlate with the complete nematode community (Korthals *et al.*, 1996; Martinez *et al.*, 2021). Martinez *et al.* (2021) suggested that the addition of organic matter in highly disturbed and contaminated mining sites may increase organic matter turnover and nutrient mineralization and, thus, plant rehabilitation. However, studies to date are missing that determine: (i) the mineralization of heavy metals by nematodes and subsequent effects on plant communities; and (ii) the heavy metal transfer to higher trophic levels (nematodes, mites, collembola, etc.) through predation.

7.3.2 Carbon and climate regulation

Soils store twice the amount of C than the atmosphere and thus play a critical role in regulating the global C cycle (Swift, 2001). However, intensive practices including land conversion from grasslands to agriculture have significantly reduced soil C stocks and have significantly contributed to global climate change (West and Post, 2002). To reverse these trends, society will need to rebuild soil C stocks through restoration efforts (Martin and Sprunger, 2022). For this reason, researchers have worked to assess the major drivers of soil organic C formation and accrual across terrestrial ecosystems. Historically, the majority view has been that chemical recalcitrance of plant inputs and accumulation of the humic soil C pool were the mechanisms that dictated soil C sequestration (Wander, 2004). However, in recent years, there has been a substantial paradigm shift, and it is now understood that soil microbial bioproducts, termed 'necromass', contribute substantially to soil C stabilization because of adsorption to mineral surfaces, leading to physically protected C in mineral-associated organic matter (MAOMC) (Cotrufo *et al.*, 2013). Given this paradigm shift, understanding how trophic-level interactions between soil fauna and microbes influence decomposition, biochemical transformations and resource allocation is critical, as such processes regulate soil C stabilization (Grandy *et al.*, 2016; Creamer *et al.*, 2022).

Nematodes are trophically diverse and exploit various niches of the soil food web, which allows them to influence soil C pools in multiple ways. Moreover, decomposition pathways such as bacterial and fungal feeding channels have contrasting impacts on soil C cycling (Malik *et al.*, 2016). For instance, decomposition pathways dominated by bacterivores are generally more associated with labile pools of C, compared to decomposition channels dominated by fungivores, which are more associated with stable pools of C (Margenot and Hodson, 2016). These dynamics can be determined by separating soil C into functionally different pools and relating such pools to fungivore dominance (Kane *et al.*, 2023) or via the Channel Index (Sprunger *et al.*, 2019). Labile pools of C persist in the soil for much less time because they are readily accessible to microbes. In contrast, stable pools of soil C are physically protected from bacteria and can lead to soil C permanence (Cotrufo and Lavallee, 2022).

In forest systems, bacterial-dominated decomposition pathways are often associated with more labile C pools that are readily accessible to bacteria, whereas fungi have been known to adapt to areas with more recalcitrant or stable C pools (Margenot and Hodson, 2016). These dynamics can be reflected within bacterivore and fungivore nematodes as well. For instance, grassland and agricultural ecosystems dominated by bacterivores are often systems that have experienced disturbance and have a less overall soil C storage (Culman *et al.*, 2010; DuPont *et al.*, 2014). An increased presence of fungivores and predators could reveal a system that has greater soil organic matter (SOM) accumulation derived from more recalcitrant materials such as perennial plant roots (Sprunger *et al.*. 2019).

To date, the link between nematodes and microbes has been understudied, which is detrimental to our fundamental understanding of soil C cycling within the soil food web. However, researchers are starting to understand key linkages between bacterivore nematodes and various bacterial taxa and how this influences nutrient cycling and plant uptake in both natural and agricultural landscapes (Jiang *et al.*, 2013; Gebremikael *et al.*, 2016). Mechanistically, there are several key soil food web processes that could drastically influence soil C cycling. For instance, nematodes directly influence SOM through predation of plant roots, which stimulates decomposition. Moreover, the top-down controls that the nematode community have on bacteria could have direct implications for soil C cycling (Martin and Sprunger, 2021). Bacterivore nematodes influence the soil metabolic quotient (i.e. rate of respiration per unit of biomass), though the effect on soil C accumulation remains contradictory (Neher, 2010). Furthermore, nematodes could contribute immensely to C stabilization through microbivory, which inhibits a microbiome's ability to degrade various soil C (Grandy *et al.*, 2016). Recent work conducted by Mielke *et al.* (2022) in grassland ecosystems using stable isotope labelling found that generally bacterivore nematodes preferentially grazed on bacteria that degrade already stored organic matter. This is a novel finding in that nematode grazing drastically alters certain taxa of the bacterial community, which ultimately dictates soil C accrual and stabilization.

There are increasing calls to further integrate microbial and soil faunal parameters with soil C metrics (Fierer *et al.*, 2021; Martin *et al.*, 2022). Recently, Martin *et al.* (2022) investigated how nematode communities integrated with other measures of soil C. The researchers found that metrics such as permanganate-oxidizable C that reflect a more processed pool of soil C tended to be positively correlated with fungivore nematodes. This is supported by other research that shows that fungal decomposition pathways are often associated with soil C stabilization processes (Hannula and Morriën, 2022). In contrast, soil respiration (i.e. mineralizable C) determined via a 24-hour assay was highly correlated with PPNs (Martin *et al.*, 2022). Herbivory on plant roots leads to quicker decomposition and a greater release of C, leading to greater overall soil C respiration (Holland *et al.*, 1996). These findings exploring the connections between nematodes and soil C are novel and could demonstrate that certain parts of the nematode community can be used as indicators of contrasting soil C pools. This could provide mechanistic insight into short- versus long-term soil C cycling in terrestrial ecosystems. However, more research is needed in this space.

7.3.3 Water regulation and purification

Nematodes are known to feed on bacteria and fungi that play a role in particle aggregation and the biological retention of toxins (Zhou *et al.*, 2012, 2022a). Therefore, it is suggested that they indirectly influence water regulation and purification through the process of microbial grazing. Many bacterial species are known to degrade or adsorb a wide range of pollutants from crude oil and heavy metals to herbicides. Some species are even used for soil remediation (Sharma *et al.*, 2022). Several studies evidenced that bacterivore nematodes were able to increase microbial degradation through microbial grazing. Zhou *et al.* (2022a) reported higher microbial activity in soil contaminated with petroleum, while Iqbal *et al.* (2020) evidenced increased lead immobilization by fungi and bacteria in the presence of nematodes. More direct evidence was presented by Zhou *et al.* (2012), who showed that inoculating with *Cephalobus* sp. significantly increased the activity of bacteria that degrade the herbicide prometryne. The mechanisms responsible for increased bacterial activity included: (i) the spread of bacteria through the migration and bioturbation of nematodes through ingestion and excretion and phoresy; and (ii) the excretion of inorganic nutrients by nematodes that were subsequently utilized by bacteria (Zhou *et al.*, 2012). Moreover, the authors reported that the presence of this *Cephalobus* sp. led to a greater herbicide removal rate when compared to soil that only contained the bacteria.

Similarly, the presence of three nematode species, namely *Caenorhabditis elegans*, *Cephalobus persegnis* and *Rhabditis marina*, influenced oil degradation in polluted soil, as well as the nematode community in general. Therefore, the magnitude of oil degradation was dependent on the nematode species (Zhou *et al.*, 2022b). In addition, Shaikhulova *et al.* (2021) suggested that *C. elegans* can survive on crude oil and is therefore capable of delivering previously ingested oil-degrading bacteria to contaminated soils. In light of these results, the use of nematodes in addition to bacteria and fungi for bioremediation of soils should be considered.

Soil nematodes can also be used as indicators of dissolved or particulate pollutants. Often single species, such as *C. elegans*, are employed as bioindicators to test the toxicity of a compound under controlled conditions (Sochová *et al.*, 2006). On the other hand, changes in the nematode community and the use of nematode-based indices can measure the effect of pollutants in the environment (Du Preez *et al.*, 2022). For example, the abundance of short-lived and fast-reproducing species is usually higher in polluted soils. Also, the Maturity Index, a nematode-based index that measures soil ecosystem disturbance, often decreases in polluted soils (Chen *et al.*, 2009; Lu *et al.*, 2020). Moreover, the abundance of specific species can be used as an indicator for stressed ecosystems, such as polluted, organically enriched or mechanically disturbed soils. For example, the bacterivore *Acrobeloides* is known to be tolerant to heavy metals and is therefore often found in heavy-metal-polluted areas (Gutiérrez *et al.*, 2016). To date, most studies have focused on the interaction between nematodes and heavy metals, while the effect of many other potential pollutants remains unknown. For further reading on the use of nematodes as bioindicators of pollution, please see Chapters 4, 5 and 6 of this volume.

7.3.4 Disease and pest regulation

Through the process of microbial dispersion, nematodes can act as vectors of beneficial bacteria and transport them to and from the rhizosphere (Knox *et al.*, 2003; Topalović and Geisen, 2023). However, harmful bacteria and viruses can also be transmitted in this way. Kanfra *et al.* (2018) determined that nematode communities interacting with microbes in apple replant disease (ARD)-infected fields resulted in a stronger negative effect on plant growth. This was observed irrespective of whether ARD microbes were added to the plants or not, confrming the potential of transmitting pathogenic microorganisms.

Another example of pest regulation is predatory nematodes consuming PPNs (Kanwar *et al.*, 2021). A comprehensive review by Kanwar *et al.* (2021) listed several mononchid, dorylaimid, diplogasterid and aphelenchid species that prey on PPNs. For example, the predator *Mononchoides gaugleri* can prey on several nematode species, including the destructive *Meloidogyne incognita* (Bilgrami *et al.*, 2005). Another mononchid example is *Clarkus papillatus*, which can kill up to 83 *Heterodera radicicola* in one day (Steiner and Heinly, 1922). This nematode was also found to predate *Aphelenchoides* sp., *Hemicriconemoides* sp., *Heterodera schachtii*, *M. hapla*, *Tylenchulus semipenetrans* and *Subanguina radicola* (Cobb, 1917; Menzel, 1920; Steiher and Heinly, 1922; Thorne, 1927). Many other examples are summarized in Kanwar *et al.* (2021).

Other mechanisms of PPN regulation have also been recorded. Studies in banana production systems listed a range of agronomic factors such as management history, the previous crop and use of cover crops, but also nematode trophic interactions, which could lead to PPN suppression (Ferris *et al.*, 2012; Chauvin *et al.*, 2015; Godefroid *et al.*, 2017). Chauvin *et al.* (2015) discovered that biochemical characteristics of litters was a critical driver of the nematode community succession, and these nematode communities influenced the infectivity of the PPN in the banana fields. Recent studies also suggest a possible interaction between PPNs and EPNs (Kenney and Eleftherianos, 2016). For example, Shapiro-Ilan *et al.* (2006) observed a 18% reduction in the egg mass production of *Meloidogyne partityla* when cadavers infected with *Steinernema riobrave* were applied to pecan tree rhizospheres. Similarly, Kepenekci *et al.* (2016) showed that dipping tomato plant roots in the cell-free supernatant derived from the EPN symbiotic bacteria fermentation reduced the egg masses produced by *M. incognita*, at the same time increasing the vigour and yield of these plants.

EPNs represent one of the best examples of pest regulation by parasitizing on insect pests. However, this group of soil nematodes is often excluded from soil biodiversity accounts and functional group analysis, mainly because the infective juvenile is the only form that naturally occurs in the soil (Stock, 2015). Their morphology also resembles other rhabditids, often resulting in misidentification. Nevertheless, EPNs are globally distributed in soils and their ability to manage insect pests is exploited to manage agroecosystems. Commercially available products containing EPNs can be applied (through augmentative or inoculative approaches) in the soil of many crops (Lacey *et al.*, 2015; Abate *et al.*, 2017). However, management practices can alter EPN interactions with the soil food web in ways that change their biocontrol potential. For example, in a citrus grove in Florida, two agricultural management systems were implemented, namely: (i) advanced citriculture production (ACPS), implemented to compensate for the damage caused by Huanglongbing (Candidatus *Liberibacter*); and (ii) conventional citriculture. The ACPS implies the use of an 'open hydroponic' method in which trees are fertigated with a specific proportion of nutrients via drip or micro-sprinker irrigation systems daily. In this approach, young trees become mature and productive earlier, ensuring the balance between economic loss and lifespan of the grove is still acceptable. The ACPS modified the soil pH and microelements in a way that reduced the native EPNs, which led to an increase in the pest–disease complex *Diaprepes–Phytophthora*. On the contrary, the conventional citriculture system maintained the presence of EPNs, which reduced the pest–disease complex also resulting in healthier soil (Campos-Herrera *et al.*, 2014).

EPNs and PPNs also contribute towards activating the resistance and defence mechanisms of plants. In the case of EPNs, Ferguson (2016) observed that tobacco plants inoculated with EPNs can activate the induced systemic resistance and be more protected against the aerial pest *Manduca sexta* (Lepidoptera: Sphingidae) and the pathogens *Peronospora tabacina* (Peronosporales: Peronosporaceae) and *Rhizoctonia solani* (Cantharellales: Ceratobasidiaceae). More recently, Kamali *et al.* (2022) demonstrated possible mechanisms behind the induced systemic resistance activation. They observed that applying EPNs in tomato roots triggered an immune response in the plant via upregulated phenylpropanoid metabolism and synthesis of protease inhibitors in their tissues. This reduced egg laying and developmental performance exhibited by above-ground herbivores such as the leaf mining insect, *Tuta absoluta*. PPNs likely play a similar role in induced systemic resistance as described for EPNs because they trigger a plant immune response when injecting their stylet or when entering the plant. Already in the 1960s, Seinhorst (1965) noted that nematode damage stimulated root production. This contributes to plant tolerance to pests and diseases and sometimes even minor damage to plant roots can result in increased yields. Mbaluto *et al.* (2021) found that resistance to *Spodoptera exigua* was induced by *M. incognita* infection through systematically stimulating responses related to jasmonic, salicylic and abscisic acids.

7.4 Concluding Nematodes as Bioindicators of Ecosystem Functions

Regarding nutrient cycling, the abundance and diversity of nematodes, particularly in agricultural systems, indicate nutrient availability and cycling, thereby enhancing primary production. By observing nematodes' relationships with soil nutrients such as N, P, K, Mg and B, we can better understand the nutrient turnover processes within ecosystems (Schmidt *et al.*, 2020). Nematodes can also potentially serve as bioindicators of soil C stabilization as an increased presence of fungivores and predators often reflects organic matter accumulation in both grassland and agricultural landscapes (Sprunger *et al.*, 2019; Martin and Sprunger, 2021). Furthermore, by using ecological toolsets such as nematode-based indices, these organisms can serve as ecological indicators of ecosystem enrichment and decomposition pathways (Du Preez *et al.*, 2022). Similarly, nematode community structure and ecological indices can be used to measure and monitor disturbance resulting from environmental pollution. Finally, nematodes can also be used as bioindicators of pest and disease regulation. Predators and omnivores require a food source, which includes PPNs (Kanwar *et al.*, 2021), while EPNs are bioindicators of insect parasitism. PPNs are also potential bioindicators of plant cover and vegetation diversity.

7.5 The Way Forward

New insights are important as they fuel the advancement of knowledge, influence policy changes and drive the evolution of best practices in various fields. We identified several areas of research which show potential for gaining a better understanding in the ecological role that nematodes fulfil. These include:

- Further research is needed to understand how specific types of nematodes, particularly fungivores, regulate soil C stabilization and influence soil C cycling. This includes a deeper understanding of the fungal decomposition pathway and its contribution to soil C stabilization.

- Comparative studies investigating the role of nematodes in ecosystem functioning covering a broader range of habitats and land-use types. Part of this exploration should focus on the potential to change and predict nematode communities through various management practices like planting of cover crops or companion planting.
- The ecological interaction between nematodes and microbes warrants further investigation, specifically in the areas of microbial grazing and dispersion. Emphasis should be placed on the potential for nematodes to disperse both beneficial and pathogenic microbes.
- The role of nematodes in water regulation (e.g. through affecting soil structure and C cycling) and purification (e.g. regulating pollutant-degrading bacteria) remains poorly studied. Further research is needed to investigate the mechanisms involved in the contribution of nematodes to water regulation and purification in various ecosystems.
- Future research needs to explore if predatory nematodes can distinguish between parasitic and beneficial nematodes. This ability could enhance their use as biological control agents in agricultural systems. Thus, future research should prioritize conducting food choice experiments among different nematode species, aiming to understand the selective feeding behaviour of predatory nematodes and its potential implications for pest management.

References

Abate, B.A., Wingfield, M.J., Slippers, B. and Hurley, B.P. (2017) Commercialisation of entomopathogenic nematodes: should import regulations be revised? *Biocontrol Science and Technology* 27, 149–168. DOI: 10.1080/09583157.2016.1278200.

Bardgett, R.D., Denton, C.S. and Cook, R. (1999) Below-ground herbivory promotes soil nutrient transfer and root growth in grassland. *Ecology Letters* 2, 357–360. DOI: 10.1046/j.1461-0248.1999.00001.x.

Bilgrami, A.L., Gaugler, R. and Brey, C.W. (2005) Prey preference and feeding behaviour of the diplogastrid predator *Mononchoides gaugleri* (Nematoda: Diplogastrida). *Nematology* 7, 333–342. DOI: 10.1163/156854105774355563.

Bünemann, E.K., Bongiorno, G., Bai, Z., Creamer, R.E., De Deyn, G. *et al.* (2018) Soil quality – a critical review. *Soil Biology and Biochemistry* 120, 105–125. DOI: 10.1016/j.soilbio.2018.01.030.

Campos-Herrera, R., El-Borai, F.E., Ebert, T.E., Schumann, A. and Duncan, L.W. (2014) Management to control citrus greening alters the soil food web and severity of a pest–disease complex. *Biological Control* 76, 41–51. DOI: 10.1016/j.biocontrol.2014.04.012.

Chauvin, C., Dorel, M., Villenave, C., Roger-Estrade, J., Thuries, L. and Risède, J.-M. (2015) Biochemical characteristics of cover crop litter affect the soil food web, organic matter decomposition, and regulation of plant-parasitic nematodes in a banana field soil. *Applied Soil Ecology* 96, 131–140. DOI: 10.1016/j.apsoil.2015.07.013.

Chen, G., Qin, J., Shi, D., Zhang, Y. and Ji, W. (2009) Diversity of soil nematodes in areas polluted with heavy metals and polycyclic aromatic hydrocarbons (PAHs) in Lanzhou, China. *Environmental Management* 44, 163–172. DOI: 10.1007/s00267-008-9268-2.

Cobb, N.A. (1917) The mononchs': a genus of free-living predatory nematodes. *Soil Science* 3, 431–486.

Cotrufo, M.F. and Lavallee, J.M. (2022) Soil organic matter formation, persistence, and functioning: a synthesis of current understanding to inform its conservation and regeneration. *Advances in Agronomy* 172, 1–66. DOI: 10.1016/bs.agron.2021.11.002.

Cotrufo, M.F., Wallenstein, M.D., Boot, C.M., Denef, K. and Paul, E. (2013) The Microbial Efficiency-Matrix Stabilization (MEMS) framework integrates plant litter decomposition with soil organic matter stabilization: do labile plant inputs form stable soil organic matter? *Global Change Biology* 19, 988–995. DOI: 10.1111/gcb.12113.

Creamer, R.E., Barel, J.M., Bongiorno, G. and Zwetsloot, M.J. (2022) The life of soils: integrating the who and how of multifunctionality. *Soil Biology and Biochemistry* 166, 108561. DOI: 10.1016/j.soilbio.2022.108561.

Creamer, R.E., Dilonardo, P., Zwetsloot, M.J., Bongiorno, G. and Barel, J.M. (2023) BIOSIS Tool. Wageningen University. Available at: https://biosisplatform.eu/ (accessed 5 May 2025).

Culman, S.W., DuPont, S.T., Glover, J.D., Buckley, D.H., Fick, G.W. *et al.* (2010) Long-term impacts of high-input annual cropping and unfertilized perennial grass production on soil properties and belowground food webs in Kansas, USA. *Agriculture, Ecosystems and Environment* 137, 13–24. DOI: 10.1016/j.agee.2009.11.008.

Du Preez, G., Daneel, M., De Goede, R., Du Toit, M.J., Ferris, H. *et al.* (2022) Nematode-based indices in soil ecology: application, utility, and future directions. *Soil Biology and Biochemistry* 169, 108640. DOI: 10.1016/j.soilbio.2022.108640.

DuPont, S.T., Beniston, J., Glover, J.D., Hodson, A., Culman, S.W. *et al.* (2014) Root traits and soil properties in harvested perennial grassland, annual wheat, and never-tilled annual wheat. *Plant and Soil* 381, 405–420. DOI: 10.1007/s11104-014-2145-2.

Ferguson, J.G. (2016) Effectiveness of entomopathogenic nematode-induced systemic resistance against pests and pathogens of tobacco and soybeans in Tennessee. Master's thesis, University of Tennessee, Knoxville, Tennessee.

Ferris, H., Pocasangre, L.E., Serrano, E., Muñoz, J., Garcia, S. *et al.* (2012) Diversity and complexity complement apparent competition: nematode assemblages in banana plantations. *Acta Oecologica* 40, 11–18. DOI: 10.1016/j.actao.2012.02.003.

Fierer, N., Wood, S.A. and De Mesquita, C.P.B. (2021) How microbes can, and cannot, be used to assess soil health. *Soil Biology and Biochemistry* 153, 108111. DOI: 10.1016/j.soilbio.2020.108111.

Freckman, D.W. and Virginia, R.A. (1997) Low-diversity Antarctic soil nematode communities: distribution and response to disturbance. *Ecology* 78, 363–369. DOI: 10.1890/0012-9658(1997)078[0363:LDASNC]2.0.CO;2.

Hannula, E.S. and Morriën, E. (2022) Will fungi solve the carbon dilemma? *Geoderma* 413, 115767. DOI: 10.1016/j.geoderma.2022.115767.

Gebremikael, M.T., Steel, H., Buchan, D., Bert, W. and De Neve, S. (2016) Nematodes enhance plant growth and nutrient uptake under C and N-rich conditions. *Scientific Reports* 6, 32862. DOI: 10.1038/srep32862.

Godefroid, M., Tixier, P., Chabrier, C., Djigal, D. and Queneherve, P. (2017) Associations of soil type and previous crop with plant-feeding nematode communities in plantain agrosystems. *Applied Soil Ecology* 113, 63–70. DOI: 10.1016/j.apsoil.2017.01.012.

Grandy, A.S., Wieder, W.R., Wickings, K. and Kyker-Snowman, E. (2016) Beyond microbes: are fauna the next frontier in soil biogeochemical models? *Soil Biology and Biochemistry* 102, 40–44. DOI: 10.1016/j.soilbio.2016.08.008.

Gutiérrez, C., Fernández, C., Escuer, M., Campos-Herrera, R., Rodríguez, M.E.B. *et al.* (2016) Effect of soil properties, heavy metals and emerging contaminants in the soil nematodes diversity. *Environmental Pollution* 213, 184–194. DOI: 10.1016/j.envpol.2016.02.012.

Holland, J.N., Cheng, W. and Crossley, D.A. (1996) Herbivore-induced changes in plant carbon allocation: assessment of below-ground C fluxes using carbon-14. *Oecologia* 107, 87–94. DOI: 10.1007/BF00582238.

Ingham, R.E., Trofymow, J.A., Ingham, E.R. and Coleman, D.C. (1985) Interactions of bacteria, fungi, and their nematode grazers: effects on nutrient cycling and plant growth. *Ecological Monographs* 55, 119–140.

Iqbal, N., Ahmed, S., Pervez, A., Nazir, R., Tang, X. and Irshad, U. (2020) The fate of lead (Pb) in multitrophic interactions among bacteria, fungi, and bacterivorous soil nematodes. *CLEAN – Soil, Air, Water* 48, 2000307. DOI: 10.1002/clen.202000307.

Jiang, Y., Sun, B., Jin, C. and Wang, F. (2013) Soil aggregate stratification of nematodes and microbial communities affects the metabolic quotient in an acid soil. *Soil Biology and Biochemistry* 60, 1–9. DOI: 10.1016/j.soilbio.2013.01.006.

Kamali, S., Javadmanesh, A., Stelinski, L.L., Kyndt, T., Seifi, A. *et al.* (2022) Beneficial worm allies warn plants of parasite attack below-ground and reduce above-ground herbivore preference and performance. *Molecular Ecology* 31, 691–712. DOI: 10.1111/mec.16254.

Kane, J.L., Kotcon, J.B., Freedman, Z.B. and Morrissey, E.M. (2023) Fungivorous nematodes drive microbial diversity and carbon cycling in soil. *Ecology* 104, e3844. DOI: 10.1002/ecy.3844.

Kanfra, X., Liu, B., Beerhues, L., Sørensen, S.J. and Heuer, H. (2018) Free-living nematodes together with associated microbes play an essential role in apple replant disease. *Frontiers in Plant Science* 9, 1666. DOI: 10.3389/fpls.2018.01666.

Kanwar, R.S., Patil, J.A. and Yadav, S. (2021) Prospects of using predatory nematodes in biological control for plant parasitic nematodes – a review. *Biological Control* 160, 104668. DOI: 10.1016/j.biocontrol.2021.104668.

Kenney, E. and Eleftherianos, I. (2016) Entomopathogenic and plant pathogenic nematodes as opposing forces in agriculture. *International Journal for Parasitology* 46, 13–19. DOI: 10.1016/j.ijpara.2015.09.005.

Kepenekci, I., Hazir, S. and Lewis, E.E. (2016) Evaluation of entomopathogenic nematodes and the supernatants of the *in vitro* culture medium of their mutualistic bacteria for the control of the root-knot nematodes *Meloidogyne incognita* and *M. arenaria*. *Pest Management Science* 72, 327–334. DOI: 10.1002/ps.3998.

Knox, O.G.G., Killham, K., Mullins, C.E. and Wilson, M.J. (2003) Nematode-enhanced microbial colonization of the wheat rhizosphere. *FEMS Microbiology Letters* 225, 227–233. DOI: 10.1016/s0378-1097(03)00517-2.

Korthals, G.W., Bongers, T., van de Leemans, T.M. and van den Brink, P.J. (1996) Long-term effects of copper and pH on the nematode community in an agroecosystem. *Environmental Toxicology and Chemistry* 15, 979–985.

Lacey, L.A., Grzywacz, D., Shapiro-Ilan, D.I., Frutos, R., Brownbridge, M. and Goettel, M.S. (2015) Insect pathogens as biological control agents: back to the future. *Journal of Invertebrate Pathology* 132, 1–41. DOI: 10.1016/j.jip.2015.07.009.

Li, G., Liu, T., Whalen, J.K. and Wei, Z. (2024) Nematodes: an overlooked tiny engineer of plant health. *Trends in Plant Science* 29, 52–63. DOI: 10.1016/j.tplants.2023.06.022.

Liu, J., Zhao, W., He, H., Kou, Y. and Liu, Q. (2022) Variations in the community patterns of soil nematodes at different soil depths across successional stages of subalpine forests. *Ecological Indicators* 136, 108624. DOI: 10.1016/j.ecolind.2022.108624.

Lu, Q., Liu, T., Wang, N., Dou, Z., Wang, K. and Zuo, Y. (2020) A review of soil nematodes as biological indicators for the assessment of soil health. *Frontiers of Agricultural Science and Engineering* 7, 275–281. DOI: 10.15302/j-fase-2020327.

Malik, A.A., Chowdhury, S., Schlager, V., Oliver, A., Puissant, J. *et al.* (2016) Soil fungal ratios are linked to altered carbon cycling. *Frontiers in Microbiology* 7, 1247. DOI: 10.3389/fmicb.2016.01247.

Margenot, A.J. and Hodson, A.K. (2016) Relationships between labile soil organic matter and nematode communities in a California oak woodland. *Nematology* 18, 1231–1245. DOI: 10.1163/15685411-00003027.

Martin, T. and Sprunger, C.D. (2021) A meta-analysis of nematode community composition across soil aggregates: implications for soil carbon dynamics. *Applied Soil Ecology* 168, 104143. DOI: 10.1016/j.apsoil.2021.104143.

Martin, T. and Sprunger, C.D. (2022) Sensitive measures of soil health reveal carbon stability across a management intensity and plant biodiversity gradient. *Frontiers in Soil Science* 2, 917885. DOI: 10.3389/fsoil.2022.917885.

Martin, T., Wade, J., Singh, P. and Sprunger, C.D. (2022) The integration of nematode communities into the soil biological health framework by factor analysis. *Ecological Indicators* 136, 108676. DOI: 10.1016/j.ecolind.2022.108676.

Martinez, J.G., Balondo, K.A., Santos, G.D., Van Ranst, E. and Moens, T. (2021) Re-establishment of nematode communities in a rehabilitated surface mining area in Sibutad, southern Philippines. *European Journal of Soil Biology* 102, 103254. DOI: 10.1016/j.ejsobi.2020.103254.

Mbaluto, C.M., Vergara, F., van Dam, N.M. and Martínez-Medina, A. (2021) Root infection by the nematode *Meloidogyne incognita* modulates leaf antiherbivore defenses and plant resistance to *Spodoptera exigua*. *Journal of Experimental Botany* 72, 7909–7926. DOI: 10.1093/jxb/erab370.

Menzel, R. (1920) Über die Nahrung der freilebenden Nematoden und die Art ihrer Aufnahme. *Verhandlungen der Naturforschenden Gesellschaft* 31, 153–188.

Mielke, L., Taubert, M., Cesarz, S., Ruess, L., Küsel, K. *et al.* (2022) Nematode grazing increases the allocation of plant-derived carbon to soil bacteria and saprophytic fungi, and activates bacterial species of the rhizosphere. *Pedobiologia* 90, 150787. DOI: 10.1016/j.pedobi.2021.150787.

Moll, J., Roy, F., Bässler, C., Heilmann-Clausen, J., Hofrichter, M. *et al.* (2021) First evidence that nematode communities in deadwood are related to tree species identity and to co-occurring fungi and prokaryotes. *Microorganisms* 9, 1454. DOI: 10.3390/microorganisms9071454.

Neher, D.A. (2010) Ecology of plant and free-living nematodes in natural and agricultural soil. *Annual Review of Phytopathology* 48, 371–394. DOI: 10.1146/annurev-phyto-073009-114439.

NRCS (Natural Resources Conservation Service) (2024) Soil Health. NCRS, US Department of Agriculture. Available at: https://www.nrcs.usda.gov/conservation-basics/natural-resource-concerns/soils/soil-health (accessed 3 March 2024).

Paterson, E., Osler, G., Dawson, L.A., Gebbing, T., Sim, A. and Ord, B. (2008) Labile and recalcitrant plant fractions are utilised by distinct microbial communities in soil: independent of the presence of roots and mycorrhizal fungi. *Soil Biology and Biochemistry* 40, 1103–1113. DOI: 10.1016/j.soilbio.2007.12.003.

Ritter, E. and Bjørnlund, L. (2005) Nitrogen availability and nematode populations in soil and litter after gap formation in a semi-natural beech-dominated forest. *Applied Soil Ecology* 28, 175–189. DOI: 10.1016/j.apsoil.2004.07.002.

Santos, P.F., Phillips, J. and Whitford, W.G. (1981) The role of mites and nematodes in early stages of buried litter decomposition in a desert. *Ecology* 62, 664–669. DOI: 10.2307/1937734.

Schmidt, J.H., Hallmann, J. and Finckh, M.R. (2020) Bacterivorous nematodes correlate with soil fertility and improved crop production in an organic minimum tillage system. *Sustainability* 12, 6730. DOI: 10.3390/su12176730.

Seinhorst, J.W. (1965) The relation between nematode density and damage to plants. *Nematologica* 11, 137–154. DOI: 10.1163/187529265X00582.

Shaikhulova, S., Fakhrullina, G., Nigamatzyanova, L., Akhatova, F. and Fakhrullin, R. (2021) Worms eat oil: *Alcanivorax borkumensis* hydrocarbonoclastic bacteria colonise *Caenorhabditis elegans* nematodes intestines as a first step towards oil spills zooremediation. *Science of the Total Environment* 761, 143209. DOI: 10.1016/j.scitotenv.2020.143209.

Shao, Y., Wang, Z., Liu, T., Kardol, P., Ma, C. *et al.* (2023) Drivers of nematode diversity in forest soils across climatic zones. *Proceedings of the Royal Society B: Biological Sciences* 290, 20230107. DOI: 10.1098/rspb.2023.0107.

Shapiro-Ilan, D.I., Nyczepir, A.P. and Lewis, E.E. (2006) Entomopathogenic nematodes and bacteria applications for control of the pecan root-knot nematode, *Meloidogyne partityla*, in the greenhouse. *Journal of Nematology* 38, 449–454.

Sharma, P., Bano, A., Singh, S.P., Dubey, N.K., Chandra, R. and Iqbal, H.M.N. (2022) Recent advancements in microbial-assisted remediation strategies for toxic contaminants. *Cleaner Chemical Engineering* 2, 100020. DOI: 10.1016/j.clce.2022.100020.

Sochová, I., Hofman, J. and Holoubek, I. (2006) Using nematodes in soil ecotoxicology. *Environment International* 32, 374–383. DOI: 10.1016/j.envint.2005.08.031.

Sprunger, C.D., Culman, S.W., Peralta, A.L., DuPont, S.T., Lennon, J.T. and Snapp, S.S. (2019) Perennial grain crop roots and nitrogen management shape soil food webs and soil carbon dynamics. *Soil Biology and Biochemistry* 137, 107573. DOI: 10.1016/j.soilbio.2019.107573.

Steiner, G. and Heinly, H. (1922) The possibility of control of *Heterodera radicicola* and other plant-injurious nemas by means of predatory nemas, especially by *Mononchus papillatus* Bastian. *Journal of the Washington Academy of Sciences* 12, 367–386. Available at: http://www.jstor.org/stable/24532726 (accessed 5 May 2025).

Stock, S.P. (2015) Diversity, biology and evolutionary relationships. In: Campos-Herrera, R. (ed.) *Nematode Pathogenesis of Insects and Other Pests*. Springer, Cham, Switzerland, pp. 3–27.

Swift, R.S. (2001) Sequestration of carbon by soil. *Soil Science* 166, 858–871. DOI: 10.1097/00010694-200111000-00010.

Thorne, G. (1927) The life history, habits and economic importance of some mononchs. *Journal of Agricultural Research of China* 34, 265–286.

Topalović, O. and Geisen, S. (2023) Nematodes as suppressors and facilitators of plant performance. *New Phytologist* 238, 2305–2312. DOI: 10.1111/nph.18925.

Topalović, O., Bredenbruch, S., Schleker, A.S.S. and Heuer, H. (2020) Microbes attaching to endoparasitic phytonematodes in soil trigger plant defense upon root penetration by the nematode. *Frontiers in Plant Science* 11, 138. DOI: 10.3389/fpls.2020.00138.

Van den Hoogen, J., Geisen, S., Routh, D., Ferris, H., Traunspurger, W. *et al.* (2019) Soil nematode abundance and functional group composition at a global scale. *Nature* 572, 194–198. DOI: 10.1038/s41586-019-1418-6.

Wall, D.H., Nielsen, U.N. and Six, J. (2015) Soil biodiversity and human health. *Nature* 528, 69–76. DOI: 10.1038/nature15744.

Wander, M.M. (2004) Soil organic matter fractions and their relevance to soil function. In: Magdoff, F. and Weil, R.R. (eds) *Soil Organic Matter in Sustainable Agriculture*. CRC Press, Boca Raton, Florida, pp. 67–102.

West, T.O. and Post, W.M. (2002) Soil organic carbon sequestration rates by tillage and crop rotation. *Soil Science Society of America Journal* 66, 1930–1946. DOI: 10.2136/sssaj2002.1930.

Whitford, W.G., Freckman, D.W., Santos, P.F., Elkins, N.Z. and Parker, L.W. (1982) The role of nematodes in decomposition in desert ecosystems. In: Freckman, D.W. (ed.) *Nematodes in Soil Ecosystems*. University of Texas Press, Austin, Texas, pp. 98–116.

Yeates, G.W. and Coleman, D.C. (1982) Role of nematodes in decomposition. In: Freckman, D.W. (ed.) *Nematodes in Soil Ecosystems*. University of Texas Press, Austin, Texas, pp. 55–81.

Zhou, J., Hu, F., Jiao, J., Liu, M. and Li, H. (2012) Effects of bacterial-feeding nematodes and prometryne-degrading bacteria on the dissipation of prometryne in contaminated soil. *Journal of Soils and Sediments* 12, 576–585. DOI: 10.1007/s11368-012-0473-5.

Zhou, J., Wang, Y., Huang, G., Zhang, C., Ai, Y. *et al.* (2022a) Variation of microbial activities and communities in petroleum-contaminated soils induced by the addition of organic materials and bacterivorous nematodes. *Ecotoxicology and Environmental Safety* 237, 113559. DOI: 10.1016/j.ecoenv.2022.113559.

Zhou, J., Xu, X., Huang, G., Li, W., Wei, Q. *et al.* (2022b) Oil degradation and variation of microbial communities in contaminated soils induced by different bacterivorous nematodes species. *Ecotoxicology and Environmental Safety* 229, 113079. DOI: 10.1016/j.ecoenv.2021.113079.

8 Case Studies Using Nematode Assemblage Analysis in Terrestrial Habitats

Péter Nagy[1]*, Thomais Kakouli-Duarte[2], Sofia Costa[3], Liliane Ruess[4] and Maria Argyropoulou[5]

[1]*Department of Zoology and Ecology, Institute for Wildlife Management and Nature Conservation, Hungarian University of Agriculture and Life Sciences, Hungary;* [2]*MENRG, enviroCORE, Department of Applied Science, South East Technological University, Ireland;* [3]*Centre of Molecular and Environmental Biology, Department of Biology, University of Minho, Campus de Gualtar, Portugal;* [4]*Institute of Biology, Ecology Group, Humboldt-Universität zu Berlin, Germany;* [5]*Department of Zoology, School of Biology, Aristotle University of Thessaloniki, Greece*

Abstract

The aim of this chapter is to illustrate the effects of various impacts of direct or indirect human origin, as revealed by nematode community parameters. We intend to give a comprehensive view on how useful these tools are to measure the effects of anthropogenic disturbances on soil ecosystems.

8.1 Agrochemicals and Plant-Based Alternatives

Research for plant-based alternatives to chemical herbicides, pesticides or fertilizers is gaining popularity. However, most studies focus on the efficacy of botanicals against plant parasites, while only a few examine their effects on free-living soil nematodes as non-target organisms. In the case studies presented herewith, soil nematode communities were studied in terms of trophic structure, nematode indices, composite metabolic footprint, diversity profiles, genera composition and interaction networks.

(1) In a greenhouse experiment, Argyropoulou *et al.* (2022) used spearmint, peppermint and rosemary as soil amendments in pots sown with tomato. After 28 days, they estimated their effect on seedling growth, soil nutrients and the soil nematode communities, while Ainalidou *et al.* (2021) studied the effects on soil microbes in the same experiments. The dynamics of the plant–soil–nematode interactions were further explored by using aromatic plants at different stages of decomposition (0, 28 and 56 days). Spearmint and peppermint directly enhanced almost all microbial groups (bacteria, fungi, protozoa), along with bacterial-feeding nematodes, raising the Enrichment Index (EI) values and the metabolic footprint. Plant feeders exhibited the reverse response, indicating nematicidal properties of the mint plants. It seems that apart from any possible nematicidal activity, botanicals offer a significant trade-off to microbial-feeding

*Email: Nagy.Peter.Istvan@uni-mate.hu

 Nematodes as Environmental Indicators: From Theory to Practice, 2nd Edition (T. Kakouli-Duarte *et al.* eds)

DOI: 10.1079/9781800624221.0008

nematodes by providing food to soil microflora. A high soil nutritional status and an enhanced plant growth were obtained, especially when mint plants were left for 28 days to decompose before sowing tomato. It seems that while aromatic plant incorporation into the soil enhanced microbes and bacterial-feeding nematodes directly, it took another month for soil enrichment with nutrients and for promotion of plant growth. At later stages of aromatic plant decomposition, their effects were still evident but far less pronounced. At that time, enrichment opportunists were replaced by general opportunists with longer life cycles. Compared with mint plants, rosemary had similar, yet less intense, effects on the soil community but on the other hand completely inhibited tomato growth.

(2) In a field experiment during the fallow period between crops, Boutsis *et al.* (2011) compared soil nematode responses to two agricultural practices, namely disinfection with metham sodium (a general biocide that is now banned in many countries) or a mixture of organic additives with nematicidal properties. The latter consisted of neem cake (a product of the neem tree *Azadirachta indica*) and a mixture of basil, marigold and clove oils. The study compared the soil nematode community before and one month after treatments. Chemical disinfection had a lethal effect on almost all nematode genera, reducing or even eliminating their populations. Only numbers of r-strategist bacterial feeders (Ba-1) were at the same level as before disinfection, probably because one month was enough for the recovery of their populations. Organic amendments favoured bacterial-feeding nematodes without affecting plant feeders, probably due to small dosage. These alterations in nematode communities were poorly reflected in changes of diversity and even less in nematode indices; EI increased, while Maturity Index (MI) and Channel Index (CI) decreased after both treatments, following the changes in the percentage contribution of Ba-1, which increased even in the chemically treated plots, where their absolute numbers were not affected. Thus, different treatments changed nematode indices almost similarly. The nematode network was smaller after chemical disinfection, poorly connected and almost entirely dependent on the dominant plant feeder, *Paratylenchus*. In the treatments with organic additives, large and more complex networks were formed, with increased interactions between genera, and probably more robust against disturbances.

(3) In the context of circular bioeconomy, Dimou *et al.* (2023) examined the potential use of olive mill wastewater (OMW), i.e. the liquid effluent from the olive oil production process, as a soil biofertilizer in pots with lettuce seedlings. OMW is rich in nutrients but is also toxic mainly due to high concentrations of phenolic compounds. Two methods of OMW detoxification were used before the experiment: one where OMW was treated with fungi and one where OMW was treated by a combination of a fungus and an alga. Non-treated OMW reduced all nematodes, and their network was the most fragmented and the least robust against disturbance. The effect was milder when OMW was detoxified either by the fungus alone or by its combination with the alga. In the latter case, plant feeders were suppressed but bacterial feeders were not affected. The most cohesive and robust nematode network was formed in the soil receiving the fungal-treated OMW, an outcome also confirmed by the high Structure Index (SI) values, which indicate more linkages in the food web.

(4) In a pot experiment, Ntalli *et al.* (2019) and Monokrousos *et al.* (2021) tested the efficacy of anise (*Pimpinella anisum*), parsley (*Petroselinum crispum*), rocket (*Eruca sativa*) and a synthetic nematicide (fluopyram) against *Meloidogyne incognita*, evaluating also their effects on soil nematode and microbial communities and on tomato growth. Fluopyram was effective against root-knot nematodes, without affecting the microbial community, but severely suppressed the non-target soil nematodes. Rocket and parsley were less efficient than fluopyram in terms of root galling and female numbers in roots, but more efficient in terms of J2 counts in soil. This means that the plant macerates are not as effective as fluopyram in reducing infectivity but affect mainly the *M. incognita* reproduction in tomato roots. All botanicals acted as soil biostimulants, enhancing enzymatic activities (urease, β-glucosidase, acid phosphatase), proliferating soil microbes and bacterial-feeding nematodes, especially the enrichment opportunists. The results were more pronounced in the case of parsley and rocket, which also enhanced plant growth.

8.2 Assessing the Ecological Impact of Recycling-Derived and Bio-Based Fertilizers in Northwest Europe

Food production is heavily reliant on sufficient soil nutrients. It is estimated that the global increasing human population will lead to the depletion of the global phosphate reserves within the next century (Cooper *et al.*, 2011). At the same time, organic wastes rich in phosphorus are increasing but discarded. Research efforts are being made towards the recovery of nutrients from waste (Mehta *et al.*, 2014) to enhance sustainable agriculture, to ensure global food production and to foster a circular economy in the fertilizer and waste management sectors. One such effort has been the European Union Interreg North-West Europe part-funded ReNu2Farm project (https://vb.nweurope.eu/projects/project-search/renu2farm-nutrient-recycling-from-pilot-production-to-farms-and-fields/, accessed 2 September 2025), aiming to increase the recycling rates of nitrogen (N), phosphorus (P) and potassium (K) in the primary food production chain in Northwest Europe. In the context of this project, field trials for the testing of recycling-derived fertilizers (RDFs) or bio-based fertilizers (BBFs) were carried out in Ireland (Karpinska *et al.*, 2021; Ryan *et al.*, 2022) and Belgium (Saju *et al.*, 2023).

In the Irish trials, (i) a non-fertilizer treatment (NF; no P, N, K or S) and (ii) a no phosphate (N, K and S applied, but no P) treatment (SP 0) served as negative controls, with mineral superphosphate (SP 40) as positive control.

A combination of cattle slurry and mineral superphosphate (CS) also served as a positive control, as this is commonly applied in agricultural practice. RDF treatments included potato waste struvite (PWS), municipal waste struvite (MWS), poultry litter ash (PLA) and sewage sludge ash (SSA); struvite is an ammonium magnesium phosphate mineral ($MgNH_4PO_4{\cdot}6H_2O$) that can be produced by crystallization and precipitation technologies (Le Corre *et al.*, 2009; Lorick *et al.*, 2020). The nutrient compositions of RDF are provided in Karpinska *et al.* (2021) and Ryan *et al.* (2022). Each treatment had five replicate plots, each 2 m × 6 m in size.

The Belgian trial focused on 'recovered nitrogen from manure' (RENURE) materials (Saju *et al.*, 2023): pig manure (PM), ammonium nitrate (AN) from stripping–scrubbing of the liquid fraction of digestate, ammonium sulfate (AS) from air scrubbing of stables and pig urine (PU). The trial included four replicate plots (6 m × 8 m) of each RENURE material. The complete nutrient profile of the treatments is detailed in Saju *et al.* (2023).

Prior to the ReNu2Farm project, there was no available literature on nematode community responses to RDF or to RENURE materials, although considerable work has been done on the effects of organic and inorganic fertilizers and composts on nematode communities (Bulluck *et al.*, 2002; Hu and Qi, 2010; Steel *et al.*, 2012, 2018). Therefore, nematode communities were used to assess the ecological impact and environmental safely of these novel fertilizers.

While nematode communities in all trials were analysed using next-generation sequencing technologies, changes in community composition, and relative abundances of nematodes, served as less detailed indicators of the soil ecosystem response to RDFs and BBFs. In the first Irish trial (Karpinska *et al.*, 2021) terrestrial nematode communities displayed significant changes under the different P fertilization treatments. Although there are several examples of dorylaimids occurring in bare sand, or heavily polluted soils, probably living on algae growing at the soil surface (Santos Dagnaisser *et al.*, 2022), in general the presence of dorylaimids reflects a mature, structured and less disturbed soil environment, as opposed to a degraded or recently disturbed ecosystem, colonized by stress-tolerant and opportunistic food nematodes (Bongers, 1990). Such disturbance occurred after the application of PLA in this P-fertilizer replacement value trial (Karpinska *et al.*, 2021); P-fertilizer replacement value is the nutritional value of any fertilizer (in this case RDF) which is used to replace mineral synthetic fertilizer of equal nutritional value. This disturbance was likely due to the high content of heavy metals in the ash derived from poultry litter waste (A. Saju, Ghent, Belgium, 2023, personal communication). It might also have been a fertilization effect, stimulating root growth, microbial activity and thus attracting different types of nematodes. The other treatment with SSA also displayed unfavourable changes to nematode diversity, and soil available P was a driving factor in shaping the nematode communities in this treatment, which also contained heavy metals. Thus, it may pose a risk to terrestrial nematode communities in long-term use and may negatively affect soil nutrient cycling and mineralization. The ash product, rich in concentrated nutrients, can certainly be used as a fertilizer additive, but it was recommended that the production process is improved to minimize the heavy metal content in the final product, to make use of the beneficial available nutrients (Karpinska *et al.*, 2021).There were fewer variations in nematode communities observed with struvites and there were no major differences between PWS and MWS and the controls in terms of ecological impact. Neither struvite fertilizer reduced the number of observed nematode species or adversely affected nematode communities. Both struvites can therefore be applied by farmers, as the main source of P, without posing a threat to terrestrial nematode communities and plant growth.

The two new RDFs, MWS and PWS, can become effective and sustainable replacements to mineral non-renewable forms of imported P in temperate Irish grasslands (Ryan *et al.*, 2022). The significant reduction of dorylaimids in the study (Ryan *et al.*, 2022) raises questions about the long-term impacts of the SSA and PLA RDF products on terrestrial nematode communities and therefore requires further investigation.

Nematode community diversity changed significantly under the various RENURE treatments (Saju *et al.*, 2023). After the application of AS, the nematode communities were more structured and less disturbed when compared to the communities in the synthetic NPK treatment.

8.3 Microelement Effects on Nematode Communities

Due to the increasing pollution from various sources, the effects of heavy metals, including microelements, on nematode communities have been attracting considerable attention since the 1980s. An overview of the relevant experiments was given in the previous edition of this book (Nagy, 2009). Here we briefly summarize the statements of that chapter illustrating how nematode communities react to this type of stress, based on a comprehensive experimental study, and we complete this with novel findings from a meta-analysis to show how nematological parameters can reveal harmful or even beneficial microelement effects.

An experimental site at Nagyhörcsök in mid-Hungary hosted a complex, long-term experiment focusing on the ways how 13 microelements affected various crop- and farm-animal-related parameters. Nematological parameters applied included: nematode density, taxon richness, colonizer–persister (*c-p*) groups and feeding groups distribution, MI, SI and diversity ordering.

The artificial pollution was applied in 1991. Nominal total values of each studied pollutant were set to 0 or 10, 30, 90 and 270 mg kg^{-1} at the beginning of the experiment. Nematode samples were taken between years 5 and 11. First, a screening of 13 microelements was performed (Nagy, 1999), then a dose–response experiment (Nagy *et al.*, 2004) and a study of long-term (5–10 years) effects were carried out (Bakonyi *et al.*, 2003). The elements involved in the study were (in alphabetic order of chemical symbols): aluminium (Al), arsenic (As), barium (Ba), cadmium (Cd), chromium (Cr), copper (Cu), mercury (Hg), molybdenum (Mo), nickel (Ni), lead (Pb), selenium (Se), strontium (Sr) and zinc (Zn). Methods applied and the most relevant results have also been summarized by Nagy (2009).

Application of the following elements (in the 'available' concentration values, as estimated by NH_4 acetate–EDTA extraction according to Lakanen and Erviö (1971), shown in brackets): Al (66.4 mg kg^{-1}), As (41.8 mg kg^{-1}), Ba (56.2 mg kg^{-1}), Hg (22.8 mg kg^{-1}), Mo (8.1 mg kg^{-1}), and Sr (132.5 mg kg^{-1}) did not significantly alter nematode abundance, taxon richness and feeding group distribution. MI showed a significant increase in Ba-, Sr- and Mo-treated plots. Therefore, the concentrations applied in this experiment were probably too low to cause toxic effects in the given calcareous chernozem soil of high buffering capacity with a pH of 7.4.

On the other hand, some elements caused significant changes in certain nematode community parameters.

Cadmium, in the 'available' concentration of 190 mg kg^{-1}, caused no detectable change in the studied community parameters, except significantly decreasing MI (Nagy, 1999). Concentration–response studies (Nagy *et al.*, 2004) showed most nematode community parameters to remain unaffected within the concentration range of 26–190 mg kg^{-1}. Only the proportions of fungal feeders and the most sensitive predatory and omnivorous taxa increased, while bacterial feeders and plant feeders decreased significantly.

In the same experiment, however, high Cd doses decreased biomass of various crop plants significantly, even 8 years after pollution. The fact that this contamination level hardly caused any measurable nematological effect underlines the tolerance of nematode assemblages to Cd. This is confirmed by Korthals *et al.* (1996) who found nematode communities relatively insensitive to Cd toxicity, although in short-term studies.

Regarding *zinc*, the maximum level of 143 mg kg^{-1} influenced nematode assemblage in several ways. Taxon richness and MI increased significantly. The proportion of fungal feeders decreased, but that of plant feeders, predators, omnivores and *c-p*3-5 nematodes doubled compared to the control plots. Although the measured available concentration was not really low, it remained below the 'B' value of 200 mg kg^{-1} as the soil contamination limit for Zn in Hungary (KvVM-EüM-FVM, 2009). However, the soil involved in this experiment was depleted in Zn. Therefore, these improved nematode community parameter values could be attributed to a microelement completion effect of the experimental treatment. This suggests that nematode assemblages are suitable to indicate beneficial microelement effects as well. Besides, from these results it may be concluded that nematode assemblages are not particularly vulnerable to Zn load under favourable soil conditions.

Chromium was shown to decrease nematode abundance, taxon richness and MI in an 'available' concentration of 1.4 mg kg^{-1} (Nagy, 1999). The proposed limit value for contaminated soils for 'available' Cr measured with the same technique was 3.0 mg kg^{-1} (Kadar, 2001). Furthermore, Cr had a clear negative concentration–response effect on nematode abundance, taxon richness, MI and SI (Nagy *et al.*, 2004). Distributions of *c-p* groups and feeding types were also significantly affected by the increasing Cr loads. The proportion of the most sensitive predatory and omnivorous taxa decreased significantly in these treatments. Diversity ordering (Tóthmérész, 1993) showed the two highest Cr treatments (approx. 0.8 and 1.4 mg kg^{-1}, respectively) to significantly decrease diversity patterns. Moreover, Bakonyi *et al.* (2003) found Cr to suppress nematode coenoses up to 10 years after pollution, even when crop yields did not reflect this type of disturbance any longer. In conclusion, nematode coenoses gave articulated reactions to Cr pollution. In particular, K-strategist dorylaimid nematodes are sensitive to this disturbance. These findings underline the sensitivity of nematode community parameters to Cr pollution. It even seems possible that changes in nematological parameters could provide an early warning signal for Cr contamination in terrestrial systems.

Selenium is a non-metallic microelement of which little is known regarding nematological effects. Nagy (1999) reported drastic negative effects of Se in an 'available' concentration of 36 mg kg^{-1} as total abundance and taxon richness decreased significantly, while it was impossible to calculate MI, due to the very low nematode numbers. It was not determined to what extent this outcome could be attributed to direct or indirect Se effects via the strong suppression of vegetation. However, Se concentration was several times higher than the contamination threshold value of 1 mg kg^{-1} (KvVM-EüM-FVM, 2009). This can explain the measured harmful Se effects. In a study of concentration-dependent effects along a nominal 'total' concentration gradient of 10, 30, 90 and 270 mg kg^{-1} (Nagy *et al.*, 2004), Se caused much more articulated effects on nematode assemblages. The corresponding 'available' Se levels, as measured in the previous year, were 0.5, 1.56, 9.32 and 36 mg kg^{-1}, respectively. Nematode density decreased significantly in the two highest pollution levels. Taxon richness values showed a continuous decline, with significant differences in the two highest pollution levels again. Maturity Index 2-5 (MI2-5) and SI values showed a slight initial increase with increasing Se concentrations that might be a consequence of the low original Se content in soil (I. Kádár, Budapest, 1998, personal communication) but then decreased sharply. Higher concentrations of Se completely eradicated predators and omnivores, resulting in increasing proportions of bacterial and fungal feeders.

In their *meta-analysis*, Chauvin *et al.* (2020) involved 15 nematological parameters and data from 37 studies. They set up four categories of: (i) no contamination, below a threshold level varying for each metal; (ii) low contamination, within the range of approx. 1.5 to 4 times the no contamination value; (iii) high contamination, from the upper limit of the previous category to its 10 times higher value; and (iv) very high contamination, above the upper limit of the high contamination category. Along the series of these contamination categories, most parameter values decreased with the increasing metal loads. Structure footprint, community footprint, trophic group abundance (except for fungal feeders) and taxon richness were identified as showing a particularly strong relationship with soil pollution levels. The SI and Shannon index (H′) decreased significantly with increasing pollution levels but did not show any difference between the high and the very high levels. MI and EI did not give a consistent response, while Nematode Channel Ratio (NCR) was not influenced by the contamination levels. Metabolic footprint patterns were influenced by pollution levels: total community footprint, structure footprint and plant-feeder footprint gave consistent reactions while enrichment footprint did not. Comparing high contamination effects to control values, structure footprint, total abundance and that of omnivores and predators, SI and NCR showed significant decrease of 20–30% on average, but confidence intervals were usually quite high. Even if this fact indicates a high variance among the various studies, such a high number of the compared data sets gives a robust indication on the usefulness of some of the studied nematological parameters.

8.4 Invasions of Alien Plants

The plant–soil feedback is probably an important factor determining the invasion success of alien plants (Kuyper and De Goede, 2013). Alien plants are assumed to experience more positive feedbacks from soil biota than their native antagonists, possibly due to mutualistic symbionts and/or lower root pathogen pressure (Callaway *et al.*, 2004; Wolfe and Klironomos, 2005; Reinhart and Callaway, 2006; van der Putten *et al.*, 2013, 2016). All case studies presented here were based on comparisons between invaded and not invaded sites or between rhizospheres of invasive versus native plants. Almost all researchers studied the effect of plant invasions on the trophic structure of nematode communities, in order to simulate the alterations of the soil food web. Some of them explored also the effects of plant invasions on the microbial communities. Most studies used nematode indices, while diversity was mostly measured by the number of nematode genera or species, the Shannon index, and sometimes diversity profiles.

8.4.1 Time since initial invasion

While alien plants are more likely to become invasive over time (Richardson and Pyšek, 2006), little is known about how time after invasion affects soil nematode communities.

(1) Belnap *et al.* (2005) studied *Bromus tectorum* invasion in grasslands of the Canyonlands National Park, comparing not invaded, recently invaded (6–7 years) and historically invaded (50+ years) sites. They found that alterations in soil communities, i.e. decrease of bacterial feeders, fungal/root feeders and of total nematode abundance, occurred mainly during the first years after invasion. Regarding the number of genera, within a few years, three of the 14 nematode genera were lost, while after 50 years, only three additional genera were lost.

(2) Karmezi *et al.* (2022) compared nematode communities in the rhizospheres of *Solanum elaeagnifolium*, one of the most prominent invasive weeds in Greece, and of its native antagonist *Cichorium intybus*, in sites where these two ruderals coexisted. These sites were along rural roadsides at margins of cultivated fields, sharing intense human activity and similar abiotic conditions (climate, seasonality, main geological substrate), but differed regarding the time after the initial invasion of *S. elaeagnifolium*, and were regarded as recently and historically invaded (10 years and 70 years after invasion, respectively). Diversity was lower in *S. elaeagnifolium* and reduced even more over time in both plant species. The nematode indices and the abundance of the bacterial feeders trophic group did not differ between the two focal plant species or between recently and historically invaded sites. Surprisingly, the alien plant was more susceptible to plant-feeder nematodes compared to its native counterpart during early invasion (10 years), but this situation reversed after its long naturalization (70 years).

8.4.2 Plant invasions in different habitats

The plant invasion effects depend not only on the characteristics of the alien plant but also on those of the invaded sites.

(3) Belnap and Phillips (2001) found that *B. tectorum* invasion in two different grassland associations led to opposite responses of soil bacteria, fungi and nematodes; for example, if any of these soil biota increased in one grassland association, it generally decreased in the other. In both systems plant feeders were almost absent, while the contrasting effects of invasion resulted from different responses of bacterial and fungal feeders to changes of their food resources.

(4) Alterations of nematode communities after *Heracleum sosnowskyi* invasion were studied in the frame of a project that covered different invaded habitats in three countries; Renčo and Baležentienė (2015) studied three habitats of Lithuania (an abandoned land, a grassland slope on a roadside and a forest edge), Renčo *et al.* (2019) studied four rural habitats of Poland with different native plant communities, while Čerevková *et al.* (2020) studied a mixed forest and a semi-natural grassland in Russia. Renčo and Baležentienė (2015) found that *H. sosnowskyi* dominance was associated with low abundances and faunal similarity of soil nematode communities of the different habitats, although several species or even trophic groups such as bacterial feeders, epidermal-cell and root-hair feeders and omnivores were not sensitive to invasion. In the study of Renčo *et al.* (2019), nematode communities responded negatively to *H. sosnowskyi* invasion, but only where it formed a monoculture. The plant's allelochemicals resulted in fewer plant feeders, fungal feeders and omnivores in those sites. A reduced abundance of plant feeders due to invasion was also recorded by Čerevková *et al.* (2020), although some plant-feeder genera responded positively to invasion, especially in the forests. The effect of *H. sosnowskyi* on microbial properties depended on habitat type and soil initial nutrient status, and this was reflected in the varied responses of bacterial feeders.

(5) Čerevková *et al.* (2019a) compared nematode communities in areas invaded and uninvaded by *Solidago gigantea* in a semi-natural grassland and a broadleaved mixed forest. Plant feeders were significantly affected. Their abundance was higher at invaded than uninvaded sites and in grassland than forest. This positive response to invasion was attributed to the well-developed root system of *S. gigantea* serving as food source and potential reservoir of plant parasites. The responses of the other tropic groups and of nematode indices were limited.

(6) Čerevková *et al.* (2019b) explored the effects of *Fallopia japonica* invasion on nematode community and soil physical and microbial properties in three natural habitats of Slovakia (forest, grassland, wetland). The impact of invasion on microbial activity varied across habitats, possibly masking differences between invaded and uninvaded plots. Despite variable responses of some nematode species, invasion had an overall negative effect on nematode communities, reducing plant feeders, fungal feeders and omnivores.

(7) Karmezi *et al.* (2023) explored the impact of *S. elaeagnifolium* invasion on soil microbial and nematode communities of Mediterranean pines and maquis. Most studied variables were affected by habitat type, while the invasion effect was different at each habitat. Compared to maquis, the soil in pines had higher silt and lower sand content, higher water content and higher organic content, supporting a much larger microbial biomass and abundance of bacterial-feeding nematodes. *S. elaeagnifolium* invasion in pines had a negative effect on soil organic content and microbial biomass, and consequently on most bacterial- and fungal-feeder nematode genera. Plant feeders were not affected. In contrast, in maquis, the soil organic content and microbial biomass responded positively to invasion, raising the few genera of enrichment opportunists and the EI. Most bacterial feeders were not affected, while plant feeders, mostly *Paratylenchus*, increased. The plants colonizing the peripheral areas in maquis probably offered a qualitative food source to microbes and root feeders, which, in pines, was not sufficient to affect the much larger microbial biomass.

8.5 Climate Change

The most important aspects of climate change that affect soil systems are temperature increase, elevated atmospheric CO_2 and altered precipitation pattern (Barreto and Lindo, 2022). Zhou *et al.* (2022a) summarized the results of 86 studies on nematodes, illustrating a considerable body of research. In particular, studies on the use of ecological indices of nematodes are discussed below.

8.5.1 Warming

Temperature can directly affect nematode life cycle and activity (Sohlenius and Boström, 1999a; Bakonyi *et al.*, 2007; Lindo, 2015), predominantly at northern latitudes. Here, nematodes play an important role in the food web, due to the lack in macrofauna in tundra soils (Procter, 1990; Boag and Yeates, 1998; Sørensen *et al.*, 2006).

An interesting approach is transplantation experiments as performed by Sohlenius and Boström (1999a,b). Undisturbed soil blocks from a tundra mire were transferred to nine sites in warmer regions in Sweden. While the nematode population density increased across sites, the MI decreased, yet only at clear-cut sites with fluctuation in temperature and moisture (Sohlenius and Boström, 1999a). Other studies used open-top chambers (OTCs) to increase temperature. Eight years of simulated climate change (+1.5°C mean summer soil temperature) at a tundra heath and a high-altitude fellfield doubled nematode population density, while again the MI declined (Ruess *et al.*, 1999a). Interestingly, at species level, warming impact was even more pronounced, pointing to direct temperature effects (Ruess *et al.*, 1999b). OTCs (+1.4°C ground surface temperature) were also used by Stark *et al.* (2023) in a long-term reindeer grazing experiment. Warming increased the abundance of fungal feeders and decreased the MI irrespective of grazing, while bacterial feeders were positively affected under heavy grazing only. This suggests that large grazers alter the strength of decomposition pathways with climate warming.

The impact of temperature investigated across grassland ecosystems revealed contrasting results. In alpine meadows, Liu *et al.* (2022) installed OTCs over seven years (+0.5, 1.2, 2.1 and 2.2°C) with the greatest temperature elevation resulting in an increase of bacterial feeders but a decrease of plant feeders, and an associated lower MI and Plant-Parasitic Index (PPI). In contrast, in steppe habitats, Li *et al.* (2013) showed a minor impact of warming, with negative trends only in fungal feeders, while Ma *et al.* (2018) report a decline in fungal and plant feeders as well as MI, PPI and SI. These warming-induced changes were mainly attributed to increased drought and associated adverse effects on vegetation production and rates of nutrient turnover. Comparably, in temperate forests (e.g. Thakur *et al.*, 2014) and crop lands (e.g. Dong *et al.*, 2013), warming solely had little effect on nematodes except when accompanied by drought. A good overview on the changes in ecological indices is given by the review of Zhou *et al.* (2022b).

8.5.2 Elevated CO_2

Atmospheric CO_2 enrichment acts predominantly indirectly via changes in vegetation, litter quality or root biomass, altering the C input below ground (Peng *et al.*, 2022). Therefore, its impact on the nematode community is mainly related to plant type and growth response as well as season (Zhou *et al.*, 2022a).

In arable soil the NCR indicated a shift towards fungi with winter wheat, yet towards bacteria with sugarbeet (Sticht *et al.*, 2009), while in forest soil the MI increased under loblolly pine but decreased under sweet gum (Neher and Weicht, 2013). The close link between nematode and plant responses to elevated CO_2 is also reflected at the level of families, genera or even species. Higher abundance of Trichodoridae and lower of Rhabditidae were reported from grassland (Yeates *et al.* (1997) and forest soil (Hoeksema *et al.*, 2000) with increased atmospheric CO_2 concentration. Other grassland studies found Anguinidae positively and Hoplolaimidae negatively affected (Ayres *et al.*, 2008). This suggests that specific combinations of soil, plant and environmental conditions at each site additionally control nematode responses.

Overall, elevated CO_2 had positive effects on nematode densities, predominantly on bacterial, fungal and plant feeders at the food web base (Zhou *et al.*, 2022a). The related bottom-up forces can affect soil ecosystem multifunctionality at higher trophic levels such as omnivores and predators (Wang *et al.*, 2023).

8.5.3 Precipitation pattern

Changes in precipitation pattern alter the duration and severity of soil water stress. As nematodes live in water-filled pore spaces and water films around soil particles, they are particularly susceptible to this climate change factor (Nielsen *et al.*, 2014).

Generally, increased precipitation improves soil conditions, resulting in higher connectivity and longer food chains in the nematode micro-food web (Zhang *et al.*, 2020). For example, Homet *et al.* (2023) reduced rainfall (−30%) in Mediterranean water-limited forests and reported consistent short- and long-term effects, i.e. a higher MI and SI, and a lower predator-to-prey ratio. Ankrom *et al.* (2022) altered precipitation in arid, semi-arid and mesic grasslands, and observed nematode community responses uniquely at each site. They identified 39 of 72 genera as characteristic of a particular precipitation regime, which were assigned an indicator value. For example, *Anaplectus* and *Panagrellus* were good indicators for precipitation reduction at the arid and semi-arid grassland, respectively, while *Mesocriconema* had the highest indicator value for increased precipitation at the mesic site.

It is noteworthy that root feeders respond differently to changes in precipitation. Ankrom *et al.* (2020) reported a low response of ecto-parasites to increased precipitation. In contrast, endoparasites increased in semi-arid and decreased in mesic grassland, which was reflected by the pattern of the herbivory index. In an arable soil, density of root feeders declined with reduced frequency, but increased with altered amount, of precipitation (Guo *et al.*, 2021). However, in the PPI, comprising all root-feeding functional groups, these effects were masked. These different reactions of plant feeders suggest various effects of altered precipitation on plant productivity.

8.5.4 Interactions of global change factors

As with other soil organisms (Rillig *et al.*, 2019), most research on nematodes has focused on the effects of individual drivers of global change, and combined factors have rarely been addressed. However, studies on multiple factors are important to understand the mechanisms involved (Peng *et al.*, 2022).

Bakonyi and Nagy (2000) compared warm dry, warm wet, cold dry and cold wet scenarios in a semi-arid grassland, and temperature was found to be the main driver of changes in nematode diversity, MI and PPI. Interactions of both factors occurred: in warm wet plots the MI was notably higher compared to warm dry plots. In contrast, Guo *et al.* (2021) report that warming and precipitation independently altered nematode communities and food web functioning. Elevated CO_2 and warming together fostered plant feeders (e.g. PPI) as well as taxa with intermediate to high *c-p* values (Mueller *et al.*, 2016; Wang *et al.*, 2019). Combining elevated CO_2 with reduced summer precipitation revealed a buffering effect of CO_2 on the impacts of low soil moisture on EI and CI (Cesarz *et al.*, 2015). That study also showed that functional guilds outperform trophic groups as indicators, highlighting the benefits of a more detailed nematode community analysis.

Manipulative studies on the three-way interaction of temperature, CO_2 and precipitation remain rare (Zhou *et al.*, 2022a). Thakur *et al.* (2019) reported that elevated CO_2 and warming along with reduced precipitation shifted the decomposition pathway in favour to bacteria, as assigned by a decrease in the CI. Mueller *et al.* (2016) showed that the effects of combined climate change variables on nematode functional composition were typically less than additive. Overall, to better predict the impacts of climate change on nematode communities more factorial studies are needed and responses should be measured for functional groups defined by trophic level and life history as well as nematode ecological indices.

8.6 Nature Protection

The importance of nematodes for nature protection goes well beyond the conservation of individual species, and they are noticeably absent from the IUCN Red List of Threatened Species (IUCN, 2023). Rather, nematodes' multiple roles in ecosystem functioning motivate efforts for conservation of their communities. The unavoidable need to include the contribution of soil biodiversity and functions to ecosystem services in conservation policies has recently been recognized by the scientific community, but severely lacks research (see Cameron *et al.*, 2018; Eisenhauer *et al.*, 2019). Soil nematode community analyses can provide relevant indicators on soil biodiversity and functions and have been proposed as descriptors for the assessment of soil ecological status in soil monitoring schemes.

8.6.1 Ecosystem restoration after intensive land use

Soil nematode communities have been used to assess restoration in sites subject to intensive agriculture in the Netherlands for many years. Nematode community development was assessed in systems where topsoil, and consequently a large part of the original soil community, was removed. Nematode recolonization was followed for five years, which allowed the appreciation of the transient abundance patterns of nematode trophic guilds. The first colonizers and dominant nematodes in the area were, expectedly, opportunists. In the absence of vegetation, plant-feeding nematodes remained in low levels, whereas the soil food web supported omnivores and predators only from the third year after soil removal. The gradual increase in the MI from as low as 1.1 up to 2.2 throughout the five years was considered indicative of ecosystem recovery (De Goede and Van Dijk, 1998). Several attempts at efficiently restoring grasslands or heathlands where agricultural topsoil was removed involved application of hay (Kardol *et al.*, 2009), biochar (Jeffery *et al.*, 2022), fresh herbage (van der Bij *et al.*, 2018), and the transference of soil (Kardol *et al.*, 2009; van der Bij *et al.*, 2018) from nearby mature, species-rich communities. The development of the soil nematode community depended strongly on resource availability and recalcitrance of the added materials. If the community in transferred soil could not be established in grasslands due to the low organic matter and erratic water availability (Kardol *et al.*, 2009), soil transfer resulted in increased abundance of all nematode trophic groups in heathland, with a corresponding 50% plant cover increase, which was considered a restoration success (van der Bij *et al.*, 2018).

Restoration of mining sites can be a difficult task not just because of topsoil removal, but also because of the inhospitable abiotic environment. In a bauxite residue disposal site in Ireland, Courtney *et al.* (2011) compared soil nematode community responses in a range of restoration practices. The authors found more taxa-diverse and mature communities in the site that had been revegetated the longest time, which suggests a long temporal scale may be required to assess restoration effects in highly disturbed systems. In China, restoration of open-pit phosphorus-mining sites was investigated in revegetation chronosequences examining tree stands up to 20 years of age (Gu *et al.*, 2019). Ten years after revegetation, nematode abundance and diversity differed between planted tree species, but differences could not be detected after 20 years, although the MI had increased significantly. Some years later, Gao *et al.* (2022) used high-throughput Illumina sequencing to unveil nematode community structure and function changes induced by restoration of an open-pit quarry and an open-pit iron mine in the desert steppe of Mongolia. Obtained operational taxonomic unit (OTU) frequencies were converted to species-level abundances so indices could be calculated on the nematode community. Those authors further tried to identify key soil nematode genera most indicative for restoration processes in the soil. These were *Cephalobus*, *Dorylaimus*, *Oxydirus*, *Paratylenchus*, *Plectus*, *Scutylenchus*, *Trichotylenchus* and *Tylencholaimus* in the open-pit quarry, and *Acrobeles*, *Cephalobus*, *Ecumenicus*, *Microdorylaimus* and *Tripylina* in the open-pit iron-mining site. Although these differed among treatments and may be context-dependent, this was an interesting attempt on the use of sentinel genera as biological monitors of restoration.

In a National Park in India, marshy areas are being replaced by terrestrial vegetation, an opposite process to restoration, where ecologically complex natural sites are reclaimed to other land uses by human action. Nematode abundance, structure and enrichment metabolic footprints were overall reduced by reclamation. Among land uses, nematode diversity was smaller in cropland compared to the original marshy land, the former having significantly smaller structure, fungal and predatory footprints, but a larger plant-feeder footprint. This is one of the few case studies where metabolic footprints were calculated and demonstrated the use of nematode community analysis to track alterations in soil ecosystems in protected areas as affected by land reclamation (Nesar *et al.*, 2023).

8.6.2 Assessment of soil biodiversity and function in protected areas

Natural, undisturbed sites in protected areas are assumed to have higher soil ecological quality, but this often lacks scientific support. Ciobanu and Popovici (2015, 2017) published relevant assessments of soil nematode communities in protected areas of Romania from samplings across Natura 2000 sites taken since the 1970s to the year 2000. Some of the most diverse communities were found in one of the Natura 2000 priority areas, that also contained more higher-trophic-group persisters, and the authors suggested nematode community analysis is useful to assess and monitor conservation (Ciobanu and Popovici, 2015) and to discriminate priority from non-priority sites (Ciobanu and Popovici, 2017). However, considering the temporal trend 1975–2000, nematode community composition did not differ between priority and non-priority sites (Ciobanu *et al.*, 2019). Functional diversity did not follow the general trend of increasing nematode taxonomic diversity with time. These findings are important for two reasons: (i) subgenera taxonomic identification is demanding and not yet widely feasible at a high-throughput level, hindering an adequate appreciation of soil-associated biodiversity; and (ii) it suggests that soil community structure and functions do not differ in areas of different conservation interest, with the implication that conservation efforts are not working to improve soil functions.

In a Nature Reserve across the Finnish–Russian border, Sushchuk and Matveeva (2021) found higher nematode diversity, lower dominance of nematode taxa and more complex food webs in the Finnish than in the Russian territories of this Nature Reserve, that were attributed to industrial activity and soil type and understorey vegetation differences among sites. The limited number of replicates (only four sampled areas) and scarce information on local drivers and pressures on the nematode communities restricted further conclusions.

To successfully make inferences on larger, widely applicable trends that can support policy development for conservation, macroecological community studies are needed. Samplings that are not only spatially but temporally explicit allow the combination of nematode community data with a range of remote observations and data sets on, for example, vegetation and climate (Guerra *et al.*, 2020). Duarte *et al.* (2024) conducted the first regional-scale study on the success of nature protection areas for forest soil biodiversity and function. In the targeted region, the North of Portugal, protected areas comprise *c.*25% of the 21,515 km^2 of the area, forming a patchwork of national parks and nature reserves aiming to promote conservation of (aboveground) species and habitats. In a total of 127 forest sites across the region, soils in protected areas had significantly higher abundance of fungal feeders and omnivores, with correspondingly larger metabolic footprints, including the structure footprint. Nematode community composition was further related to edaphoclimatic factors, with the abundance of omnivores and predators being positively related to soil organic carbon content and moisture. This suggests these groups are more sensitive to climate change effects (but see Section 8.5 above).

8.7 Conclusions

This chapter has presented an overview on the application of nematode community-related parameters as a tool to reflect perturbation in terrestrial systems. The examples given cover a broad range of man-made disturbances from agrochemicals, heavy metals, plant invasions and climate change to nature conservation.

Plant-based alternatives to agrochemicals fuel the soil food web, enhance the microbial-feeding nematodes, especially the bacterial feeders, and increase the nematode metabolic footprint. Botanicals with nematicidal activity are primarily targeted against plant feeders, since by providing food for soil microflora, they offer a significant trade-off to microbial-feeding nematodes.

Nematode assemblages and nematode-based indices are useful tools to evaluate the impact of recycling-derived and bio-based fertilizers in the soil environment. Soil nutritional profiles

Table 8.1. Overview of the major human impacts on nematode assemblage in terrestrial habitats. The respective case studies are given in the main text. RENURE = recovered nitrogen from manure, MI = Maturity Index, PPI = Plant-Parasitic Index, SI = Structure Index. (Author's own table.)

Type of disturbance/treatment	Expected environmental impact	Major nematode community response
General biocide (metham sodium), synthetic nematicide (fluopyram)		Severe suppression of all nematode populations. Small, poorly connected and unstable nematode networks
Spearmint, peppermint, rosemary Neem and mixture of basil, marigold and clove oils Anise, parsley, rocket Detoxified olive mill wastewater		Enhancement of microbial-feeding nematodes, especially of bacterial feeders. Increased metabolic footprint. Mild but significant, perhaps dosage-dependent, reduction of plant feeders
Bio-based fertilizer Potato waste struvite, municipal waste struvite, poultry litter ash, sewage sludge ash, pig manure, RENURE materials from pig waste	Changes in nematode community composition and biodiversity	Changes in nematode abundance. Ashes negatively affected dorylaimids. Nematode communities more structured and less disturbed after struvite and RENURE application
Microelement pollution Cadmium, chromium, selenium, zinc	Various consequences, from mostly toxic effects to revealing completion of essential elements	Cd: no consistent response, probably low sensitivity of nematode communities Cr: decrease of various parameters, even at low levels. Long-term effects (after 10 years). Concentration-dependent toxicity Se: drastic toxic effects at high concentrations Zn: increasing parameter values in low concentrations
Invasion of alien plants	Biodiversity and productivity decrease, crop loss, alteration of native plant communities, degradation of habitats and ecosystem functions	Alterations of nematological parameters depend on time since initial invasion, habitat type (native plant community, soil parameters) and the chemical nature of the invasive plant
Warming	Altered life cycle and activity of biota	Increased population density and lower MI (tundra), decrease of plant feeders and lower PPI (grassland)
Elevated CO_2	Indirect impact via changes in vegetation, litter quality or root biomass	Impact linked to plant responses, predominantly acts on root feeders, food web bottom-up forces
Precipitation pattern	Effects on activity of the semi-aquatic soil fauna	Higher connectivity and longer food chains of the nematode micro-food web
Ecosystem restoration	Recolonization and increased food web complexity	Increasing abundance, diversity and MI
Protected sites	Conservation of above-ground ecosystems	Higher diversity, larger abundance of persister groups, higher SI

create distinct conditions, and these are reflected in nematode feeding groups as well as key taxa and nematode-based indices. As such, environmental nematology can play an important role in soil fertility management programmes.

For microelement pollutions and additions to soil, some nematode community parameters are particularly usable to indicate toxic conditions, while others proved suitable to point out favourable effects, especially in case of a zinc treatment. Of course, the characteristics and forms of the chemical element itself, and soil conditions, largely influence the effects on nematode community structure, yet index values are definitely worthy to consider for biological indication when such issues emerge.

Regarding invasions of alien plants, results from different case studies often contradict to each other, as alterations of nematological parameters depend on time since initial plant invasion, habitat type (native plant community, soil parameters) and the chemical nature of the invasive plant.

The response of the nematode fauna to global change factors is rapid and offers good indicator potential to assign impacts on food web structure and function as well as on soil conditions and health. Generally, nematode ecological indices were more affected in the long term than the short term, and responses varied due to ecosystem and ambient climate conditions. The current state of research highlights the importance of functional guild- or even genera-level studies to address nematode community responses, as well as the need for more multifactor experiments for a comprehensive ecological interpretation of global change factors.

Until now, in nature conservation, soil nematode community analyses remained largely underexploited. Relevant indicators on soil biodiversity and functions have been proposed as descriptors for the assessment of soil ecological status in soil monitoring schemes. However, nematode-related parameters should also be taken into account when conservation steps are considered on a broader scale. Considering the important role of nematodes as regulators of soil nutrient turnover, they act both as a driver and an indicator of the related processes.

An overview of nematode community responses in the light of various disturbances or treatments is given in Table 8.1.

References

Ainalidou, A., Bouzoukla, F., Menkissoglu-Spiroudi, U., Vokou, D. and Karamanoli, K. (2021) Impacts of decaying aromatic plants on the soil microbial community and on tomato seedling growth and metabolism: suppression or stimulation? *Plants* 10, 1848. DOI: 10.3390/plants10091848.

Ankrom, K., Franco, A.L.C., Fonte, S.J., Gherardi, L.A., de Tomasel, C.M. *et al.* (2020) Ecto- and endoparasitic nematodes respond differently across sites to changes in precipitation. *Oecologia* 193, 761–771. DOI: 10.1007/s00442-020-04708-7.

Ankrom, K., Franco, A.L.C., Fonte, S.J., Gherardi, L.A., de Tomasel, C.M. *et al.* (2022) Ecological maturity and stability of nematode communities in response to precipitation manipulations in grasslands. *Applied Soil Ecology* 170, 104263. DOI: 10.1016/j.apsoil.2021.104263.

Argyropoulou, M.D., Karmezi, M., Tsiafouli, M., Chalkos, D., Bountla, A. *et al.* (2022) Soil amendments with spearmint, peppermint and rosemary enhance the community of free-living nematodes and improve soil quality, while having strikingly different effects on plant growth. *Life* 12, 1121. DOI: 10.3390/life12081121.

Ayres, E., Wall, D.H., Simmons, B.L., Field, C.B., Milchunas, D.G. *et al.* (2008) Belowground nematode herbivores are resistant to elevated atmospheric CO_2 concentrations in grassland ecosystems. *Soil Biology and Biochemistry* 40, 978–985. DOI: 10.1016/j.soilbio.2007.11.018.

Bakonyi, G. and Nagy, P. (2000) Temperature- and moisture induced changes in the structure of the nematode fauna of a semiarid grassland – patterns and mechanisms. *Global Change Biology* 6, 697–707. DOI: 10.1046/j.1365-2486.2000.00354.x.

Bakonyi, G., Nagy, P. and Kádár, I. (2003) Long-term effects of heavy metals and microelements on nematode assemblage. *Toxicology Letters* 140–141, 391–401. DOI: 10.1016/S0378-4274(03)00035-3.

Bakonyi, G., Nagy, P., Kovács-Láng, E., Kovács, E., Barabás, S. *et al.* (2007) Soil nematode community structure as affected by temperature and moisture in a temperate semiarid shrubland. *Applied Soil Ecology* 37, 31–40. DOI: 10.1016/j.apsoil.2007.03.008.

Barreto, C. and Lindo, Z. (2022) Response of soil biodiversity to global change. *Pedobiologia* 90, 150792. DOI: 10.1016/j.pedobi.2022.150792.

Belnap, J. and Phillips, S.L. (2001) Soil biota in an ungrazed grassland: response to annual grass (*Bromus tectorum*) invasion. *Ecological Applications* 11, 1261–1275. DOI: 10.1890/1051-0761(2001)011[1261:SBIAUG]2.0.CO;2.

Belnap, J., Phillips, S.L., Sherrod, S.K. and Moldenke, A. (2005) Soil biota can change after exotic plant invasion: does this affect ecosystem processes? *Ecology* 86, 3007–3017. DOI: 10.1890/05-0333.

Boag, B. and Yeates, G.W. (1998) Soil nematode biodiversity in terrestrial ecosystems. *Biodiversity and Conservation* 7, 617–630. DOI: 10.1023/A:1008852301349.

Bongers, T. (1990) The maturity index: an ecological measure of environmental disturbance based on nematode species composition. *Oecologia* 83, 14–19. DOI: 10.1007/BF00324627.

Boutsis, G., Stamou, G.P. and Argyropoulou, M.D. (2011) Short term effects of soil disinfection with metham sodium and organic alternatives on nematode communities. *Community Ecology* 12, 161–170. DOI: 10.1556/ComEc.12.2011.2.3.

Bulluck, L.R., Barker, K.R. and Ristaino, J.B. (2002) Influences of organic and synthetic soil fertility amendments on nematode trophic groups and community dynamics under tomatoes. *Applied Soil Ecology* 21, 233–250. DOI: 10.1016/S0929-1393(02)00089-6.

Callaway, R., Thelen, G., Rodriguez, A. and Holben, W.E. (2004) Soil biota and exotic plant invasion. *Nature* 427, 731–733. DOI: 10.1038/nature02322.

Cameron, E.K., Martins, I.S., Lavelle, P., Mathieu, J., Tedersoo, L. *et al.* (2018) Global gaps in soil biodiversity data. *Nature Ecology and Evolution* 2, 1042–1043. DOI: 10.1038/s41559-018-0573-8.

Čerevková, A., Miklisová, D., Bobuľská, L. and Renčo, M. (2019a) Impact of the invasive plant *Solidago gigantea* on soil nematodes in a semi-natural grassland and a temperate broadleaved mixed forest. *Journal of Helminthology* 94, e51. DOI: 10.1017/S0022149X19000324.

Čerevková, A., Bobuľská, L., Miklisová, D. and Renčo, M. (2019b) A case study of soil food web components affected by *Fallopia japonica* (Polygonaceae) in three natural habitats in Central Europe. *Journal of Nematology* 51, 1–16. DOI: 10.21307/jofnem-2019-042.

Čerevková, A., Ivashchenko, K., Miklisová, D., Ananyeva, N. and Renčo, M. (2020) Influence of invasion by Sosnowsky's hogweed on nematode communities and microbial activity in forest and grassland ecosystems. *Global Ecology and Conservation* 21, e00851. DOI: 10.1016/j.gecco.2019.e00851.

Cesarz, S., Reich, P.B., Scheu, S., Ruess, L., Schaefer, M. *et al.* (2015) Nematode functional guilds, not trophic groups, reflect shifts in soil food webs and processes in response to interacting global change factors. *Pedobiologia* 58, 23–32. DOI: 10.1016/j.pedobi.2015.01.001.

Chauvin, C., Trambolho, M., Hedde, M., Makowski, D., Cérémonie, H. *et al.* (2020) Soil nematodes as indicators of heavy metal pollution: a meta-analysis. *Open Journal of Soil Science* 10, 579–601. DOI: 10.4236/ojss.2020.1012028.

Ciobanu, M. and Popovici, I. (2015) Soil nematode communities in three Natura 2000 sites of the Trascău Mountains (Romania). *Annals of Forest Research* 58, 311–322. DOI: 10.15287/afr.2015.363.

Ciobanu, M. and Popovici, I. (2017) Structural and functional diversity of nematode fauna associated with habitats located in the Natura 2000 site Apuseni (Romania). *Russian Journal of Nematology* 25, 23–36. DOI: 10.24411/0869-6918-2017-10010.

Ciobanu, M., Eisenhauer, N., Stoica, I.A. and Cesarz, S. (2019) Natura 2000 priority and non-priority habitats do not differ in soil nematode diversity. *Applied Soil Ecology* 135, 166–173. DOI: 10.1016/j.apsoil.2018.12.009.

Cooper, J., Lombardi, R., Boardman, D. and Carliell-Marquet, C. (2011) The future distribution and production of global phosphate rock reserves. *Resources Conservation and Recycling* 57, 78–86. DOI: 10.1016/j.resconrec.2011.09.009.

Courtney, R., Keith, A.M. and Harrington, T. (2011) Nematode assemblages in bauxite residue with different restoration histories. *Restoration Ecology* 19, 758–764. DOI: 10.1111/j.1526-100X.2010.00734.x.

De Goede, R.G.M. and Van Dijk, T.S. (1998) Establishment of carabid beetle and nematode populations in a nature restoration project after the abandonment of arable land. *Applied Soil Ecology* 9, 355–360. DOI: 10.1016/S0929-1393(98)00089-4.

Dimou, M.D., Monokrousos, N., Katapodis, P., Diamantopoulou, P.A., Argyropoulou, M.D. *et al.* (2023) Use of microbially treated olive mill wastewaters as soil organic amendments; their short-term effects on the soil nematode community. *Diversity* 15, 497. DOI: 10.3390/d15040497.

Dong, Z., Hou, R., Chen, Q., Ouyang, Z. and Ge, F. (2013) Response of soil nematodes to elevated temperature in conventional and no-tillage cropland systems. *Plant and Soil* 373, 907–918. DOI: 10.1007/s11104-013-1846-2.

Duarte, A.C., Guerra, C.A., Cano-Diaz, C., Zeiss, R., Carvalho-Santos, C. *et al.* (2024) Effects of protected areas on soil nematode communities in forests of the North of Portugal. *Biodiversity and Conservation* 33, 73–89. DOI: 10.1007/s10531-023-02732-6.

Eisenhauer, N., Bonn, A. and A. Guerra, C. (2019) Recognizing the quiet extinction of invertebrates. *Nature Communications* 10, 50. DOI: 10.1038/s41467-018-07916-1.

Gao, Z., Liu, Z., Wu, Z., Huang, J., Zhang, L. *et al.* (2022) Soil nematode communities and functions: ecological package restoration responses in damaged desert steppe open-pit mining ecosystems. *Land Degradation and Development* 33, 1015–1028. DOI: 10.1002/ldr.4208.

Gu, L.P., Kong, J.J., Chen, K. and Guo, Y.Q. (2019) Monitoring soil biological properties during the restoration of a phosphate mine under different tree species and plantation types. *Ecotoxicology and Environmental Safety* 180, 130–138. DOI: 10.1016/j.ecoenv.2019.05.001.

Guerra, C.A., Heintz-Buschart, A., Sikorski, J., Chatzinotas, A., Guerrero-Ramírez, N. *et al.* (2020) Blind spots in global soil biodiversity and ecosystem function research. *Nature Communications* 11, 3870. DOI: 10.1038/s41467-020-17688-2.

Guo, X., Endler, A., Poll, C., Marhan, S. and Ruess, L. (2021) Independent effects of warming and altered precipitation pattern on nematode community structure in an arable field. *Agriculture, Ecosystems and Environment* 316, 107467. DOI: 10.1016/j.agee.2021.107467.

Hoeksema, J.D., Lussenhop, J. and Teeri, J.A. (2000) Soil nematodes indicate food web response of elevated atmospheric CO_2. *Pedobiologia* 44, 725–735. DOI: 10.1078/S0031-4056(04)70085-2.

Homet, P., Ourcival, J.-M., Gutiérrez, E., Domínguez-Begines, J., Matías, L. *et al.* (2023) Short- and long-term responses of nematode communities to predicted rainfall reduction in Mediterranean forests. *Soil Biology and Biochemistry* 179, 108974. DOI: 10.1016/j.soilbio.2023.108974.

Hu, C. and Qi, Y. (2010) Effect of compost and chemical fertilizer on soil nematode community in a Chinese maize field. *European Journal of Soil Biology* 46, 230–236. DOI: 10.1016/j.ejsobi.2010.04.002.

IUCN (2023) The IUCN Red List of Threatened Species. Available at: www.iucnredlist.org (accessed 14 April 2023).

Jeffery, S., van de Voorde, T.F.J., Harris, W.E., Mommer, L., Van Groenigen, J.W. *et al.* (2022) Biochar application differentially affects soil micro-, meso-macro-fauna and plant productivity within a nature restoration grassland. *Soil Biology and Biochemistry* 174, 108789. DOI: 10.1016/j.soilbio.2022.108789.

Kadar, I. (2001) *Considerations of Soil Properties in Application of Soil Contamination Limit Values II. Mineral Elements* (in Hungarian). Report for Ministry of Environmental Protection, Budapest, pp. 1–30.

Kardol, P., Bezemer, T.M. and van der Putten, W.H. (2009) Soil organism and plant introductions in restoration of species-rich grassland communities. *Restoration Ecology* 17, 258–269. DOI: 10.1111/j.1526-100X.2007.00351.x.

Karmezi, M., Krigas, N. and Argyropoulou, M.D. (2022) The invasion and long naturalization of *Solanum elaeagnifolium* affects the soil nematode community: evidence from a comparative study. *Agronomy* 12, 2346. DOI: 10.3390/agronomy12102346.

Karmezi, M., Krigas, N., Papatheodorou, E.M. and Argyropoulou, M.D. (2023) The invasion of alien populations of *Solanum elaeagnifolium* in two Mediterranean habitats modifies the soil communities in different ways. *Plants* 12, 2193. DOI: 10.3390/plants12112193.

Karpinska, A., Ryan, D., Germaine, K., Dowling, D., Forrestal, P. *et al.* (2021) Soil microbial and nematode community response to the field application of recycled bio-based fertiliser in Irish grassland. *Sustainability* 13, 12342. DOI: 10.3390/su132212342.

Korthals, G.W., van der Ende, A., van Megen, H., Lexmond, T.M., Kammenga, J.E. *et al.* (1996) Short-term effects of cadmium, copper, nickel and zinc on soil nematodes from different feeding and life-history strategy groups. *Applied Soil Ecology* 4, 107–117. DOI: 10.1016/0929-1393(96)00113-8.

Kuyper, T.W. and De Goede, R.G.M. (2013) Interactions between higher plants and soil-dwelling organisms. In: van der Maarel, E. and Franklin, J. (eds) *Vegetation Ecology*. Wiley, Chichester, UK, pp. 260–284. DOI: 10.1002/9781118452592.ch9.

KvVM-EüM-FVM (2009) (IV. 14) KvVM-EüM-FVM joint ministerial decree on the threshold values for the protection of geological medium and groundwater bodies from contamination and on the measurements of contaminations (in Hungarian). Available at: https://net.jogtar.hu/jogszabaly?docid=a0900006.kvv (accessed 3 September 2025).

Lakanen, E. and Erviö, R. (1971) A comparison of eight extractants for the determination of plant available micronutrients in soils. *Acta Agronomica Fennica* 123, 223–232.

Le Corre, K.S., Valsami-Jones, E., Hobbs, P. and Parsons, S.A. (2009) Phosphorus recovery from wastewater by struvite crystallization: a review. *Critical Reviews in Environmental Science and Technology* 39(6), 433–477. DOI: 10.1080/10643380701640573.

Li, Q., Bai, H., Liang, W., Xia, J., Wan, S. *et al.* (2013) Nitrogen addition and warming independently influence the belowground micro-food web in a temperate steppe. *PLoS ONE* 8, e60441. DOI: 10.1371/journal.pone.0060441.

Lindo, Z. (2015) Warming favours small-bodied organisms through enhanced reproduction and compositional shifts in belowground systems. *Soil Biology and Biochemistry* 91, 271–278. DOI: j.soilbio.2015.09.003.

Liu, Y., Wang, W., Liu, P., Zhou, H., Chen, Z. *et al.* (2022) Plant–soil mediated effects of long-term warming on soil nematodes of alpine meadows on the Qinghai–Tibetan plateau. *Biology* 11, 1596. DOI: 10.3390/biology11111596.

Lorick, D., Macura, B., Ahlström, M., Grimvall, A. and Harder, R. (2020) Effectiveness of struvite precipitation and ammonia stripping for recovery of phosphorus and nitrogen from anaerobic digestate: a systematic review. *Environmental Evidence* 9, 27. DOI: 10.1186/s13750-020-00211-x.

Ma, Q., Yu, H., Liu, X., Xu, Z., Zhou, G. *et al.* (2018) Climatic warming shifts the soil nematode community in a desert steppe. *Climatic Change* 150, 243–258. DOI: 10.1007/s10584-018-2277-0.

Mehta, C.M., Khunjar, W.O., Nguyen, V., Tait, S. and Batstone, D. (2014) Technologies to recover nutrients from waste streams: a critical review. *Critical Reviews in Environmental Science and Technology* 45, 385–427. DOI: 10.1080/10643389.2013.866621.

Monokrousos, N., Argyropoulou, M.D., Tzani, K., Menkissoglou-Spiroudi, U., Boutsis, G. *et al.* (2021) The effect of botanicals with nematicidal activity on the structural and functional characteristics of the soil nematode community. *Agriculture* 11, 326. DOI: 10.3390/agriculture11040326.

Mueller, K.E., Blumenthal, D.M., Carrillo, Y., Cesarz, S., Ciobanu, M. *et al.* (2016) Elevated CO_2 and warming shift the functional composition of soil nematode communities in a semiarid grassland. *Soil Biology and Biochemistry* 103, 46–51. DOI: 10.1016/j.soilbio.2016.08.005.

Nagy, P. (1999) Effects of an artificial metal pollution on nematode assemblage of a calcareous loamy chernozem soil. *Plant and Soil* 212, 35–47. DOI: 10.1023/A:1004657924496.

Nagy, P. (2009) Case studies using nematode assemblage analysis in terrestrial habitats. In: Wilson, M.J. and Kakouli-Duarte, T. (eds) *Nematodes as Environmental Indicators.* CAB International, Wallingford, UK, pp. 172–187.

Nagy, P., Bakonyi, G., Bongers, T., Kádár, I., Fábián, M. *et al.* (2004) Effects of microelements on soil nematode assemblages seven years after contaminating an agricultural field. *Science of The Total Environment* 320, 131–143. DOI: 10.1016/j.scitotenv.2003.08.006.

Neher, D.A. and Weicht, T.R. (2013) Nematode genera in forest soil respond differentially to elevated CO_2. *Journal of Nematology* 45, 214–222.

Nesar, H., Afzal, S., Imran, Z. and Ahmad, W. (2023) Impact of marshy area reclamation by various vegetations on soil–nematode community structure in Dachigam National Park. *Soil Ecology Letters* 5, 220166. DOI: 10.1007/s42832-022-0166-y.

Nielsen, U.N., Ayres, E., Wall, D.H., Li, G., Bardgett, R.D. *et al.* (2014) Global-scale patterns of assemblage structure of soil nematodes in relation to climate and ecosystem properties. *Global Ecology and Biogeography* 23, 968–978. DOI: 10.1111/geb.12177.

Ntalli, N., Zioga, D., Argyropoulou, M.D., Papatheodorou, E.M., Menkissoglu-Spiroudi, U. *et al.* (2019) Anise, parsley and rocket as nematicidal soil amendments and their impact on non-target soil organisms. *Applied Soil Ecology* 143, 17–25. DOI: 10.1016/j.apsoil.2019.05.024.

Peng, Y., Peñuelas J., Vesterdal, L., Yue, K., Puguero, G. *et al.* (2022) Responses of soil fauna communities to the individual and combined effects of multiple global change factors. *Ecology Letters* 25, 1961–1973. DOI: 10.1111/ele.14068.

Procter, D.L. (1990) Global overview of the functional roles of soil-living nematodes in terrestrial communities and ecosystems. *Journal of Nematology* 22, 1–7.

Reinhart, K.O. and Callaway, R.M. (2006) Soil biota and invasive plants. *New Phytologist* 170, 445–457. DOI: 10.1111/j.1469-8137.2006.01715.x.

Renčo, M. and Baležentiené, L. (2015) An analysis of soil free-living and plant-parasitic nematode communities in three habitats invaded by *Heracleum sosnowskyi* in central Lithuania. *Biological Invasions* 17, 1025–1039. DOI: 10.1007/s10530-014-0773-3.

Renčo, M., Kornobis, F.W., Domaradzki, K., Jakubska-Busse, A., Jurová, J. *et al.* (2019) How does an invasive *Heracleum sosnowskyi* affect soil nematode communities in natural conditions? *Nematology* 21, 71–89. DOI: 10.1163/15685411-00003196.

Richardson, D.M. and Pyšek, P. (2006) Plant invasions: merging the concepts of species invasiveness and community invasibility. *Progress in Physical Geography: Earth and Environment* 30, 409–431. DOI: 10.1191/0309133306pp490pr.

Rillig, M.C., Lehmann, A., Aguilar-Trigueros, C.A., Buchert, S., Wulf, A. *et al.* (2019) The role of multiple global change factors in driving soil functions and microbial diversity. *Science* 366, 886–890. DOI: 10.1126/science.aay2832.

Ruess, L., Michelsen, A., Schmidt, I.K. and Jonasson, S. (1999a) Simulated climate change affecting microorganisms, nematode density and biodiversity in subarctic soils. *Plant and Soil* 212, 63–73. DOI: 10.1023/A:1004567816355.

Ruess, L., Michelsen, A. and Jonasson, S. (1999b) Simulated climate change in subarctic soils: responses in nematode species composition and dominance structure. *Nematology* 1(5), 513–526. DOI: 10.1163/156854199508513.

Ryan, D., Karpinska, A., Forrestal, P.J., Ashekuzzaman, S.M., Kakouli-Duarte, T. *et al.* (2022) The impact of bio-based fertilizer integration into conventional grassland fertilization programmes on soil bacterial, fungal, and nematode communities. *Frontiers in Sustainable Food Systems* 6, 832841, DOI: 10.3389/fsufs.2022.832841.

Saju, A., Van De Sande, T., Ryan, D., Karpinska, A., Sigurnjak, I. *et al.* (2023) Exploring the short-term in-field performance of Recovered Nitrogen from Manure (RENURE) materials to substitute synthetic nitrogen fertilisers. *Cleaner and Circular Bioeconomy* 5, 100043. DOI: 10.1016/j.clcb.2023.100043.

Santos Dagnaisser, L., Granato Barbosa dos Santos, M., Vieira Santa Rita, A., Chaves Cardoso, J., Fonseca de Carvalho, D. and Vieira de Mendonça, H. (2022) Microalgae as bio-fertilizer: a new strategy for advancing modern agriculture, wastewater bioremediation, and atmospheric carbon mitigation. *Water, Air, & Soil Pollution* 233, 477. DOI: 10.1007/s11270-022-05917-x.

Sohlenius, B. and Boström, S. (1999a) Effects of global warming on nematode diversity in a Swedish tundra soil – a soil transplantation experiment. *Nematology* 1, 695–709. DOI: 10.1163/156854199508720.

Sohlenius, B. and Boström, S. (1999b) Effects of climate change on soil factors and metazoan microfauna (nematodes, tardigrades and rotifers) in a Swedish tundra soil – a soil transplantation experiment. *Applied Soil Ecology* 12, 113–128. DOI: 10.1016/S0929-1393(98)00168-1.

Sørensen, L.I., Holmstrup, M., Maraldo, K., Christensen, S. and Christensen, B. (2006) Soil fauna communities and microbial respiration in high Arctic tundra soils at Zackenberg, Northeast Greenland. *Polar Biology* 29, 189–195. DOI: 10.1007/s00300-005-0038-9.

Stark, S., Väisänen, M., Männistö, M.K., Kühn, J. and Ruess, L. (2023) Long-term grazing intensity by reindeer alters the response of the soil micro-food web to simulated climate change in subarctic tundra. *Oikos* 2023(10), e09855. DOI: 10.1111/oik.09855.

Steel, H., Vandecasteele, B., Willekens, K., Sabbe, K., Moens, T. and Bert, W. (2012) Nematode communities and macronutrients in composts and compost-amended soils as affected by feedstock composition. *Applied Soil Ecology* 61, 100–112. DOI: 10.1016/j.apsoil.2012.05.004.

Steel, H., Moens, T., Vandecasteele, B., Hendrickx, F., De Neve, S., Neher, D.A. *et al.* (2018) Factors influencing the nematode community during composting and nematode-based criteria for compost maturity. *Ecological Indicators* 85, 409–421. DOI: 10.1016/j.ecolind.2017.10.039.

Sticht, C., Schrader, S., Giesemann, A. and Weigel, H.-J. (2009) Sensitivity of nematode feeding types in arable soil to free air CO_2 enrichment (FACE) is crop specific. *Pedobiologia* 52, 337–349. DOI: 10.1016/j.pedobi.2008.12.001.

Sushchuk, A.A. and Matveeva, E.M. (2021) Soil nematodes of coniferous forests in the Finnish–Russian friendship nature reserve. *Nature Conservation Research* 6, 76–88. DOI: 10.24189/ncr.2021.015.

Thakur, M.P., Reich, P.B., Fisichelli, N.A., Stefanski, A., Cesarz, S. *et al.* (2014) Nematode community shifts in response to experimental warming and canopy conditions are associated with plant community changes in a temperate-boreal forest ecotone. *Oecologia* 175, 713–723. DOI: 10.1007/s00442-014-2927-5.

Thakur, M.P., Del Real, I.M., Cesarz, S., Steinauer, K., Reich, P.B. *et al.* (2019) Soil microbial, nematode, and enzymatic responses to elevated CO_2, N fertilization, warming and reduced precipitation. *Soil Biology and Biochemistry* 135, 184–193. DOI: 10.1016/j.soilbio.2019.04.020.

Tóthmérész, B. (1993) DivOrd 1.50. A program for diversity ordering. *Tiscia* 27, 33–44.

van der Bij, A.U., Weijters, M.J., Bobbink, R., Harris, J.A., Pawlett, M. *et al.* (2018) Facilitating ecosystem assembly: plant–soil interactions as a restoration tool. *Biological Conservation* 220, 272–279. DOI: 10.1016/j.biocon.2018.02.010.

van der Putten, W.H., Bardgett, R.D., Bever, J.D., Bezemer, T.M., Casper, B.B. *et al.* (2013) Plant–soil feedbacks: the past, the present and future challenges. *Journal of Ecology* 101, 265–276. DOI: 10.1111/1365-2745.12054.

van der Putten, W.H., Bradford, M.A., Brinkman, E.P., van de Voorde, T.F.J. and Veen, G.F. (2016) Where, when and how plant–soil feedback matters in a changing world. *Functional Ecology* 30, 1109–1121. DOI: 10.1111/1365-2435.12657.

Wang, J., Li, M., Zhang, X., Liu, X., Li, L. *et al.* (2019) Changes in soil nematode abundance and composition under elevated CO_2 and canopy warming in a rice paddy field. *Plant and Soil* 445, 425–437. DOI: 10.1007/s11104-019-04330-4.

Wang, J., Shi, X., Lucas-Borja, M.E., Guo, Q., Mao, J. *et al.* (2023) Soil nematode abundances drive agroecosystem multifunctionality under short-tern elevated CO_2 and O_3. *Global Change Biology* 29, 1618–1627. DOI: 10.1111/gcb.16546.

Wolfe, B.E. and Klironomos, J.N. (2005) Breaking new ground: soil communities and exotic plant invasion. *BioScience* 55, 477–487. DOI: 10.1641/0006-3568(2005)055[0477:BNGSCA]2.0.CO;2.

Yeates, G.W., Tate, K.R. and Newton, P.C.D. (1997) Response of the fauna of a grassland soil to doubling of atmospheric carbon dioxide concentration. *Biology and Fertility of Soils* 25, 307–315.

Zhang, G., Sui, X., Li, Y., Jia, M., Wang, Z. *et al.* (2020) The response of soil nematode fauna to climate drying and warming in *Stipa breviflora* desert steppe in Inner Mongolia, China. *Journal of Soils and Sediments* 20, 2166–2180. DOI: 10.1007/s11368-019-02555-5.

Zhou, J., Wu, J., Huang, J., Sheng, X., Dou, X. *et al.* (2022a) A synthesis of soil nematode responses to global change factors. *Soil Biology and Biochemistry* 165, 108538. DOI: 10.1016/j.soilbio.2021.108538.

Zhou, J., Lu, M., Sheng, X. and Wu, J. (2022b) Effects of nitrogen addition and warming on nematode ecological indices: a meta-analysis. *European Journal of Soil Biology* 110, 103407. DOI: 10.1016/j.ejsobi.2022.103407.

9 Case Studies Using Nematode Assemblage Analysis in Aquatic Habitats

Cristina Gambi[1*‡], Sebastian Höss[2‡], Francesca Ape[3], Roberto Danovaro[1], Simone Mirto[4], Walter Traunspurger[5] and Aldo Zullini[6]

[1]Department of Life and Environmental Sciences, Polytechnic University of Marche, Ancona, Italy; [2]Ecossa, Starnberg, Germany; [3]Institute of Marine Sciences, National Research Council (ISMAR-CNR), Bologna, Italy; [4]Institute of Anthropic Impacts and Sustainability in Marine Environment, National Research Council (IAS-CNR), Palermo, Italy; [5]Animal Ecology, Bielefeld University, Germany; [6]Dipartimento di Biotecnologie & Bioscienze, Università di Milano-Bicocca, Italy

Abstract

Both marine and freshwater habitats are threatened by the activity of mankind, with industrial discharges, sewage disposal and oil spills all representing major threats exacerbated by global changes. Among meiofaunal taxa, nematodes are the most abundant multicellular organisms on Earth, including systems with extremely challenging conditions, accounting for about two-thirds of all metazoans. Nematodes are abundant and diverse in both marine and freshwater habitats (see Chapter 2, this volume). They are not only ubiquitous but also characterized by a high species richness, covering a broad range of sensitivity to various anthropogenic disturbances. Nematodes have the potential to be used as bioindicators, and they also display short resilience times after the end of disturbances. Here we report case studies relating to the use of nematode assemblage analysis in both marine and freshwater habitats. Further examples can also be found in other chapters of this volume.

9.1 Marine Habitats

Free-living nematodes, thanks to their biodiversity and ecological role, can be considered an effective tool to assess the impacts of multiple natural and anthropogenic disturbances on marine benthic ecosystems. Nematodes' diversity and their life strategies are widely used as proxies for health assessment of environmental ecosystems and to detect the effect of different impacts on marine ecosystems ranging from coastal to deep sea (Ferris and Bongers, 2006; Fraschetti *et al.*, 2006, 2016; Moreno *et al.*, 2008, 2011; Danovaro *et al.*, 2009; Gambi *et al.*, 2009, 2020a,b; Goodsell *et al.*, 2009; Semprucci and Balsamo, 2012; Losi *et al.*, 2013; Pusceddu *et al.*, 2014; Zeppilli *et al.*, 2015; Semprucci *et al.*, 2016; Schratzberger and Somerfield, 2020). Moreover, the analysis of their functional traits, such as buccal apparatus, allows gathering important insights on their functional diversity (Giere, 2009). Here we present an update on the status of marine ecosystems in response to several historical (hydrocarbon and heavy metal impacts,

*Email: c.gambi@staff.univpm.it

‡Equally contributed to first authorship.

 Nematodes as Environmental Indicators: From Theory to Practice, 2nd Edition (T. Kakouli-Duarte *et al.* eds)

DOI: 10.1079/9781800624221.0009

organic enrichment) and emerging threats (microplastics, climate change, oxygen depletion) and disturbances (trawling and deep-sea mining). We report a synthesis of the most recent literature after Wilson and Kakouli-Duarte (2009) on the use of nematodes as indicators of the effect of different types of anthropogenic disturbance in marine ecosystems.

9.1.1 Marine nematode response to hydrocarbon and heavy metal pollution

In marine as well as freshwater ecosystems, sediments are a major repository for contaminants, which can alter benthic biodiversity, food web structure and ecosystem functioning. The tolerance to metal pollutants has been shown to vary widely among nematode species in both marine and freshwater habitats (Hägerbäumer *et al.*, 2015; Gambi *et al.*, 2020a), and it has been observed that a long exposure to metals can lead to the selection of tolerant nematode species (Bastami *et al.*, 2017; Gambi *et al.*, 2020a,b), mainly with a K-strategy, able to persist in sediments chronically impacted by high contamination levels (Gambi *et al.*, 2020a). Nematode assemblages can show a lower variability in species composition when compared to macrofaunal assemblages (Fraschetti *et al.*, 2016; Gambi *et al.*, 2020b). The dominant nematode taxa in contaminated sediments typically include the genera *Sabatieria*, *Spilophorella*, *Desmodora* and *Pseudochromadora* (Schratzberger *et al.*, 2006; Moreno *et al.*, 2009; Fraschetti *et al.*, 2016). On the other hand, nematode assemblages in less polluted and pristine sediments are often dominated by species belonging to the genera *Richtersia* and *Desmoscolex*, sensitive to various types of disturbance (Moreno *et al.*, 2011). The effects of heavy metals contamination on nematode assemblages can vary with food availability (Ryu *et al.*, 2011). Indeed, 'colonizers' often dominate in highly contaminated sediments with low food availability, while sediments with low levels of metal concentration and higher food availability are often dominated by 'persister' species (*sensu* Bongers and Ferris, 1999; Giere, 2009; Semprucci *et al.*, 2010; Semprucci and Balsamo, 2012). Laboratory investigations using acute copper toxicity tests on *Halomonhystera disjuncta* associated to different environmental conditions revealed that this species exposed to high hydrostatic pressure significantly increased its sensitivity to copper, whereas lower temperatures resulted in an apparently increased copper tolerance, possibly because of a slower metabolism under low temperatures (Mevenkamp *et al.*, 2017). Hydrostatic pressure and temperature can thus significantly affect metal toxicity for nematodes and therefore need to be considered in toxicity assessments also in the case of deep-sea habitats affected by heavy metal contamination (Mevenkamp *et al.*, 2017). The effect of phenanthrene, one of the most common polycyclic aromatic hydrocarbons, can be exacerbated with increasing temperature as documented in experiments with *Diplolaimelloides delyi* leading to a significantly reduced development and population growth rate, no reproduction, or even 100% mortality (Pereira Pontes *et al.*, 2021). Thermally induced toxicity effects on marine populations can be significant, and current climate change and warming may have significant implications for known and emerging pollutants affecting marine diversity, food webs and ecosystem functions.

9.1.2 Marine nematode response to organic enrichment

Organic and nutrient enrichment represents a threat to marine ecosystems, and an example is provided by aquaculture activities and their biodeposition (i.e. faeces, pseudofaeces and uneaten food). Organic enrichment results in changes in sediment composition, biogeochemistry and habitat heterogeneity that can impact benthic communities, including meiofauna (Sutherland *et al.*, 2007; Grego *et al.*, 2009; Holmer, 2010; Mirto *et al.*, 2010, 2012; Dal Zotto *et al.*, 2016). Meiofaunal abundances may respond differently (positively, negatively or no response) to biodeposition from aquaculture activities, depending on the site and farm characteristics. The effects of the organic enrichment on total nematode abundance, biomass (including the body size) and diversity are generally not consistent but site-specific, and structural changes in the assemblages result from the replacement of sensitive species by tolerant species (Sutherland *et al.*, 2007; Netto and Valgas, 2010; Mirto *et al.*, 2010, 2012; Mirto *et al.*, 2014; Dal Zotto *et al.*, 2016; Venekey and Melo, 2016; Lacoste *et al.*, 2020; Pinto *et al.*, 2021; Ridall and Ingels, 2021). Mirto *et al.* (2014) evidenced the importance of life strategy as the Maturity Index and the turnover diversity of nematodes for detection of the organic pollution produced by fish farms in different habitats (*Posidonia oceanica* seagrass meadows, and non-vegetated sediments) in the Mediterranean. The genera *Daptonema* and *Prochromadorella* were more abundant in fish-farm-impacted sediments, while species of *Richtersia, Desmoscolex* and *Halalaimus*, generally associated with pristine conditions (Moreno *et al.*, 2011), have been found to be sensitive to organic enrichment due to aquaculture activities. These findings agree with other studies, which showed, in the sediments affected by fish (Vezzulli *et al.*, 2008; Abebe *et al.*, 2012) and mussel (Netto and Valgas, 2010) farms and shrimp ponds in the Ca Mau mangrove forests (Tran *et al.*, 2022), also the dominance of the genera *Terschellingia* and *Sabatieria* which, like *Daptonema*, are known for their tolerance to organic pollution. A study conducted at a fish farm located in a tropical area (Pinto *et al.*, 2021) showed fast changes in dominance and losses of sensitive genera below the cages during an eight-month rearing cycle. Moreover, a different trophic structure of nematode assemblages was found in fish-farm sediments from that of the control site, with lower values of the trophic diversity index and the dominance of selective deposit feeders at the impacted site (Giere, 2009; Semprucci *et al.*, 2010). The decrease of trophic diversity and the dominance of opportunistic, deposit-feeding (selective and non-selective) nematodes, observed also in shellfish-farm sediments (Lacoste *et al.*, 2020), are related to the organic enrichment and the enhancement of microbial abundance in aquaculture-impacted sites (Ape *et al.*, 2019). Changes in the nematode trophic structure have been also observed in a study conducted in Hong Kong Bay where a shift from the carnivorous to herbivorous feeding type was observed in the organically polluted sites due to the oxygen stress, prey limitation, and increased abundance and nutritional quality of fresh organic matter released in this eutrophic area (Xu *et al.*, 2022). Microcosm experiments of organic enrichment also showed that nematode abundance and diversity decreased, with major changes also for the taxonomic composition and trophic structure of the assemblages (Armenteros *et al.*, 2010). These patterns were driven by chemical stressors such as ammonia and hydrogen sulfide in which only tolerant species resulted able to survive in these hypoxic conditions (Armenteros *et al.*, 2010).

9.1.3 Marine nematode response to microplastics

Microplastics (<5 mm) contamination is of increasing concern for benthic fauna, including nematodes, and is now widespread at all latitudes and even in the most remote oceanic regions (Tirelli *et al.*, 2022). There is very limited information on the *in situ* relationship between microplastic concentrations and meiofauna (nematode) diversity (Haegerbaeumer *et al.*, 2019), nor are there studies on how microplastic particles elicit changes in the species diversity of an ecosystem. A manipulative experiment was set up to assess the relative effect of microplastic pollution on nematode diversity and trophic composition (Corinaldesi *et al.*, 2022). The abundance of epistrate feeders (species belonging to *Desmodora*) increased at the expense of predators (*Viscosia* spp.) and non-selective deposit feeders (species belonging to *Daptonema* and *Axonolaimus*). The reduction of predatory and non-selective deposit-feeding nematodes could be due to: (i) ingestion of microplastic particles covered in organic matter, followed by feeding impairment (Galloway *et al.*, 2017; Corinaldesi *et al.*, 2021); or (ii) physical damage due to adhesion of microplastic particles on the tissues of marine organisms, thus causing feeding impairment and stress/abrasions as observed in some marine organisms (Wright *et al.*, 2013; Corinaldesi *et al.*, 2021). Another laboratory experiment with microcosms assessed the multifaceted effects of polyvinyl chloride and chrysene (alone or combined) on free-living marine nematodes (Hedfi *et al.*, 2022). The results highlighted a significant decrease in the abundance of nematodes as well as the disappearance of sensitive nematode species and their replacement with the most tolerant groups.

9.1.4 Marine nematode response to trawling

The most consistent response of nematodes to varying intensities and frequencies of *in situ* bottom fishing was change in abundances of species and genera, mediated via fishing-induced modifications of the sediment (Schratzberger and Somerfield, 2020). Indeed, mixing of sediments following the passage of fishing gear releases buried organic material and benefits those species that can use this surplus either directly or via increased primary production (Hinz *et al.*, 2008; Liu *et al.*, 2011). Similarly, increased organic loading of the sediment can lead to a shift towards microbial-dominated, anaerobic food chains, resulting in the proliferation of the few species that can survive in oxygen-poor sediments and are able to exploit microbial food sources (Franco *et al.*, 2008). The consequences of mine tailings disposal and associated metals in several marine coastal systems worldwide (Schaanning *et al.*, 2019) are still largely unknown for the benthic fauna. Field and laboratory experiments on the subsequent effects of deposition of (artificially resuspended) contaminated or organic-rich sediments revealed that sediment resuspension and plume deposition could result in a prompt significant decrease of nematode species richness with significant changes also for the species composition (Ingels *et al.*, 2014; Pusceddu *et al.*, 2014; Gambi *et al.*, 2022; Tinlin-Mackenzie *et al.*, 2023). Changes in species composition can also indicate consequences on nematode life strategies and functional diversity due to the increasing relevance of early colonizers (opportunistic) species as observed in both coastal and deep-sea ecosystems (Liu *et al.*, 2009). Nematode assemblages in sediments affected by the plume deposition were of a lower functional diversity resulting from the loss of a specific trophic guild (predators), and this might have additional consequences on benthic food webs and ecosystem functions in areas affected by trawling activities. Severe consequences of chronically trawled sediments have been documented along the continental slope where a significant decrease in nematode species richness (25%) has been reported (Pusceddu *et al.*, 2014). Moreover, nematode assemblages in trawled sediments were characterized by more opportunistic species than those observed in non-trawled sediments, as reported also from shallow-water studies (Liu *et al.*, 2009). This observation, coupled with the significantly higher mean individual biomass of nematodes in trawled sediments, suggests that the disturbance exerted by bottom trawling removed preferentially small individuals, more susceptible to be resuspended, while leaving on site larger and heavier organisms.

9.1.5 Marine nematode response to deep-sea mining

Seabed mining is expected to determine habitat modifications and species mortality due to mechanical disturbances (Van Dover, 2014; Jones *et al.*, 2017) and impacts due to the release and spread of toxin-laden water from the generation of plumes (Boschen *et al.*, 2013). Deep-sea sediments impacted by simulated mining activities may take decades to recover. While faunal abundance recovers more quickly than diversity, the latter can remain low even after multiple decades (Gollner *et al.*, 2017; van Dover *et al.*, 2017; Hauquier *et al.*, 2019; Simon-Lledó *et al.*, 2019). Nematode abundance was significantly lower 26 years after the experimental disturbance carried out in the Clarion–Clipperton Fracture Zone (Pacific Ocean) (Miljutin *et al.*, 2011) while the number of genera was higher post-disturbance (70 genera) than pre-disturbance (50 genera) (Mahatma, 2009). Also, nematode community composition differed significantly when monitored after 26 years of disturbance (Miljutin *et al.*, 2011). When nematodes were compared among different licence areas of the Clarion–Clipperton Fracture Zone, almost half of the assemblages consisted of rare genera, each contributing less than 5% to the overall abundance and displaying a distribution which was usually restricted to a single licence area (Hauquier *et al.*, 2019). Most of the rare nematode genera and species (many singletons and doubletons; Pape *et al.*, 2021) could be particularly vulnerable to mining-induced changes (Vanreusel *et al.*, 2016; Pape *et al.*, 2017; Vanreusel *et al.*, 2017; Macheriotou *et al.*, 2020). Conversely, nematode abundance, diversity and community composition were similar between a test mine site and a reference mining site in the abyssal nodule fields of the Central Indian Ocean (Singh *et al.*, 2019). Both areas were dominated by Monhysteridae, Chromadoridae, Xylidae and Leptolaimidae and by cosmopolitan genera including *Acantholaimus*, *Halalaimus*, *Thalassomonhystera* and *Leptolaimus* which were common with other nodule fields such as the Clarion–Clipperton Fracture Zone (Miljutina *et al.*, 2010) and the Peru Basin in the tropical South-East Pacific. The differences in nematode assemblage composition between the test mining and control sites in the Clarion–Clipperton Fracture Zone and in the Central Indian Ocean can be the result of the different approach used to simulate the experimental mining activities.

9.1.6 Marine nematode response to climate change

Global average temperature, frequency of temperature extremes and amplitude of daily fluctuations are all expected to change under predicted scenarios of climate change. Across habitats, nematodes' response to temperature shift revealed that abundance generally decreased under both elevated constant and fluctuating temperature treatments (Gingold *et al.*, 2013; Mevenkamp *et al.*, 2018; Vafeiadou *et al.*, 2018a; Schratzberger *et al.*, 2023) and under increasing exposure time to extreme temperatures (Ape *et al.*, 2018). The same variability has been also reported for other meiofaunal taxa although their low abundances often precluded unequivocal conclusions on the effects of changing temperature (Meadows *et al.*, 2015; Schratzberger *et al.*, 2023). Changes in the daily temperature fluctuations as well as in the frequency of extreme values may influence nematode species' interactions, with consequences on the structure of the nematode assemblages and ecosystem functioning (Vafeiadou *et al.*, 2018a). The abundances of a few opportunistic species in tropical subtidal assemblages increased when the effects of elevated temperature were combined with a reduction in pH (Lee *et al.*, 2017). Consistently observed shifts in dominance patterns of nematode species resulted from species-specific physiological tolerance to changes in temperature and pH (Mevenkamp *et al.*, 2018). De Meester *et al.* (2015) and Vafeiadou *et al.* (2018b) demonstrated that specific nematode species can alter their relative behaviour in response to elevated temperature and such shifts might trigger changes in abundance of temperature-tolerant species with relevant consequences for assemblage composition and trophic structure. Some nematode families generally considered as colonizers and resistant to the effects of climate change (Chromadoridae, Rhabditidae and Xyalidae; Meadows *et al.*, 2015; Lee *et al.*, 2017) did not increase in abundance with increasing temperature (Colossi Brustolin *et al.*, 2019). This might result from changes in fitness of individual species and indirect effects through changes in food availability (Lee *et al.*, 2017; Ullah *et al.*, 2018). The effect of temperature increase on nematode assemblages is also exacerbated when warming is combined with acidification, leading to increases of fast-growing opportunistic genera (e.g. *Anticoma*) which feed on detritus and bacterial mats (Schratzberger *et al.*, 2007). The fast response of colonizers together with the changes in food quantity and quality can determine relevant effects on the nematode assemblages under future scenarios. A decrease in nematode trophic diversity, mainly related to the loss of predatory nematodes, can cause cascading effects on the dynamics of the benthic food web (Gingold *et al.*, 2013) and affect its resilience to other types of disturbances (Teixidó *et al.*, 2018). Changing temperatures can have differential effects on cryptic species (Derycke *et al.*, 2007). Some cryptic species of *H. disjuncta* can tolerate a wide range of temperatures. Individuals remain active at low temperatures (<3°C) and show high fitness at higher temperatures in the range of 10–16°C when compared to other cryptic species (Van Campenhout *et al.*, 2014). Other species belonging to the Monhysteridae family that commonly co-occur at the same locations, *Diplolaimelloides oschei* and *Diplolaimelloides meyli*, tend to be more impacted by low temperatures but appear to be tolerant to much more elevated temperatures (up to 35°C) than *H. disjuncta* (Vafeiadou *et al.*, 2018a). Competitive interactions among co-occurring species of Monhysteridae are common and can strongly impact population abundances, thus species of this family can be used as relevant indicators of environmental change and of ecological interactions (dos Santos *et al.*, 2009). Microcosm studies have also highlighted that the effects of changing temperature along with inter- and intraspecific competition might affect life-history traits and population development of nematode species in different ways: higher fitness under daily fluctuating temperatures and lower fitness under elevated constant temperature, with severe consequences on species coexistence and thus assemblage composition (Vafeiadou and Moens, 2021).

9.1.7 Marine nematode response to extreme environmental conditions

Global ocean deoxygenation is predicted to further extend oxygen minimum zones (OMZs) whose expansion has been documented since the 1960s (Breitburg *et al.*, 2018), furthering the need to understand these regions (Long *et al.*, 2021). Nematodes are among the metazoans most tolerant of low-oxygen conditions, and they thus play major roles in sea-floor ecosystem processes at OMZs. A total of 110 species belonging to 24 families were found along the bathymetric transect in the OMZ of the Indian Ocean (Singh and Ingole, 2016). The dominant species were *Terschellingia longicaudata*, *Desmodora* sp.1, *Sphaerolaimus gracilis* and *Theristus ensifer*, and seven species were exclusively reported in the OMZ: *Pselionema* sp.1, *Choanolaimus* sp.2, *Halichoanolaimus* sp.1, *Cobbia dentata*, *Daptonema* sp.1, *Trissonchulus* sp.1 and *Minolaimus* sp.1. A new species, *Spinonema gracilispiculum* sp. n., was described from the OMZ of the eastern Arabian Sea (216 m) margin (Jacob *et al.*, 2020). Only few individuals of this species were found exclusively in sandy low-oxygen (i.e. dissolved oxygen 0.35 ml l^{-1}) sediments, despite the analysis of 288 sediment samples from an extensive survey along the eastern Arabian Sea OMZ region. A recent investigation conducted along the oxic/anoxic interface in the Istanbul Strait's outlet area of the Black Sea detected the presence of 137 species from 63 genera and 30 families at the depth ranges (75–300 m) studied (Sergeeva *et al.*, 2021). A total of 34 species were reported for the first time from Turkey, suggesting that free-living nematodes of the Black Sea are much more diverse than is currently known and requiring further taxonomic investigations. Also, in the OMZ off Baja California a new nematode species *Glochinema spinithorni* has been discovered (Neira *et al.*, 2005), suggesting that OMZs function as isolated habitats by promoting endemism at low taxonomic levels. Nematode assemblages were studied in the OMZ (122–972 m) off Concepción, Central Chile, recording 33 genera (belonging to 17 families), with higher abundances at low-oxygen sites (122–364 m) and the lowest diversity at the most oxygen-depleted site (122 m), despite the high availability of organic carbon (Neira *et al.*, 2013). A field experiment designed to assess the effects of induced hypoxic–anoxic conditions on nematode genus diversity in *P. oceanica* meadows in Spain (Western Mediterranean) revealed that, as a phylum, nematodes tolerated hypoxic–anoxic conditions, as their species richness did not change when compared to the oxygenated sediments (Gambi *et al.*, 2009). *Desmoscolex* and *Bolbolaimus* were replaced by *Chromadorella*, *Sabatiera* and *Polysigma*, more tolerant to the anoxic conditions, and this change in genus composition reflected also changes in the trophic composition. Selective deposit feeders and predators decreased and were replaced by non-selective deposit feeders and epistrate feeders. Laboratory experiments showed that short-term suboxic and anoxic/partially sulfidic conditions had similar structuring effects on nematode assemblages, resulting from species-specific survival in suboxic and anoxic/partially sulfidic conditions (Steyaert *et al.*, 2007). *Daptonema setosum*, *Daptonema tenuispiculum* and *Chromadora macrolaima* were dominant under oxic conditions but disappeared when oxygen levels were reduced. The dominant species *Sabatieria pulchra* was slightly reduced, as was *Terschellingia communis*, while *Metachromadora vivipara* increased under suboxic and anoxic conditions.

9.2 Freshwater Habitats

Analysis of macrobenthic invertebrates is an important component in determinations of the ecological status of freshwater systems within the European Union Water Framework Directive (EU WFD) (European Community, 2000; Sandin and Hering, 2004), with different indices for the evaluation of various kinds of anthropogenic stress (Hering *et al.*, 2004). However, meiobenthic invertebrates, such as nematodes, are not included in routine monitoring programmes, although they are key players in benthic ecosystems (Schratzberger and Ingels, 2018; Traunspurger, 2021) and, in some freshwater habitats (fine sediments; biofilms), represent the prevailing fauna, both in terms of abundance and diversity

(Peters and Traunspurger, 2005; Wolfram *et al.*, 2010; Patrício *et al.*, 2012). Since the publication of Wilson and Kakouli-Duarte (2009), substantial developments have been made in the field of nematode-based tools for assessing: (i) the quality of freshwater ecosystems; and (ii) the response of freshwater nematode communities to chemical pollution (see Chapter 6 this volume; also Häegerbäeumer *et al.*, 2015; Höss and Traunspurger, 2021; Höss *et al.*, 2022 for other reviews). These developments require a fundamental update of this chapter, by adding new field studies addressing: (i) various types of anthropogenic stress, such as chemical pollution, eutrophication and climate change; and (ii) the use of SPEAR (SPEcies At Risk) indices, namely the NemaSPEAR[%]-index (see Chapter 6, this volume).

9.2.1 Contaminated samples across different river basins

Barbuto and Zullini (2005) compared various Italian rivers in terms of their nematode community structure. These authors applied the Taxonomic Distinctness Index (Δ+), which is defined by the average taxonomic path length between any two randomly chosen species, traced through a Linnean phylogenetic classification of the full set of species involved (Clarke and Warwick, 1998). Based on literature data, the Δ+ was calculated for ten samplings from seven different Italian rivers (Fig. 9.1A). All the rivers fell into the 95% confidence funnel, except the highly polluted Lambro in which very polluted stretches diverged considerably from the normal situation. Habitats outside the confidence funnel are subject to environmental stresses, usually pollution, but other factors can also affect this index. It is noteworthy that the best river environments (Oglio, Adige, Seveso U), namely those that were unpolluted, were at or above the average expected value.

Wolfram *et al.* (2012) used a weight-of-evidence (WoE) approach to evaluate the sediment quality of rivers in three different European catchments: Elbe (Czech Republic), Scheldt (Belgium) and Llobregat (Spain). Among various lines of evidence (LoE) (chemical contamination, sediment toxicity testing, *in situ* benthic fauna), nematode fauna was investigated along with routinely monitored macrozoobenthos (MZB) to link biological impacts with chemical contamination. Distinct benthic communities (both MZB and nematodes) were recorded in the various catchments but also at sites with lower versus higher chemical contamination. Moreover, the

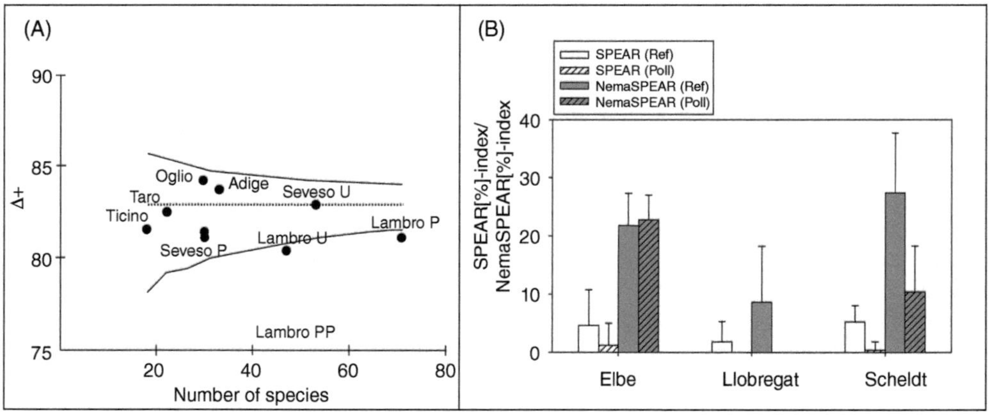

Fig. 9.1. (A) Taxonomic Distinctness Index (Δ+) values from seven Italian rivers. The data were plotted against the number of species as points on the simulated 95% confidence funnel based on random selections from the total Italian species list. U = unpolluted; P = polluted; PP = very polluted. (Reprinted with permission from Barbuto and Zullini, 2005.) (B) SPEAR[%]- and NemaSPEAR[%]-index calculated for macroinvertebrates and nematodes, respectively, at polluted sites (Poll) and the respective reference site with lower pollution (Ref) in different European river catchments. SPEAR = SPEcies At Risk. (Author's own image.)

NemaSPEAR[%]-index (Höss *et al.*, 2011; see also Chapter 6, this volume) showed comparable results compared to the macroinvertebrate-based SPEAR[%]-index (Liess and Von der Ohe, 2005). Both indices showed relatively low values for reference and polluted sites, which is supported by the relatively high toxic potentials of some of the reference sites (Wolfram *et al.*, 2012). However, at reference sites, the NemaSPEAR[%] values were substantially higher than the SPEAR[%] values, indicating a better suitability of the nematode-based index in fine sediments (Fig. 9.1B).

9.2.2 Pollution gradient approach

Investigating pollution gradients within a single river system is one step towards a better understanding of cause-and-effect relationships. Depending on the length of the investigated river stretch, hydro-morphological factors are relatively constant, while a pollution input (e.g. wastewater or industrial effluents) will increase chemical pollution from a point source. More than 30 years ago, Zullini (1976) published a study from the river Seveso, a small Italian river less than 50 km long. This river has a steep pollution gradient, since it crosses one of the most populated and industrialized areas of northern Italy. As part of the study, five stations were sampled at monthly intervals during a one-year period (site 1, near source = slightly polluted; site 5, on the northern outskirts of Milan = very polluted).

In Zullini's study, the quantity and quality of pollution were not directly measured as contaminant concentrations in the sediment, but rather by three water-chemical measurements that are indicative of pollution: (i) dissolved oxygen concentration, which indicates the amount of oxygen left in the water (and available to resident organisms) after its depletion due to organic pollution; (ii) chemical oxygen demand (COD), which reflects the amount of dissolved organic matter that originated from sewage, industrial and agricultural waste; and (iii) electrical conductivity (μS cm^{-1}), which is proportional to the salt content of the river and indicates the amount of inorganic, mainly industrial, pollution. These measurements suggested that pollution increased strongly along the Seveso River. Dissolved oxygen dropped from 95% (average at site 1) to only 35% (average at site 5) while COD and conductivity increased from 12 to 235 mg O_2 l^{-1} and from 255 to 881 μS cm^{-1}, respectively (Fig. 9.2A).

Species belonging to the Rhabditida played an important role in discriminating among the various sites. Zullini (1976) considered Rhabditida (referred to as Secernentea) as indicators of mainly organic pollution, since these nematodes are bacterial feeders and typical inhabitants of rotten matter and dung. Diplogasteromorpha peaked in the medium–heavily polluted stretch of the river, whereas Rhabditomorpha peaked in the terminal, most polluted stretch. If the percentage of Rhabditida (excluding *Fictor fictor* as an indicator of clean habitats; Zullini, 1976) is chosen as an indicator of water pollution, then increasing values of pollution were evident along the river course, from 12% at site 1 to almost 100% at site 3 (Fig. 9.2B). These values reflected the increasing degree of pollution as determined by the above-mentioned chemical parameters (Fig. 9.2A). In addition, the Maturity Index (see Chapter 5, this volume) correlated well with the pollution levels at the different sites, dropping from 2.82 at site 1 to 1.21 at site 3 (Fig. 9.2B). The two proxies, percentage of Rhabditida and Maturity Index, were negatively correlated, as rhabditid species are usually categorized as colonizers, with a colonizer–persister (*c-p*) value of 1. Consistent with the above findings, *F. fictor* should be assigned a higher *c-p* score (*c-p*3 or *c-p*4) rather than considered as a typical colonizer, with a *c-p* value of 1. If a higher *c-p* score had been applied to *F. fictor*, the difference between site 1 and all other sites would have been greater, as this species only occurred at site 1.

Sonne *et al.* (2018) investigated a stream located in western Denmark (Grindsted) where both agricultural and urban activities are present (including two large, contaminated sites), investigating the ecological status in relation to the complex chemical contamination dynamics documented in this stream (Sonne *et al.*, 2017). It was possible to link the meiobenthic and nematode community structure, sampled in the bed sediment along a contamination gradient, with a contaminated groundwater plume that likely has influenced redox conditions in the groundwater such that toxic elements (e.g. arsenic) released from the altered redox conditions subsequently affected the meiofauna community.

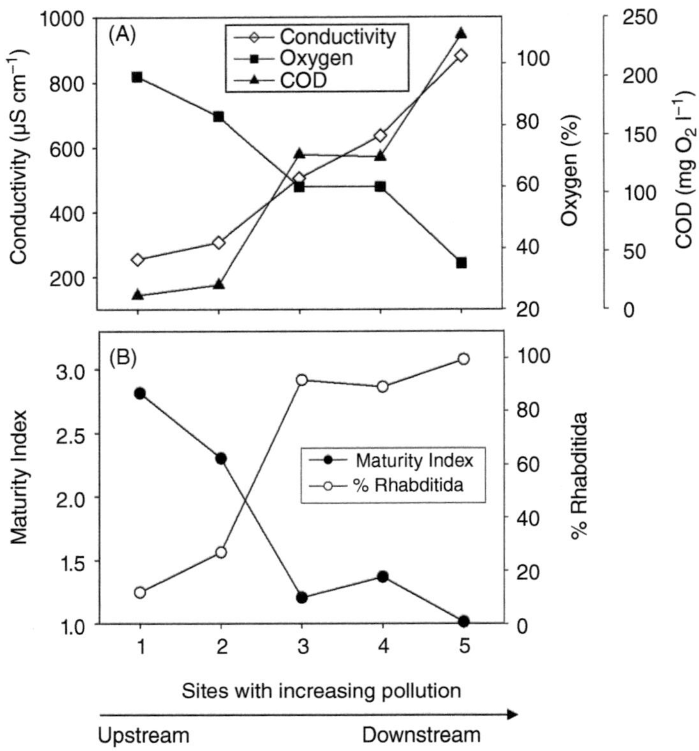

Fig. 9.2. (A) Electrical conductivity (µS cm^{-1}), oxygen saturation (%) and chemical oxygen demand (COD, in mg O_2 l^{-1}) and (B) Maturity Index and percentage of Rhabditida (% Rhabditida) at increasingly polluted sites from the Seveso River (northern Italy). (Author's own image using data taken from Zullini, 1976.)

In that study, the NemaSPEAR[%]$_{genus}$ (Höss *et al.*, 2017; see Chapter 6, this volume) helped to distinguish chemical-induced effects on meiofauna from other in-stream parameters that can influence meiofauna composition (Sonne *et al.*, 2018). The study of Sonne *et al.* (2018) represents one of the first providing evidence of ecological consequences resulting from the interaction of organic and inorganic chemical stressors, released in groundwater–surface water environments, where the meiobenthic community could be linked to the more stable biogeochemical water quality parameters in the hyporheic zone.

9.2.3 Eutrophication

Ristau *et al.* (2015) evaluated the suitability of nematodes as indicator organisms for assessing ecosystem deteriorations induced by eutrophication. A total of 15 different lakes located in Germany and Sweden, characterized by various trophic states ranging from oligo- to eutrophic (trophic state index, TSI: 39.7–71.3; Carlson, 1977), were studied. Besides the taxonomic composition, nematodes were categorized by several life-cycle, feeding and morphological traits. Both composition of nematode species and traits could clearly be distinguished between oligo-, meso- and trophic lakes. However, in contrast to macrobenthos, nematode traits did not show a higher bioindication potential than taxonomic composition.

While *Eumonhystera filiformis*, *Monhystera paludicola*, *Monhystera* cf. *lemani*, *Dorylaimus stagnalis*, *Brevitobrilus stefanskii* and *Theristus agilis* benefited from eutrophication, *Ironus ignavus*, *Trischostoma monohystera*, *Prismatolaimus intermedius*, *Euteratocephalus palustris*, *Rhabdolaimus terrestris* and *Rhabdolaimus aquaticus* were more dominant in oligotrophic lakes. Only the two trait modalities '5–10 offspring per reproductive cycle' and 'suction feeder' showed a positive relationship with increased levels of eutrophication, whereas five trait modalities ('chewer', 'slender body shape', 'very slender body shape', 'obligate asexual reproduction' and 'body length < 500 μm') were related to lower TSI values (Fig. 9.3A; Ristau *et al.*, 2015). Interestingly, the NemaSPEAR[%]-index (calculated from the data of Ristau *et al.*, 2015), although indicating good to high ecological status at all sites, was negatively correlated with trophic status (Fig. 9.3B; NemaSPEAR[%]-index at oligotrophic, mesotrophic and eutrophic lakes: 69.1 ± 3.0, 47.5 ± 6.6 and 40.1 ± 6.4, respectively).

These results suggest that benthic nematode communities can be responsive to the trophic state of lakes by changing the species composition, maybe driven by beneficial conditions for certain life-cycle, morphological or feeding traits. The response of nematodes to nutrient enrichment had previously been evidenced in experimental studies (Ristau *et al.*, 2012). However, future research must increase the knowledge on bioecological trait data for nematode species, in order to use a trait-based bioindication to assess the impact of eutrophication on sediment quality. Additionally, the NemaSPEAR[%]-index, developed for indicating pollution-induced effects on benthic communities, might be a promising measure for eutrophication.

9.2.4 Climate change – effects of drought and temperature

In the view of global climate change and its severe consequences for the environment, drought and temperature stress in aquatic ecosystems have to be addressed as part of a multi-stress situation for aquatic organisms. Dry conditions promote water stress, direct effects of solar radiation and high stream-bed temperatures, which progressively affect aquatic organisms (Lake, 2000). Majdi *et al.* (2020a) investigated the meiofauna of intermittent streams in Spain with various patterns of duration and frequency of non-flow events. Due to their trophic versatility and high colonization ability, meiofauna plays a key role for the restoration of desiccated stream beds after flow returns. Nematode communities responded to alternations of flow and drought conditions with higher densities and number of species (Fig. 9.4) in intermittent compared to permanent streams, as in dry periods typical soil nematode species could easily colonize the stream beds (Majdi *et al.*, 2020a). Moreover, nematode density was significantly correlated with the frequency of drought periods.

In a field study, Majdi *et al.* (2020b) investigated, among other organisms, biofilm-inhabiting

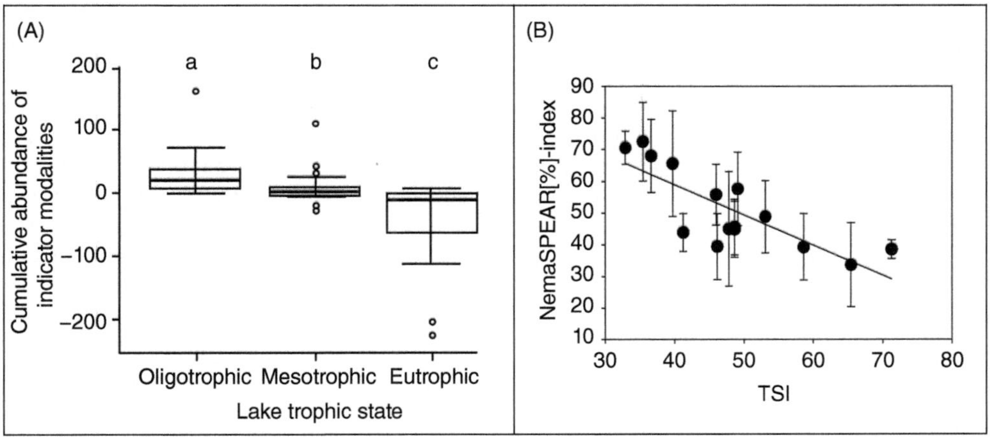

Fig. 9.3. (A) Cumulative abundances, weighted by CCA (canonical correspondence analysis) scores, of indicator trait modalities for nematodes sampled at sites of German and Swedish lakes characterized by different trophic states (oligotrophic sites: $n = 30$, mesotrophic sites: $n = 41$, eutrophic sites: $n = 29$); different letters above the box plots indicate significant differences (Kruskal–Wallis test with pairwise Holm-corrected Wilcoxon tests, $P < 0.001$). (Modified from Ristau *et al.*, 2015.) (B) NemaSPEAR[%]-index (calculated for nematode species data of Ristau *et al.*, 2015; mean values with their standard deviation represented by error bars) plotted against the trophic state index (TSI); linear regression: $r^2 = 0.64$, $P < 0.001$.

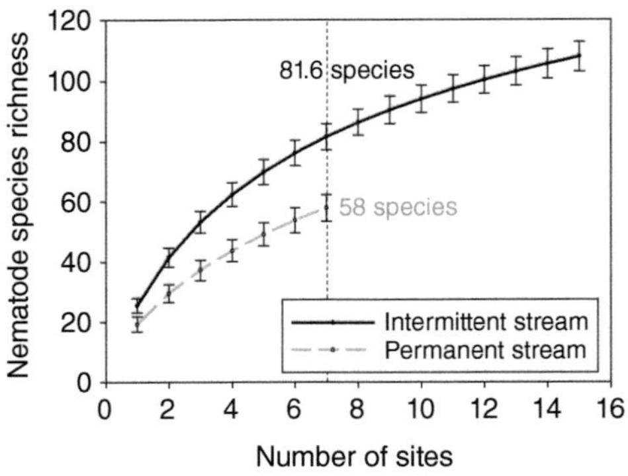

Fig. 9.4. Sample-based rarefaction curves for nematode species richness measured in 15 intermittent streams (solid black line) and seven permanent streams (dashed grey line). (From Majdi *et al.*, 2020a. CC BY 4.0.)

nematode communities at various sites of a stream that were differently impacted by thermal effluents of a nuclear power plant. The effluents caused a significant average temperature increase of 1.5°C at the impacted site, compared to the unimpacted control site. Moreover, the daily temperature fluctuations were higher at the heated sites. Interestingly, nematode communities were more diverse and showed a different species composition at the heated compared to the control sites. At the heated sites, nematode feeding types were distributed more homogeneously, with higher proportions of epistrate feeders, predators and omnivores. Heat-induced changes to the trophic structure of nematodes will also have consequences on food web structure and energy transfer, and thus on the functions of biofilms for the aquatic ecosystem. The distinct response of nematodes to drought events and changes in temperature makes them potential candidates as indicators for climate change-driven disturbances in ecosystems.

9.3 Conclusions

Nematodes can be considered among the most suitable bioindicators due to their worldwide distributions across habitats and ecosystems, high abundances, and genus- and species-specific responses to several types of disturbance. Marine and freshwater nematodes respond to various types of anthropogenic stress (chemical pollution, droughts, eutrophication, microplastics, climate change and oxygen depletion) and disturbances (trawling and deep-sea mining) with shifts in the composition of species and functional traits. Although the response of nematode assemblages to anthropogenic pressures is complex (and impact-specific), literature available suggests that severe disturbance leads to the dominance of opportunistic/colonizer species (regardless of the nature or type of disturbance) to the detriment of persistent species. The analysis of the nematode life strategy may be recommended in both marine and freshwater ecosystems. Although the mechanistic reasons for differences in vulnerability between nematodes species to certain stress factors are still not well understood, there are promising approaches to use and further develop nematode-based bioindices (e.g. NemaSPEAR[%]-index for freshwater nematodes; see also Chapter 6, this volume). Vulnerability to stress is always governed by the complex interactions: (i) between single species and the stress factor (sensitivity or tolerance); (ii) between the species, the stress factor and the environment (exposure, uptake, multiple stress); and (iii) within the food web (indirect effects: competition, food availability and quality, etc.). Thus, trait-based bioindication (such as the Maturity Index for soil nematodes) is a promising concept largely adopted in marine ecosystems and applied to assess the response to different types of disturbance. However, there are still large knowledge gaps regarding the responsiveness of traits to environmental factors, which makes a customized development for the indication of certain types of stress difficult. Therefore, empirically derived indices, such as the NemaSPEAR[%]-index, might be a good alternative for indicating ecosystem stress. However, although it could be shown that the NemaSPEAR[%]-index is responding not only to chemical contamination but also to eutrophication, it might be necessary to adapt the index according to the requirements for other types of stress. For this, all information on the response of nematode species or traits from the above-mentioned case studies can be helpful for any refinements. Although the number of investigations based on the use of nematodes to assess several types of disturbance has increased in the last decade, most work remains to improve our knowledge on free-living nematodes' diversity as biological indicators showing features and identifiable responses in both marine and freshwater environments.

References

Abebe, E., Joseph, M. and Bert, W. (2012) Temporal dynamics in nematode biodiversity and community structure at an experimental open ocean aquaculture site, Gulf of Maine, USA. *Russian Journal of Nematology* 20(2), 127–140.

Ape, F., Sarà, G., Airoldi, L., Mancuso, F.P. and Mirto, S. (2018) Influence of environmental factors and biogenic habitats on intertidal meiofauna. *Hydrobiologia* 807, 349–366.

Ape, F., Manini, E., Quero, G.M., Luna, G.M., Sarà, G. *et al.* (2019) Biostimulation of *in situ* microbial degradation processes in organically-enriched sediments mitigates the impact of aquaculture. *Chemosphere* 226, 715–725.

Armenteros, M., Pérez-García, J.A., Ruiz-Abierno, A., Díaz-Asencio, L., Helguera, Y. *et al.* (2010) Effects of organic enrichment on nematode assemblages in a microcosm experiment. *Marine Environmental Research* 70(5), 374–382.

Barbuto, M. and Zullini, A. (2005) The nematode community of two Italian rivers (Taro and Ticino). *Nematology* 7, 667–675.

Bastami, K.D., Taheri, M., Foshtomi, M.Y., Haghparast, S., Hamzehpou, A. *et al.* (2017) Nematode community structure in relation to metals in the southern of Caspian Sea. *Acta Oceanologica Sinica* 36, 79–86.

Bongers, T. and Ferris, H. (1999) Nematode community structure as a bioindicator in environmental monitoring. *Trends in Ecology & Evolution* 14, 224–228.

Boschen, R.E., Rowden, A.A., Clark, M.R. and Gardner, J.P.A. (2013) Mining of deep-sea seafloor massive sulfides: a review of the deposits, their benthic communities, impacts from mining, regulatory frameworks and management strategies. *Ocean Coastal Management* 84, 54–67.

Breitburg, D., Levin, L.A., Oschlies, A., Grégoire, M., Chavez, F.P. *et al.* (2018) Declining oxygen in the global ocean and coastal waters. *Science* 359, eaam7240.

Carlson, R.E. (1977) A trophic state index for lakes. *Limnology and Oceanography* 22(2), 361–369.

Clarke, K.R. and Warwick, R.M. (1998) A taxonomic distinctness index and its statistical properties. *Journal of Applied Ecology* 35, 523–531.

Colossi Brustolin, M., Nagelkerken, I., Moitinho Ferreira, C., Goldernberg, S.U., Ullah, H. and Fonseca, G. (2019) Future ocean climate homogenizes communities across habitats through diversity loss and rise of generalist species. *Global Change Biology* 25, 3539–3548.

Corinaldesi, C., Canensi, S., Dell'Anno, A., Tangherlini, M., Di Capua, I. *et al.* (2021) Multiple impacts of microplastics can threaten marine habitat-forming species. *Communications Biology* 4, 431.

Corinaldesi, C., Canensi, S., Carugati, L., Lo Martire, M., Marcellini, F. *et al.* (2022) Organic enrichment can increase the impact of microplastics on meiofaunal assemblages in tropical beach systems. *Environmental Pollution* 292(B), 118415.

Dal Zotto, M., Santulli, A., Simonini, R. and Todaro, M.A. (2016) Organic enrichment effects on a marine meiofauna community, with focus on Kinorhyncha. *Zoologischer Anzeiger – A Journal of Comparative Zoology* 265, 127–140.

Danovaro, R., Gambi, C., Höss, S., Mirto, S., Traunspurger, W. and Zullini, A. (2009) Case studies using nematode assemblage analysis in aquatic habitats. In: Wilson, M.J. and Kakouli-Duarte, T. (eds) *Nematodes as Environmental Indicators*. CAB International, Wallingford, UK, pp. 146–171.

De Meester, N., Dos Santos, G.A.P., Rigaux, A., Valdes, Y., Derycke, S. and Moens, T. (2015) Daily temperature fluctuations alter interactions between closely related species of marine nematodes. *PLoS One* 10(7), e0131625.

Derycke, S., Hendrickx, F., Backeljau, T., D'Hondt, S., Camphijn, L. *et al.* (2007) Effects of sublethal abiotic stressors on population growth and genetic diversity of *Pellioditis marina* (Nematoda) from the Westerschelde estuary. *Aquatic Toxicology* 82(2), 110–119.

dos Santos, G.A.P., Derycke, S., Genevois, V.G.F., Coelho, L.C.B.B., Correia, M.T.S. and Moens, T. (2009) Interactions among bacterial-feeding nematode species at different levels of food availability. *Marine Biology* 156, 629–640.

European Community (2000) Directive 2000/60/EC of the European Parliament and the Council of 23 October 2000 establishing a framework for Community action in the field of water policy. *Official Journal of the European Community* L327, 1–73.

Ferris, H. and Bongers, T. (2006) Nematode indicators of organic enrichment. *Journal of Nematology* 38, 3–12.

Franco, M., Steyaert, M., Cabral, H.N., Tenreiro, R., Chambel, L. *et al.* (2008) Impact of discards of beam trawl fishing on the nematode community from the Tagus estuary (Portugal). *Marine Pollution Bulletin* 56(10), 1728–1736.

Fraschetti, S., Gambi, C., Giangrande, A., Musco, L., Terlizzi, A. and Danovaro, R. (2006) Structural and functional response of meiofauna rocky assemblages to sewage pollution. *Marine Pollution Bulletin* 52, 540–548.

Fraschetti, S., Guarnieri, G., Gambi, C., Bevilacqua, S., Terlizzi, A. and Danovaro, R. (2016) Impact of offshore gas platforms on the structural and functional biodiversity of nematodes. *Marine Environmental Research* 115, 56–64.

Galloway, T.S., Cole, M. and Lewis, C. (2017) Interactions of microplastic debris throughout the marine ecosystem. *Nature Ecology and Evolution* 1, 116.

Gambi, C., Bianchelli, S., Pérez, M., Invers, O., Ruiz, J.M. and Danovaro, R. (2009) Biodiversity response to experimental induced hypoxic–anoxic conditions in seagrass sediments. *Biodiversity and Conservation* 18, 33–54.

Gambi, C., Dell'Anno, A., Corinaldesi, C., Lo Martire, M., Musco, L. *et al.* (2020a) Impact of historical contamination on meiofaunal assemblages: the case study of the Bagnoli-Coroglio Bay (southern Tyrrhenian Sea). *Marine Environmental Research* 156, 104907.

Gambi, C., Canals, M., Corinaldesi, C., Dell'Anno, A., Manea, E. *et al.* (2020b) Impact of historical sulfide mine tailings discharge on meiofaunal assemblages (Portmán Bay, Mediterranean Sea). *Science of The Total Environment* 736, 139641.

Gambi, C., Canals, M., Corinaldesi, C., Dell'Anno, A., Manea, E. *et al.* (2022) Impact of resuspended mine tailings on benthic biodiversity and ecosystem processes: the case study of Portmán Bay, Western Mediterranean Sea, Spain. *Environmental Pollution* 301, 119021.

Giere, O. (2009) *Meiobenthology: The Microscopic Motile Fauna of Aquatic Sediments*, 2nd edn. Springer, Berlin.

Gingold, R., Moens, T. and Rocha-Olivares, A. (2013) Assessing the response of nematode communities to climate change-driven warming: a microcosm experiment. *PLoS One* 8(6), e66653.

Gollner, S., Kaiser, S., Menzel, L., Jones, D.O.B., Brown, A. *et al.* (2017) Resilience of benthic deep-sea fauna to mining activities. *Marine Environmental Research* 129, 76–101.

Goodsell, P.J., Underwood, A.J. and Chapman, M.G., (2009) Evidence necessary for taxa to be reliable indicators of environmental conditions or impacts. *Marine Pollution Bulletin* 58, 323–331.

Grego, M., De Troch, M., Forte, J. and Malej, A. (2009) Main meiofauna taxa as an indicator for assessing the spatial and seasonal impact of fish farming. *Marine Pollution Bulletin* 58(8), 1178–1186.

Haegerbaeumer, A., Mueller, M.-T., Fueser, H. and Traunspurger, W. (2019) Impacts of micro-and nano-sized plastic particles on benthic invertebrates: a literature review and gap analysis. *Frontiers in Environmental Science* 7, 17.

Hägerbäumer, A., Hoss, S., Heininger, P. and Traunspurger, W. (2015) Experimental studies with nematodes in ecotoxicology: an overview. *Journal of Nematology* 47(1), 11–27.

Hauquier, F., Macheriotou, L., Bezerra, T.N., Egho, G., Martínez Arbizu, P. and Vanreusel, A. (2019) Distribution of free-living marine nematodes in the Clarion–Clipperton Zone: implications for future deep-sea mining scenarios. *Biogeosciences* 16, 3475–3489.

Hedfi, A., Ali, M.B., Korkobi, M., Allouche, M., Harrath, A.H. *et al.* (2022) The exposure to polyvinyl chloride microplastics and chrysene induces multiple changes in the structure and functionality of marine meiobenthic communities. *Journal of Hazardous Materials* 436, 129161.

Hering, D., Moog, O., Sandelin, L. and Verdonschot, P.F.M. (2004) Overview and application of the AQEM assessment system. *Hydrobiologia* 516, 1–20.

Hinz, H., Hiddink, J.G., Forde, J. and Kaiser, M.J. (2008) Large-scale responses of nematode communities to chronic otter-trawl disturbance. *Canadian Journal of Fisheries and Aquatic Sciences* 65, 723–732.

Holmer, M. (2010) Environmental issues of fish farming in offshore waters: perspectives, concerns and research needs. *Aquaculture Environment Interactions* 1, 57–70.

Höss, S. and Traunspurger, W. (2021) Freshwater nematodes as bioindicators in field studies – the Nema-SPEAR[%]-index. In: Traunspurger, W. (ed.) *Ecology of Freshwater Nematodes*. CAB International, Wallingford, UK, pp. 323–340.

Höss, S., Claus, E., Von der Ohe, P.C., Brinke, M., Güde, H. *et al.* (2011) Nematode species at risk – a metric to assess pollution in soft sediments of freshwaters. *Environment International* 37, 940–949.

Höss, S., Heininger, P., Claus, E., Möhlenkamp, C., Brinke, M. *et al.* (2017) Validating the Nema-SPEAR[%]-index for assessing sediment quality regarding chemical-induced effects on benthic communities in rivers. *Ecological Indicators* 73, 52–60.

Höss, S., Gansfort, B. and Traunspurger, W. (2022) Nematodes as ecological and environmental indicators. In: Glazer, I., Shapiro-Ilan, D. and Sternberg, P.W. (eds) *Nematodes as Model Organisms*. CAB International, Wallingford, UK, pp. 165–195.

Ingels, J., Dashfield, S., Somerfield, P., Widdicombe, S. and Austen, M. (2014) Interactions between multiple large macrofauna species and nematode communities – mechanisms for indirect impacts of trawling disturbance. *Journal of Experimental Marine Biology and Ecology* 456, 41–49.
Jacob, J., Rayaroth, A.P., Useph, A.J.K., Philip, R. and Damodaran, R. (2020) A novel species *Spinonema gracilispiculum* sp. n. (Nematoda: Desmodoridae) from the oxygen minimum zone of Eastern Arabian Sea margin. *Zootaxa* 4869(4), zootaxa.4869.4.6.
Jones, D.O.B., Kaiser, S., Sweetman, A. K., Smith, C.R., Menot, L. *et al.* (2017) Biological responses to disturbance from simulated deep-sea polymetallic nodule mining. *PLoS One* 12(2), e0171750.
Lacoste, É., Boufahja, F., Pelaprat, C., Le Gall, P., Berteaux, T. *et al.* (2020) First simultaneous assessment of macro-and meiobenthic community response to juvenile shellfish culture in a Mediterranean coastal lagoon (Thau, France). *Ecological Indicators* 115, 106462.
Lake, P.S. (2000) Disturbance, patchiness, and diversity in streams. *Journal of the North American Benthological Society* 19, 573–592.
Lee, M.R., Torres, R. and Manríquez, P.H. (2017) The combined effects of ocean warming and acidification on shallow-water meiofaunal assemblages. *Marine Environmental Research* 131, 1–9.
Liess, M. and Von der Ohe, P.C. (2005) Analyzing effects of pesticides on invertebrates in streams. *Environmental Toxicology and Chemistry* 24, 954–965.
Liu, X.-S., Cheung, S.G. and Shin, P.K.S. (2009) Meiofauna with special reference to nematodes in trawling ground of subtropical Hong Kong. *Marine Pollution Bulletin* 58(4), 607–615.
Liu, X.S., Xu, W.Z., Cheung, S.G. and Shin, P.K.S. (2011) Response of meiofaunal community with special reference to nematodes upon deployment of artificial reefs and cessation of bottom trawling in subtropical waters, Hong Kong. *Marine Pollution Bulletin* 63(5–12), 376–384.
Long, A.M., Jurgensen, S.K., Petchel, A.R., Savoie, E.R. and Brum, J.R. (2021) Microbial ecology of oxygen minimum zones amidst ocean deoxygenation. *Frontiers in Microbiology* 12, 748961.
Losi, V., Moreno, M., Gaozza, L., Vezzulli, L., Fabiano, M. and Albertelli, G. (2013) Nematode biomass and allometric attributes as indicators of environmental quality in a Mediterranean harbour (Ligurian Sea, Italy). *Ecological Indicators* 30, 80–89.
Macheriotou, L., Rigaux, A., Derycke, S. and Vanreusel, A. (2020) Phylogenetic clustering and rarity imply risk of local species extinction in prospective deep-sea mining areas of the Clarion–Clipperton Fracture Zone. *Proceedings of the Royal Society Part B* 287, 20192666.
Mahatma, R. (2009) Meiofauna communities of the Pacific Nodule Province: abundance, diversity and community structure. PhD thesis, Universität Oldenburg, Germany.
Majdi, N., Colls, M., Weiss, L., Acuña, V., Sabater, S. and Traunspurger, W. (2020a) Duration and frequency of non-flow periods affect the abundance and diversity of stream meiofauna. *Freshwater Biology* 65, 1906–1922.
Majdi, N., Uthoff, J., Traunspurger, W., Laffaille, P. and Maire, A. (2020b) Effect of water warming on the structure of biofilm-dwelling communities. *Ecological Indicators* 117, 106622.
Meadows, A.S., Ingels, J., Widdicombe, S., Hale, R. and Rundle, S.D. (2015) Effects of elevated CO_2 and temperature on an intertidal meiobenthic community. *Journal of Experimental Marine Biology and Ecology* 469, 44–56.
Mevenkamp, L., Brown, A., Hauton, C., Kordas, A., Thatje, S. and Vanreusel, A. (2017) Hydrostatic pressure and temperature affect the tolerance of the free-living marine nematode *Halomonhystera disjuncta* to acute copper exposure. *Aquatic Toxicology* 192, 178–183.
Mevenkamp, L., Ong, E.Z., Van Colen, C., Vanreusel, A. and Guilini, K. (2018) Combined, short-term exposure to reduced seawater pH and elevated temperature induces community shifts in an intertidal meiobenthic assemblage. *Marine Environmental Research* 133, 32–44.
Miljutin, D.M., Miljutina, M.A., Martinez Arbizu, P. and Galeron, J. (2011) Deep-sea nematode assemblage has not recovered 26 years after experimental mining of polymetallic nodules (Clarion–Clipperton Fracture Zone, Tropical Eastern Pacific). *Deep Sea Research Part I: Oceanographic Research Papers* 58, 885–897.
Miljutina, M.A., Miljutin, D.M., Mahatma, R. and Galéron, J. (2010) Deep-sea nematode assemblages of the Clarion–Clipperton Nodule Province (Tropical North-Eastern Pacific). *Marine Biodiversity* 40, 1–15.
Mirto, S., Bianchelli, S., Gambi, C., Krzelj, M., Pusceddu, A. *et al.* (2010) Fish-farm impact on metazoan meiofauna in the Mediterranean Sea: analysis of regional vs. habitat effects. *Marine Environmental Research* 69(1), 38–47.
Mirto, S., Gristina, M., Sinopoli, M., Maricchiolo, G., Genovese, L. *et al.* (2012) Meiofauna as an indicator for assessing the impact of fish farming at an exposed marine site. *Ecological Indicators* 18, 468–476.

Mirto, S., Arigò, C., Genovese, L., Pusceddu, A., Gambi, C. and Danovaro, R. (2014) Nematode assemblage response to fish-farm impact in vegetated (*Posidonia oceanica*) and non-vegetated habitats. *Aquaculture Environment Interactions* 5, 17–28.

Moreno, M., Ferrero, T.J., Gallizia, I., Vezzulli, L., Albertelli, G. and Fabiano, M. (2008) An assessment of the spatial heterogeneity of environmental disturbance within an enclosed harbour through the analysis of meiofauna and nematode assemblage. *Estuarine, Coastal and Shelf Science* 77, 565–576.

Moreno, M., Vezzulli, L., Marin, V., Laconi, P., Albertelli, G. and Fabiano, M. (2009) The use of meiofauna diversity as an indicator of pollution in harbours. *ICES Journal of Marine Science* 65, 1428–1435.

Moreno, M., Semprucci, F., Vezzulli, L., Balsamo, M., Fabiano, M. and Albertelli, G. (2011) The use of nematodes in assessing ecological quality status in the Mediterranean coastal ecosystems. *Ecological Indicators* 11, 328–336.

Neira, C., Decraemer, W. and Backeljau, T. (2005) A new species of *Glochinema* (Epsilonematidae: Nematoda) from the oxygen minimum zone off Baja California, NE Pacific and phylogenetic relationships at species level within the family. *Cahiers de Biologie Marine* 46, 105–126.

Neira, C., King, I., Mendoza, G., Sellanes, J., De Ley, P. and Levin, L.A. (2013) Nematode community structure along a central Chile margin transect influenced by the oxygen minimum zone. *Deep Sea Research Part I: Oceanographic Research Papers* 78, 1–15.

Netto, S.A. and Valgas, I. (2010) The response of nematode assemblages to intensive mussel farming in coastal sediments (Southern Brazil). *Environmental Monitoring and Assessment* 162, 81–93.

Pape, E., Bezerra, T.N., Hauquier, F. and Vanreusel, A. (2017) Limited spatial and temporal variability in meiofauna and nematode communities at distant but environmentally similar sites in an area of interest for deep-sea mining. *Frontiers in Marine Science* 4, 205.

Pape, E., Bezerra, T.N., Gheerardyn, H., Buydens, M., Kieswetter, A. and Vanreusel, A. (2021) Potential impacts of polymetallic nodule removal on deep-sea meiofauna. *Scientific Reports* 11, 19996.

Patrício, J., Adão, H., Neto, J.M., Alves, A.S., Traunspurger, W. and Marques, J.C. (2012) Do nematode and macrofauna assemblages provide similar ecological assessment information? *Ecological Indicators* 14, 124–137.

Pereira Pontes, L., Vafeiadou, A.-M., Lobato de Franca, F.J., Cavalcante, R.A., de Araújo França, D.A. *et al.* (2021) Toxic effects of phenanthrene intensify with an increase of temperature for the populations of a free-living nematode. *Ecological Indicators* 120, 106868.

Peters, L. and Traunspurger, W. (2005) Species distribution of free-living nematodes and other meiofauna in littoral periphyton communities of lakes. *Nematology* 7, 267–280.

Pinto, T.K., Nascimento, R.L., Ferreira, R.C., dos Santos, W.A., Hamilton, S. and Cavalli, R.O. (2021) Impact of cage aquaculture of cobia (*Rachycentron canadum*) on nematode assemblages in the tropics. *Regional Studies in Marine Science* 48, 102056.

Pusceddu, A., Bianchelli, S., Martín, J., Puig, P., Palanques, A. *et al.* (2014) Chronic and intensive bottom trawling impairs deep-sea biodiversity and ecosystem functioning. *Proceedings of the National Academy of Sciences USA* 111(24), 8861–8866.

Ridall, A. and Ingels, J. (2021) Suitability of free-living marine nematodes as bioindicators: status and future considerations. *Frontiers in Marine Science* 8, 685327.

Ristau, K., Faupel, M. and Traunspurger, W. (2012) The effects of nutrient enrichment on a freshwater meiofaunal assemblage. *Freshwater Biology* 57, 824–834.

Ristau, K., Spann, N. and Traunspurger, W. (2015) Species and trait compositions of freshwater nematodes as indicative descriptors of lake eutrophication. *Ecological Indicators* 53, 196–205.

Ryu, J., Khim, J.S., Kang, S.-G., Kang, D., Lee, C.-H. and Koh, C.-H. (2011) The impact of heavy metal pollution gradients in sediments on benthic macrofauna at population and community levels. *Environmental Pollution* 159, 2622–2629.

Sandin, L. and Hering, D. (2004) Comparing macroinvertebrate indices to detect organic pollution across Europe: a contribution to the EC Water Framework Directive intercalibration. *Hydrobiologia* 516, 55–68.

Schaanning, M.T., Trannum, H.C., Øxnevad, S. and Ndungu, K. (2019) Benthic community status and mobilization of Ni, Cu and Co at abandoned sea deposits for mine tailings in SW Norway. *Marine Pollution Bulletin* 141, 318–331.

Schratzberger, M. and Ingels, J. (2018) Meiofauna matters: the roles of meiofauna in benthic ecosystems. *Journal of Experimental Marine Biology and Ecology* 502, 12–25.

Schratzberger, M. and Somerfield, P.J. (2020) Effects of widespread human disturbances in the marine environment suggest a new agenda for meiofauna research is needed. *Science of The Total Environment* 728, 138435.

Schratzberger, M., Warr, K. and Rogers, S.I. (2006) Patterns of nematode populations in the southwestern North Sea and their link to other components of the benthic fauna. *Journal of Sea Research* 55, 113–127.

Schratzberger, M., Warr, K. and Rogers, S.I. (2007) Functional diversity of nematode communities in the southwestern North Sea. *Marine Environmental Research* 63, 368–389.

Schratzberger, M., Danovaro, R., Ingels, J., Montagna, P.A., Rohal, L.M. *et al.* (2023) Hidden players – meiofauna mediate ecosystem effects of anthropogenic disturbances in the ocean. In: Giere, O. and Schratzberger, M. (eds) *New Horizons in Meiobenthos Research: Profiles, Patterns and Potential*. Springer, Cham, Switzerland, pp. 175–226.

Semprucci, F. and Balsamo, M. (2012) Free-living marine nematodes as bioindicators: past, present and future perspectives. *Environmental Research* 6(1), 17–35.

Semprucci, F., Boi, P., Manti, A., Covazzi-Harriague, A., Rocchi, M. *et al.* (2010) Benthic communities along a littoral of the Central Adriatic Sea (Italy). *Helgoland Marine Research* 64, 101–115.

Semprucci, F., Colantoni, P. and Balsamo, M. (2016) Is maturity index an efficient tool to assess the effects of the physical disturbance on the marine nematode assemblages? – A critical interpretation of disturbance-induced maturity successions in some study cases in Maldives. *Acta Oceanologica Sinologica* 35, 89–98.

Sergeeva, N.G., Urkmez, D. and Revkova, T. (2021) Meiobenthic nematodes at the deep oxic/anoxic boundary of the Black Sea (Istanbul Strait Outlet Area) with new records for Turkey. *Regional Studies in Marine Science* 46, 101904.

Simon-Lledó, E., Bett, B.J., Huvenne, V.A.I., Köser, K., Schoening, T. *et al.* (2019) Biological effects 26 years after simulated deep-sea mining. *Scientific Reports* 9, 8040.

Singh, R. and Ingole, B.S. (2016) Structure and function of nematode communities across the Indian western continental margin and its oxygen minimum zone. *Biogeosciences* 13, 191–209.

Singh, R., Sautya, S. and Ingole, B.S. (2019) The community structure of the deep-sea nematode community associated with polymetallic nodules in the Central Indian Ocean Basin. *Deep Sea Research Part II: Topical Studies in Oceanography* 161, 16–28.

Sonne, A.T., McKnight, U.S., Rønde, V. and Bjerg, P.L. (2017) Assessing the chemical contamination dynamics in a mixed land use stream system. *Water Research* 125, 141–151.

Sonne, A.T., Rasmussen, J.J., Höss, S., Traunspurger, W., Bjerg, P.L. and McKnight, U.S. (2018) Linking ecological health to co-occurring organic and inorganic chemical stressors in a groundwater-fed stream system. *Science of The Total Environment* 642, 1153–1162.

Steyaert, M., Moodley, L., Nadong, T., Moens, T., Soetaert, K. and Vincx, M. (2007) Responses of intertidal nematodes to short-term anoxic events. *Journal of Experimental Marine Biology and Ecology* 345, 175–184.

Sutherland, T.F., Levings, C.D., Petersen, S.A., Poon, P. and Piercey, B. (2007) The use of meiofauna as an indicator of benthic organic enrichment associated with salmonid aquaculture. *Marine Pollution Bulletin* 54(8), 1249–1261.

Teixidó, N., Gambi, M.C., Parravacini, V., Kroeker, K., Micheli, F. *et al.* (2018) Functional biodiversity loss along natural CO_2 gradients. *Nature Communications* 9(1), 5149.

Tinlin-Mackenzie, A., Sugden, H., Scott, C.L., Kennedy, R. and Fitzsimmons, C. (2023) Trawling for evidence: an ecosystem-based multi-method trawling impact assessment. *Fisheries Research* 268, 106858.

Tirelli, V., Suaria, G. and Lusher, A.L. (2022) Microplastics in polar samples. In: Rocha-Santos, T., Costa, M.F. and Mouneyrac, C. (eds) *Handbook of Microplastics in the Environment*. Springer, Cham, Switzerland, pp. 281–322. DOI: 10.1007/978-3-030-39041-9_4.

Tran, T.T., Yen, N.T.M., Quang, N.X., Veettil, B.K., Nguyen, T. *et al.* (2022) Maturity index of nematode communities and its potential use for measuring environmental disturbances with special emphasis on effects of organic enrichment in mangrove ecosystems. *Thalassas* 38, 355–366.

Traunspurger, W. (2021) *Ecology of Freshwater Nematodes*. CAB International, Wallingford, UK.

Ullah, H., Nagelkerken, I., Goldenberg, S.U. and Fordham, D.A. (2018) Climate change could drive marine food web collapse through altered trophic flows and cyanobacterial proliferation. *PLoS Biology* 16(1), e2003446.

Vafeiadou, A.M. and Moens, T. (2021) Effects of temperature and interspecific competition on population fitness of free-living marine nematodes. *Ecological Indicators* 120, 106958.

Vafeiadou, A.M., Bretaña, B.L.P., Van Colen, C., dos Santos, G.A.P. and Moens, T. (2018a) Global warming-induced temperature effects to intertidal tropical and temperate meiobenthic communities. *Marine Environmental Research* 142, 163–177.

Vafeiadou, A.M., Chintiroglou, C.C. and Moens, T. (2018b) Effects of an increased temperature regime on the population dynamics and species interactions of marine nematodes. *Journal of Experimental Marine Biology and Ecology* 502, 142–152.

Van Campenhout, J., Derycke, S., Moens, T. and Vanreusel, A. (2014) Differences in life histories refute ecological equivalence of cryptic species and provide clues to the origin of bathyal Halomonhystera (Nematoda). *PLoS One* 9, e111889.

Van Dover, C.L. (2014) Impacts of anthropogenic disturbances at deep-sea hydrothermal vent ecosystems: a review. *Marine Environmental Resources* 102, 59–72.

van Dover, C.L., Ardron, J.A., Escobar, E., Gianni M., Gjerde, K.M. *et al.* (2017) Biodiversity loss from deep-sea mining. *Nature Geoscience* 10, 464–465.

Vanreusel, A., Hilario, A., Ribeiro, P.A., Menot, L. and Arbizu, P.M. (2016) Threatened by mining, polymetallic nodules are required to preserve abyssal epifauna. *Scientific Reports* 6, 26808.

Venekey, V. and Melo, T.P.G.D. (2016) Nematodes as indicators of shrimp farm impact on an Amazonian estuary (Curuçá, Pará, Brazil). *Brazilian Journal of Oceanography* 64, 75–87.

Vezzulli, L., Moreno, M., Marin, V., Pezzati, E., Bartoli, M. and Fabiano, M. (2008) Organic waste impact of capture-based Atlantic bluefin tuna aquaculture at an exposed site in the Mediterranean Sea. *Estuarine, Coastal and Shelf Science* 78(2), 369–384.

Wilson, M.J. and Kakouli-Duarte, T. (eds) (2009) *Nematodes as Environmental Indicators*. CAB International, Wallingford, UK.

Wolfram, G., Orendt, C., Höss, S., Großschartner, M., Adamek, Z. *et al.* (2010) The macroinvertebrate and nematode community from soft sediments in impounded sections of the river Elbe near Pardubice, Czech Republic. *Lauterbornia* 69, 87–105.

Wolfram, G., Höss, S., Orendt, C., Schmitt, C., Adamek, Z. *et al.* (2012) Assessing the impact of chemical pollution on benthic invertebrates from three different European rivers using a weight-of-evidence approach. *Science of The Total Environment* 438, 498–509.

Wright, S.L., Thompson, R.C. and Galloway, T.S. (2013) The physical impacts of microplastics on marine organisms: a review. *Environmental Pollution* 178, 483–492.

Xu, W., Shin, P.K.S. and Sun, J. (2022) Organic enrichment induces shifts in the trophic position of infauna in a subtropical benthic food web, Hong Kong. *Frontiers in Marine Science* 9, 937477. DOI: 10.3389/fmars.2022.937477.

Zeppilli, D., Sarrazin, J., Leduc, D., Martinez Arbizu, P., Fontaneto, D. *et al.* (2015) Is the meiofauna a good indicator for climate change and anthropogenic impacts? *Marine Biodiversity* 45, 505–535.

Zullini, A. (1976) Nematodes as indicators of river pollution. *Nematologia Mediterranea* 4, 13–22.

10 Nematodes in Extreme Environments

Xia Xue[1]*, Diana H. Wall[†2] and Byron J. Adams[3]
[1]*The Fifth Affiliated Hospital of Zhengzhou University, Zhengzhou, China;* [2]*Formerly Department of Biology and School of Global Environmental Sustainability, Fort Collins, Colorado, USA;* [3]*Department of Biology, Brigham Young University, and Monte L. Bean Life Science Museum, Provo, Utah, USA*

Abstract

Nematodes inhabit all extreme ecosystems on Earth. They tolerate multiple extreme environmental stressors, including extreme temperatures, low and high pH, salinity, desiccation, freezing, hypoxia and ultraviolet radiation. They present a number of adaptations that allow them to survive under such conditions, so a number of endemic nematode species inhabit harsh ecosystems such as hot and cold deserts, Arctic and alpine regions, deep-sea sediments, and alkaline and hypersaline lakes. The nematode *Scottnema lindsayae*, for example, is the most abundant metazoan in the dry deserts of Antarctica, while bacterivores and some plant parasites are abundant in the Sahara, Mojave and Negev deserts. The study of the diversity of adaptations of extremotolerant nematode species, such as anhydrobiosis, cryoprotectant and antifreeze proteins synthesis, and DNA repair pathways, provide fundamental insight into the adaptation of life to environmental variation.

10.1 Introduction

Extreme environments are common on Earth, often characterized by challenging physicochemical conditions like extreme temperatures, limited water, and high or low pressures (Amils *et al.*, 2007; Wharton, 2007; Block *et al.*, 2009). These conditions also introduce challenges to sampling, culturing and studying life in these habitats. However, many organisms, including nematodes, have evolved adaptations to survive in extreme environments (Amils *et al.*, 2007; Wharton, 2007). Nematodes are essential to the functioning of most ecosystems and play important roles in the biosphere (Yeates, 2003; Ferris, 2010; Nielsen *et al.*, 2014). They are found in a wide range of environments (Schratzberger *et al.*, 2019; Norton and Niblack 2020), including polar, Antarctic, high-elevation and desert environments (McSorley, 2003; Sapir, 2021). They can tolerate or even thrive under extreme environmental stressors (Sapir, 2021; Ingels *et al.*, 2023). Most extreme environments have more than one extreme condition, for instance polar deserts, which are both extremely cold and dry (Campbell *et al.*, 1998; Wingfield *et al.*, 2011). Nematodes isolated from these habitats can tolerate multiple stressors, including extreme temperatures, pH, salinity, desiccation, radiation and hypoxia (Treonis and Wall, 2005; Yan *et al.*, 2018). In parallel, nematodes are sensitive to environmental changes and can survive climate-induced

*Email: xue-xiasophia@hotmail.com

 Nematodes as Environmental Indicators: From Theory to Practice, 2nd Edition (T. Kakouli-Duarte *et al.* eds)
DOI: 10.1079/9781800624221.0010

environmental changes via acclimatization, adaptation or migration (Callaghan *et al.*, 2004; Hauquier and Vanreusel, 2012).

10.2 Nematodes in Extreme Environments

Abiotic factors, such as temperature and soil chemistry, greatly influence the distribution ecology of free-living nematodes (Stuart *et al.*, 2015; Sánchez-Moreno and Ferris, 2018; Liu *et al.*, 2019; van den Hoogen *et al.*, 2019), especially in extreme environments (Aanderud *et al.*, 2018; Caruso *et al.*, 2019; Franco *et al.*, 2022). These nematodes have evolved various strategies to adapt to harsh conditions, notably anhydrobiosis, a state of suspended animation induced by desiccation (Crowe and Madin, 1974; Demeure *et al.*, 1979; McGill *et al.*, 2015). While extreme environments can limit nematode distribution, they can also exert strong, directional selection pressures on nematode populations to better survive extreme environmental stressors (Wall and Viginia, 1999; Zullini, 2018; Xue *et al.*, 2021).

For example, in the harsh environment of the Antarctic McMurdo Dry Valleys (MDVs), life thrives in suitable soil habitats (Wall and Viginia, 1999), where extreme temperatures above the thermal optimum for nematodes can expose populations to the risk of heat or freezing, which is common in polar and high-altitude regions (Wharton, 2007; Block *et al.*, 2009). Extremotolerant nematodes have also been found in alkaline, hypersaline and arsenic-rich lake sediments (Shih *et al.*, 2019; Kanzaki *et al.*, 2021; Jung *et al.*, 2024), soda lakes, deep-sea brine pools, thermal springs, heavy-metal-polluted soils, and hot and polar desert soils. The predominance of nematodes in these extreme habitats suggests their widespread exaptation to such conditions (Adhikari *et al.*, 2009; Rebecchi *et al.*, 2020; Robinson *et al.*, 2023).

Several inherent characteristics contribute to the successful adaptation of nematodes to extreme habitats, such as their conserved cylindrical body, lack of appendages, tolerance to low-oxygen conditions, simple energy requirements, multiple strategies of suspended animation and exaptations for cross-tolerance to multiple stressors (Sapir, 2021). These characteristics, acting additively or synergistically, enable nematodes to thrive in a wide range of environments. Understanding the distribution patterns and adaptation of nematodes is crucial, as they play an important role in how terrestrial ecosystems respond to global environmental change (Wall, 2007; Nielsen and Wall, 2013; Siebert *et al.*, 2020; Khanal and Land, 2023).

The distribution pattern of nematodes is a potential indicator of future environmental change. However, nematodes evolving from standing genetic variation may face challenges to adapting to rapid environmental changes, as the most suitable genotypes for extreme environments may be lost through genetic drift or founder effects.

10.2.1 Cold deserts

In contrast to the heat and aridity of hot deserts, cold deserts present nematodes with challenges of extreme cold and desiccation, necessitating different but equally remarkable adaptations to survive. The MDVs are the largest ice-free region of Antarctica and perhaps the quintessential example of a cold desert. Here, nematodes are primarily found in mineral soils, and in sediments of ephemeral glacial melt streams and permanently ice-covered lakes (Wall and Viginia, 1999; Wharton, 2003; Wharton and Brown, 2012). Five nematode genera are endemic in MDV soils, including *Plectus*, *Scottnema*, *Panagrolaimus*, *Geomonhystera* and *Eudorylaimus* (Sohlenius *et al.*, 1996; Andrassy, 1998; Simmons *et al.*, 2009; Adams *et al.*, 2014). MDV nematode species are distributed heterogeneously in this unique ecosystem, mainly due to variability in soil properties such as organic matter content, availability of liquid water and salinity (Courtright *et al.*, 2001; Adams *et al.*, 2014). This means that the nematodes are not evenly distributed throughout the MDVs but rather are found in certain areas where the soil is more conducive to their survival. The Antarctic desert is one of the driest and coldest soils on Earth, and Antarctic nematodes, such as *Scottnema lindsayae*, *Panagrolaimus* sp. DAW1 and *Plectus murrayi*, can enter anhydrobiosis to protect themselves under dehydration conditions (Crowe *et al.*, 1977; Demeure *et al.*, 1979; Treonis *et al.*, 2000; Wharton and Raymond, 2015).

The distribution ecology of nematodes in the Antarctic is still being studied, but it is clear that nematodes play an important role in the ecological functioning of this unique ecosystem (Adams *et al.*, 2006; Velasco-Castrillón *et al.*, 2014). Of the 54 nematode species thus far identified in Antarctica, 22 are endemic to the ice-free soil environment (Terauds and Lee, 2016). Because of the extreme cold temperatures, desiccated soils, and cold, dark and long winters (Freckman and Virginia, 1997), nematodes are highly specialized but differ in their suitability to various habitats across the landscape, such as adaptative responses to freezing (Nkem *et al.*, 2005; Adhikari *et al.*, 2010a; McGill *et al.*, 2015; Wharton and Raymond, 2015) and dehydration (Adhikari *et al.*, 2010a; Wharton, 2015). *S. lindsayae* is the dominant nematode species in abundance and distribution around MDV soils (Zawierucha *et al.*, 2019). This species has a broad ecological amplitude or niche space, demonstrating high tolerance for desiccation and high salinity as compared to other Antarctic nematodes (Caruso *et al.*, 2019). Thus, *Scottnema* persists across the widest range of soil habitats in Antarctica (Adams *et al.*, 2006, 2014). For *Plectus* and *Eudorylaimus*, soils with higher moisture and more organic soil carbon are preferred. Thus, most *Plectus* are found near the wetted margins of streams and lakes, while *Eudorylaimus* are found in the soil with lower salinity (Adams *et al.*, 2006, 2014). The investigation so far remains limited, therefore the distribution of some nematodes is less understood, such as *Geomonhystera* (Adams *et al.*, 2006, 2014). Notably, the spatial distribution of soil nematode populations in Antarctic terrestrial ecosystems is tightly regulated by environmental factors, making them highly sensitive to changes in soil properties (Robinson *et al.*, 2023; Santos *et al.*, 2023).

10.2.2 Hot deserts

Hot deserts comprise a diverse group of soil habitats that are characterized by limited moisture and reduced plant biomass. These conditions make it challenging for carbon, nitrogen and other nutrients to be stored or cycled through the soil food web (Hadley and Szarek, 1981). In desert ecosystems, nematodes play important roles in nutrient cycling and decomposition (Freckman *et al.*, 1987). Like other extreme ecosystems, soil temperature, moisture and nutrient levels in desert soils are the main factors that determine nematode distribution (Freckman and Virginia, 1997; Zhang *et al.*, 2020), but biotic drivers are also important (Marais *et al.*, 2020). Nematodes in hot desert soils are mostly found in the upper soil layer (around 10 cm depth from the top) and mainly feeding on microorganisms (Whitford, 1996; Hu *et al.*, 2020; Treonis *et al.*, 2022). Several strategies related to stress tolerance are implicated in nematode survival and ecosystem functioning in hot desert soils, such as cryptobiosis and the production of heat-shock proteins that help protect cellular structures and enzymes from heat-induced damage (Treonis *et al.*, 2000; Wharton *et al.*, 2005; McGill *et al.*, 2015; Raymond and Wharton, 2016; Treonis *et al.*, 2022).

When plants are present, common nematodes of hot deserts include *Pratylenchus thornei* and *Helicotylenchus dihystera*, root-feeding parasites, both of them present in the Sonoran and Saharan deserts (Bohra and Sultana, 2012). In the Mojave Desert, *Mesodorylaimus obscurus*, *Pararhyssocolpus desertus* and *Tylenchus* sp. are the dominant metazoans that play important roles in soil nutrient cycling and decomposition (Whitford *et al.*, 1982; Treonis *et al.*, 2012). In the Negev, *Tylenchorhynchus*, *Acrobeles*, *Acrobeloides* and *Cervidellus* are the most abundant taxa. Freckman and Virginia (1989) investigated the spatial and seasonal distribution of plant-feeding nematodes and the potential significance of herbivory in four mesquite (*Prosopis* spp.) communities with differing rooting depths and distributions. They found that the majority of nematodes and roots in the playa system were located below a soil depth of 0.5 m, while in the creosote bush (*Larrea tridentata*) system, most roots and nematodes were recovered in the upper 0.5 m of soil. Pen-Mouratov *et al.* discovered that the upper 0–10 cm of soil contained the highest abundance of soil free-living nematodes (Pen-Mouratov *et al.*, 2008a,b). They also found that the fungivore/bacterivore (F/B) ratio was a useful ecological index for reflecting the vertical distribution of nematode communities under different plant species and soil layers.

10.2.3 Arctic and alpine regions

Nematode biodiversity and abundance are much higher in Arctic regions than in Antarctica (Procter, 1984). In most Arctic and alpine regions, nematodes are primarily found in soils with relatively high moisture and productivity (Kergunteuil *et al.*, 2016; Semeraro *et al.*, 2022). Meanwhile, environmental factors including temperature, soil nutrient levels and biotic interactions are important drivers for nematode distribution in alpine soil ecosystems as well (Porazinska *et al.*, 2021). Even at high latitudes, nematodes tend to be more abundant at lower elevations and warmer temperatures, where soil and vegetation are more plentiful (Bardgett, 2005). However, several species of nematodes have been found at higher altitudes, including alpine meadows and tundra habitats (Kergunteuil *et al.*, 2016; Semeraro *et al.*, 2022). As with Antarctic nematodes, those from Arctic and alpine soil ecosystems have also evolved stress response mechanisms for dealing with cold and dry conditions that are likely similar to those of Antarctic species (Mulvey, 1969). Holovachov (2014) provides a comprehensive overview of nematodes reported from the entire Arctic region. Not surprisingly, species endemic to Antarctica are congeners of those in the MDVs, including *Plectus*, *Eudorylaimus* and *Geomonhystera*. *Scottnema* is a monospecific genus found only in Antarctica, but the closely related *Cephalobus* genus is found in Arctic deserts (Nadler *et al.*, 2006, Holovachov 2014).

Moving to Northern Tibet, which experiences a dry and cold climate, the alpine meadows in the region are of great ecological importance. Liu *et al.* (2022) identified three representative plant communities (*Potentilla* spp., *Kobresia pygmaea* and *Kobresia littledalei*) and studied the soil nematode communities associated with each community. The study area had seasonally frozen soil, and soil nematodes were collected using the shallow basin method. The researchers analysed various soil chemical properties and examined the trophic composition, colonizer–persister (*c-p*) group structure, functional structure index, and their relationship with soil properties. They collected a total of 33,038 nematodes belonging to different taxonomic groups and found variations in nematode community composition across different plant communities in the alpine meadows, with herbivorous and bacterivorous nematodes being the dominant groups. Azzoni *et al.* (2015) conducted a study analysing the nematode and rotifer fauna on the Miage Glacier and the Belvedere Glacier, which are the two largest debris-covered glaciers in the Italian Alps (Azzoni *et al.*, 2015). They found that the dominant genus of nematodes was *Plectus*, which is (as mentioned above) commonly found in habitats at high latitudes and altitudes.

Some studies have reported that soil pH is another key factor driving the diversity of nematodes in high-altitude and alpine areas due to its impacts on the trophic structure of soils (Kashyap *et al.*, 2022), while soil nitrogen and carbon are more tightly linked to nematode abundance in the Arctic ecosystem. For example, bacterivorous nematodes are abundant in soils enriched in nitrogen, while nematodes are loosly linked to carbon cycling in high-mountain dry shrubs at altitude (Savin *et al.*, 2001; Postma-Blaauw *et al.*, 2005; Afzal *et al.*, 2021). Hofmann *et al.* investigated the occurrence of methane-cycling microorganisms in high-alpine soils in the Austrian Alps (Hofmann *et al.*, 2016). They assessed the spatial distribution of these microorganisms along an altitudinal gradient and found that methanogens, which are involved in methane production, can persist in high-alpine cold soils. This research helps in understanding the potential impact of climate warming on high-altitude environments. Kouser *et al.* (2021) proposed expanding the use of nematode-derived ecological indices to link soil and climate properties with soil food webs and ecosystem processes along elevation gradients. They confirmed that the functional contribution of soil nematodes to below-ground ecosystem processes, including carbon and energy flow, is stronger at high elevations.

The diversity and abundance of soil nematodes decrease along elevation gradients in high-altitude regions, such as Gangotri National Park, the Banihal Pass of the Pir Panjal mountain range, and high-altitude areas of China (Kergunteuil *et al.*, 2016; Devetter *et al.*, 2017; Afzal *et al.*, 2021). In the European Alps, populations of nematodes increase in habitats that are less rocky, such as grasslands and meadows (Nagy and Grabherr, 2009; Renčo, 2012). Furthermore, season and interannual variation in weather also drive patterns of alpine nematode population structure (Jenkins *et al.*, 2006; Bhusal *et al.*, 2015). Here, nematode distribution patterns are highly related to biotic and abiotic conditions, especially from high-elevation alpine soil where nematode diversity can be mainly driven by local physicochemical factors as they occur along elevation gradients (Kergunteuil *et al.*, 2016; Devetter *et al.*, 2017).

10.3 Nematode Adaptation to Other Extreme Environments

The broad ecological amplitude of nematodes allows them to be compared across habitat types to study the evolution of survival strategies in response to diverse environmental stressors (Wharton, 2004; Yeates *et al.*, 2009; Norton and Niblack, 2020). Extreme environments pose significant threats to nematodes, since freezing temperatures, dehydration and intense pressures can directly damage nematode cells and disrupt their vital physiological processes (Wharton, 2007). When resources are limited, nematodes slow their metabolic processes and activities (e.g. reproduction) to maintain necessary biological processes (Wharton, 2007). Even though stress-tolerant nematodes have evolved several strategies to survive in harsh environments, their abundance and diversity are lower than those of nematode communities and populations in more hospitable habitats.

Nematodes are frequently the most abundant animal taxon found in extreme environments, including the deep-subsurface and deep-sea sediments (McSorley, 2003; Hodda *et al.*, 2006; Sapir, 2021). Although the reasons for the abundance of nematodes in extreme habitats remain unknown, a unique combination of characteristics of nematodes may explain their successful adaptation to extreme habitats.

10.3.1 pH

Extreme pH, including acidic and alkaline environments, threatens nematodes and other organisms living in such habitats (Šalamún *et al.*, 2014). Nematodes maintain their internal pH balance through unique physiological and biochemical bio-strategies, such as producing protective cuticles and altering ion exchange across cell membranes (Petronijevic and Rogers, 1987; Sims *et al.*, 1994; Cong *et al.*, 2020). Besides actively moving away from extreme pH areas, nematodes can produce a variety of enzymes and compounds that enable them to neutralize extreme pH in their immediate proximity (Stringham and Candido, 1994). It has been found that nematode adaptations to mitigate environments with extreme pH reveal fundamental mechanics of cellular physiology, biochemistry and behaviours that are involved in tolerating extremely harsh environments.

Halicephalobus mephisto has been reported from the deep subsurface of Earth where the soil pH is highly alkaline. At pH of 9.9, *H. mephisto* tolerates this harsh alkaline environment by way of a protective cuticle and enhanced ability to repair damaged DNA (Borgonie *et al.*, 2011; Weinstein 2019). *Plectus aquatilis*, a species found in alkaline lakes at the base of the Adobe Hills in California, can alter its ion regulation to balance its internal pH by producing ion transporters as well as a set of enzymes, such as alkaline proteases and chitinases, that break down organic matter and use it as resources (Winitsky, 2020). Steward *et al.* investigated the diversity of nitrogenase genes in Mono Lake, California, which is an alkaline and moderately hypersaline lake (Steward *et al.*, 2004). They found that the nematodes in Mono Lake play a role in nitrogen dynamics in this nitrogen-limited lake (Shih *et al.*, 2019; Sapir, 2021). Both protective cuticles and unique enzymes are essential for tolerating extremely alkaline environments.

Nematodes can also detect chemical or environmental cues to escape extreme acidic conditions (Stamps and Linit, 2001). To cope with an acidic environment, nematodes can alter their acid tolerance-related genes and protein expression, as well as regulate their cellular metabolism (Grewal *et al.*, 2011; Dhal and Pal, 2022). The model nematode *Caenorhabditis elegans* has been shown to acclimatize to extremely acidic environments (pH as low as 2.5) by maintaining cell-membrane pH homeostasis and preventing DNA damage (Parida, 2022). Additionally, nematodes have been shown to regulate their acid-tolerant metabolic pathways by adjusting the levels of calcium and potassium ions within their bodies (Butterworth and Barrett, 1985). Most importantly, the protective cuticle of most nematodes can shield them from the acidic threats of most environments by avoiding tissue damage (Bird, 1980).

10.3.2 Freeze tolerance

Freeze tolerance in nematodes was first demonstrated by Asahina (1959), who showed that *Aphelenchoides ritzemabosi* could survive freezing when subjected to a gradual, slow freeze. Following this, Sayre (1964) applied classification methods for freeze tolerance from arthropod research to nematodes, suggesting various strategies for cold survival. Nematodes employ freeze tolerance and avoidance strategies, enabling them to survive extreme cold, as demonstrated in both laboratory and field studies (Wharton, 2003; Smith *et al.*, 2008). For example, *Panagrolaimus* sp. DAW1 avoids freezing by establishing a supercooling point (Raymond and Wharton, 2016), while *Trichostrongylus colubriformis* larvae use outer sheaths for protection (Abbas and Hildreth, 2019).

Mechanistically, nematodes use cryoprotectants, dehydration, rapid cold hardening and antifreeze proteins to survive freezing temperatures (Wharton, 2003). The phenotypic plasticity of these strategies allows for enhanced freeze tolerance after acclimatization to sublethal temperatures (Mallard *et al.*, 2020; Leung *et al.*, 2023). The evolution of phenotypic plasticity likely involves multilevel selection, including DNA methylation, gene expression and cell morphology. In high-stress environments, freezing also induces osmotic pressure changes, further triggering dehydration, which prevents ice recrystallization during thawing, and facilitating anhydrobiosis (Wharton, 2003). Advances in molecular biology tools continue to uncover the evolutionary strategies these nematodes employ to cope with freezing conditions.

For normal development in temperate and frigid regions, nematodes have evolved the ability to survive freezing conditions (Adhikari *et al.*, 2010b; Wharton and Raymond, 2015). Freeze tolerance is widespread among nematodes, but not all nematode species tolerate freezing (Wharton and Block, 1993; Treonis and Wall, 2005; Wharton and Raymond, 2015). Freezing tolerance strategies have been divided into freeze tolerance and avoidance (Wharton, 2003; Smith *et al.*, 2008). Numerous laboratory and field studies have shown that nematodes can withstand the low temperatures they may experience under natural conditions (McSorley, 2003; Wharton, 2003). For example, the infective larvae of the vertebrate parasite *T. colubriformis* can survive soil surface temperatures as low as –10°C (Abbas and Hildreth, 2019). The soil temperature where plant-parasitic northern root-knot nematodes (*Meloidogyne hapla*) survive in winter can reach –16°C (Wu *et al.*, 2018). Moreover, free-living nematodes from the Antarctic dry valleys, where soil surface temperatures frequently fall below –60°C, show a high tolerance to freezing temperatures as their populations are frequently the focus of ecological and evolutionary studies.

A. ritzemabosi, a plant-parasitic nematode that primarily affects the leaves of ornamental plants, has also evolved freezing tolerance (Asahina, 1959). One way by which nematodes protect their bodies from freezing and gain resistance to low temperatures is by avoiding freezing through protective dehydration (Wharton *et al.*, 2005). After removing surface water, *P. davidi* does not freeze in a low-temperature air medium (Wharton *et al.*, 2005). Its resistance strategy shifts to avoiding freezing, by establishing a supercooling point of at least –60°C (Raymond and Wharton, 2016). By combining cryoprotectants, dehydration-induced tolerance, acclimatization, rapid cold hardening and repair mechanisms, *P. davidi* has developed a remarkable strategy for withstanding freezing conditions and maintaining its viability (Wharton and Raymond, 2015; Raymond and Wharton, 2016). Although the third-instar larvae of *T. colubriformis* have protective outer sheaths, some individuals are still subject to inoculation-induced freezing, known as ice nucleation or ice nucleation activity (INA), which refers to the process by which certain microorganisms or biological substances can initiate the formation of ice crystals at temperatures above the freezing point of water (Wharton *et al.*, 2005). Moreover, *Panagrolaimus* sp. DAW1 can resist low-temperature stress by simultaneously adopting production of antifreeze proteins (e.g. LEA) (Seybold *et al.*, 2017; Wang *et al.*, 2020). It has been shown that the freezing tolerance of nematodes has certain phenotypic plasticity and can take optimal cold tolerance countermeasures according to environmental changes. Nematodes using freeze tolerance rely on ice active proteins (anti-freezing proteins), primarily functioning *in vivo* as recrystallization inhibitors, to prevent damaging ice crystal growth during exposure to sublethal low temperatures (Adhikari *et al.*, 2010b; Wharton and Raymond, 2015). Freezing tolerance of certain nematode species has been shown to be enhanced after acclimatization to sublethal cold temperatures (Asahina, 1959; Adhikari *et al.*, 2010b; Wharton and Raymond, 2015). When nematodes are subjected to low-temperature stress, it is also accompanied by other stresses (Suzuki *et al.*, 2014). For example, when cytosolic water begins to freeze, salt ions are continuously expelled from the growing ice crystals, resulting in a significant increase in the osmotic concentration of the non-frozen portion (Wharton, 2003, 2004). Thus, nematodes are dehydrated under the stress of high osmotic pressure, eventually leading to the body avoiding inoculation ice and improving its cold tolerance (Charwat *et al.*, 2002). With advances in laboratory and analytical tools for molecular biology and 'omics, the study of nematode cold tolerance is expected to continue to reveal physiological and evolutionary insights into these unique adaptive strategies (Adhikari *et al.*, 2010b).

10.3.3 Heat tolerance and thermophily

To cope with the heat and high-temperature risks that limit survival and distribution, nematodes can shorten and thicken their body shape to maintain body water content (Chen *et al.*, 2020). Nematodes can reduce heat stress by elongating the body size to increase surface area to more efficiently dissipate heat (Wharton and Brown, 2012). The evolution of specialized physical structures, like large mouthparts and gut linings, enables nematodes to graze more efficiently under heat threats (Mukuka *et al.*, 2010; Gingold *et al.*, 2013). Under heat stress, the larger mouthparts can provide an advantage by allowing nematodes to more effectively puncture plant cells and access nutrients, while the specialized gut lining composed of cuticle layers helps them prevent desiccation (dehydration) by reducing water loss. The cuticle may have adaptations to resist damage from high temperatures, ensuring that the nematode's digestive system remains functional (Wharton, 2012). Additionally, nematodes are highly motile and can move towards cooler areas to avoid heat from direct sunlight, such as moving deeper underground or under vegetation (Wharton, 2012). To conserve water and energy under heat stress, nematodes slow their metabolism, alter their cell-membrane composition and upregulate heat-shock proteins (Wharton, 2004; Tahseen, 2012).

Thermophilic nematodes are those that prefer high temperatures, and their distribution ecology is strongly linked to environmental temperatures (Garrity *et al.*, 2010; Hoffmann *et al.*, 2013). Unlike most nematodes which typically inhabit soils habitats with a wide range of temperatures, thermophilic nematodes thrive in habitats with consistent, extremely high temperatures (Jana, 1973; Liu *et al.*, 2016; Bellec *et al.*, 2020). Moreover, they play essential roles in nutrient cycling in such ecosystems by being involved in the decomposition and cycling of organic matter under high temperatures (Steel *et al.*, 2013; Nisa *et al.*, 2021). However, the distribution of thermophilic nematodes is limited by the availability of suitable habitats, which are often localized around geothermal features.

Anhydrobiosis is an additional strategy to survive desiccation, freezing, heat or aridity. When in the state of anhydrobiosis, nematodes lose most of their body water and shut down their metabolic processes. In this state, nematodes can survive extremely high (or low) temperatures or desiccated conditions for extended periods of time (Crowe and Madin, 1974; Crowe *et al.*, 1977; Demeure *et al.*, 1979). To enter anhydrobiosis, nematodes produce special LEA (late embryogenesis abundant) proteins, which protect their cells from damage during dehydration (Hibshman and Goldstein, 2021). Nematodes also produce trehalose, a type of sugar that helps to stabilize their membranes and prevent damage from freezing or drying (Wharton *et al.*, 2005). In addition to entering anhydrobiosis, nematodes can burrow into soil or other substrates to avoid extreme conditions on the surface.

10.3.4 Ultraviolet radiation

Ultraviolet (UV) radiation threatens life on Earth by breaking DNA and proteins. It causes irreversible damage and limits organisms' distribution strongly in different habitats (Oarga, 2009). Nematodes are recognized as highly resistant to radiation compared with other animals by developing different adaptive strategies (Alicata, 1956; Offutt and Riggs, 1970; Yanase *et al.*, 1999). Nematodes have evolved pigments that act as natural UV absorbers, which are referred to as 'sunscreen pigments' (Schulze and Schierenberg, 2008). These pigments are distributed throughout the nematode's body, especially in the cuticle, and they have the capacity to absorb and dissipate UV radiation to prevent the radiation from penetrating deeper into the nematode's tissues and causing damage. In addition to pigments, nematodes possess specialized cuticular structures that aid in UV resistance, including reflective layers or thickened cuticles. Reflective layers can bounce back UV radiation, reducing its penetration into the nematode's body. Thickened cuticles provide an additional physical barrier that shields internal tissues from UV exposure. In parallel, by studying *C. elegans*, it has been found that nematodes can activate DNA repair pathways to cope with radiation damage, including base excision repair, nucleotide excision repair and double-strand-break repair (Dhakal *et al.*, 2021). These DNA repair pathways of nematodes can be triggered in response to radiation exposure to maintain genetic stability and enable them to survive under these threats. Nematodes can also produce protective molecules, including antioxidants and heat-shock proteins, to adapt to radiation stress (Zhao and Wang, 2012; Farahani *et al.*, 2020). Notably, nematodes are capable of entering anhydrobiosis during development, as staying in a state of suspended animation could help them avoid radiation damage (Watanabe *et al.*, 2007; Gusev *et al.*, 2010).

10.3.5 Global change

Climate change can have a significant impact on nematode distribution (Jenkins *et al.*, 2006; Nielsen *et al.*, 2011, 2014; Nielsen and Wall, 2013; Nielsen and King, 2015; Mueller *et al.*, 2016; Andriuzzi *et al.*, 2018; Siebert *et al.*, 2020). As temperatures rise, nematodes may expand their ranges to include previously inhospitable habitats (Jenkins *et al.*, 2006; Simmons *et al.*, 2009; Andriuzzi *et al.*, 2018). Additionally, changes in precipitation patterns can alter soil moisture content, affecting the survival and distribution of nematodes that rely on specific moisture conditions for survival. We must continue to monitor and study these impacts to understand better how climate change may affect nematode populations.

Some nematode species expand the ranges of their habitats as environmental temperatures rise, while others contract, leading to changes in the diversity, structure and functioning of their communities (Simmons *et al.*, 2009; Gingold *et al.*, 2013; Andriuzzi *et al.*, 2018). For example, *P. murrayi* in MDVs might be dominant instead of *S. lindsayae* due to its preference for relatively warmer and wetter conditions (Kowalewski *et al.*, 2006; Simmons *et al.*, 2009). Soil moisture conditions are one of the most important drivers of nematode distribution. Since most nematode species have evolved to narrow ranges of soil moisture tolerance, climate warming that results in changed precipitation regimes would produce changes in habitat suitability, resulting in changes in community structure (Zhou *et al.*, 2022). In addition, warming can indirectly change the distribution patterns of nematodes by affecting the distribution and abundance of other biota, both below- and above-ground (Somasekhar and Prasad, 2011). Altered biotic interactions and food web disruptions, such as changes in primary food sources, can drive changes in nematode population dynamics (Liu *et al.*, 2022; Gao *et al.*, 2023). A more detailed report on the effects of climate change on soil nematode communities can be read in Chapter 11 of this volume.

10.4 Conclusion

This chapter underscores the extraordinary adaptability of nematodes, highlighting their ecological significance and survival strategies in extreme environments. These insights not only enhance our understanding of life on Earth but also offer critical perspectives on ecological resilience and planetary protection. Research on these resilient organisms not only deepens our understanding of the boundaries of life on Earth but also expands our quest for life beyond our planet. The study of nematodes in extreme environments provides vital insights for planetary protection protocols, as their resilience mirrors potential life forms on other planets. Additionally, understanding their role in soil ecosystems can inform strategies for managing the impacts of climate change on global biodiversity and nutrient cycling. Furthermore, studying nematodes' adaptations provides critical insights into soil ecosystem resilience and its capacity to mitigate the impacts of climate change. As we unravel the secrets of nematodes' resilience, we gain a glimpse into the intricate web of life's adaptations, underscoring the importance of these diminutive yet formidable creatures in shaping the future of our planet. In this chapter, we delve into the significance of comprehending these seemingly insignificant yet resilient creatures and their potential implications for the future of terrestrial ecosystems.

References

Aanderud, Z.T., Saurey, S., Ball, B.A., Wall, D.H., Barrett, J.E. *et al.* (2018) Stoichiometric shifts in soil C:N:P promote bacterial taxa dominance, maintain biodiversity, and deconstruct community assemblages. *Frontiers in Microbiology* 9, 1401. DOI: 10.3389/fmicb.2018.01401.

Abbas, I. and Hildreth, M. (2019) Egg autofluorescence and options for detecting peanut agglutinin binding for the identification of *Haemonchus contortus* eggs in fecal samples. *Veterinary Parasitology* 267, 69–74.

Adams, B.J., Bardgett, R.D., Ayres, E., Wall, D.H., Aislabie, J. *et al.* (2006) Diversity and distribution of Victoria Land biota. *Soil Biology and Biochemistry* 38, 3003–3018. DOI: 10.1016/j.soilbio.2006.04.030.

Adams, B., Wall, D., Virginia, R., Broos, E. and Knox, M. (2014) Ecological biogeography of the terrestrial nematodes of Victoria Land, Antarctica. *ZooKeys* 419, 29–71. DOI: 10.3897/zookeys.419.7180.

Adhikari, B.N., Wall, D.H. and Adams, B.J. (2009) Desiccation survival in an Antarctic nematode: molecular analysis using expressed sequenced tags. *BMC Genomics* 10, 69. DOI: 10.1186/1471-2164-10-69.

Adhikari, B.N., Tomasel, C.M., Li, G., Wall, D.H. and Adams, B.J. (2010a) The Antarctic nematode *Plectus murrayi*: an emerging model to study multiple stress survival. *Cold Spring Harbor Protocols* 2010, 142. DOI: 10.1101/pdb.emo142.

Adhikari, B.N., Wall, D.H. and Adams, B.J. (2010b) Effect of slow desiccation and freezing on gene transcription and stress survival of an Antarctic nematode. *Journal of Experimental Biology* 213, 1803–1812. DOI: 10.1242/jeb.032268.

Afzal, S., Nesar, H., Imran, Z. and Ahmad, W. (2021) Altitudinal gradient affect abundance, diversity and metabolic footprint of soil nematodes in Banihal-Pass of Pir-Panjal mountain range. *Scientific Reports* 11, 16214.

Alicata, J.E. (1956) Observations on the possibility of developing a strain of *Trichinella spiralis* resistant to radiation. *Journal of Parasitology* 42, 656–657.

Amils, R., Ellis-Evans, C. and Hinghofer-Szalkay, H.G. (2007) *Life in Extreme Environments*. Springer, Dordrecht, The Netherlands.

Andrassy, I. (1998) Nematodes in the sixth continent. *Journal of Nematode Morphology and Systematics* 1, 107–186.

Andriuzzi, W.S., Adams, B.J., Barrett, J.E., Virginia, R.A. and Wall, D.H. (2018) Observed trends of soil fauna in the Antarctic Dry Valleys: early signs oswf shifts predicted under climate change. *Ecology* 99, 312–321. DOI: 10.1002/ecy.2090.

Asahina, E. (1959) Prefreezing as a method enabling animals to survive freezing at an extremely low temperature. *Nature* 184, 1003–1004.

Azzoni, R., Franzetti, A., Fontaneto, D., Zullini, A. and Ambrosini, R. (2015) Nematodes and rotifers on two Alpine debris-covered glaciers. *Italian Journal of Zoology* 82, 616–623.

Bardgett, R. (2005) *The Biology of Soil: A Community and Ecosystem Approach*. Oxford University Press, Oxford.

Bellec, L., Cambon-Bonavita, M.A., Durand, L., Aube, J., Gayet, N. *et al.* (2020) Microbial communities of the shallow-water hydrothermal vent near Naples, Italy, and chemosynthetic symbionts associated with a free-living marine nematode. *Frontiers in Microbiology* 11, 2023.

Bhusal, D.R., Tsiafouli, M.A. and Sgardelis, S.P. (2015) Temperature-based bioclimatic parameters can predict nematode metabolic footprints. *Oecologia* 179, 187–199.

Bird, A.F. (1980) The nematode cuticle and its surface. *Aging and Other Model Systems* 2, 213–236.

Block, W., Lewis Smith, R. and Kennedy, A. (2009) Strategies of survival and resource exploitation in the Antarctic fellfield ecosystem. *Biological Reviews* 84, 449–484.

Bohra, P. and Sultana, R. (2012) *Nematodes of Thar Desert, Rajasthan*. Occasional Paper No. 334. Zoological Survey of India, Kolkata, India.

Borgonie, G., García-Moyano, A., Litthauer, D., Bert, W., Bester, A. *et al.* (2011) Nematoda from the terrestrial deep subsurface of South Africa. *Nature* 474, 79–82.

Butterworth, P.E. and Barrett, J. (1985) Anaerobic metabolism in the free-living nematode *Panagrellus redivivus*. *Physiological Zoology* 58, 9–17.

Callaghan, T.V., Björn, L.O., Chernov, Y., Chapin, T., Christensen, T.R. *et al.* (2004) Biodiversity, distributions and adaptations of Arctic species in the context of environmental change. *AMBIO: A Journal of the Human Environment* 33, 404–417.

Campbell, I.B., Claridge, G.G.C., Campbell, D.I. and Balks, M.R. (1998) The soil environment of the McMurdo Dry Valleys, Antarctica. In: Priscu, J.C. (ed.) *Ecosystem Dynamics in a Polar Desert: the McMurdo Dry Valleys, Antarctica*. American Geophysical Union, Washington, DC, pp. 297–322. DOI: 10.1029/AR072p0297.

Caruso, T., Hogg, I.D., Nielsen, U.N., Bottos, E.M., Lee, C.K. *et al.* (2019) Nematodes in a polar desert reveal the relative role of biotic interactions in the coexistence of soil animals. *Communications Biology* 2, 63. DOI: 10.1038/s42003-018-0260-y.

Charwat, S.M., Fisher, J.M. and Wyss, U. (2002) The effect of osmotic stress on desiccation survival and water content of four nematode species. *Nematology* 4, 89–97.

Chen, X., Xue, W., Xue, J., Griffiths, B.S. and Liu, M. (2020) Contribution of bacterivorous nematodes to soil resistance and resilience under copper or heat stress. *Soil Ecology Letters* 2, 220–229.

Cong, Y., Yang, H., Zhang, P., Xie, Y., Cao, X. and Zhang, L. (2020) Transcriptome analysis of the nematode *Caenorhabditis elegans* in acidic stress environments. *Frontiers in Physiology* 11, 1107.

Courtright, E.M., Wall, D.H. and Virginia, R.A. (2001) Determining habitat suitability for soil invertebrates in an extreme environment: the McMurdo Dry Valleys, Antarctica. *Antarctic Science* 13, 9–17.

Crowe, J.H. and Madin, K.A. (1974) Anhydrobiosis in tardigrades and nematodes. *Transactions of the American Microscopical Society* 93, 513–524. DOI: 10.2307/3225155.

Crowe, J.H., Madin, K.A.C. and Loomis, S.H. (1977) Anhydrobiosis in nematodes: metabolism during resumption of activity. *Journal of Experimental Zoology* 201, 57–64. DOI: 10.1002/jez.1402010107.

Demeure, Y., Freckman, D.W. and Van Gundy, S.D. (1979) Anhydrobiotic coiling of nematodes in soil. *Journal of Nematology* 11, 189–195. Available at: https://pmc.ncbi.nlm.nih.gov/articles/PMC2617952/ (accessed 14 August 2025).

Devetter, M., Háněl, L., Řeháková, K. and Doležal, J. (2017) Diversity and feeding strategies of soil microfauna along elevation gradients in Himalayan cold deserts. *PLoS One* 12, e0187646.

Dhakal, R., Yosofvand, M., Yavari, M., Abdulrahman, R., Schurr, R. *et al.* (2021) Review of biological effects of acute and chronic radiation exposure on *Caenorhabditis elegans*. *Cells* 10, 1966.

Dhal, S. and Pal, H. (2022) Regulation of glycolysis and Krebs cycle during biotic and abiotic stresses. In: Roychoudhury, A. (ed.) *Photosynthesis and Respiratory Cycles during Environmental Stress Response in Plants*. Apple Academic Press, New York, pp. 263–308.

Farahani, S., Bandani, A.R., Alizadeh, H., Goldansaz, S.H. and Whyard, S. (2020) Differential expression of heat shock proteins and antioxidant enzymes in response to temperature, starvation, and parasitism in the carob moth larvae, *Ectomyelois ceratoniae* (Lepidoptera: Pyralidae). *PLoS One* 15, e0228104.

Ferris, H. (2010) Contribution of nematodes to the structure and function of the soil food web. *Journal of Nematology* 42, 63–67. Available at: https://pubmed.ncbi.nlm.nih.gov/22736838 (accessed 14 August 2025).

Franco, A.L., Adams, B.J., Diaz, M.A., Lemoine, N.P., Dragone, N.B. *et al.* (2022) Response of Antarctic soil fauna to climate-driven changes since the Last Glacial Maximum. *Global Change Biology* 28, 644–653.

Freckman, D.W. and Virginia, R.A. (1989) Plant-feeding nematodes in deep-rooting desert ecosystems. *Ecology* 70, 1665–1678.

Freckman, D.W. and Virginia, R.A. (1997) Low-diversity Antarctic soil nematode communities: distribution and response to disturbance. *Ecology* 78, 363–369.

Freckman, D.W., Whitford, W.G. and Steinberger, Y. (1987) Effect of irrigation on nematode population dynamics and activity in desert soils. *Biology and Fertility of Soils* 3, 3–10.

Gao, Z., Huang, J., Gao, W., Jia, M., Li, X., Han, G. and Zhang, G. (2023) Exploring the effects of warming and nitrogen deposition on desert steppe based on soil nematodes. *Land Degradation and Development* 34, 682–697.

Garrity, P.A., Goodman, M.B., Samuel, A.D. and Sengupta, P. (2010) Running hot and cold: behavioral strategies, neural circuits, and the molecular machinery for thermotaxis in *C. elegans* and *Drosophila*. *Genes and Development* 24, 2365–2382.

Gingold, R., Moens, T. and Rocha-Olivares, A. (2013) Assessing the response of nematode communities to climate change-driven warming: a microcosm experiment. *PLoS ONE* 8, e66653.

Grewal, P.S., Bai, X. and Jagdale, G.B. (2011) Longevity and stress tolerance of entomopathogenic nematodes. In: Perry, R.N. and Wharton, D.A. (eds) *Molecular and Physiological Basis of Nematode Survival*. CAB International, Wallingford, UK, pp. 157–181.

Gusev, O., Nakahara, Y., Vanyagina, V., Malutina, L., Cornette, R. *et al.* (2010) Anhydrobiosis-associated nuclear DNA damage and repair in the sleeping chironomid: linkage with radioresistance. *PLoS ONE* 5, e14008.

Hadley, N.F. and Szarek, S.R. (1981) Productivity of desert ecosystems. *BioScience* 31, 747–753.

Hauquier, F. and Vanreusel, A. (2012) Nematode (re)colonisation after climate-change induced disturbance events in the Antarctic. In: *Conference Abstracts, XXXII SCAR Open Science Conference: Antarctic Science and Policy Advice in a Changing World, 16–19 July 2012, Portland, Oregon, USA*. Available at: https://scar.org/~documents/route%3A/download/3339 (accessed 19 August 2025).

Hibshman, J.D. and Goldstein, B. (2021) LEA motifs promote desiccation tolerance *in vivo*. *BMC Biology* 19, 263.

Hodda, M., Ocana, A. and Traunspurger, W. (2006) Nematodes from extreme freshwater habitats. In: Abebe, E., Traunspurger, W. and Andrássy, I. (eds) *Freshwater Nematodes: Ecology and Taxonomy*. CAB International, Wallingford, UK, pp. 179–210.

Hoffmann, A.A., Chown, S.L. and Clusella-Trullas, S. (2013) Upper thermal limits in terrestrial ectotherms: how constrained are they? *Functional Ecology* 27, 934–949.

Hofmann, K., Pauli, H., Praeg, N., Wagner, A.O. and Illmer, P. (2016) Methane-cycling microorganisms in soils of a high-alpine altitudinal gradient. *FEMS Microbiology Ecology* 92, fiw009.

Holovachov, O. (2014) Nematodes from terrestrial and freshwater habitats in the Arctic. *Biodiversity Data Journal* 19, e1165.

Hu, C., Pen-Mouratov, S. and Steinberger, Y. (2020) Vertical distribution of soil free-living nematode in a playa habitat in the North-Western Negev desert, Israel. *Helminthologia* 57, 361–375.

Ingels, J., Zeppilli, D. and Giere, O. (2023) Meiofauna – adapted to life at the limits. In: Giere, O. and Schratzberger, M. (eds) *New Horizons in Meiobenthos Research: Profiles, Patterns and Potentials*. Springer, Cham, Switzerland, pp. 363–400.

Jana, B. (1973) The thermal springs of Bakreswar, India: physico-chemical conditions, flora and fauna. *Hydrobiologia* 41, 291–307.

Jenkins, E., Veitch, A., Kutz, S., Hoberg, E. and Polley, L. (2006) Climate change and the epidemiology of protostrongylid nematodes in northern ecosystems: *Parelaphostrongylus odocoilei* and *Protostrongylus stilesi* in Dall's sheep (*Ovis d. dalli*). *Parasitology* 132, 387–401.

Jung, J., Loschko, T., Reich, S., Rassoul-Agha, M. and Werner, M.S. (2024) Newly identified nematodes from the Great Salt Lake are associated with microbialites and specially adapted to hypersaline conditions. *Proceedings of the Royal Society B: Biological Sciences* 291, 20232653.

Kanzaki, N., Yamashita, T., Lee, J.S., Shih, P.-Y., Ragsdale, E.J. and Shinya, R. (2021) *Tokorhabditis* n. gen. (Rhabditida, Rhabditidae), a comparative nematode model for extremophilic living. *Scientific Reports* 11, 16470.

Kashyap, P., Afzal, S., Rizvi, A.N., Ahmad, W., Uniyal, V.P. and Banerjee, D. (2022) Nematode community structure along elevation gradient in high altitude vegetation cover of Gangotri National Park (Uttarakhand), India. *Scientific Reports* 12, 1428. DOI: 10.1038/s41598-022-05472-9.

Kergunteuil, A., Campos-Herrera, R., Sánchez-Moreno, S., Vittoz, P. and Rasmann, S. (2016) The abundance, diversity, and metabolic footprint of soil nematodes is highest in high elevation alpine grasslands. *Frontiers in Ecology and Evolution* 4, 84.

Khanal, C. and Land, J. (2023) Study on two nematode species suggests climate change will inflict greater crop damage. *Scientific Reports* 13, 14185.

Kouser, Y., Shah, A.A. and Rasmann, S. (2021) The functional role and diversity of soil nematodes are stronger at high elevation in the lesser Himalayan Mountain ranges. *Ecology and Evolution* 11, 13793–13804.

Kowalewski, D.E., Marchant, D.R., Levy, J.S. and Head, J.W. (2006) Quantifying low rates of summertime sublimation for buried glacier ice in Beacon Valley, Antarctica. *Antarctic Science* 18, 421–428. DOI: 10.1017/S0954102006000460.

Leung, C., Grulois, D., Quadrana, L. and Chevin, L.M. (2023) Phenotypic plasticity evolves at multiple biological levels in response to environmental predictability in a long-term experiment with a halotolerant microalga. *PLoS Biology* 21, e3001895.

Liu, L., Salam, N., Jiao, J.-Y., Jiang, H.-C., Zhou, E.-M. *et al.* (2016) Diversity of culturable thermophilic actinobacteria in hot springs in Tengchong, China and studies of their biosynthetic gene profiles. *Microbial Ecology* 72, 150–162.

Liu, T., Hu, F. and Li, H. (2019) Spatial ecology of soil nematodes: perspectives from global to micro scales. *Soil Biology and Biochemistry* 137, 107565.

Liu, Y., Wang, W., Liu, P., Zhou, H., Chen, Z. and Suonan, J. (2022) Plant–soil mediated effects of long-term warming on soil nematodes of alpine meadows on the Qinghai–Tibetan Plateau. *Biology* 11, 1596.

Mallard, F., Nolte, V. and Schlötterer, C. (2020) The evolution of phenotypic plasticity in response to temperature stress. *Genome Biology and Evolution* 12, 2429–2440.

Marais, E., Maggs-Kölling, G., Sherman, C., Doniger, T., Liu, R. *et al.* (2020) Profiling soil free-living nematodes in the Namib Desert, Namibia. *Journal of Arid Land* 12, 130–143.

McGill, L.M., Shannon, A.J., Pisani, D., Félix, M.-A., Ramløv, H. *et al.* (2015) Anhydrobiosis and freezing-tolerance: adaptations that facilitate the establishment of *Panagrolaimus* nematodes in polar habitats. *PLoS ONE* 10, e0116084. DOI: 10.1371/journal.pone.0116084.

McSorley, R. (2003) Adaptations of nematodes to environmental extremes. *Florida Entomologist* 86, 138–142.

Mueller, K.E., Blumenthal, D.M., Carrillo, Y., Cesarz, S., Ciobanu, M. *et al.* (2016) Elevated CO_2 and warming shift the functional composition of soil nematode communities in a semiarid grassland. *Soil Biology and Biochemistry* 103, 46–51.

Mukuka, J., Strauch, O., Waeyenberge, L., Viaene, N., Moens, M. and Ehlers, R.-U. (2010) Heat tolerance among different strains of the entomopathogenic nematode *Heterorhabditis bacteriophora*. *BioControl* 55, 423–434.

Mulvey, R. (1969) Soil-inhabiting nematodes of the orders Araeolaimida, Chromadorida, Enoplida, and Monhysterida from the Canadian high Arctic. *Canadian Journal of Zoology* 47, 365–382.

Nadler, S.A., De Ley, P., Mundo-Ocampo, M., Smythe, A.B., Stock, S.P. *et al.* (2006) Phylogeny of Cephalobina (Nematoda): molecular evidence for recurrent evolution of probolae and incongruence with traditional classifications. *Molecular Phylogenetics and Evolution* 40(3), 696–711.

Nagy, L. and Grabherr, G. (2009) *The Biology of Alpine Habitats*. Oxford University Press, Oxford.

Nielsen, U.N. and King, C.K. (2015) Abundance and diversity of soil invertebrates in the Windmill Islands region, East Antarctica. *Polar Biology* 38, 1391–1400. DOI: 10.1007/s00300-015-1703-2.

Nielsen, U.N. and Wall, D.H. (2013) The future of soil invertebrate communities in polar regions: different climate change responses in the Arctic and Antarctic? *Ecology Letters* 16, 409–419.

Nielsen, U.N., Wall, D.H., Adams, B.J. and Virginia, R.A. (2011) Antarctic nematode communities: observed and predicted responses to climate change. *Polar Biology* 34, 1701–1711.

Nielsen, U.N., Ayres, E., Wall, D.H., Li, G., Bardgett, R.D. *et al.* (2014) Global-scale patterns of assemblage structure of soil nematodes in relation to climate and ecosystem properties. *Global Ecology and Biogeography* 23, 968–978.

Nisa, R.U., Tantray, A.Y., Kouser, N., Allie, K.A., Wani, S.M. *et al.* (2021) Influence of ecological and edaphic factors on biodiversity of soil nematodes. *Saudi Journal of Biological Sciences* 28, 3049–3059.

Nkem, J.N., Virginia, R.A., Barrett, J.E., Wall, D.H. and Li, G. (2005) Salt tolerance and survival thresholds for two species of Antarctic soil nematodes. *Polar Biology* 29, 643–651. DOI: 10.1007/s00300-005-0101-6.

Norton, D.C. and Niblack, T.L. (2020) Biology and ecology of nematodes. In: Nickle, W.R. (ed.) *Manual of Agricultural Nematology.* CRC Press, Boca Raton, Florida, pp. 47–72.

Oarga, A. (2009) Life in extreme environments. *Revista de Biologia e Ciencias da Terra* 9, 1–10.

Offutt, M. and Riggs, R. (1970) Radiation-induced resistance to root-knot nematodes in Korean Lespedeza. *Crop Science* 10, 49–50.

Parida, L. (2022) The locomotory characteristics of *Caenorhabditis elegans* in various external environments: a review. *Applied Animal Behaviour Science* 255, 105741.

Pen-Mouratov, S., Shukurov, N., Plakht, J. and Steinberger, Y. (2008a) Soil microbial activity and free-living nematode community in the upper soil layer of the anticline erosional cirque, Makhtesh Ramon, Israel. *European Journal of Soil Biology* 44(1), 71–79.

Pen-Mouratov, S., Shukurov, N. and Steinberger, Y. (2008b) Influence of industrial heavy metal pollution on soil free-living nematode population. *Environmental Pollution* 152, 172–183.

Petronijevic, T. and Rogers, W. (1987) The physiology of infection with nematodes: the role of intracellular pH in the development of the early parasitic stage. *Comparative Biochemistry and Physiology Part A: Physiology* 88, 207–212.

Porazinska, D.L., de Mesquita, C.P.B., Farrer, E.C., Spasojevic, M.J., Suding, K.N. and Schmidt, S.K. (2021) Nematode community diversity and function across an alpine landscape undergoing plant colonization of previously unvegetated soils. *Soil Biology and Biochemistry* 161, 108380.

Postma-Blaauw, M., de Vries, F.T., De Goede, R., Bloem, J., Faber, J. and Brussaard, L. (2005) Within-trophic group interactions of bacterivorous nematode species and their effects on the bacterial community and nitrogen mineralization. *Oecologia* 142, 428–439.

Procter, D. (1984) Towards a biogeography of free-living soil nematodes. I. Changing species richness, diversity and densities with changing latitude. *Journal of Biogeography* 11, 103–117.

Raymond, M.R. and Wharton, D.A. (2016) The ability to survive intracellular freezing in nematodes is related to the pattern and distribution of ice formed. *Journal of Experimental Biology* 219, 2060. DOI: 10.1242/jeb.137190.

Rebecchi, L., Boschetti, C. and Nelson, D.R. (2020) Extreme-tolerance mechanisms in meiofaunal organisms: a case study with tardigrades, rotifers and nematodes. *Hydrobiologia* 847, 2779–2799.

Renčo, M. (2012) The soil nematodes in natural and semi-natural grasslands and their use as bioindicators. In: Zhang, W.J. (ed.) *Grasslands: Types, Biodiversity and Impacts.* Nova Science Publishers, Hauppauge, New York, pp. 119–145.

Robinson, C.M., Hansen, L.D., Xue, X. and Adams, B.J. (2023) Temperature response of metabolic activity of an Antarctic nematode. *Biology* 12, 109.

Šalamún, P., Kucanová, E., Brázová, T., Miklisová, D., Renčo, M. and Hanzelová, V. (2014) Diversity and food web structure of nematode communities under high soil salinity and alkaline pH. *Ecotoxicology* 23, 1367–1376.

Sánchez-Moreno, S. and Ferris, H. (2018) Nematode ecology and soil health. In: Sikora, R.A., Coyne, D., Hallmann, J. and Timper, P. (eds) *Plant-Parasitic Nematodes in Subtropical and Tropical Agriculture.* CAB International, Wallingford, UK, pp. 62–86.

Santos, A., Gómez-Espinoza, O., Núñez-Montero, K., Zárate, A., Andreote, F.D. *et al.* (2023) Measuring the effect of climate change in Antarctic microbial communities: toward novel experimental approaches. *Current Opinion in Biotechnology* 81, 102918.

Sapir, A. (2021) Why are nematodes so successful extremophiles? *Communicative and Integrative Biology* 14, 24–26.

Savin, M., Görres, J., Neher, D. and Amador, J. (2001) Biogeophysical factors influencing soil respiration and mineral nitrogen content in an old field soil. *Soil Biology and Biochemistry* 33, 429–438.

Sayre, R. (1964) Cold-hardiness of nematodes. I. Effects of rapid freezing on the eggs and larvae of *Meloidogyne incognita* and *M. hapla*. *Nematologica* 10, 168–179.

Schratzberger, M., Holterman, M., van Oevelen, D. and Helder, J. (2019) A worm's world: ecological flexibility pays off for free-living nematodes in sediments and soils. *BioScience* 69, 867–876.

Schulze, J. and Schierenberg, E. (2008) Cellular pattern formation, establishment of polarity and segregation of colored cytoplasm in embryos of the nematode *Romanomermis culicivorax*. *Developmental Biology* 315, 426–436.

Semeraro, S., Kergunteuil, A., Sánchez-Moreno, S., Puissant, J., Goodall, T. *et al.* (2022) Relative contribution of high and low elevation soil microbes and nematodes to ecosystem functioning. *Functional Ecology* 36, 974–986.

Seybold, A.C., Wharton, D.A., Thorne, M.A. and Marshall, C.J. (2017) Investigating trehalose synthesis genes after cold acclimation in the Antarctic nematode *Panagrolaimus* sp. DAW1. *Biology Open* 6, 1953–1959.

Shih, P.-Y., Lee, J.S., Shinya, R., Kanzaki, N., Pires-daSilva, A. *et al.* (2019) Newly identified nematodes from Mono Lake exhibit extreme arsenic resistance. *Current Biology* 29, 3339–3344 e4.

Siebert, J., Ciobanu, M., Schädler, M. and Eisenhauer, N. (2020) Climate change and land use induce functional shifts in soil nematode communities. *Oecologia* 192(1), 281–294.

Simmons, B.L., Wall, D.H., Adams, B.J., Ayres, E., Barrett, J.E. and Virginia, R.A. (2009) Long-term experimental warming reduces soil nematode populations in the McMurdo Dry Valleys, Antarctica. *Soil Biology and Biochemistry* 41, 2052–2060. DOI: 10.1016/j.soilbio.2009.07.009.

Sims, S., Ho, N., Magas, L., Geary, T., Barsuhn, C. and Thompson, D. (1994) Biophysical model of the transcuticular excretion of organic acids, cuticle pH and buffer capacity in gastrointestinal nematodes. *Journal of Drug Targeting* 2, 1–8.

Smith, T., Wharton, D. and Marshall, C. (2008) Cold tolerance of an Antarctic nematode that survives intracellular freezing: comparisons with other nematode species. *Journal of Comparative Physiology B* 178, 93–100.

Sohlenius, B., Boström, S. and Hirschfelder, A. (1996) Distribution patterns of microfauna (nematodes, rotifers and tardigrades) on nunataks in Dronning Maud Land, East Antarctica. *Polar Biology* 16, 191–200. DOI: 10.1007/BF02329207.

Somasekhar, N. and Prasad, J.S. (2011) Plant–nematode interactions: consequences of climate change. In: Venkateswarlu, B., Shanker, A., Shanker, C. and Maheswari, M. (eds) *Crop Stress and Its Management: Perspectives and Strategies*. Springer, Dordrecht, The Netherlands, pp. 547–564. DOI: 10.1007/978-94-007-2220-0_17.

Stamps, W.T. and Linit, M.J. (2001) Interaction of intrinsic and extrinsic chemical cues in the behaviour of *Bursaphelenchus xylophilus* (Aphelenchida: Aphelenchoididae) in relation to its beetle vectors. *Nematology* 3, 295–301.

Steel, H., Buchan, D., De Neve, S., Couvreur, M., Moens, T. and Bert, W. (2013) Nematode and microbial communities in a rapidly changing compost environment: how nematode assemblages reflect composting phases. *European Journal of Soil Biology* 56, 1–10.

Steward, G., Zehr, J., Jellison, R., Montoya, J. and Hollibaugh, J. (2004) Vertical distribution of nitrogen-fixing phylotypes in a meromictic, hypersaline lake. *Microbial Ecology* 47, 30–40.

Stringham, E.G. and Candido, E.P.M. (1994) Transgenic hsp 16-*Lacz* strains of the soil nematode *Caenorhabditis elegans* as biological monitors of environmental stress. *Environmental Toxicology and Chemistry* 13, 1211–1220.

Stuart, R.J., Barbercheck, M.E. and Grewal, P.S. (2015) Entomopathogenic nematodes in the soil environment: distributions, interactions and the influence of biotic and abiotic factors. In: Campos-Herrera, R. (ed.) *Nematode Pathogenesis of Insects and Other Pests. Sustainability in Plant and Crop Protection*. Springer, Cham, Switzerland, pp. 97–137. DOI: 10.1007/978-3-319-18266-7_4.

Suzuki, N., Rivero, R.M., Shulaev, V., Blumwald, E. and Mittler, R. (2014) Abiotic and biotic stress combinations. *New Phytologist* 203, 32–43.

Tahseen, Q. (2012) Nematodes in aquatic environments: adaptations and survival strategies. *Biodiversity Journal* 3, 13–40.

Terauds, A. and Lee, J.R. (2016) Antarctic biogeography revisited: updating the Antarctic Conservation Biogeographic Regions. *Diversity and Distributions* 22, 836–840. DOI: 10.1111/ddi.12453.

Treonis, A.M. and Wall, D.H. (2005) Soil nematodes and desiccation survival in the extreme arid environment of the Antarctic Dry Valleys. *Integrative and Comparative Biology* 45, 741–750.

Treonis, A., Sutton, K., Kavanaughf, B., Narla, A., McLlarky, T. *et al.* (2012) Soil nematodes and their prokaryotic prey along an elevation gradient in the Mojave Desert (Death Valley National Park, California, USA). *Diversity* 4, 363–374.

Treonis, A.M., Wall, D.H. and Virginia, R.A. (2000) The use of anhydrobiosis by soil nematodes in the Antarctic Dry Valleys. *Functional Ecology* 14, 460–467. DOI: 10.1046/j.1365-2435.2000.00442.x.

Treonis, A.M., Marais, E. and Maggs-Kölling, G. (2022) Nematode communities indicate diverse soil functioning across a fog gradient in the Namib Desert gravel plains. *Ecology and Evolution* 12, e9013.

van den Hoogen, J., Geisen, S., Routh, D., Ferris, H., Traunspurger, W. *et al.* (2019) Soil nematode abundance and functional group composition at a global scale. *Nature* 572, 194–198. DOI: 10.1038/s41586-019-1418-6.

Velasco-Castrillón, A., Schultz, M.B., Colombo, F., Gibson, J.A.E., Davies, K.A. *et al.* (2014) Distribution and diversity of soil microfauna from East Antarctica: assessing the link between biotic and abiotic factors. *PLoS One* 9(1), e87529. DOI: 10.1371/journal.pone.0087529.

Wall, D.H. (2007) Global change tipping points: above- and below-ground biotic interactions in a low diversity ecosystem. *Philosophical Transactions of the Royal Society B: Biological Sciences* 362, 2291–2306.

Wall, D.H. and Viginia, R.A. (1999) Controls on soil biodiversity: insights from extreme environments. *Applied Soil Ecology* 13, 137–150. DOI: 10.1016/S0929-1393(99)00029-3.

Wang, Y., Liu, C., Ren, P. and Li, Y. (2020) Pre-cold acclimation promotes cryoprotectant accumulation and enhances freezing tolerance of *Meloidogyne incognita*. *Journal of Phytopathology* 168, 337–341.

Watanabe, M., Nakahara, Y., Sakashita, T., Kikawada, T., Fujita, A. *et al.* (2007) Physiological changes leading to anhydrobiosis improve radiation tolerance in *Polypedilum vanderplanki* larvae. *Journal of Insect Physiology* 53, 573–579.

Weinstein, D.J. (2019) Genome-wide analysis of gene expression in *Halicephalobus mephisto* (the devil worm). MS thesis, American University, Washington, DC. ProQuest Dissertations & Theses, 13811979.

Wharton, D.A. (2003) The environmental physiology of Antarctic terrestrial nematodes: a review. *Journal of Comparative Physiology B* 173, 621–628. DOI: 10.1007/s00360-003-0378-0.

Wharton, D.A. (2004) Survival strategies. In: Gaugler, R. and Bilgrami, A.L. (eds) *Nematode Behaviour*. CAB International, Wallingford, UK, pp. 371–399.

Wharton, D.A. (2007) *Life at the Limits: Organisms in Extreme Environments*. Cambridge University Press, Cambridge.

Wharton, D.A. (2012) *A Functional Biology of Nematodes*. Springer and Johns Hopkins University Press, Baltimore, Maryland.

Wharton, D.A. (2015) Anhydrobiosis. *Current Biology* 25, R1114–R1116.

Wharton, D.A. and Block, W. (1993) Freezing tolerance in some Antarctic nematodes. *Functional Ecology* 7, 578–584.

Wharton, D.A. and Brown, I.M. (2012) A survey of terrestrial nematodes from the McMurdo Sound region, Antarctica. *New Zealand Journal of Zoology* 16, 467–470. DOI: 10.1080/03014223.1989.10422914.

Wharton, D.A. and Raymond, M.R. (2015) Cold tolerance of the Antarctic nematodes *Plectus murrayi* and *Scottnema lindsayae*. *Journal of Comparative Physiology B* 185(3), 281–289. DOI: 10.1007/s00360-014-0884-2.

Wharton, D.A., Downes, M.F., Goodall, G. and Marshall, C.J. (2005) Freezing and cryoprotective dehydration in an Antarctic nematode (*Panagrolaimus davidi*) visualised using a freeze substitution technique. *Cryobiology* 50(1), 21–28. DOI: 10.1016/j.cryobiol.2004.09.004.

Whitford, W.G. (1996) The importance of the biodiversity of soil biota in arid ecosystems *Biodiversity and Conservation* 5, 185–195.

Whitford, W., Freckman, D.W., Santos, P., Elkins, N. and Parker, L. (1982) The role of nematodes in decomposition in desert ecosystems. In: Freckman, D.W. (ed.) *Nematodes in Soil Ecosystems*. University of Texas Press, Austin, Texas, pp. 98–116.

Wingfield, J.C., Kelley, J.P. and Angelier, F. (2011) What are extreme environmental conditions and how do organisms cope with them? *Current Zoology* 57, 363–374.

Winitsky, S.E. (2020) Vascular flora of adobe valley and surrounding hills, Mono County, California. *Aliso: A Journal of Systematic and Floristic Botany* 38, 32–61.

Wu, X., Zhu, X., Wang, Y., Liu, X., Chen, L. and Duan, Y. (2018) The cold tolerance of the northern root-knot nematode, *Meloidogyne hapla*. *PLoS One* 13, e0190531.

Xue, X., Suvorov, A., Fujimoto, S., Dilman, A.R. and Adams, B.J. (2021) Genome analysis of *Plectus murrayi*, a nematode from continental Antarctica. *G3 Genes|Genomes|Genetics* 11, jkaa045. DOI: 10.1093/g3journal/jkaa045.

Yan, D., Yan, D., Song, X., Yu, Z., Peng, D. *et al.* (2018) Community structure of soil nematodes under different drought conditions. *Geoderma* 325, 110–116.

Yanase, S., Hartman, P.S., Ito, A. and Ishii, N. (1999) Oxidative stress pretreatment increases the X-radiation resistance of the nematode *Caenorhabditis elegans*. *Mutation Research/Fundamental and Molecular Mechanisms of Mutagenesis* 426, 31–39.

Yeates, G.W. (2003) Nematodes as soil indicators: functional and biodiversity aspects. *Biology and Fertility of Soils* 37, 199–210.

Yeates, G.W., Scott, M.B., Chown, S.L. and Sinclair, B.J. (2009) Changes in soil nematode populations indicate an annual life cycle at Cape Hallett, Antarctica. *Pedobiologia* 52, 375–386. DOI: 10.1016/j.pedobi.2009.01.001.

Zawierucha, K., Marshall, C.J., Wharton, D. and Janko, K. (2019) A nematode in the mist: *Scottnema lindsayae* is the only soil metazoan in remote Antarctic deserts, at greater densities with altitude. *Polar Research* 38, 1–12. DOI: 10.33265/polar.v38.3494.

Zhang, G., Sui, X., Li, Y., Jia, M., Wang, Z. *et al.* (2020) The response of soil nematode fauna to climate drying and warming in *Stipa breviflora* desert steppe in Inner Mongolia, China. *Journal of Soils and Sediments* 20, 2166–2180.

Zhao, Y.-L. and Wang, D.-Y. (2012) Formation and regulation of adaptive response in nematode *Caenorhabditis elegans*. *Oxidative Medicine and Cellular Longevity* 2012, 564093.

Zhou, J., Wu, J., Huang, J., Sheng, X., Dou, X. and Lu, M. (2022) A synthesis of soil nematode responses to global change factors. *Soil Biology and Biochemistry* 165, 108538.

Zullini, A. (2018) Cosmopolitism and endemism in free-living nematodes. *Biogeographia – The Journal of Integrative Biogeography* 33, 33–39.

11 Nematodes as Indicators of Climate Change and Environmental Restoration

Qi Li[1]*, Wenju Liang[1] and Sara Sánchez-Moreno[2]
[1]*CAS Key Laboratory of Forest Ecology and Silviculture, Institute of Applied Ecology, Chinese Academy of Sciences, Shenyang, China;* [2]*Department of Biodiversity and Evolutionary Biology, National Museum of Natural Sciences (MNCN-CSIC), Madrid, Spain*

Abstract
Climate change is affecting multiple aspects of life on Earth. Soil biota, as one of the main drivers of soil resilience against climate stress, plays a key role in mitigating climate change effects on ecosystems. As nematodes respond to climatic and environmental conditions, they are often used as indicators of climate change impacts in agricultural systems and to evaluate the success of restoration of degraded ecosystems. The effects of climate change on soil nematodes are variable, with CO_2 often stimulating nematodes due to its positive effects on plant biomass. Other climate change factors, such as warming, might induce negative responses on nematode communities, depending on regional climatic conditions and local factors. To reverse soil degradation, traditional restoration techniques have been primarily focused on restoring plant diversity, but new methods aiming to restore soil functions might effectively protect nematode diversity under global changes.

11.1 Introduction

Climate change is one of the most serious threats that humankind is facing nowadays. The 6th Assessment Report published by the Intergovernmental Panel on Climate Change (IPCC) lists the numerous effects climate change has already caused in ecosystems and on human societies (IPCC, 2022). Climate change has impacted ecosystems, people, settlements and infrastructure, increasing the frequency of extreme climatic events (heat extremes, heavy floods, droughts, wildfires). Such impacts affect human health and degrade ecosystems, producing increasingly irreversible losses in terrestrial, freshwater and marine ecosystems.

Negative effects of climate change on agricultural systems are also expected (IPCC, 2022). In the future, climate change will continue weakening soil health and multiple ecosystem services, increasing pest pressure and undermining food productivity. Other effects of climate and global change on soils include salinization, soil acidification and biodiversity loss. Warming, nitrogen deposition, soil carbon loss, and other global changes, impact on soils and alter the composition and abundance of soil communities, their density, biomass and trophic interactions, affecting soil food web structure and derived ecosystem services (Ott *et al.*, 2012; Yin *et al.*, 2020). Under this context, protecting soil biological communities is imperative, since

*Email: liq@iae.ac.cn

DOI: 10.1079/9781800624221.0011

they determine the resilience of the whole soil system to climate change (De Vries *et al.*, 2012). As one of the most abundant soil invertebrates, soil nematodes are useful indicators of environmental conditions and ecosystem functioning (van den Hoogen *et al.*, 2019). Nematode communities are widely used as bioindicators to assess the effects of climate change in agroecosystems, and their response to climatic conditions helps us understand the effects of elevated CO_2, warming and drought on terrestrial ecosystems and agriculture sustainability.

In the soil, multiple biotic and abiotic factors determine the composition and abundances of nematode communities. Nematodes in low trophic levels (mainly bacterivores and fungivores) are affected by microbial communities, which in turn rely on water, carbon and nitrogen availability, affected by root exudates, litter and plant communities. Such assemblages in lower positions are resources for nematodes at higher trophic levels, so bottom-up (resource availability) and top-down (predation pressure) forces do occur. These systems are affected by environmental factors and climatic stress (Fig. 11.1).

11.2 Effects of Climate Change on Soil Nematode Communities

11.2.1 Elevated atmospheric CO_2

In terrestrial ecosystems, elevated CO_2 (eCO_2) might increase plant photosynthesis and net primary production, thereby promoting carbon input into the soil, which can indirectly stimulate soil biota through changes in plant litter and the exudation of nutrients into the rhizosphere (Wang *et al.*, 2021). Due to these plant-mediated processes, the response of soil nematodes to climate change is often more obvious in rhizosphere soil than in root-free or bulk soil. This is likely due to the increase of root biomass and rhizo-deposition associated to increased biomass/activity of soil microorganisms under eCO_2 (Wang *et al.*, 2021). For instance, an increase in the abundance of herbivores under eCO_2 is linked to increases in root biomass (Ayres *et al.*, 2008) and leads to high Plant-Parasitic Index (PPI) values (Wang *et al.*, 2019), which is an indicator of complex and diverse assemblages of plant-parasitic nematodes. In addition, eCO_2

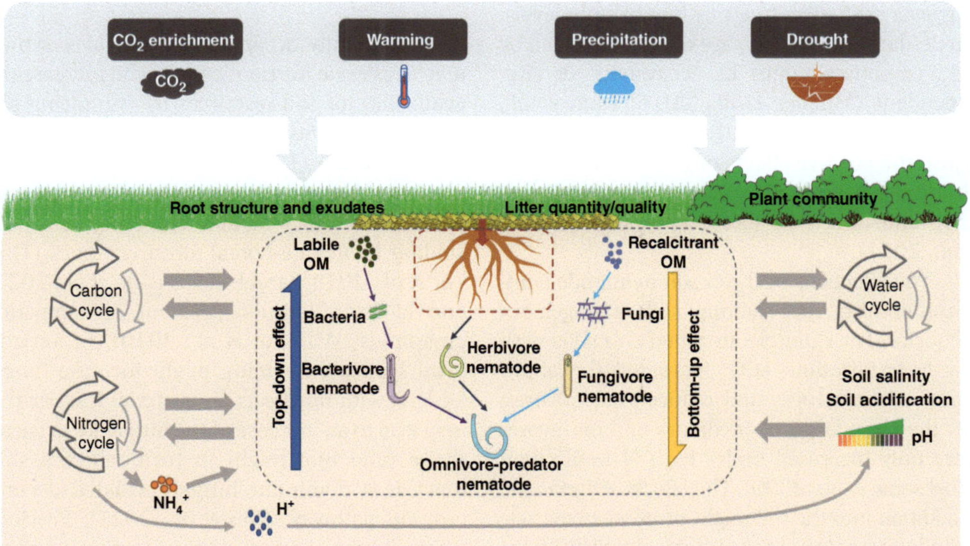

Fig. 11.1. Individual and combined effects of multiple factors of climate change on soil nematode communities and food webs. Climate change factors (CO_2 enrichment, warming, changes in precipitation regimes, drought) induce changes in the composition of plant communities, root structure and exudation, and litter and organic matter (OM) quality and quantity, which influence nematode food webs through their effects on the carbon, nitrogen and water cycles (bottom-up effects). Changes in soil chemical parameters, including soil pH, salinity and mineral forms of nitrogen, may also act as toxicants for soil organisms. (Author's own image.)

might promote plant growth, cause increases in C:N ratios in plant tissues and disrupt the allocation of plant resources between growth and defence, reducing plant defences against plant-parasitic nematodes (Sun *et al.*, 2010). Commonly, the effects of eCO_2 on nematode communities are weaker under field conditions than in controlled laboratory or greenhouse conditions, probably due to the higher variations of field environment than the controlled conditions in laboratory experiments. Effects of eCO_2 on low nematode trophic guilds are usually more evident than those on higher ones. In croplands, eCO_2 increases herbivores, but often has no noticeable effects on omnivores and predators (Wang *et al.*, 2019). In grazed pastures with naturally elevated CO_2 levels in New Zealand, total nematode abundances and richness increased with increasing CO_2 levels. In such CO_2-enriched sites, nematode abundances and diversity decreased, and only bacterivore nematodes increased with increasing CO_2 concentrations (Yeates *et al.*, 1999). Oppositely, in a free-air CO_2 enrichment experiment developed in a grazed pasture, all nematode trophic groups were promoted by CO_2 enrichment (Yeates *et al*, 2003). Although multiple evidence suggests that nematodes in low trophic levels, particularly fungivores, are often stimulated by CO_2, responses might be ecosystem- or site-dependent (Mueller *et al.*, 2016; Dam *et al.*, 2017; Song *et al.*, 2020). Among ecosystem types, the positive effects of eCO_2 on nematode abundances are more obvious in croplands and grasslands than in forest ecosystems (Zhou *et al.*, 2022).

The effects of eCO_2 on soil nematode communities also vary among soil management scenarios. For example, in aspen tree soils with low N fertilization, eCO_2 increased the abundance of herbivores and reduced bacterivores, but the abundance of predators and omnivores were only increased under high N fertilization (Hoeksema *et al.*, 2000). The effects of increased N addition into the soil might promote nematode abundances at the expense of decreased diversity, especially driven by the increase of some opportunistic bacterivores (e.g. Rhabditidae) (Shaw *et al.*, 2019). As moist and warm conditions can facilitate N uptake by plants and soil biota, the positive effects of low N inputs may be further enhanced under eCO_2 (Xing *et al.*, 2022). Oppositely, high N inputs can decrease nematode abundance through soil acidification and ammonium toxicity (Wei *et al.*, 2012; Chen *et al.*, 2015). Nitrogen fertilization can also indirectly change the diversity of soil nematodes by changing plant community composition and soil chemical conditions, which further influence the structure and stability of soil food webs through above- and below-ground interactions (Bardgett and Wardle, 2010; Li *et al.*, 2013; Neher and Weicht, 2013). Finally, the characteristics of the plant community might also modulate the response of nematodes to eCO_2. For instance, eCO_2 exerted opposite effects on the abundance of plant-parasitic nematodes under two different invasive plant species (Xiao *et al.*, 2013). In a rice–wheat rotation system, fertilization rates and residue management determined nematode responses to eCO_2. Herbivore nematodes, for example, increased under eCO_2 and high fertilization levels. However, they increased under eCO_2 and plant residue incorporation but decreased under eCO_2 and no residue incorporation (Li *et al.*, 2009).

11.2.2 Warming

It is commonly accepted that global warming may accelerate nutrient cycling and increase the availability of soil resources by enhancing soil microbial activity and organic matter decomposition rates (Thakur *et al.*, 2018). Global warming has been reported to increase soil nematode density in different ecosystems, including temperate-boreal forest ecotones (Thakur *et al.*, 2014), arable fields (Guo *et al.*, 2021) and along a latitude gradient of temperate grasslands (Wilschut *et al.*, 2019). In agricultural systems, warming might increase fungal feeders, with no effects on bacterial feeders; this may lead to an increase of the fungivore to bacterivore ratio that might in turn reflect a shift towards a dominant fungal-mediated decomposition pathway (Guo *et al.*, 2021). Previous studies have also reported that warming increases microbivore nematodes, compared to herbivores, under sufficient soil water availability. This is due to plant-increased allocation of photosynthesis products below ground, that stimulates microbial biomass and activity (Thakur *et al.*, 2014, 2018).

In other agroecosystems, such as in paddy soils, warming might reduce overall nematode abundances due to a decline in nematode reproduction rates (Okada *et al.*, 2014). Contrasting nematode responses in experimental studies may be explained by the variety in warming methods applied, ecosystem types or other specific factors, such as local climatic conditions and the plant community composition. For example, nematode diversity and fungivore, omnivore, predator and plant-parasitic nematodes were negatively affected by experimental warming in a desert steppe ecosystem (Ma *et al.*, 2018). Increases in the fungivore to bacterivore ratio occurred in relatively cold regions (Nielsen *et al.*, 2014). Experimental warming increased nematode diversity in diverse plant communities but decreased it in monocultures, along an experimental gradient of plant community richness. This is attributed to the heterogeneity in the soil environment under a diverse plant community, which allows coexistence of more nematode species due to greater resource availability (Thakur *et al.*, 2017). Nematodes, as drivers of plant performance, may also shape plant communities, and shifts in nematode communities due to warming may cause further changes in plant–soil interactions (Wilschut *et al.*, 2019). The effects of warming might also be modulated by other climatic factors. Indeed, some studies found that synergistic effects of warming and eCO_2 reduce omnivores and predators, simplifying the soil food web, and increasing the prevalence of opportunistic nematodes of colonizer–persister (*c-p*) groups *c-p*1 and *c-p*2 (Carrillo *et al.*, 2016).

11.2.3 Reduced precipitation and drought

Increasing aridity associated with climate change exerts a major impact on the structure and functioning of terrestrial ecosystems (IPCC, 2013). As soil nematodes rely on the water film that surrounds soil particles or water-filled soil pores to move, they are sensitive to changes in soil water content (Franco *et al.*, 2022). However, nematode communities, especially some large nematodes, may not always follow pore size distributions but can also relate to soil aggregates, as aggregates may preserve water while the surrounding soils remain dry (Neher *et al.*, 1999). Altered moisture conditions, induced by precipitation changes, might affect nematode activity, reproduction and community species composition (Blankinship *et al.*, 2011), while drought reduces the abundance and the taxonomic richness of soil nematode communities (Nielsen *et al.*, 2014; Franco *et al.*, 2019). Increasing aridity affects higher trophic guilds (i.e. predators) more obviously than lower trophic positions (i.e. fungivores and herbivores) (Xiong *et al.*, 2021) due to their longer generation times and lower reproductive rates (Ferris *et al.*, 2001). Besides, the effects of drought may also lead to species-specific responses due to the differences in the physiological tolerance of individual species (Xiong *et al.*, 2020). Thus, increasing aridity might lead to shifts in nematode communities within and between trophic levels (Andriuzzi *et al.*, 2018).

In addition to direct effects on nematode reproduction and development, soil water content may indirectly affect nematodes by affecting microorganisms and plants (Bardgett and Wardle, 2010). Thus, current and future changes in precipitation regimes may differentially influence soil nematode communities, depending on their composition, ecosystem type, and the direction and strength of precipitation change (Bristol *et al.*, 2023). For example, in mesic grasslands, water pulses decrease total nematode abundances (Franco *et al.*, 2019). In desert soils, however, nematode communities might resist changes in the frequency of summer precipitations, surviving in hot and dry soils, but not in hot and moist soils (Darby *et al.*, 2011).

In temperate forests, precipitation maximizes the negative effects of nitrogen deposition on soil nematodes, decreasing the abundance of omnivores and predators (Sun *et al.*, 2013), so reduced precipitation might mitigate the negative effects of nitrogen deposition in soils. In general, reductions in rainfall are more likely to decrease soil nematode abundances in grasslands or deserts, with cascading effects on soil processes (Bristol *et al.*, 2023). Decreased precipitation frequency also decreased the diversity and abundance of soil nematodes in degraded grasslands, due to reduced soil moisture and increased variation (temporal, spatial) of water availability (Chen *et al.*, 2021). Fungivore and bacterivore nematodes respond differently to changes in rainfall frequency, probably because fungi are more resistant to desiccation than bacteria (Barnard *et al.*, 2013). Oppositely, positive effects of increased rainfall on total nematode abundances and herbivores have been mainly observed in long-term experiments via increased plant productivity (Bristol *et al.*, 2023).

11.2.4 The interactive effects of multiple factors of climate change

Anthropogenic activities have intensified global changes (Rillig *et al.*, 2019; Peng *et al.*, 2022), and the responses of soil nematodes to such changes vary in relation to global change factors and their interactions. When several climate change factors co-occur, synergistic effects may appear, so multifactorial studies are necessary to assess the potential effects of interactions among multiple aspects of climate change (Eisenhauer *et al.*, 2012). Warming and reduced precipitation can synergistically decrease soil resource inputs and nutrient turnover (Thakur *et al.*, 2018), which may in turn induce nutrient-depleted soil conditions and enhance fungivores, due to their relatively high environmental stress tolerance (Song *et al.*, 2016). In a multifactor experiment, Kardol *et al.* (2010) found that water shortage and its effects in the plant communities were the main drivers of the interactive effects of climate change factors on soil nematodes, reducing the abundances of fungivores and omnivores. They showed that, due to the complexity of the biological interactions in soil food webs, assessing the effects of global change on soil nematodes requires intensive research on multiple trophic levels in the soil food web, including plants, microorganisms and soil fauna. Together with climatic and edaphic drivers, biological interactions also play a role in shaping nematode communities (see Chapter 3, this volume).

11.3 Nematodes as Indicators of Ecological Restoration

11.3.1 Restoring nematode communities in perturbed soil

Soil restoration aims to recover full soil functioning and diversity after soil perturbation, commonly resulting from land-use change and anthropogenic activities that result in soil pollution, erosion, desertification and soil organic carbon depletion (Fig. 11.2). These associated changes in land cover and soil processes are

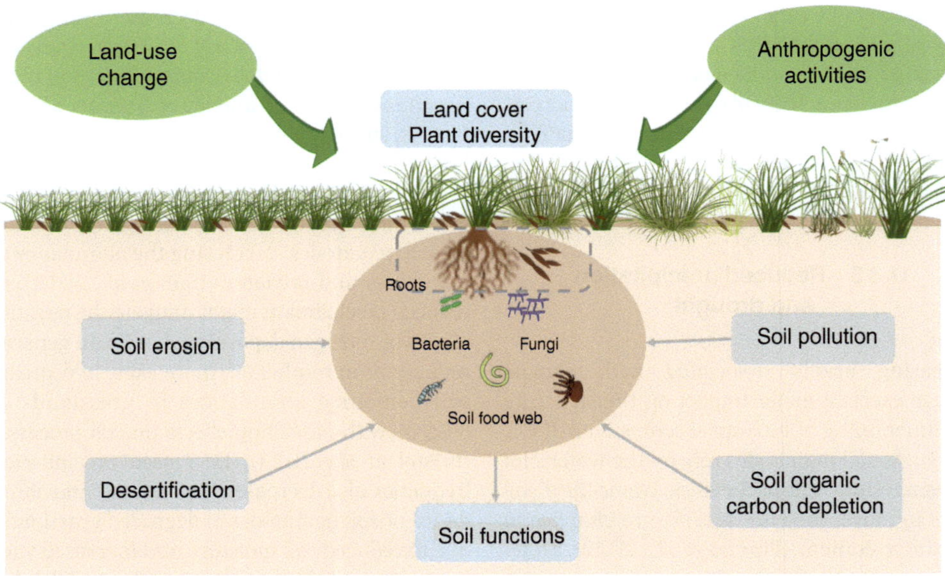

Fig. 11.2. Land-use change and anthropogenic activities often lead to changes in land cover and plant diversity, soil erosion, desertification, soil pollution and soil organic carbon depletion. All these factors affect directly and indirectly the soil microbiota and the soil food web, with consequences in soil functioning. (Author's own image.)

recognized as one of the most important components of global change (Sala *et al.*, 2000; Todd *et al.*, 2006). Ecosystem and soil restoration traditionally aimed to recover above-ground biodiversity, while restoration effects on soil diversity were commonly considered a side effect of successful above-ground-aimed restoration. However, restoring soil biodiversity globally is essential to ensure ecosystem health in the future (De Deyn and Kooistra, 2021; Averill *et al.*, 2022).

Given the impacts of different restoration strategies on soil and plant properties in degraded systems (Chen *et al.*, 2013), restoration activities directly and indirectly influence soil nematode communities. In Northern China, livestock exclosure and nutrient addition are widely used to restore degraded grasslands (Bai *et al.*, 2010; Ren *et al.*, 2018). A case study found that livestock exclosure positively influenced omnivore and predator nematodes, and this in turn increased the Maturity Index (MI), which is an indicator of ecological succession (see Chapter 5, this volume). Conversely, nutrient addition weakened these effects by reducing soil pH and enhancing soil ammonium concentration (Zhang *et al.*, 2019). In natural semi-arid grassland, moderate perturbation might enhance nematode abundances due to increase of bacterivores and herbivores, while severe degradation leads to simplified nematode food webs with opportunistic nematodes dominating the community (Yang *et al.*, 2021).

Effects of ecological restoration on soil nematode communities are also mediated by environmental factors such as nutrient availability and root architecture. In intact northern prairies, nematode food webs are modulated by the organic matter decomposition process and litter inputs. When subjected to restoration after relevant degradation, however, nematode communities are determined by the resources provided by roots (Otfinowski and Coffey, 2020). In contrast to natural ecosystems, the replacement of tallgrass prairie with agricultural systems is generally characterized by increased nutrient inputs and physical soil disturbance, which negatively affect soil biodiversity and ecosystem processes (Neher, 1999). For instance, tillage and nitrogen fertilization increase the relative abundance of bacterivores and result in a shift towards bacterial-dominated pathways in decomposition (Bardgett and McAlister, 1999; Neher, 2001). While restoration is also achieved by re-establishing the native plant community, other actions, such as topsoil removal, may be also used after extreme perturbations. An experiment was done in central Europe in a degraded grassland due to intensive agricultural management, where topsoil was removed aiming to withdraw excessive nutrients and opportunistic organisms. This intervention proved to successfully improve both plant and nematode communities in the long term, especially if accompanied by the introduction of propagules of the target vegetation (Resch *et al.*, 2019). In a subtropical agroecosystem, biochar amendment promoted the energy flux through fungivores and omnivore-carnivores at the expense of carbon flux through the herbivore channel, promoting restoration and multifunctionality in the system (Zhu *et al.*, 2023).

Mining activities have often caused serious degradation of above-ground vegetation, soil erosion and chemical pollution (Gwenzi, 2021). Post-mining sites often obviously differ from the original pre-disturbance sites in terms of soil properties, structure and functioning (Harris, 2009; Gwenzi, 2021). Due to their key role in the maintenance of ecosystem functions, soil organisms contribute to soil restoration by enhancing plant growth and, for example, reducing the content of heavy metals (Lu *et al.*, 2021). Such heavy metal pollution in soils commonly decreases the abundance and diversity of soil biota, which might ameliorate by revegetation restoration, especially in shrub-dominated revegetated soils (Lu *et al.*, 2021). Nematodes might even disappear from mining soils due to heavy perturbation, but might re-establish with appropriate remediation (Courtney *et al.*, 2011). In an open-pit iron mine wasteland, the annual removal of the dominant plant species, aiming to increase vegetation diversity, resulted in positive effects on the complexity of the nematode community, with taxa such as the predators *Oxydirus* and *Tripylina* and the bacterivore *Cephalobus* increasing their abundances after restoration management in the desert steppe (Gao *et al.*, 2022). Mine tailings are commonly stored in dams that on occasions trigger ecological accidents, polluting extensive land surfaces with heavy metals. In Southern Spain, copper, lead and zinc from a pyrite mine dam breakage negatively affected nematode community diversity and structure two years after the accident (Sánchez-Moreno *et al.*, 2006). In Brazil, iron tailings released to the environment decreased soil pH, soil organic carbon, microbial biomass and bacterivore nematodes (Ribeiro *et al.*, 2023). In extremely acidic soils, nematode abundances might remain extremely low several years after restoration (Martínez *et al.*, 2021).

11.4 Conclusion

Global change, including climate change, land-use changes and extensive soil degradation at global scales, is the main threat to soil biodiversity nowadays. In such context, ecological restoration plays a relevant role to reverse soil diversity loss, especially in agricultural and polluted soils. The diversity, structure and functioning of nematode communities, closely linked to soil physical, chemical and biological properties, are successfully used to assess the effects of disturbance in soil systems, as well as to assess the success of restoration activities.

References

Andriuzzi, W.S., Adams, B.J., Barrett, J.E., Virginia, R.A. and Wall, D.H. (2018) Observed trends of soil fauna in the Antarctic Dry Valleys: early signs of shifts predicted under climate change. *Ecology* 99, 312–321. DOI: 10.1002/ecy.2090.

Averill, C., Anthony, M.A., Baldrian, P., Finkbeiner, F., van den Hoogen, J. *et al.* (2022) Defending Earth's terrestrial microbiome. *Nature Microbiology* 7, 1717–1725. DOI: 10.1038/s41564-022-01228-3.

Ayres, E., Wall, D.H., Simmons, B.L., Field, C.B., Milchunas, D.G. *et al.* (2008) Belowground nematode herbivores are resistant to elevated atmospheric CO_2 concentrations in grassland ecosystems. *Soil Biology and Biochemistry* 40, 978–985. DOI: 10.1016/j.soilbio.2007.11.018.

Bai, Y.F., Wu, J.G., Clark, C.M., Naeem, S., Pan, Q.M. *et al.* (2010) Tradeoffs and thresholds in the effects of nitrogen addition on biodiversity and ecosystem functioning: evidence from inner Mongolia Grasslands. *Global Change Biology* 16, 358–372. DOI: 10.1111/j.1365-2486.2009.02142.x.

Bardgett, R.D. and McAlister, E. (1999) The measurement of soil fungal:bacterial biomass ratios as an indicator of ecosystem self-regulation in temperate meadow grasslands. *Biology and Fertility of Soils* 29, 282–290.

Bardgett, R.D. and Wardle, D.A. (2010) *Aboveground–Belowground Linkages: Biotic Interactions, Ecosystem Processes, and Global Change*. Oxford University Press, Oxford.

Barnard, R.L., Osborne, C.A. and Firestone, M.K. (2013) Responses of soil bacterial and fungal communities to extreme desiccation and rewetting. *The ISME Journal* 7, 2229–2241. DOI: 10.1038/ismej.2013.104.

Blankinship, J.C., Niklaus, P.A. and Hungate, B.A. (2011) A meta-analysis of responses of soil biota to global change. *Oecologia* 165, 553–565. DOI: 10.1007/s00442-011-1909-0.

Bristol, D., Hassan, K., Blankinship, J.C. and Nielsen, U.N. (2023) Responses of nematode abundances to increased and reduced rainfall under field conditions: a meta-analysis. *Ecosphere* 14, e4364. DOI: 10.1002/ecs2.4364.

Carrillo, Y., Dijkstra, F.A., LeCain, D. and Pendall, E. (2016) Mediation of soil C decomposition by arbuscular mycorrizhal fungi in grass rhizospheres under elevated CO_2. *Biogeochemistry* 127, 45–55. DOI: 10.1007/s10533-015-0159-3.

Chen, D.M., Zheng, S.X., Shan, Y.M., Taube, F. and Bai, Y.F. (2013) Vertebrate herbivore-induced changes in plants and soils: linkages to ecosystem functioning in a semi-arid steppe. *Functional Ecology* 27, 273–281. DOI: 10.1111/1365-2435.12027.

Chen, D.M., Lan, Z.C., Hu, S.J. and Bai, Y.F. (2015) Effects of nitrogen enrichment on belowground communities in grassland: relative role of soil nitrogen availability vs. soil acidification. *Soil Biology and Biochemistry* 89, 99–108. DOI: 10.1016/j.soilbio.2015.06.028.

Chen, J., Yao, Y., Wang, G., Zhong, X.Y., Yang, T.X. and Sun, W. (2021) Decreased precipitation frequency altered abundance, but not community structure, of soil nematodes in degraded grasslands. *Ecological Indicators* 131, 108184. DOI: 10.1016/j.ecolind.2021.108184.

Courtney, R., Keith, A.M. and Harrington, T. (2011) Nematode assemblages in bauxite residue with different restoration histories. *Restoration Ecology* 19, 758–764. DOI: 10.1111/j.1526-100X.2010.00734.x.

Dam, M., Bergmark, L. and Vestergard, M. (2017) Elevated CO_2 increases fungal-based micro-foodwebs in soils of contrasting plant species. *Plant and Soil* 415, 549–561. DOI: 10.1007/s11104-017-3191-3.

Darby, B.J., Neher, D.A., Housman, D.C. and Belnap, J. (2011) Few apparent short-term effects of elevated soil temperature and increased frequency of summer precipitation on the abundance and taxonomic diversity of desert soil micro- and meso-fauna. *Soil Biology and Biochemistry* 43, 1474–1481. DOI: 10.1016/j.soilbio.2011.03.020.

De Deyn, G.B. and Kooistra, L. (2021) The role of soils in habitat creation, maintenance and restoration. *Philosophical Transactions of the Royal Society B: Biological Sciences* 376, 20200170. DOI: 10.1098/rstb.2020.0170.

De Vries, F.T., Liiri, M.E., Bjørnlund, L., Setälä, H.M., Christensen, S. and Bardgett, R.D. (2012) Legacy effects of drought on plant growth and the soil food web. *Oecologia* 170(3), 821–833. DOI: 10.1007/s00442-012-2331-y.

Eisenhauer, N., Cesarz, S., Koller, R., Worm, K. and Reich, P.B. (2012) Global change belowground: impacts of elevated CO_2, nitrogen, and summer drought on soil food webs and biodiversity. *Global Change Biology* 18, 435–447. DOI: 10.1111/j.1365-2486.2011.02555.x.

Ferris, H., Bongers, T. and de Goede, R.G.M. (2001) A framework for soil food web diagnostics: extension of the nematode faunal analysis concept. *Applied Soil Ecology* 18, 13–29. DOI: 10.1016/S0929-1393(01)00152-4.

Franco, A.L.C., Gherardi, L.A., de Tomasel, C.M., Andriuzzi, W.S., Ankrom, K.E. *et al.* (2019) Drought suppresses soil predators and promotes root herbivores in mesic, but not in xeric grasslands. *Proceedings of the National Academy of Sciences USA* 116, 12883–12888. DOI: 10.1073/pnas.1900572116.

Franco, A.L.C., Guan, P.T., Cui, S.Y., de Tomasel, C.M., Gherardi, L.A. *et al.* (2022) Precipitation effects on nematode diversity and carbon footprint across grasslands. *Global Change Biology* 28, 2124–2132. DOI: 10.1111/gcb.16055.

Gao, Z.W., Liu, Z.Y., Wu, Z.H., Huang, J., Zhang, L. *et al.* (2022) Soil nematode communities and functions: ecological package restoration responses in damaged desert steppe open-pit mining ecosystems. *Land Degradation and Development* 33, 1015–1028. DOI: 10.1002/ldr.4208.

Guo, X.H., Endler, A., Poll, C., Marhan, S. and Ruess, L. (2021) Independent effects of warming and altered precipitation pattern on nematode community structure in an arable field. *Agriculture, Ecosystems and Environment* 316, 107467. DOI: 10.1016/j.agee.2021.107467.

Gwenzi, W. (2021) Rethinking restoration indicators and end-points for post-mining landscapes in light of novel ecosystems. *Geoderma* 387, 114944. DOI: 10.1016/j.geoderma.2021.114944.

Harris, J. (2009) Soil microbial communities and restoration ecology: facilitators or followers? *Science* 325, 573–574. DOI: 10.1126/science.1172975.

Hoeksema, J.D., Lussenhop, J. and Teeri, J.A. (2000) Soil nematodes indicate food web responses to elevated atmospheric CO_2. *Pedobiologia* 44, 725–735. DOI: 10.1078/S0031-4056(04)70085-2.

IPCC (Intergovernmental Panel on Climate Change) (2013) *Climate Change 2013: The Physical Science Basis: Working Group I Contribution to the Fifth Assessment Report of the Intergovernmental Panel on Climate Change* [Stocker, T.F., Qin, D., Plattner, G.-K., Tignor, M., Allen, S.K. *et al.* (eds)]. Cambridge University Press, Cambridge and New York.

IPCC (Intergovernmental Panel on Climate Change) (2022) Summary for Policymakers [Pörtner, H.-O., Roberts, D.C., Poloczanska, E.S., Mintenbeck, K., Tignor, M. *et al.* (eds)]. In: *Climate Change 2022: Impacts, Adaptation, and Vulnerability. Contribution of Working Group II to the Sixth Assessment Report of the Intergovernmental Panel on Climate Change* [Pörtner, H.-O., Roberts, D.C., Tignor, M., Poloczanska, E.S., Mintenbeck, K. *et al.* (eds)]. Cambridge University Press, Cambridge and New York, pp. 3–33. DOI: 10.1017/9781009325844.001.

Kardol, P., Cregger, M.A., Campany, C.E. and Classen, A.T. (2010) Soil ecosystem functioning under climate change: plant species and community effects. *Ecology* 91, 767–781. DOI: 10.1890/09-0135.1.

Li, Q., Xu, C.G., Liang, W.J., Zhong, S., Zheng, X.H. and Zhu, J.G. (2009) Residue incorporation and N fertilization affect the response of soil nematodes to the elevated CO_2 in a Chinese wheat field. *Soil Biology and Biochemistry* 41, 1497–1503. DOI: 10.1016/j.soilbio.2009.04.006.

Li, Q., Bai, H.H., Liang, W.J., Xia, J.Y., Wan, S.Q. and van der Putten, W.H. (2013) Nitrogen addition and warming independently influence the belowground micro-food web in a temperate steppe. *PLoS One* 8, e60441. DOI: 10.1371/journal.pone.0060441.

Lu, P., Tan, Y.J., Dai, N.H., Di, M.X., Weng, X.D. *et al.* (2021) Composition and structure of soil fauna communities and their relationships with environmental factors in copper mine waste rock after re-vegetation. *Global Ecology and Conservation* 31, e01889. DOI: 10.1016/j.gecco.2021.e01889.

Ma, Q.H., Yu, H.Y., Liu, X.D., Xu, Z.Z., Zhou, G.S. and Shi, Y.H. (2018) Climatic warming shifts the soil nematode community in a desert steppe. *Climatic Change* 150, 243–258. DOI: 10.1007/s10584-018-2277-0.

Martínez, J.G., Balondo, K.A., dos Santos, G., Van Ranst, E. and Moens, T. (2021) Re-establishment of nematode communities in a rehabilitated surface mining area in Sibutad, southern Philippines. *European Journal of Soil Biology* 102, 103254. DOI: 10.1016/j.ejsobi.2020.103254.

Mueller, K.E., Blumenthal, D.M., Carrillo, Y., Cesarz, S., Ciobanu, M. *et al.* (2016) Elevated CO_2 and warming shift the functional composition of soil nematode communities in a semiarid grassland. *Soil Biology and Biochemistry* 103, 46–51 DOI: 10.1016/j.soilbio.2016.08.005.

Neher, D.A. (1999) Soil community composition and ecosystem processes: comparing agricultural ecosystems with natural ecosystems. *Agroforestry Systems* 45, 159–185.

Neher, D.A. (2001) Role of nematodes in soil health and their use as indicators. *Journal of Nematology* 33, 161–168.

Neher, D.A. and Weicht, T.R. (2013) Nematode genera in forest soil respond differentially to elevated CO_2. *Journal of Nematology* 45(3), 214–222.

Neher, D.A., Weicht, T.R., Savin, M., Görres, J.H. and Amador, J.A. (1999) Grazing in a porous environment. 2. Nematode community structure. *Plant and Soil* 212, 85–99.

Nielsen, U.N., Ayres, E., Wall, D.H., Li, G., Bardgett, R.D. *et al.* (2014) Global-scale patterns of assemblage structure of soil nematodes in relation to climate and ecosystem properties. *Global Ecology and Biogeography* 23, 968–978. DOI: 10.1111/geb.12177.

Okada, H., Sakai, H., Tokida, T., Usui, Y., Nakamura, H. and Hasegawa, T. (2014) Elevated temperature has stronger effects on the soil food web of a flooded paddy than does CO_2. *Soil Biology and Biochemistry* 70, 166–175. DOI: 10.1016/j.soilbio.2013.12.029.

Otfinowski, R. and Coffey, V. (2020) Can root traits predict communities of soil nematodes in restored northern prairies? *Plant and Soil* 453, 459–471. DOI: 10.1007/s11104-020-04624-y.

Ott, D., Rall, B.C. and Brose, U. (2012) Climate change effects on macrofaunal litter decomposition: The interplay of temperature, body masses and stoichiometry. *Philosophical Transactions of the Royal Society B: Biological Sciences* 367, 3025–3032. DOI: 10.1098/rstb.2012.0240.

Peng, Y., Penuelas, J., Vesterdal, L., Yue, K., Peguero, G. *et al.* (2022) Responses of soil fauna communities to the individual and combined effects of multiple global change factors. *Ecology Letters* 25, 1961–1973. DOI: 10.1111/ele.14068.

Ren, G.H., Wang, C.X., Dong, K.H., Zhu, H.S., Wang, Y.C. and Zhao, X. (2018) Effects of grazing exclusion on soil–vegetation relationships in a semiarid grassland on the Loess Plateau, China. *Land Degradation and Development* 29, 4071–4079. DOI: 10.1002/ldr.3164.

Resch, M.C., Schütz, M., Graf, U., Wagenaar, R., van der Putten, W.H. and Risch, A.C. (2019) Does topsoil removal in grassland restoration benefit both soil nematode and plant communities? *Journal of Applied Ecology* 56, 1782–1793. DOI: 10.1111/1365-2664.13400.

Ribeiro, L.G., Puerari, H.H., Silva, A.O., Vaz, K.A., dos Santos, J.V. *et al.* (2023) Structure and composition of the nematode community in a restoration area affected by iron tailings. *Pedobiologia* 97–98, 150864. DOI: 10.1016/j.pedobi.2023.150864.

Rillig, M.C., Ryo, M., Lehmann, A., Aguilar-Trigueros, C.A., Buchert, S. *et al.* (2019) The role of multiple global change factors in driving soil functions and microbial biodiversity. *Science* 366, 886–890. DOI: 10.1126/science.aay2832.

Sala, O.E., Chapin, F.S. III, Armesto, J.J., Berlow, E., Bloomfield, J. *et al.* (2000) Global biodiversity scenarios for the year 2100. *Science* 287, 1770–1774.

Sanchez-Moreno, S., Camargo, J.A. and Navas, A. (2006) Ecotoxicological assessment of the impact of residual heavy metals on soil nematodes in the Guadiamar River Basin (Southern Spain). *Environmental Monitoring and Assessment* 116, 245–262. DOI: 10.1007/s10661-006-7398-7.

Shaw, E.A., Boot, C.M., Moore, J.C., Wall, D.H. and Baron, J.S. (2019) Long-term nitrogen addition shifts the soil nematode community to bacterivore-dominated and reduces its ecological maturity in a subalpine forest. *Soil Biology and Biochemistry* 130, 177–184. DOI: 10.1016/j.soilbio.2018.12.007.

Song, M., Li, X.M., Jing, S.S., Lei, L.J., Wang, J.L. and Wan, S.Q. (2016) Responses of soil nematodes to water and nitrogen additions in an old-field grassland. *Applied Soil Ecology* 102, 53–60. DOI: 10.1016/j.apsoil.2016.02.011.

Song, Y., Liu, J. and Chen, F. (2020) Elevated CO_2 not increased temperature has specific effects on soil nematode community either with planting of transgenic *Bt* rice or non-*Bt* rice. *Peer J* 8, e8547. DOI: 10.7717/peerj.8547.

Sun, Y., Cao, H., Yin, J., Kang, L. and Ge, F. (2010) Elevated CO_2 changes the interactions between nematode and tomato genotypes differing in the JA pathway. *Plant Cell and Environment* 33, 729–739. DOI: 10.1111/j.1365-3040.2009.02098.x.

Sun, X.M., Zhang, X.K., Zhang, S.X., Dai, G.H., Han, S.J. and Liang, W.J. (2013) Soil nematode responses to increases in nitrogen deposition and precipitation in a temperate forest. *PLoS One* 8, e82468. DOI: 10.1371/journal.pone.0082468.

Thakur, M.P., Reich, P.B., Fisichelli, N.A., Stefanski, A., Cesarz, S. *et al.* (2014) Nematode community shifts in response to experimental warming and canopy conditions are associated with plant community changes in the temperate-boreal forest ecotone. *Oecologia* 175, 713–723. DOI: 10.1007/s00442-014-2927-5.

Thakur, M.P., Tilman, D., Purshcke, O., Ciobanu, M., Cowles, J. *et al.* (2017) Climate warming promotes species diversity, but with greater taxonomic redundancy in complex environments. *Science Advances* 3, e1700866. DOI: 10.1126/sciadv.1700866.

Thakur, M.P., Reich, P.B., Hobbie, S.E., Stefanski, A., Rich, R. *et al.* (2018) Reduced feeding activity of soil detritivores under warmer and drier conditions. *Nature Climate Change* 8, 75–78. DOI: 10.1038/s41558-017-0032-6.

Todd, T.C., Powers, T.O. and Mullin, P.G. (2006) Sentinel nematodes of land-use change and restoration in tallgrass prairie. *Journal of Nematology* 38(1), 20–27.

van den Hoogen, J., Geisen, S., Routh, D., Ferris, H., Traunspurger, W. *et al.* (2019) Soil nematode abundance and functional group composition at a global scale. *Nature* 572, 194–198. DOI: 10.1038/s41586-019-1418-6.

Wang, J.Q., Li, M., Zhang, X.H., Liu, X.J., Li, L.Q. *et al.* (2019) Changes in soil nematode abundance and composition under elevated [CO_2] and canopy warming in a rice paddy field. *Plant and Soil* 445, 425–437. DOI: 10.1007/s11104-019-04330-4.

Wang, J.Q., Shi, X.Z., Li, L.Q. and Zhang, X.H. (2021) Changes in soil nematodes in rhizosphere and non-rhizosphere soils following combined elevated [CO_2] and canopy warming in a winter wheat field. *Geoderma* 386, 114907. DOI: 10.1016/j.geoderma.2020.114907.

Wei, C.Z., Zheng, H.W., Li, Q., Lü, X.T., Yu, Q. *et al.* (2012) Nitrogen addition regulates soil nematode community composition through ammonium suppression. *PLoS One* 7, e43384. DOI: 10.1371/journal.pone.0043384.

Wilschut, R.A., Geisen, S., Martens, H., Kostenko, O., de Hollander, M. *et al.* (2019) Latitudinal variation in soil nematode communities under climate warming-related range-expanding and native plants. *Global Change Biology* 25, 2714–2726. DOI: 10.1111/gcb.14657.

Xiao, H.F., Schaefer, D.A., Lei, Y.B., Zhen, Y.L., Li, Y.P. *et al.* (2013) Influence of invasive plants on nematode communities under simulated CO_2 enrichment. *European Journal of Soil Biology* 58, 91–97. DOI: 10.1016/j.ejsobi.2013.07.002.

Xing, W., Lu, X.M., Niu, S.L., Chen, D.M., Wang, J.S. *et al.* (2022) Global patterns and drivers of soil nematodes in response to nitrogen enrichment. *Catena* 213, 106235. DOI: 10.1016/j.catena.2022.106235.

Xiong, D., Wei, C.Z., Wubs, E.R.J., Veen, G.F., Liang, W.J. *et al.* (2020) Nonlinear responses of soil nematode community composition to increasing aridity. *Global Ecology and Biogeography* 29, 117–126. DOI: 10.1111/geb.13013.

Xiong, D., Wei, C.Z., Wang, X.G., Lü, X.T., Fang, S. *et al.* (2021) Spatial patterns and ecological drivers of soil nematode diversity in natural grasslands vary among vegetation types and trophic position. *Journal of Animal Ecology* 90, 1367–1378.

Yang, J.J., Wu, X.F., Chen, Y., Yang, Z.B., Liu, J.S. *et al.* (2021) Combined attributes of soil nematode communities as indicators of grassland degradation. *Ecological Indicators* 131, 108215. DOI: 10.1016/j.ecolind.2021.108215.

Yeates, G.W., Newton, P.C.D. and Ross, D.J. (1999) Response of soil nematode fauna to naturally elevated CO_2 levels influenced by soil pattern. *Nematology* 1(3), 285–293. DOI: 10.1163/156854199508289.

Yeates, G.W., Newton, P.C.D. and Ross, D.J. (2003) Significant changes in soil microfauna in grazed pasture under elevated carbon dioxide. *Biology and Fertility of Soils* 38, 319–326. DOI: 10.1007/s00374-003-0659-5.

Yin, R., Siebert, J., Eisenhauer, N. and Schaedler, M. (2020) Climate change and intensive land use reduce soil animal biomass via dissimilar pathways. *eLife* 9, e54749. DOI: 10.7554/eLife.54749.

Zhang, Z.W., Li, Q., Zhang, H.Y., Hu, Y.Y., Hou, S.L. *et al.* (2019) The impacts of nutrient addition and livestock exclosure on the soil nematode community in a degraded grassland. *Land Degradation and Development* 30, 1574–1583. DOI: 10.1002/ldr.3344.

Zhou, J., Wu, J.P., Huang, J.X., Sheng, X.J., Dou, X.L. and Lu, M. (2022) A synthesis of soil nematode responses to global change factors. *Soil Biology and Biochemistry* 165, 108538. DOI: 10.1016/j.soilbio.2021.108538.

Zhu, B., Wan, B., Liu, T., Zhang, C., Cheng, L. *et al.* (2023) Biochar enhances multifunctionality by increasing the uniformity of energy flow through a soil nematode food web. *Soil Biology and Biochemistry* 183, 109056. DOI: 10.1016/j.soilbio.2023.109056.

12 Soil Nematode Communities Across Large Spatial Scales

Stefan Geisen[1]*, Rutger A. Wilschut[1], Manqiang Liu[2], Diana H. Wall[†3] and Sara Sánchez-Moreno[4]

[1]Laboratory of Nematology, Wageningen University and Research, The Netherlands; [2]Centre for Grassland Microbiome, State Key Laboratory of Herbage Improvement and Grassland Agro-Ecosystems, Lanzhou, China; [3]Formerly Department of Biology and School of Global Environmental Sustainability, Fort Collins, Colorado, USA; [4]Department of Biodiversity and Evolutionary Biology, National Museum of Natural Sciences (MNCN-CSIC), Madrid, Spain

Abstract

Nematode communities are shaped by a number of factors including soil properties, biome characteristics, climate, agricultural management and other anthropogenic perturbations. Also, biotic factors such as plants, microbes and other animals drive nematode community structure. At large spatial scales, effects of land-use intensification and climate drive nematode responses to agricultural management. Environmental filters such as soil pH, texture, vegetation or aridity may determine soil diversity at regional, continental and global scales. While recent studies show that soil nematode abundances are highest in tundra and boreal forests, large-scale patterns of nematode diversity remain elusive. Large-scale samplings, designed to answer specific questions and performed at high taxonomical resolution, are needed to disentangle nematode biogeographical patterns and to enhance our understanding on the roles of and threats to nematode diversity. Coordinated efforts to monitor soil diversity at large scales should be prioritized to ensure soil health and the global benefits provided by soil biodiversity.

12.1 Diversity of Nematodes, Including Taxonomic Richness and Community Composition, is Not Constant but Depend on Edaphic and Climatic Properties

12.1.1 Main drivers of nematode communities

Soil is a complex habitat that varies in dozens, if not hundreds, of often locally fluctuating physicochemical properties. Many of these properties, determined by both natural and anthropogenic factors, shape nematode abundances, community composition and community dynamics (Song *et al*., 2017; van den Hoogen *et al*., 2019) (Fig. 12.1). The most important factors influencing nematode abundances are soil characteristics, particularly texture and organic carbon content (van den Hoogen *et al*., 2019). As nematodes are water-bound, soil moisture, and thus precipitation, also play a key role in structuring nematode communities: nematode abundances and community diversity decrease with increasing soil aridity (Xiong *et al*., 2020). Also other soil

*Email: stefan.geisen@wur.nl

DOI: 10.1079/9781800624221.0012

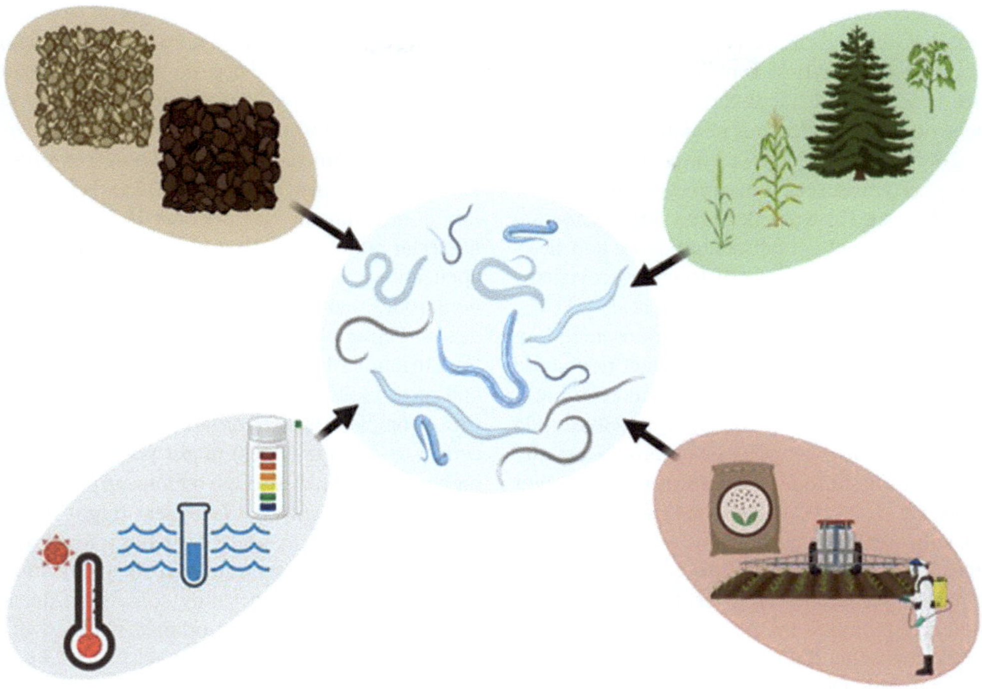

Fig. 12.1. Nematode communities are affected by multiple drivers that might be classified into four main groups: (i) soil properties, including parent material, soil texture, soil carbon and soil nutrients (top left); (ii) biome characteristics and vegetation diversity (top right); (iii) climate and climate change stresses, including soil temperature, water availability, erosion, desertification and acidification (bottom left); and (iv) agricultural soil management and other anthropogenic impacts such as changes in land use and soil sealing (bottom right). (Author's own image.)

characteristics and climatic factors affect nematode communities (van den Hoogen *et al.*, 2019), showing the complexity of nematode responses to the environment that partly explain changes observed across latitudinal ranges (Song *et al.*, 2017). A part of these complex responses can be explained by the fact that nematode taxa differ in their responses to (changes in) their environment, which allows soil ecologists to monitor soil disturbance on the basis of nematode community composition (see Chapters 4–6, this volume).

The composition of nematode communities is also tightly connected to other groups in the biological community. Plants are highlighted as important regulators of soil nematode populations, abundance and growth, as their provision of carbon in the form of exudates or litter is a major energy source into soils, driving the composition of microbial communities and, as prey, microbivorous nematodes. Moreover, plants can be hosts to different types of root-feeding nematodes, which can be host-specialists or host-generalists. Especially in monocultures of suitable host plants, root feeders can become a dominant component of the soil nematode community (Atandi *et al.*, 2017; Song *et al.*, 2017). Soil nematodes are commonly reported to increase with increasing plant species richness. In particular, Scherber *et al.* (2010) highlighted that both nematode abundances and especially species richness increased with an increase in plant species, the latter more than for other soil biota. Multiple mechanisms may jointly determine such a positive relationship between plant and nematode communities. Firstly, increasing plant species diversity may result in an increased number of root-feeding nematode taxa in the community, as considerable numbers of root-feeding nematode taxa have a limited host range (Topalovic and Geisen, 2023). Moreover, an increased plant diversity may also lead to concurrent increases in fungal and bacterial abundances and diversity, for example, due to a more diverse input of carbon sources through root exudates and litter. Indeed, plants accumulate species-specific root-associated nematode communities. Legumes, for instance, support the growth of rapidly growing bacterivore nematodes (Rhabditida), while forbs foster the growth of fungivorous nematodes (Viketoft *et al.*, 2009). Finally, nematode diversity (here used as a synonym for species richness) seems also to be strongly shaped by vegetation type, with temperate trees promoting the highest abundance and diversity of nematodes compared to other ecosystems like grasslands or trophic sites (Song *et al.*, 2017).

Besides plants, other biotic groups including microbes and animals in soil impact nematode communities. As aforementioned, microbes, and especially bacteria, strongly affect nematode abundances, as bacterivore nematodes usually dominate nematode communities (van den Hoogen *et al.*, 2019). Bacteria are usually increasing with available carbon and, therefore, with increasing plant inputs into the soil system. In turn, nematodes may buffer increases in bacterial biomass by increasing in numbers (Landesman *et al.*, 2011). As their main consumers, higher-trophic-level nematodes often follow abundance patterns of bacterivores, while changes in general substrate availability similarly affect nematode feeding groups (van den Hoogen *et al.*, 2019). A last nematode group to be highlighted are entomopathogenic nematodes, which depend on the availability of insect hosts. More than 120 spp. of entomopathogenic nematodes have been described, and some are commercially used as biocontrol organisms (Bhat *et al.*, 2020).

12.2 Anthropogenic Impacts on Soil Nematode Communities

Human activities are deeply altering soil conditions worldwide. At global scales, a number of anthropogenic impacts, including chemical and plastic pollution, erosion, salinization and soil sealing, threaten soils. Additionally, agricultural practices that include large chemical inputs, heavy tilling and reduced plant diversity also degrade soil condition. Such threats modify the physicochemical niches in which nematodes live, and thus directly influence their establishment and persistence.

In agricultural systems, fertilization, tillage and habitat simplification are the main threats to soil diversity. Fertilization effects on soil biota depend on the organismal group, type of fertilizer, habitat (e.g. soil type) and climatic conditions. In general, organic and organo-mineral fertilization have positive general effects on soil fauna abundance due to the concomitant increase of soil carbon and nitrogen that boost basal resources for soil organisms (Zhu *et al.*, 2023). Fertilizers might thus favour bacterivores by enhancing bacterial populations that bloom after fertilizer applications, which might increase opportunistic nematodes, particularly in the family Rhabditidae or Cephalobidae, leading to an increase in the Enrichment Index (Puissant *et al.*, 2021), an indicator of soil fertility (see Chapter 5, this volume). Other complex forms of organic matter (manure, composts, crop residues) might positively affect higher trophic links through bottom-up processes, and, oppositely, mineral fertilizers might negatively affect soil diversity. Synthetic fertilizers and nitrogen deposition induce soil acidification in nutrient-poor, often sandy soils, with detrimental effects on the abundance (Xing *et al.*, 2023) and diversity (Puissant *et al.*, 2021) of soil organisms.

When combining results of multiple studies from different geographic regions, contrasting responses of soil nematodes to different types of organic amendments become evident. Nitrogen additions increase total nematode abundances but reduce richness, while carbon additions foster nematode diversity (Liu *et al.*, 2016). The purported overall positive effect of nitrogen addition on nematode abundances might be primarily due to the increased abundances of herbivorous nematodes (Liu *et al.*, 2016), which are enhanced by the positive effects of fertilizers on plant biomass, and of some bacterivores. Observed negative effects of nitrogen addition on nematode diversity can be explained by toxic effects of excess nitrogen on predators and omnivores, reflected in the decrease of the Structure and Maturity Indices (Liu *et al.*, 2016), indicators of soil food web complexity and connectance (see Chapter 5, this volume). Similar negative impacts of nitrogen additions on nematode community composition can also be found in natural systems (Eisenhauer *et al.*, 2012).

In agricultural systems, tillage also affects soil diversity by disturbing soil aggregation and by altering the physical matrix in which soil organisms move. Often, tillage is most detrimental to large soil organisms such as earthworms but also omnivorous/predatory nematodes that need large soil pores to move through the soil matrix. These large nematodes, such as many taxa of Dorylaimidae, are more sensitive to both chemical and physical perturbations, and tillage therefore especially affects omnivore and predatory nematodes, reducing nematode richness and the Maturity Index at global scales (Puissant *et al.*, 2021). In contrast, reducing tillage increases omnivore nematodes, although such effect varies with depth, climate, soil texture, tillage method and frequency, and concurrent fertilization and herbicide application (Betancur-Corredor *et al.*, 2023). Standard tillage, however, possesses some benefits for farmers, and helps controlling plant-parasitic nematodes. Indeed, at the national level, adoption of no-tillage practices increased the abundance of herbivorous nematodes in cropping systems in Germany (van Capelle *et al.*, 2012).

Overall, the effects of land-use intensification on soil nematodes greatly depend on the type of ecosystem and other environmental and climatic drivers. At global scales, interactive effects of land-use intensification and soil organic carbon determine the abundances of fungivores, herbivores and omnivores, while its interaction with annual mean temperature and vegetation determine bacterivore and omnivore abundances, respectively (Li *et al.*, 2022). Land-use intensification reduces nematode diversity, an effect that is stronger in forests than in grasslands, croplands or urban soils (Pothula *et al.*, 2019). Due to the bottom-up effects of organic enrichment already described, total nematode abundances might increase in managed compared to unmanaged systems, specifically in undisturbed natural habitats, secondary vegetation habitats and urban areas (Li *et al.*, 2022). Such increase is due to the enhancement of bacterivorous and herbivorous nematodes in managed lands, while sensitive omnivores and predators commonly decline in such scenarios (Li *et al.*, 2022). Despite the reduction in both diversity and richness of nematodes through agricultural management, the evolutionary diversification rates may be greater in agricultural than in natural habitats (Gong *et al.*, 2021). This phenomenon may be attributed to shifts in nematode community compositions towards rapidly evolving taxa (r-strategists) in agricultural habitats (Gong *et al.*, 2021). Land-use intensification is also commonly accompanied by further simplifications of the agroecosystem. It is widely acknowledged that monocultures increase pests and diseases in crops, and that plant diversification either in space (intercropping, polycultures) or in time (rotations) greatly enhances plant health and soil fertility. At global scales, plant-feeding nematodes are reduced by crop rotations, and such effect amplifies with time (Puissant *et al.*, 2021).

In both natural and agricultural ecosystems, extreme weather events can have a profound impact on soil nematode communities. Given their dependency on soil water, drought and flooding events strongly affect nematode communities. Typically, increased precipitation enhances nematode abundances, while severe drought can lead to a virtual loss of soil nematodes (Landesman *et al.*, 2011; Yan *et al.*, 2018). Landesman *et al.* (2011) and Yan *et al.* (2018) also reported differential sensitivity of nematode taxa to changes in soil water availability leading to changes in nematode communities. For example, the genus *Acrobeloides* within the family Cephalobidae exhibited higher relative abundance under drought conditions compared to the ambient soil moisture treatment (Lu *et al.*, 2023). In the same study, drought also led to a shift in nematode communities towards smaller-sized species, a reduced body size at the population scale, and a decrease in size-at-age of *Acrobeloides* at the individual level (Lu *et al.*, 2023). These findings suggest that large-sized nematodes are more sensitive to drought than small-sized nematodes and are in line with multiple observational studies on the size structure of nematode communities along aridity gradients, showing generally smaller nematodes with increasing aridity (Andriuzzi *et al.*, 2020; Li *et al.*, 2023). Many of these driving factors are substantially affected by climate change, which has profound effects on nematode communities (see Chapter 11, this volume).

Other global anthropogenic impacts that threaten soil organisms include different types of pollution as well as salinization. Microplastics are now widely present in all ecosystems on Earth, but their impact on soil animal communities is still poorly understood. So far, evidence shows that nematodes can directly ingest microplastics, at least in marine habitats (Fueser *et al.*, 2020). Microplastic toxicity might derive, however, from chemicals released by plastic remains in the soil, such as wire tyres (Kim *et al.*, 2021), exerting negative effects on nematode growth and reproduction (Liu *et al.*, 2023). The effects of other anthropogenic impacts are even less known. As an example, salinization reduces nematode density, especially in arid and semi-arid regions (see, e.g., Su *et al.*, 2016; de Sá *et al.*, 2021). Erosion transports and redistributes nematodes and changes the physicochemical soil environment, affecting soil nematode density and diversity (Baxter *et al.*, 2013). However, the impacts of these and other anthropogenic factors (soil compaction, introduction of invasive plant species, use of genetically modified organisms, soil sealing) on nematode communities at global scales have not been assessed. Before it is too late, it is extremely relevant to deepen our knowledge on the global drivers of soil diversity and reverse the current patterns of global decline of nematode and other soil organisms due to climate change and the biodiversity crisis (see Chapter 11, this volume).

12.3 Nematodes Across Larger Spatial Scales

12.3.1 At the (sub)continental scale

Recent advancements have greatly enhanced our understanding of the macroecology and biogeography of soil nematodes at the (sub)continental or regional scale. It has been observed that nematodes, along with fungi and protists, are primarily shaped by environmental selection-based deterministic processes during community assembly, in contrast to smaller microorganisms like bacteria that are more influenced by dispersal-based stochastic processes (Luan *et al.*, 2020). This emphasizes the importance of ecological or environmental filtering effects in the assembly of nematode communities. Current evidence suggests that the diversity, distribution and evolutionary dynamics of soil nematodes are largely controlled by bottom-up forces such as the characteristics of plant species and environmental conditions at the regional scale. For example, Wu *et al.* (2016) found that in coastal China, mean annual temperature variation and sediment pH were more important for nematode community composition than vegetation type, whereas a study in Europe suggested that the composition of nematode communities along latitudinal transects was primarily shaped by latitude and plant species identity rather than soil characteristics (Wilschut *et al.*, 2019). In natural habitats, soil nematodes exhibit the highest diversity and species richness at around 30°N latitude (Gong *et al.*, 2021). However, agricultural habitats show no significant variation in nematode diversity across latitudes, and there is a reduction in both diversity and richness, particularly in subtropical zones (Gong *et al.*, 2021), a pattern that may be explained as a result of biotic homogenization driven by agricultural practices (Li *et al.*, 2020).

In addition, climatic factors, such as precipitation and aridity, have been identified as key drivers of nematode communities on a regional scale, and may be the underlying drivers of latitudinal and altitudinal shifts in nematode communities observed in some large-scale sampling studies (Wilschut *et al.*, 2019). Precipitation explains the variation in nematode abundance, trophic groups, and the ratio of bacterial-feeding to fungal-feeding nematodes (Chen *et al.*, 2015). Experimental studies simulating increased rainfall of 30% during the growing season have shown that nematodes are more responsive than microbes, resulting in an increase in below-ground biomass or density across a broad range of arid and semi-arid grasslands (Chen *et al.*, 2015). Conversely, aridity exerts a pronounced effect on nematode communities. As aridity increases, there is a decline in nematode abundance, richness and similarity, indicating the sensitivity of these communities to aridity rather than the plant community and environmental characteristics (Xiong *et al.*, 2020). Notably, bacterial-feeding nematodes exhibit greater resilience to aridity compared to fungal-feeding nematodes, suggesting differences in their adaptations to this environmental stress (Xiong *et al.*, 2020). Surprisingly, however, an investigation of two ecosystems on the Tibetan Plateau showed that climatic factors contributed little to changes in nematode communities (Chen *et al.*, 2021). Although climatic factors are considered to be the most important factors in structuring below-ground communities (Nielsen *et al.*, 2014), numerous studies have shown that soil nematodes are not always sensitive to changes in mean annual precipitation and mean annual temperature (Darby *et al.*, 2011; Eisenhauer *et al.*, 2012). These apparent findings suggest that changes in soil nematode community composition are highly ecosystem- and spatial-scale-dependent (Liu *et al.*, 2019). In short, regional-scale variation in soil resources, plant species characteristics and climate may be the main factors explaining changes in soil nematode diversity, but nematode communities respond differently, or even oppositely, to biotic and abiotic drivers (Liu *et al.*, 2019). The relative contribution of mutually superimposed environmental factors to the spatial distribution patterns of soil nematode diversity needs to be further investigated.

More importantly, much of the variation in soil nematode community composition remains unexplained, possibly because some of the important factors driving nematode community composition are still not adequately measured. This unexplained variation commonly exceeds 50% (Chen *et al.*, 2015; Wilschut *et al.*, 2019). Viketoft (2013) also suggested that measuring more microorganisms could potentially reduce unexplained variation in nematode community composition. Traditionally, climate and soil characteristics have been generally considered to be key in driving species distributions at large scales, while biotic interactions only played a role at small scales (Pearson and Dawson, 2003). Currently, however, a growing body of research suggests that biotic interactions are equally critical for species distribution at large scales (Araújo and Luoto, 2007; Cazelles *et al.*, 2016), suggesting that more attention should be paid in the future to the important role of trophic interactions on nematode community variation at large scales. In summary, there are still many challenges in understanding the spatial distribution and drivers of nematodes at the regional scale, and in deciphering how regional-scale drivers interact to shape nematode communities at the global scale (Fig. 12.2).

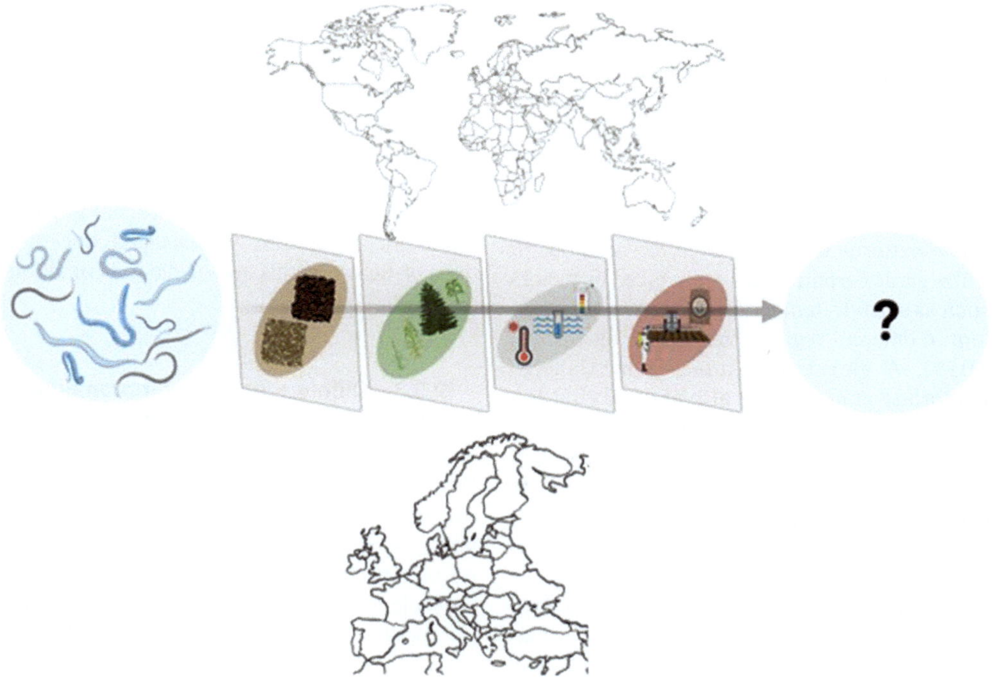

Fig. 12.2. Nematode communities are affected by multiple drivers at local and regional scales, including soil properties, vegetation type, climate and soil management. Multiple drivers interact in complex ways, determining the composition and abundance of nematode communities at global scales through processes that are still unexplored. (Author's own image.)

12.3.2 At the global scale

Soil nematode abundance and richness at the global scale display a clear latitudinal pattern, with increasing nematode abundances and diversity at higher latitudes, but a decrease in polar regions (Nielsen *et al.*, 2014). This pattern was supported by a global meta-analysis by Song *et al.* (2017), which found higher nematode abundance and genus richness in temperate forests compared to tropical forests. The meta-analysis also identified peak values observed at latitudes between 30° and 55°. In contrast, a recent study using molecular methods found that species richness in the Arctic highland tundra might be surprisingly high, with alpha diversity comparable to that observed in Malaysian tropical rainforests (Kerfahi *et al.*, 2016). These inconsistent conclusions may be due to large gaps in the coverage of existing data sets, highlighting the need for greater taxonomic resolution to draw more reliable conclusions. However, recent research conducted by van den Hoogen *et al.* (2019) has expanded our understanding of the spatial distribution of soil nematodes at a global scale. Analysing over 6700 georeferenced samples, the study revealed distinct patterns across different ecosystems. Nematode abundances were highest in tundra regions, followed by boreal forests and temperate broadleaf forests, while Mediterranean forests, Antarctic sites and hot deserts had the lowest nematode abundances. These patterns were attributed to variations in soil organic carbon stocks, influenced by factors such as altitude, temperature and moisture content in different regions (van den Hoogen *et al.*, 2019). A global meta-analysis by Gallego-Zamorano *et al.* (2023) also showed that the response of nematode abundance to nitrogen enrichment depended on mean annual precipitation. In arid areas, nematode abundance tended to decrease with nitrogen enrichment, while in wet areas, it exhibited an increase, with variations observed among different feeding guilds.

Furthermore, in addition to natural factors, increasing anthropogenic factors also have an important influence on the distribution of soil nematodes. Pothula *et al.* (2019) evaluated the effect of anthropogenic interventions on nematode richness and abundance by comparing urban, agricultural, disturbed grassland, natural grassland and forest ecosystems. The results showed that the overall richness was higher in forests than in all other ecosystem. In contrast, overall nematode abundance was highest in disturbed grassland, agricultural and forest ecosystems. These meta-analyses support the concept that human intervention can adversely affect nematode richness and abundance. In contrast, Li *et al.* (2022) conducted a reanalysis of the global data set derived from van den Hoogen *et al.* (2019) and investigated the impact of anthropogenic drivers, such as management practices and urbanization. The study observed that nematode abundance in urban areas exhibited greater variability and was less influenced by environmental factors compared to other land-use types. This finding challenges the conventional notion that human land use universally leads to a decline in animal abundance across trophic groups. Instead, urbanization is found to have unique effects on nematode communities, indicating that urban areas may possess unique ecological dynamics that influence nematode abundance.

Given the natural and anthropogenic pressures, the species-specific adaptation capability plays a crucial role in shaping the global distribution of soil nematodes. Cephalobidae, a cosmopolitan family, demonstrates adaptability to various soil habitats globally, including ponds, polar, tundra, mine tailing and deserts (Liu *et al.*, 2019). Similarly, fungivorous nematodes of the Aphelenchoidae family exhibit cosmopolitan distribution and resilience to stressful environments, including those with elevated concentrations of heavy metals and acidic soil conditions (Liu *et al.*, 2019).

12.3.3 Differences and commonalities between large- and local-scale studies

Why do we need to make clear distinctions between results obtained from local-scale and large-scale studies? Among different reasons, it is relevant to clearly determine the study focus. Local-scale studies are often performed to disentangle questions related to distinct local treatments, such as the assessment of diverse agricultural practices. If such studies are addressed at the global scale, confounding factors such as climate-related ones are enormous. For example, seasonal differences between hemispheres do not allow sampling at the same time, while extreme comparisons including tropics to tundra will have major effects on studied patterns. Only with sampling efforts of potentially hundreds of thousands of samples, with intense sampling at local scales of interest, will the differences in the biogeography of nematode taxa be observable and interpretable. As this sampling effort has never been made, large-scale studies have resulted in observation of large-scale biogeographical patterns. At the regional to continental level, for example, sampling sites are often chosen following gradients of aridity or under specific vegetation or biome structure. These approaches allow reducing the complexity of confounding factors linked to a specific question such as deciphering the effects of drought on soil nematode communities (Lu *et al.*, 2023). To disentangle the effects of global factors on soil organisms, research should be performed at the meta-scale ensuring that they are not influenced by confounding local-scale differences (Fig. 12.3). Relevant open questions are, for example, the role of the most important physicochemical drivers on nematode abundances or the extent of latitudinal differences in community assembly (van den Hoogen *et al.*, 2019; Luan *et al.*, 2020; Aslani *et al.*, 2022). To answer many of the research questions addressed in local to global studies, follow-up experiments in controlled conditions are often advisable (Berlinches de Gea *et al.*, 2023). Only these studies allow manipulating and thereby confirming the causal role of the observed patterns, such as physicochemical drivers on nematode abundances and community composition.

12.4 Limitations and Benefits of Large-Scale Studies with a Guide for Future Approaches

12.4.1 Methods to synchronize and how to proceed

Despite their inherent difficulties and high costs, an increasing number of studies focus on large-scale soil biodiversity patterns, many performed

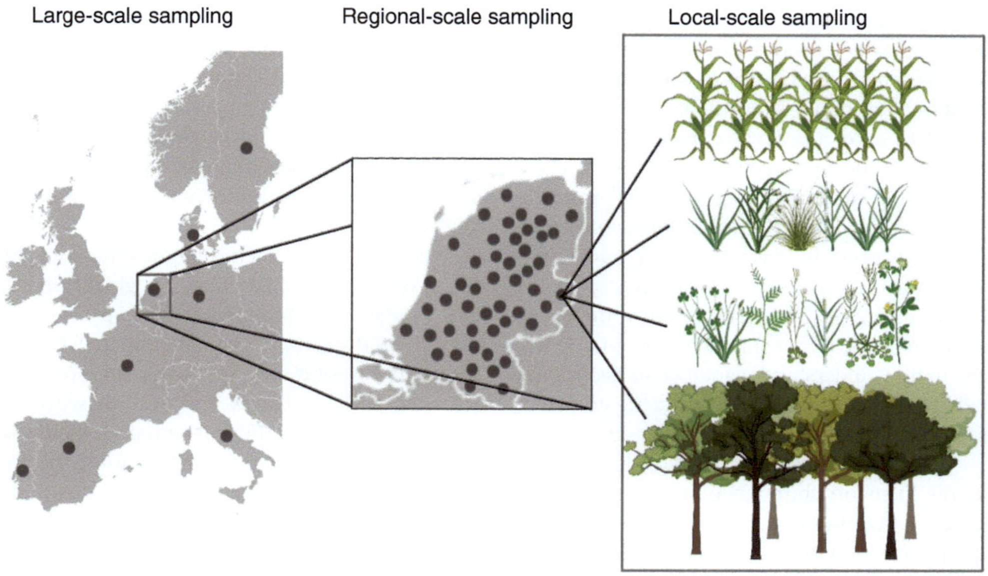

Fig. 12.3. Differences in sampling scale illustrating the need to design study-specific questions. Large-scale sampling is illustrated here with samples taken across Europe that might reveal latitudinal differences in nematode communities as driven by profound climatic differences. For a better understanding of regional processes, more samples need to be taken within a comparable climate, such as shown for The Netherlands. Fundamental differences exist even at the regional scale, as illustrated for major differences in vegetation structure across forests, natural and managed grassland, and cropland. Note that, if large-scale or regional-scale samples differ in factors such as vegetation, such factors might overrule other factors of interest such as climatic factors. This shows the need for a thorough definition of the research question, including the scale, to address the question and a matching sampling design. (Author's own image.)

at the global scale. They focus on fungi and bacteria (Tedersoo *et al.*, 2014; Thompson *et al.*, 2017; Bahram *et al.*, 2018; Delgado-Baquerizo *et al.*, 2018), protists (Oliverio *et al.*, 2020), collembola (Potapov *et al.*, 2023), earthworms (Phillips *et al.*, 2019) or eukaryotes in general (Aslani *et al.*, 2022). What have these studies in common and how do they differ from global studies performed on soil nematodes (Song *et al.*, 2017; van den Hoogen *et al.*, 2019)? Can we learn from studies on other soil organisms to optimize global studies on soil nematodes?

Arguably all global studies fall short in sample numbers. Luckily, sampling scales on global soil biodiversity analyses are increasing rapidly, and have come from only 11 locations to map global soil animal distribution (Wu *et al.*, 2011) to, more recently, 83 sampling points (Bastida *et al.*, 2020). Currently, above-mentioned studies on other soil organisms include sample numbers generally in the hundreds, and a few in the thousands. Although high, such numbers are still too limited to allow accurate assessments of soil communities to cover accurately Earth soil diversity. Large-scale sampling efforts like the LUCAS (Land Use and Coverage Area frame Survey) project are needed to better cover these questions, as many thousands of samples are needed to cover the environmental heterogeneity even at the continental scale (Orgiazzi *et al.*, 2022). Our understanding of global soil biodiversity would benefit from large increased sample numbers to cover especially understudied regions like the Arctic and large parts of Africa and South America. Also improved modelling approaches (Ludwig *et al.*, 2023) will help to fully understand nematode biogeography, their functional importance, and the contribution of nematodes to global biodiversity.

Available nematode studies at large scales mainly focus on abundance and biomass data, limited in taxonomic resolution to the functional group level (van den Hoogen *et al.*, 2019), and more rarely to family or genus level (Song *et al.*, 2017). Oppositely, global studies on microorganisms often provide more in-depth taxonomic information equalling roughly genera and sometimes species, thanks to molecular sequencing approaches (Geisen *et al.*, 2019a). Large-scale nematode studies at higher taxonomical resolution, to the genus, family, species or even (sub)species level, would greatly enhance our understanding of nematode species distributions and biogeography, and would permit inferences on the dynamics of species distribution and to identify local extinctions and invasive patterns (Veresoglou *et al.*, 2015; Geisen *et al.*, 2019b; Thakur *et al.*, 2019).

An intrinsic limitation of many large-scale studies is that nematode data are gathered from diverse sources, including different sampling schemes and techniques, with different nematode extraction methods, and by various scientific teams, with no common standardized methodologies. Studies developed with common protocols on soil sampling, sample processing and data analyses (e.g. Bahram *et al.*, 2018; Delgado-Baquerizo *et al.*, 2018) reduce confounding factors that might influence, and obscure, the obtained data. Efforts that try to standardize scientific methodologies to obtaining homogeneous data are pivotal to better compare samples and allow more accurate and informative global studies, as described in Potapov *et al.* (2022). It is important to note that sampling should be optimized for nematode analysis, as the sample size used for microorganisms (>1 g usually) is not sufficient to study nematode diversity in soil (>100 g recommended) (Wiesel *et al.*, 2015). Studies performed on minute soil samples to characterize nematode diversity based on DNA sequencing should be treated with care and re-evaluated with more suitable methods.

Molecular approaches might solve some of the caveats mentioned above on global nematode studies (like higher sampling processing, the potential for resolving individuals to a higher taxonomic resolution). Yet, these methods are also not without biases and cannot serve as stand-alone tools to study all nematode biodiversity facets (e.g. are limited in determining species abundances, biomasses and activities), as detailed in Chapter 15 of this volume. Also, data obtained from molecular methods cannot directly be compared to morphological information as there are biases associated with both (Griffiths *et al.*, 2018). Calibrations are needed to allow merging molecular and morphological-based identification methods. Such calibration, and the joint use of both, is a pivotal task to increase our understanding on the role of soil nematodes in ecosystems and to make nematode studies at the global scale more feasible.

12.4.2 Links between scientific values and societal relevance of nematode large-scale studies

As explained in Chapter 1 (this volume), soil animals such as nematodes are crucial to guarantee functioning of terrestrial ecosystems through their contribution to carbon and nitrogen cycling, water purification and plant production. With accelerating multifaceted global challenges/threats (Rillig *et al.*, 2019) that are impacting soils and peoples beyond borders (van der Putten *et al.*, 2023), there is uncertainty on whether soil biodiversity and ecosystem functions will be maintained in the future. This adds to an urgency, and perhaps an opportunity, for soil biodiversity scientists to provide not only solid integrated syntheses and research results useful to society, and particularly to policy makers, but also to re-evaluate our own practices and collaborations to assure that global change impacts on soils, biodiversity, ecosystems and people are minimized.

To achieve this goal, soil biodiversity has to be inventoried and monitored. It is important to know how and where 'hotspots' of soil nematode biodiversity are to allow policy makers to prioritize adequate management options to preserve soil biodiversity, to support sustainable plant pathogen control, and to predict, and prevent, soil biodiversity losses (Guerra *et al.*, 2020). Soil samplings at large spatial scales (regional, continental) and/or across different ecosystems and management types are increasing, but an increased sampling effort is particularly necessary in understudied areas such as Africa, South America and the Arctic.

We, as nematologists, together with other soil scientists, are moving towards potentially new global analyses of the distribution and functional significance of individual taxa, populations and communities of the different components of soil biodiversity (animals, microbes, protists, bacteria and fungi). Coordinating such efforts may help setting newer and higher standards on soil research and enhance our comprehension of current and past soil biodiversity research. Could we, as soil researchers, unite with others studying soil organisms at a global scale, and establish a code of conduct, a common framework and a shared goal for global soil biodiversity research? Below, we build on key points derived from ongoing regional and global projects, emphasizing research on nematodes and other soil biota for consideration as we move forward to fill global knowledge gaps. We note that we recognize, commend and encourage all global efforts and field research to be safe from harassment, and statements and actions should support diversity, equity and inclusion.

There is an urgency. The world is changing. Soil habitats and species are changing due to multi-coexisting global challenges (from climate change to soil pollution). Soil biodiversity underpins soil health and is influential in addressing land management options. Policy makers need us to prioritize, at global and regional scales, conservation efforts to sustain biodiversity and soil health. Since nematodes and their interactions are an integral part of the soil food web and present ubiquitous distributions, global surveys could be achieved such as those performed by LUCAS (https://ec.europa.eu/eurostat/web/lucas/overview, accessed 15 August 2025) and Soil BON (Soil Biodiversity Observation Network) (https://monitor.soilbon.org/, accessed 15 August 2025). Such initiatives involve usually many collaborators, coordination and cooperation among scientists from many disciplines and relevant funding unless everyone volunteers (Potapov *et al.*, 2022).

The Ecological Society of America, for example, includes in their Code of Conduct, 'Ecologists will conduct their research so as to avoid or minimize adverse effects on ecosystems and human communities'. To obtain large-scale data, large numbers of local soil samples, some including plant roots, and/or soil monoliths, have to be collected, each differing in their impact on soil conservation. Coordinating and optimizing the number of samples needed for global studies of nematodes and other soil biota, prior to soil sampling, may reduce sampling impact on multiple ecosystems. Will we train tomorrow's students from multiple disciplines to minimize soil disturbance, standardize soil biodiversity studies and integrate local human communities? Avoiding helicopter research by inclusion of local collaborators and partners who are historically marginalized is part of global science ethics (Nature's Editorial Team, 2022) and is, for example, part of Soil BON (Guerra *et al.*, 2021). Acknowledgement, when possible, of the current and/or past landowners where soil samples are taken, as well as contributions from citizen scientists and other partners, is also necessary in papers and open access data.

Nematology has advanced and soil biodiversity is recognized as part of the global Earth's biodiversity. In December 2022, the United Nations (UN) Convention on Biological Diversity (CBD) invited several soil-related organizations such as the Intergovernmental Technical Panel on Soils (Global Soil Partnership), the UN Convention to Combat Desertification, the Global Soil Biodiversity Initiative and soil scientists to support the post-2020 Framework on Biodiversity. In Europe, a new Proposal for a Directive on Soil Monitoring and Resilience is being discussed, as a result of the societal and policy concerns on soil sustainability, soil health, and the global benefits provided by soil biodiversity.

The extreme diversity of soils is driven by a number of environmental factors that occur at large scale, and shaped by anthropogenic factors such as soil management, pollution and erosion. Soil nematode studies at large spatial scales provide relevant insight into the evolutionary, ecological and biogeographical history of soil diversity, and help understanding of the effects of environmental threats such as climate change in ecosystems. Increasing our knowledge on the global distribution of soil nematodes is crucial, and common global scientific efforts to assess global soil nematode diversity should be promoted to assess and protect soil biodiversity worldwide.

References

Andriuzzi, W.S., Franco, A.L.C., Ankrom, K.E., Cui, S., de Tomasel, C.M. *et al.* (2020) Body size structure of soil fauna along geographic and temporal gradients of precipitation in grasslands. *Soil Biology and Biochemistry* 140, 107638.

Araújo, M.B. and Luoto, M. (2007) The importance of biotic interactions for modelling species distributions under climate change. *Global Ecology and Biogeography* 16, 743–753. DOI: 10.1111/j.1466-8238.2007.00359.x.

Aslani, F., Geisen, S., Ning, D., Tedersoo, L. and Bahram, M. (2022) Towards revealing the global diversity and community assembly of soil eukaryotes. *Ecology Letters* 25, 65–76.

Atandi, J.G., Haukeland, S., Kariuki, G.M., Coyne, D.L., Karanja, E.N. *et al.* (2017) Organic farming provides improved management of plant parasitic nematodes in maize and bean cropping systems. *Agriculture, Ecosystems and Environment* 247, 265–272.

Bahram, M., Hildebrand, F., Forslund, S.K., Anderson, J.L., Soudzilovskaia, N.A. *et al.* (2018) Structure and function of the global topsoil microbiome. *Nature* 560, 233–237.

Bastida, F., Eldridge, D.J., Abades, S., Alfaro, F.D., Gallardo, A. *et al.* (2020) Climatic vulnerabilities and ecological preferences of soil invertebrates across biomes. *Molecular Ecology* 29, 752–761.

Baxter, C., Rowan, J.S., McKenzie, B.M. and Neilson, R. (2013) Understanding soil erosion impacts in temperate agroecosystems: bridging the gap between geomorphology and soil ecology using nematodes as a model organism. *Biogeosciences* 10, 7133–7145. DOI: 10.5194/bg-10-7133-2013.

Berlinches de Gea, A., Hautier, Y. and Geisen, S. (2023) Interactive effects of global change drivers as determinants of the link between soil biodiversity and ecosystem functioning. *Global Change Biology* 29, 296–307.

Betancur-Corredor, B., Lang, B. and Russell, D.J. (2023) Organic nitrogen fertilization benefits selected soil fauna in global agroecosystems. *Biology and Fertility of Soils* 59, 1–16. DOI: 10.1007/s00374-022-01677-2.

Bhat, A.H., Chaubey, A.K. and Askary, T.H. (2020) Global distribution of entomopathogenic nematodes, *Steinernema* and *Heterorhabditis*. *Egyptian Journal of Biological Pest Control* 30, 31. DOI: 10.1186/s41938-020-0212-y.

Cazelles, K., Mouquet, N., Mouillot, D. and Gravel, D. (2016) On the integration of biotic interaction and environmental constraints at the biogeographical scale. *Ecography* 39, 921–931. DOI: 10.1111/ecog.01714.

Chen, D., Cheng, J., Chu, P., Hu, S., Xie, Y., Tuvshintogtokh, I. and Bai, Y. (2015) Regional-scale patterns of soil microbes and nematodes across grasslands on the Mongolian plateau: relationships with climate, soil, and plants. *Ecography* 38, 622–631. DOI: 10.1111/ecog.01226.

Chen, H., Luo, S., Li, G., Jiang, W., Qi, W. *et al.* (2021) Large-scale patterns of soil nematodes across grasslands on the Tibetan Plateau: relationships with climate, soil and plants. *Diversity* 13, 369. DOI: 10.3390/d13080369.

Darby, B.J., Neher, D.A., Housman, D.C. and Belnap, J. (2011) Few apparent short-term effects of elevated soil temperature and increased frequency of summer precipitation on the abundance and taxonomic diversity of desert soil micro- and meso-fauna. *Soil Biology and Biochemistry* 43, 1474–1481. DOI: 10.1016/j.soilbio.2011.03.020.

de Sá, C.S.B., Shiosaki, R.K., dos Santos, A.M. and da Silva Campos, M.A. (2021) Salinization causes abrupt reduction in soil nematode abundance in the Caatinga area of the Submedio San Francisco Valley, Brazilian semiarid region. *Pedobiologia* 85–86, 150729. DOI: 10.1016/j.pedobi.2021.150729.

Delgado-Baquerizo, M., Oliverio, A.M., Brewer, T.E., Benavent-González, A., Eldridge, D.J. *et al.* (2018) A global atlas of the dominant bacteria found in soil. *Science* 359, 320–325.

Eisenhauer, N., Cesarz, S., Koller, R., Worm, K. and Reich, P.B. (2012) Global change belowground: impacts of elevated CO_2, nitrogen, and summer drought on soil food webs and biodiversity. *Global Change Biology* 18, 435–447. DOI: 10.1111/j.1365-2486.2011.02555.x.

Fueser, H., Mueller, M.-T. and Traunspurger, W. (2020) Ingestion of microplastics by meiobenthic communities in small-scale microcosm experiments. *Science of The Total Environment* 746, 141276. DOI: 10.1016/j.scitotenv.2020.141276.

Gallego-Zamorano, J., de Jonge, M.M.J., Runge, K., Huls, S.H., Wang, J. *et al.* (2023) Context-dependent responses of terrestrial invertebrates to anthropogenic nitrogen enrichment: a meta-analysis. *Global Change Biology* 29, 4161–4173. DOI: 10.1111/gcb.16717.

Geisen, S., Briones, M.J.I., Gan, H., Behan-Pelletier, V.M., Friman, V.-P. *et al.* (2019a) A methodological framework to embrace soil biodiversity. *Soil Biology and Biochemistry* 136, 107536.

Geisen, S., Wall, D.H. and van der Putten, W.H. (2019b) Challenges and opportunities for soil biodiversity in the Anthropocene. *Current Biology* 29, R1036–R1044.

Gong, X., Chen, X., Geisen, S., Zhang, J., Zhu, H. *et al.* (2021) Agricultural habitats are dominated by rapidly evolving nematodes revealed through phylogenetic comparative methods. *Soil Biology and Biochemistry* 155, 108183. DOI: 10.1016/j.soilbio.2021.108183.

Griffiths, B.S., de Groot, G.A., Laros, I., Stone, D. and Geisen, S. (2018) The need for standardisation: exemplified by a description of the diversity, community structure and ecological indices of soil nematodes. *Ecological Indicators* 87, 43–46.

Guerra, C.A., Heintz-Buschart, A., Sikorski, J., Chatzinotas, A., Guerrero-Ramírez, N. *et al.* (2020) Blind spots in global soil biodiversity and ecosystem function research. *Nature Communications* 11, 3870.

Guerra, C.A., Wall, D.H. and Eisenhauer, N. (2021) Unearthing soil ecological observations: see supporting information as supplementary material. *Soil Organisms* 93, 79–81.

Kerfahi, D., Tripathi, B.M., Porazinska, D.L., Park, J., Go, R. and Adams, J.M. (2016) Do tropical rain forest soils have greater nematode diversity than High Arctic tundra? A metagenetic comparison of Malaysia and Svalbard. *Global Ecology and Biogeography* 25, 716–728.

Kim, S.W., Liang, Y., Lozano, Y.M. and Rillig, M.C. (2021) Microplastics reduce the negative effects of litter-derived plant secondary metabolites on nematodes in soil. *Frontiers in Environmental Science* 9, 790560. DOI: 10.3389/fenvs.2021.790560.

Landesman, W.J., Treonis, A.M. and Dighton, J. (2011) Effects of a one-year rainfall manipulation on soil nematode abundances and community composition. *Pedobiologia* 54, 87–91.

Li, G., Wilschut, R.A., Luo, S., Chen, H., Wang, X. *et al.* (2023) Nematode biomass changes along an elevational gradient are trophic group dependent but independent of body size. *Global Change Biology* 29, 4898–4909.

Li, X., Zhu, H., Geisen, S., Bellard, C., Hu, F. *et al.* (2020) Agriculture erases climate constraints on soil nematode communities across large spatial scales. *Global Change Biology* 26, 919–930.

Li, X., Liu, T., Li, H., Geisen, S., Hu, F. and Liu, M. (2022) Management effects on soil nematode abundance differ among functional groups and land-use types at a global scale. *Journal of Animal Ecology* 91, 1770–1780. DOI: 10.1111/1365-2656.13744.

Liu, M., Feng, J., Shen, Y. and Zhu, B. (2023) Microplastics effects on soil biota are dependent on their properties: a meta-analysis. *Soil Biology and Biochemistry* 178, 108940.

Liu, T., Chen, X.Y., Hu, F., Ran, W., Shen, Q.R. *et al.* (2016) Carbon-rich organic fertilizers to increase soil biodiversity: evidence from a meta-analysis of nematode communities. *Agriculture, Ecosystems and Environment* 232, 199–207. DOI: 10.1016/j.agee.2016.07.015.

Liu, T., Hu, F. and Li, H. (2019) Spatial ecology of soil nematodes: perspectives from global to micro scales. *Soil Biology and Biochemistry* 137, 107565. DOI: 10.1016/j.soilbio.2019.107565.

Lu, L., Li, G., He, N., Li, H., Liu, T. *et al.* (2023) Drought shifts soil nematodes to smaller size across biological scales. *Soil Biology and Biochemistry* 184, 109099.

Luan, L., Jiang, Y., Cheng, M., Dini-Andreote, F., Sui, Y. *et al.* (2020) Organism body size structures the soil microbial and nematode community assembly at a continental and global scale. *Nature Communications* 11, 6406.

Ludwig, M., Moreno-Martinez, A., Hölzel, N., Pebesma, E. and Meyer, H. (2023) Assessing and improving the transferability of current global spatial prediction models. *Global Ecology and Biogeography* 32, 356–368.

Nature's Editorial Team (2022) *Nature* addresses helicopter research and ethics dumping. *Nature* 606, 7.

Nielsen, U.N., Ayres, E., Wall, D.H., Li, G., Bardgett, R.D. *et al.* (2014) Global-scale patterns of assemblage structure of soil nematodes in relation to climate and ecosystem properties. *Global Ecology and Biogeography* 23, 968–978. DOI: 10.1111/geb.12177.

Oliverio, A.M., Geisen, S., Delgado-Baquerizo, M., Maestre, F.T., Turner, B.L. and Fierer, N. (2020) The global-scale distributions of soil protists and their contributions to belowground systems. *Science Advances* 6, eaax8787.

Orgiazzi, A., Panagos, P., Fernández-Ugalde, O., Wojda, P., Labouyrie, M. *et al.* (2022) LUCAS Soil Biodiversity and LUCAS Soil Pesticides, new tools for research and policy development. *European Journal of Soil Science* 73, e13299.

Pearson, R.G. and Dawson, T.P. (2003) Predicting the impacts of climate change on the distribution of species: are bioclimate envelope models useful? *Global Ecology and Biogeography* 12, 361–371. DOI: 10.1046/j.1466-822X.2003.00042.x.

Phillips, H.R.P., Guerra, C.A., Bartz, M.L.C., Briones, M.J.I., Brown, G. *et al.* (2019) Global distribution of earthworm diversity. *Science* 366, 480–485.

Potapov, A.M., Sun, X., Barnes, A.D., Briones, M.J., Brown, G.G. *et al.* (2022) Global monitoring of soil animal communities using a common methodology. *Soil Organisms* 94, 55–68.

Potapov, A.M., Guerra, C.A., van den Hoogen, J., Babenko, A., Bellini, M.P. *et al.* (2023) Globally invariant metabolism but density-diversity mismatch in springtails. *Nature Communications* 14, 674.

Pothula, S.K., Grewal, P.S., Auge, R.M., Saxton, A.M. and Bernard, E.C. (2019) Agricultural intensification and urbanization negatively impact soil nematode richness and abundance: a meta-analysis. *Journal of Nematology* 51, 1–17. DOI: 10.21307/jofnem-2019-011.

Puissant, J., Villenave, C., Chauvin, C., Plassard, C., Blanchart, E. and Trap, J. (2021) Quantification of the global impact of agricultural practices on soil nematodes: a meta-analysis. *Soil Biology and Biochemistry* 161, 108383. DOI: 10.1016/j.soilbio.2021.108383.

Rillig, M.C., Ryo, M., Lehmann, A., Aguilar-Trigueros, C.A., Buchert, S. *et al.* (2019) The role of multiple global change factors in driving soil functions and microbial biodiversity. *Science* 366, 886–890.

Scherber, C., Eisenhauer, N., Weisser, W.W., Schmid, B., Voigt, W. *et al.* (2010) Bottom-up effects of plant diversity on multitrophic interactions in a biodiversity experiment. *Nature* 468, 553–556.

Song, D., Pan, K., Tariq, A., Sun, F., Li, Z. *et al.* (2017) Large-scale patterns of distribution and diversity of terrestrial nematodes. *Applied Soil Ecology* 114, 161–169.

Su, Y., Liu, T., Wang, X. and Yang, R. (2016) Salinity effects on soil organic carbon and its labile fractions, and nematode communities in irrigated farmlands in an arid region, northwestern China. *Sciences in Cold and Arid Regions* 8, 46–53. DOI: 10.3724/SP.J.1226.2016.00046.

Tedersoo, L., Bahram, M., Põlme, S., Kõljalg, U., Yorou, N.S. *et al.* (2014) Fungal biogeography. Global diversity and geography of soil fungi. *Science* 346, 1256688.

Thakur, M.P., van der Putten, W.H., Cobben, M.M.P., van Kleunen, M. and Geisen, S. (2019) Microbial invasions in terrestrial ecosystems. *Nature Reviews Microbiology* 17(10), S.621–631.

Thompson, L.R., Sanders, J.G., McDonald, D., Amir, A., Ladau, J. *et al.* (2017) A communal catalogue reveals Earth's multiscale microbial diversity. *Nature* 551, 457–463.

Topalovic, O. and Geisen, S. (2023) Nematodes as suppressors and facilitators of plant performance. *New Phytologist* 238, 2305–2312.

van Capelle, C., Schrader, S. and Brunotte, J. (2012) Tillage-induced changes in the functional diversity of soil biota – a review with a focus on German data. *European Journal of Soil Biology* 50, 165–181. DOI: 10.1016/j.ejsobi.2012.02.005.

van den Hoogen, J., Geisen, S., Routh, D., Ferris, H., Traunspurger, W. *et al.* (2019) Soil nematode abundance and functional group composition at a global scale. *Nature* 572, 194–198.

van der Putten, W.H., Bardgett, R.D., Farfan, M., Montanarella, L., Six, J. and Wall, D.H. (2023) Soil biodiversity needs policy without borders. *Science* 379, 32–34.

Veresoglou, S.D., Halley, J.M. and Rillig, M.C. (2015) Extinction risk of soil biota. *Nature Communications* 6, 8862.

Viketoft, M. (2013) Determinants of small-scale spatial patterns: importance of space, plants and abiotics for soil nematodes. *Soil Biology and Biochemistry* 62, 92–98. DOI: 10.1016/j.soilbio.2013.03.012.

Viketoft, M., Bengtsson, J., Sohlenius, B., Berg, M.P., Petchey, O. *et al.* (2009) Long-term effects of plant diversity and composition on soil nematode communities in model grasslands. *Ecology* 90, 90–99.

Wiesel, L., Daniell, T.J., King, D. and Neilson, R. (2015) Determination of the optimal soil sample size to accurately characterise nematode communities in soil. *Soil Biology and Biochemistry* 80, 89–91.

Wilschut, R.A., Geisen, S., Martens, H., Kostenko, O., de Hollander, M. *et al.* (2019) Latitudinal variation in soil nematode communities under climate warming-related range-expanding and native plants. *Global Change Biology* 25, 2714–2726. DOI: 10.1111/gcb.14657.

Wu, J., Chen, H. and Zhang, Y. (2016) Latitudinal variation in nematode diversity and ecological roles along the Chinese coast. *Ecology and Evolution* 6, 8018–8027. DOI: 10.1002/ece3.2538.

Wu, T., Ayres, E., Bardgett, R.D., Wall, D.H. and Garey, J.R. (2011) Molecular study of worldwide distribution and diversity of soil animals. *Proceedings of the National Academy of Sciences USA* 108, 17720–17725.

Xing, W., Lu, X., Geng, S., Ding, J. and Bai, Y. (2023) Mechanisms underlying the negative effects of nitrogen addition on soil nematode communities in global grassland ecosystems. *Geoderma* 436, 116564. DOI: 10.1016/j.geoderma.2023.116564.

Xiong, D., Wei, C., Wubs, E.R.J., Veen, G.J., Liang, W. *et al.* (2020) Nonlinear responses of soil nematode community composition to increasing aridity. *Global Ecology and Biogeography* 29, 117–126. DOI: 10.1111/geb.13013.

Yan, D., Yan, D., Song, X., Yu, Z., Peng, D. *et al.* (2018) Community structure of soil nematodes under different drought conditions. *Geoderma* 325, 110–116.

Zhu, Y., Bian, H., Ju, C., Xu, C., Zhou, Y. *et al.* (2023) Fertilization alters the abundance but not the diversity of soil fauna: a meta-analysis. *Global Ecology and Biogeography* 32, 482–494. DOI: 10.1111/geb.13641.

13 Ecotoxicity Testing with Nematodes

Sebastian Höss[1]* and Phillip L. Williams[2]
[1]*Ecossa, Starnberg, Germany;* [2]*College of Public Health, University of Georgia, Athens, Georgia, USA*

Abstract
Nematodes are suitable organisms for testing the toxicity of chemicals in water, sediment and soil, with *Caenorhabditis elegans* being the most frequently used nematode species in toxicity testing. Moreover, also multi-species test systems are available for assessing effects of chemicals at higher ecological levels (community/food web level). *C. elegans* has successfully been used for testing the toxicity of field samples from soils and freshwater sediments, with standardized methods helping to achieve reliable data (ISO 10872:2020; ASTM E2172-22). Additionally, *C. elegans* is an excellent model organism for studying toxicity mechanisms *in vivo*, as single chemicals or chemical mixtures can be tested in various substrates, such as water, sediments and soil, by using a large range of toxicity end points, exposure methods and high-throughput systems. A wide variety of toxicants were tested with *C. elegans*, where this species turned out to be particularly suitable for testing toxic particles (e.g. microplastics, nanoparticles).

13.1 Introduction

This chapter gives an overview on the use of free-living (non-parasitic) nematodes in experimental and applied ecotoxicology. Since the publication of *Nematodes as Environmental Bioindicators* (Wilson and Kakouli-Duarte, 2009), research on nematodes as toxicity test organisms has made considerable progress resulting in several review articles and book chapters on the use of nematodes in experimental ecotoxicology (Haegerbaeumer *et al.*, 2015; Höss, 2021; Höss *et al.*, 2022a), especially on the model nematode species *Caenorhabditis elegans* (e.g. Leung *et al.*, 2008; Tejeda-Benitez and Olivero-Verbel, 2016; Hunt, 2017; Chen *et al.*, 2019). Test protocols for toxicity tests in water, sediments and soils have been further standardized or revised by international organizations including the International Organization for Standardization (ISO) (ISO, 2010, 2020) and the American Society for Testing and Materials (ASTM) (ASTM, 2022). Moreover, multi-species test systems (e.g. using microcosms) have been developed for assessing the toxicity of chemicals on the native nematode fauna in soils and sediments (e.g. Brinke *et al.*, 2010; Höss *et al.*, 2014).

13.2 Toxicity Testing in Ecotoxicology

Anthropogenic pollutants can strongly influence the functions of an ecosystem, such as nutrient

*Email: hoess@ecossa.de

 Nematodes as Environmental Indicators: From Theory to Practice, 2nd Edition (T. Kakouli-Duarte *et al.* eds)

DOI: 10.1079/9781800624221.0013

cycling and overall habitat quality. Moreover, pollutants may accumulate in food webs or migrate to groundwater, thus also affecting human health via contaminated food or drinking water. The fact that the hazard of countless anthropogenic substances that are released into the environment can no longer be assessed solely by chemical analysis, has forced scientists and regulators to develop biological methods that allow for the estimation of the bioavailability and toxicity of environmental pollutants (Rand, 2003).

In vivo toxicity tests (exposing living organism) provide information about the effects of single substances, as well as of known or unknown chemical mixtures, on organisms under standardized laboratory conditions. Depending on the type of organism, various exposure scenarios can be chosen to simulate different uptake routes of substances into the organism and thus consider their bioavailability. Additionally, the use of various toxicological end points provides information about the mode of action of certain toxicants.

In ecotoxicology, bioassays with organisms of a single species or with assemblages of multiple species have two major aims: (i) toxicity assessment of environmental samples containing at least partly unknown mixtures of chemicals, integrating all factors that influence the interaction between a potential toxicant and the organism; and (ii) prediction of hazardous effects by studying the bioavailability, toxicity thresholds and modes of action of known substances or substance mixtures on various test organisms in different substrates.

In vivo toxicity tests attempt to provide a balance between interpretability and practicability on one hand and ecological relevance on the other hand (Fig. 13.1). The whole organism represents a black box where processes that determine the response to chemicals such as metabolization, detoxification and elimination, as well as the mode of action, appear as an effect on a certain toxicological end point. Microcosm toxicity testing, considering multi-species exposure with native benthic or soil communities, provides a more realistic exposure scenario, allowing indirect effects (e.g. via the food web) to be considered.

Sediment and soil toxicity testing is a challenge for ecotoxicologists. In these complex matrices, the interaction between toxicants, organisms and the environment is not easy to understand, which makes it difficult to interpret toxicity data. A variety of sediment- and soil-dwelling organisms have been used for assessments of solid phases (Tuikka *et al.*, 2011; Feiler *et al.*, 2013; Diepens *et al.*, 2014; Alves and Cardoso, 2016; Simpson *et al.*, 2016). Although nematodes are one of the most important and abundant organisms in sediments and soils, the use of this animal phylum has always been under-represented in toxicological assessments (see Table 13.1).

13.3 Suitability of Nematodes as Test Organisms

For more than five decades, nematodes have been used as test organisms for laboratory bioassays (e.g. Boroditsky and Samoiloff, 1973; for an overview see Haegerbaeumer *et al.*, 2015 and Höss *et al.*, 2022a). Mainly free-living (i.e. non-parasitic), bacterivorous nematodes have been used as test organisms: besides the marine *Monhystera disjuncta*, the soil nematodes *Pristionchus pacificus*, *Plectus acuminatus*, species of the genus *Panagrellus* and *Diploscapter* sp. (see Haegerbaeumer *et al.*, 2015 and Höss *et al.*, 2022a and references herein), ecotoxicology with nematodes has focused on the intensively studied *C. elegans* (Fig. 13.2). This species that predominantly

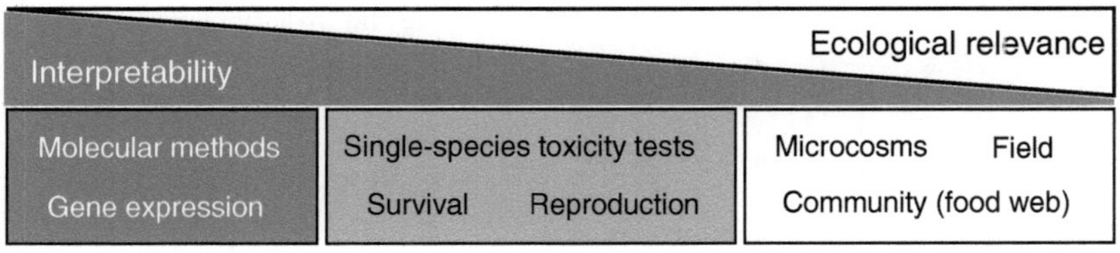

Fig. 13.1. Classification of various ecotoxicological methods regarding two important criteria, interpretability and ecological relevance. (Author's own image.)

Table 13.1. Standards for sediment and soil toxicity tests using invertebrates as test organisms; standards can be found at https://www.astm.org (American Society for Testing and Materials International), https://www.iso.org (International Organization for Standardization) and https://www.oecd.org (Organisation for Economic Co-operation and Development) (all accessed 20 August 2025). (Author's own table.)

Standard	Test organism	Substrate
ASTM E1367-03	*Ampelisca abdita*, *Eohaustorius estuarius*, *Leptocheirus plumulosus*, *Rhepoxynius abronius* (Amphipoda)	Estuarine and marine sediments
ASTM E1611-21	*Neanthes arenaceodentata* (Annelida)	Estuarine and marine sediments
ASTM E1676-12	*Eisenia fetida*, *Enchytraeus albidus* (Annelida)	Soil
ASTM E1706-20	*Hyalella azteca* (Amphipoda), *Chironomus tentans* (Insecta)	Freshwater sediment
ASTM E2172-22	*Caenorhabditis elegans* (Nematoda)	Soil
ISO 10872	*C. elegans* (Nematoda)	Freshwater sediment and soil
ISO 11267	*Folsomia candida* (Collembola)	Soil
ISO 11268-1/2	*E. fetida*, *Eisenia andrei* and other earthworm species (Annelida)	Soil
ISO 14371	*Heterocypris incongruens* (Ostracoda)	Freshwater sediment
ISO 15952	*Helix aspersa* (Mollusca)	Soil
ISO 16303	*H. azteca* (Amphipoda)	Freshwater sediment
ISO 16387	*E. albidus* (Annelida)	Soil
ISO 16712	*Corophium volutator* (Amphipoda)	Estuarine and marine sediments
ISO 17512-1/2	*E. fetida*, *E. andrei* (Annelida), *F. candida* (Collembola)	Soil
ISO 20963	*Oxythyrea funesta* (Coleoptera)	Soil
ISO 21285	*Hypoaspis aculeifer* (Acari)	Soil
ISO 23266	*Oppia nitens* (Acari)	Soil
OECD 207	*E. fetida* (Annelida)	Soil
OECD 218/219/233	*Chironomus riparius* (Insecta)	Freshwater sediment
OECD 220	*E. albidus* (Annelida)	Soil
OECD 222	*E. fetida* (Annelida)	Soil
OECD 225	*Lumbriculus variegatus* (Annelida)	Freshwater sediment
OECD 226	*H. aculeifer* (Acari)	Soil
OECD 228	*Musca autumnalis*, *Scathophaga stercoraria* (Diptera)	Soil, dung

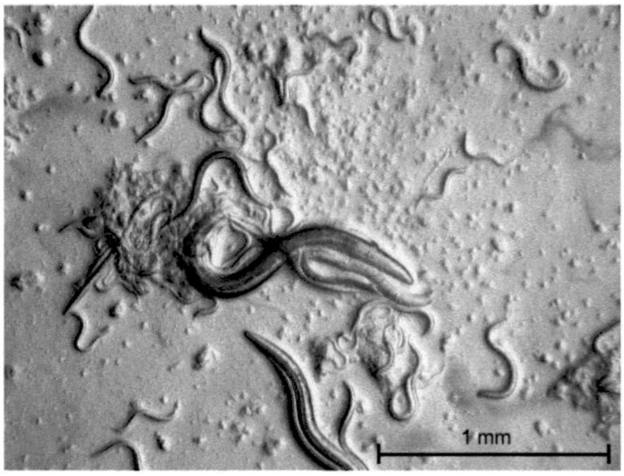

Fig. 13.2. Photograph of *Caenorhabditis elegans*: hermaphrodite (centre) surrounded by offspring in various juvenile stages crawling on agar. (Author's own image.)

occurs in microbe-rich habitats, such as decaying plant material (e.g. compost) (Félix and Braendle, 2010), shows a relatively wide tolerance to environmental factors such as pH, salinity, water hardness and particle grain size (Khanna *et al.*, 1997; Höss *et al.*, 1999) and, thus, has proven to be an adequate test organism for various substrates. Methods have been developed for assessing the toxicity of wastewater (Hitchcock *et al.*, 1997), sediments (Traunspurger *et al.*, 1997; Höss *et al.*, 2012; Schertzinger *et al.*, 2019), soils (Freeman *et al.*, 2000; Höss *et al.*, 2009, 2012; Huguier *et al.*, 2013) and waste (Höss and Römbke, 2019). For these purposes, a variety of toxicological end points were studied for *C. elegans*: lethality (e.g. Williams and Dusenbery, 1990a), growth (e.g. Traunspurger *et al.*, 1997;

Saleem *et al.*, 2023), reproduction (e.g. Traunspurger *et al.*, 1997) and behaviour (e.g. Williams and Dusenbery, 1990b). Besides classical toxicity tests, *C. elegans* also has been used as a test organism for uptake, bioconcentration and bioaccumulation studies (e.g. Haitzer *et al.*, 1999; Jackson *et al.*, 2005; Spann *et al.*, 2015; Fueser *et al.*, 2019). Moreover, due to their short generation time, continuous reproduction and high densities, nematodes are also suitable for multi-species test systems, such as microcosms or other model ecosystems (Haegerbaeumer *et al.*, 2015). Furthermore, nematodes represent different trophic levels, with species feeding on detritus, bacteria, algae, fungi and higher plants, besides omnivorous and predatory species (see Chapters 1 and 2, this volume), allowing to assess chemical effects within food webs.

In the following, we provide an overview of the use of nematodes in ecotoxicity testing. To do this, we reviewed exemplary studies that investigated the toxicity of single substances or contaminated field samples on nematodes, regarding different nematode species, toxicity end points and test substrates. The focus is on the most frequently used species, *C. elegans*, by presenting examples and case studies where this test organism has been used in ecotoxicological assessments. Also, multi-species systems are considered.

13.4 Toxicity of Different Single Substances or Chemical Mixtures on Nematodes

One of the major tasks of toxicity tests is the assessment of the impact of potentially harmful chemicals, either as a single substance or combined with other contaminants in a chemical mixture. Data on the susceptibility of single species towards certain chemicals is necessary for the authorization of chemicals, as well as for a better interpretation of toxicity data of contaminated field samples. However, it is very important to consider the differences in sensitivity between various species and toxicity end points, as well as the different exposure routes in various substrates, such as water, sediment or soil. Therefore, this section gives examples for the toxicity assessment of a variety of dissolved and particulate substances with nematodes and highlights innovative methods available for nematode toxicity testing.

13.4.1 *C. elegans* used for testing a large variety of chemical substances

Dissolved and particulate chemical substances have been tested for their effects on *C. elegans* at levels ranging from the molecular to the whole organism. Studies have been carried out for endocrine disruptors (reviewed in Höss and Weltje, 2007), pesticides (reviewed in Meyer and Williams, 2014), metals (e.g. Höss *et al.*, 2011b), polycyclic aromatic hydrocarbons (PAHs) (e.g. Spann *et al.*, 2015; Fischer *et al.*, 2016), nanomaterials (reviewed in Wu *et al.*, 2019), microplastics (Mueller *et al.*, 2020; Kim *et al.*, 2023), natural toxins (e.g. Wei *et al.*, 2003) and the joint toxicity of chemical mixtures (e.g. Wittkowski *et al.*, 2019). To investigate ecotoxicological effects on a more relevant ecological level, toxicodynamic approaches, in which life-cycle traits such as population growth rates are considered, have been applied (Alda Álvarez *et al.*, 2005; Brinke *et al.*, 2013). To study toxicological mechanisms, effects of chemicals and field samples on *C. elegans* have been explored at the genome or transcriptome level (Hunt, 2017; Chapter 14, this volume). However, in spite of the suitability and relevance of nematodes in ecotoxicology, nematodes are still rarely used in routine risk assessments of chemicals in sediments (e.g. within the framework of the REACH (Registration, Evaluation, Authorisation and Restriction of Chemicals) Regulation; Cesnaitis *et al.*, 2014). This is due to: (i) the general preference to use macroscopic organisms for bioassays; (ii) myths of a generally low sensitivity of nematodes to toxic chemicals; and (iii) the generally very slow process of including new methods in existing regulatory routines. This can only be changed by persistently spreading facts on the suitability of nematodes in the field of ecotoxicology.

13.4.2 Innovations for testing chemicals with *C. elegans*

Passive dosing

Testing substances that are scarcely soluble in water (hydrophobic chemicals) is challenging, as these substances easily bind to any particulate matter (food, organic debris) in the medium or to the wall of the test vials. This leads to losses of the tested chemical substance from the water and to unstable exposure concentrations. This issue also exists for volatile substances, where chemicals evaporate from the water to the air. Therefore, passive dosing systems have been developed, where these losses are compensated by partitioning from a reservoir loaded with the hydrophobic chemical (e.g. silicone materials, oils; Mayer *et al.*, 1999; Smith *et al.*, 2010; Trac *et al.*, 2018). This innovative exposure technique also has been successfully applied for toxicity testing with *C. elegans* (Kwon *et al.*, 2011; Fischer *et al.*, 2016; Mueller *et al.*, 2020). Fischer *et al.* (2016) could show that, in contrast to active spiking with solvent, exposure concentrations of several PAHs could be maintained constant. Consequently, the half-maximal effective concentration (EC_{50}) values achieved for *C. elegans'* reproduction after exposure via passive dosing were considerably lower than those for active spiking (phenanthrene: 45 versus 115 μg l^{-1}; pyrene: 8 versus 31 μg l^{-1}, respectively). In a study on the toxicity of polystyrene microbeads of different sizes for *C. elegans*, it could be ruled out that traces of the monomer styrene were responsible for the polystyrene effects (Mueller *et al.*, 2020). By using headspace passive dosing of the volatile substance styrene, exact threshold concentrations could be determined for the monomer, which were then compared to styrene concentrations in polystyrene bead suspensions, measured by headspace passive sampling.

Automated testing approaches

Automated platforms that will rapidly test *C. elegans* for a variety of toxicological end points following exposure to chemicals or other environmental stressors have become commercially available. The most useful of these high-throughput systems have resulted from the advancements in the use of large particle flow cytometry. These systems are capable of quickly analysing high numbers of organisms in an automated system equipped with software that can determine a variety of toxicological end points including feeding, lethality, growth, reproduction and behaviour. Depending on the system, it can count the number of organisms, detect changes in fluorescence (using transgenic strains containing green fluorescent protein (GFP)), and determine changes in growth and behaviour. The approach eliminates manual sorting processes and reduces human error while allowing for large-scale screens that were not previously possible. The organisms can be exposed by any number of standard approaches and then transferred to a multi-well plate for processing within the automated platform system. Dr Jonathon Freedman's laboratory led the effort to adapt this technology for toxicity exposures using *C. elegans* and published many of the early studies (Boyd *et al.*, 2010a,b).

Molecular and genetic testing

As the first multicellular organism to have its DNA fully sequenced, *C. elegans* provides many advantages for the use of genetic markers and transgenic strains to explore toxicological outcomes and molecular changes following environmental exposures. Transgenic strains, some with GFP labelling (Chalfie *et al.*, 1994), are readily available and have been used by many researchers. For example, Ma *et al.* (2009) developed a transgenic strain of *C. elegans* with GFP attached to the promotor region of metallothionein (a family of cytoplasmatic metal-binding proteins) that in the presence of specific metals (Cd, Hg, Cu and Zn) in the soil resulted in increasing GFP fluorescence intensity with the increase in metal concentration. Other researchers have developed GFP-tagged specific neurons that allow their neural activity to be assessed with environmental exposures. Using a high-throughput system (see above), Tang *et al.* (2020) compared wild-type *C. elegans* to a number of transgenic strains and was able to determine both the mechanisms of action and detoxification for *C. elegans* exposed to Ni in the environment.

Reactive oxygen species (ROS)

Reactive oxygen species (ROS) can be potentially damaging by-products of aerobic metabolism but, within normal physiological levels, they play a role in cell signalling. It has been challenging to understand the differences between the beneficial and harmful roles of ROS. *C. elegans* is now being used as a model for understanding the ROS function in physiology and disease outcomes including ageing. With these advances, ROS is being measured in many environmental studies using *C. elegans* and the results used to help determine toxicological effects (Miranda-Vizuete and Veal, 2017).

13.5 Comparison of Different Invertebrate Species

The extrapolation of toxicity data from lower-tier studies, such as single-species toxicity tests, to higher ecological levels is a crucial issue in environmental risk assessment (Forbes and Calow, 2002). Although it is known that the toxicity of chemicals to various species can vary by several orders of magnitude, from a practical standpoint, only a limited number of test organisms can be chosen to be used in routine application of toxicity testing. Since the most sensitive relevant species is seldom known, a test organism usually is chosen based on ease of handling and accessibility.

It is difficult to compare the toxicity data from different studies, as the test conditions (exposure, food, test duration, temperature, etc.) can vary considerably and these differences can significantly influence the outcome of a toxicity test. Thus, only studies that compared the toxicity of chemicals on various species under the same experimental conditions are considered here.

13.5.1 Comparison of various nematode species

Kammenga *et al.* (1994) investigated the acute toxicity of Cd and pentachlorophenol (PCP) on 12 different terrestrial nematode species belonging to different taxonomic and ecological groups: bacterial feeders, a fungal feeder, a plant feeder, predators and omnivores. These species exhibited large differences in sensitivity (median lethal dose (LC_{50}) values: 0.5 to > 34.5 µmol PCP l^{-1}; 29 to > 800 µmol Cd l^{-1}). The differences in sensitivity were only partly related to the taxonomic or ecological groups. Species of the group of Chromadorea (*sensu* De Ley and Blaxter, 2002) were less sensitive to PCP than species of the group of Enoplea (*sensu* De Ley and Blaxter, 2002), while no differences were observed for Cd. There appeared to be a relationship between feeding groups and PCP sensitivity. Carnivorous, omnivorous and plant-feeding nematodes were relatively sensitive, whereas bacterial and fungal feeders were more tolerant.

Boyd and Williams (2003) compared the sensitivity of three bacterial-feeding nematodes, all belonging to the group of Rhabditida (*sensu* De Ley and Blaxter, 2002): *C. elegans* (Rhabditidae), *Panagrellus redivivus* (Panagrolaimidae) and *P. pacificus* (Neodiplogasteridae), in terms of their sensitivity towards Cu in water and soil. The LC_{50} showed a relatively high variation between the three species, with 19 mg l^{-1} (48 mg kg^{-1}) for *P. pacificus*, 85 mg l^{-1} (179 mg kg^{-1}) for *C. elegans* and 160 mg l^{-1} (251 mg kg^{-1}) for *P. redivivus* for water (soil). Although survival of *C. elegans* was affected at considerably higher concentrations of Cu than the survival of *P. pacificus* (factor 4.5 and 3.7 for water and soil, respectively), sublethal toxicity end points, such as reproduction and movement, showed no difference in sensitivity between the two species (EC_{50} reproduction: 2.0 and 2.2 mg l^{-1} for *C. elegans* and *P. pacificus*, respectively) or even a higher sensitivity of *C. elegans* (EC_{50} movement: 2.1 and 8.1 mg l^{-1} for *C. elegans* and *P. pacificus*, respectively). The authors speculated that the Cu-induced changes in movement of *C. elegans* might have protected it from lethal effects, explaining the lower sensitivity of this species in the acute test.

Comparing the sensitivity of *C. elegans* to metals and PAHs with nematode species extracted from freshwater sediments, Haegerbaeumer *et al.* (2018) revealed that *C. elegans* was well within the range of sensitivities of freshwater nematode species, being slightly more sensitive to metals, but less sensitive to PAHs.

13.5.2 Comparison of nematodes with other standard test organisms

Bezchlebová *et al.* (2007) compared the toxicity of the polychlorinated insecticide toxaphene on various soil invertebrates, including the nematode *C. elegans*, the collembolan *Folsomia candida*, and the oligocaetes *Eisenia fetida*, *Enchytraeus albidus* and *Enchytraeus crypticus*. *F. candida* showed to be most sensitive with a lowest observed effect concentration (LOEC) of 6.2 mg kg^{-1}, followed by *C. elegans*, *E. fetida* and the *Enchytraeus* species (no observed effect concentration (NOEC): 69, 496 and >600 mg kg^{-1}, respectively). However, it should be noted that, at 48 hours, the exposure period for *C. elegans* was much lower than for *E. fetida* and *Enchytraeus* (28 and 42 days, respectively). In a study with aquatic sediments, linear alkylbenzene sulfonates (LAS) had a similar toxicity to *Lumbriculus variegatus* and *C. elegans* in spiked sediments, with LOECs of 82 and 100 mg kg^{-1}, respectively (Comber *et al.*, 2006). Using soils spiked with five metallic salts, Peredney and Williams (2000) compared lethality data with *C. elegans* to lethality data with the earthworm *E. fetida*. The *C. elegans* exposure was for 24 hours compared to earthworm exposure of 14 days. *C. elegans* appeared to be more sensitive than the earthworm for two of the five metals, Pb and Cd, and less sensitive to Cu, Ni and Zn. Comparing the toxicity of various engineered nanomaterials among a variety of aquatic species (algae, bacteria, invertebrates, fish) using the species sensitivity distribution (SSD) approach, *C. elegans* appeared to be the least sensitive species for TiO_2 (NOEC > 100 mg l^{-1}) but in the middle range of sensitivities for Ag and graphene (NOEC < 10 mg l^{-1}; Zhao *et al.*, 2021; Fig. 13.3). The SSD approach was also used by Kim *et al.* (2023) comparing the sensitivity of various soil organisms to microplastics. They found *C. elegans* to be one of the most sensitive test organisms (Fig. 13.3).

13.6 Comparison of Different Toxicity End Points

In toxicity tests with nematodes and especially with *C. elegans*, various toxicity end points were used. Besides survival, sublethal end points include growth, fertility, reproduction, feeding and movement behaviour. Toxicity end points can be characterized by their sensitivity (sublethal versus lethal), by their ecological relevance (e.g. reproduction as a measure for population growth has a higher ecological relevance than body length), by their specificity in response to certain types of toxicants (e.g. behaviour for neurotoxicants; reproduction for endocrine disruptors), or simply by their reliability and practicability.

Although survival is less sensitive than sublethal toxicity end points, such as reproduction and behaviour, the discrepancy in sensitivity can vary from species to species. Boyd and Williams (2003) compared the toxicity of Cu on survival, reproduction and movement of two different bacterial-feeding species, *C. elegans* and *P. pacificus*. For *C. elegans*' reproduction and movement, acute-to-chronic ratios (ACR = LC_{50}/EC_{50}) were considerably higher (ACR: 41.9 and 41.4, respectively) than for *P. pacificus* (ACR: 8.6 and 2.3, respectively). Generally, ACRs for *C. elegans* were considerably higher for metals ($ACR_{reproduction}$: 76) compared to carbamate pesticides ($ACR_{reproduction}$: 12) and narcotics ($ACR_{reproduction}$: 7) (Ristau *et al.*, 2015).

Anderson *et al.* (2004) showed that movement behaviour might be a specific toxicity end

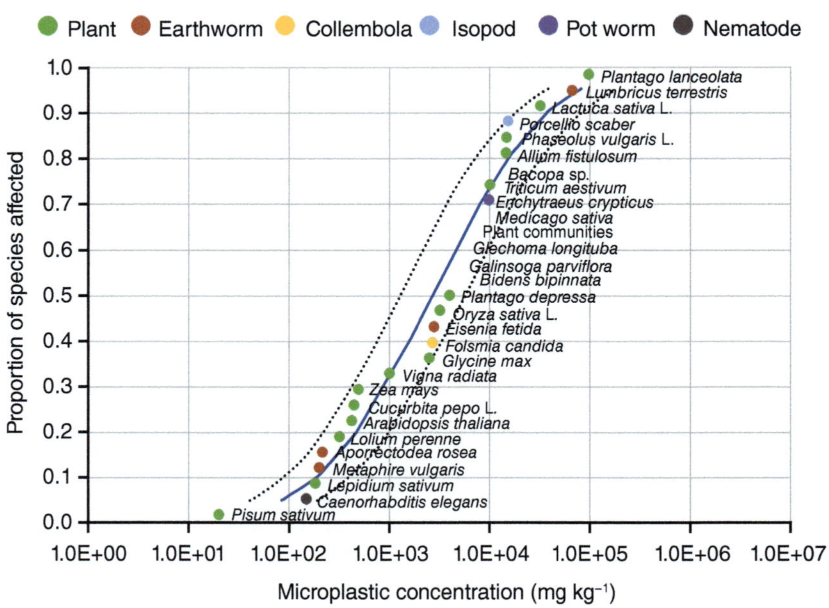

Fig. 13.3. Species sensitivity distributions (SSDs; dotted lines: 95% confidence intervals) for microplastic toxicity in soil ecosystems. (Reproduced from Kim *et al.*, 2023 with permission of Elsevier.)

point for neural toxicants. In 4-hour toxicity tests with *C. elegans*, six chemicals of three different chemical classes with known neural toxicity (heavy metals: Pb and Al; organic solvents: ethanol and acetone; organic pesticides: levamisol and chlorpyrifos) had a considerably higher toxicity on locomotion of the nematodes, compared to toxicants of the respective classes without neural toxicity (Cu; dimethylsulfoxide (DMSO); mebendazol; Fig. 13.4). The specificity of behavioural toxicity end points was already indicated in two earlier studies (Williams and Dusenbery, 1990b; Anderson *et al.*, 2001). However, to detect neural toxicity, the exposure time had to be short (4 hours), so that other symptoms of general toxicity, such as starving, could not mask the specific effect on nematode behaviour.

Endocrine disruption is also a very specific mode of action that has attracted much attention in the last 30 years. Höss and Weltje (2007) reviewed the literature that dealt with effects of endocrine disruptors on nematodes, and demonstrated, with their own data, that endocrine disruptors were able to stimulate the reproduction of *C. elegans*. Although it was not proved that endocrine disruption caused the stimulating effects, studies at the molecular level suggest that nematode reproduction can be influenced by endocrine-active substances (Höss and Weltje, 2007).

13.7 Comparison of Different Test Substrates – The Role of Chemical Uptake Routes

A contaminant can only exhibit a toxic effect to an organism if it is bioavailable. The bioavailability of compounds in a complex matrix, such as soils and sediments, is on one hand determined by the partitioning of chemicals between the particulate and dissolved phase. On the other hand, the major interfaces for chemical exchange with the surrounding medium are the cuticle and the gut, making the movement and feeding behaviour of the test organism crucial factors for chemical exposure.

It is known that the cuticle of *C. elegans* is nearly impermeable to metals and organic chemicals (Jackson *et al.*, 2005; Sandhu *et al.*, 2021), so that the uptake of toxicants mainly occurs after ingestion via the gut (Höss *et al.*, 2011b; Spann *et al.*, 2015). Unfortunately, little is known about permeability for free-living aquatic species. For dorylaims, a high variability of the cuticle within this group has been reported (Peña Santiago, 2020). For particulate contaminants (e.g. microplastic), Fueser *et al.* (2019) showed that the uptake of polystyrene beads for bacterial-feeding nematodes is dependent on the ratio of bead diameter to size of the buccal cavity, whereas stylet-feeding nematodes

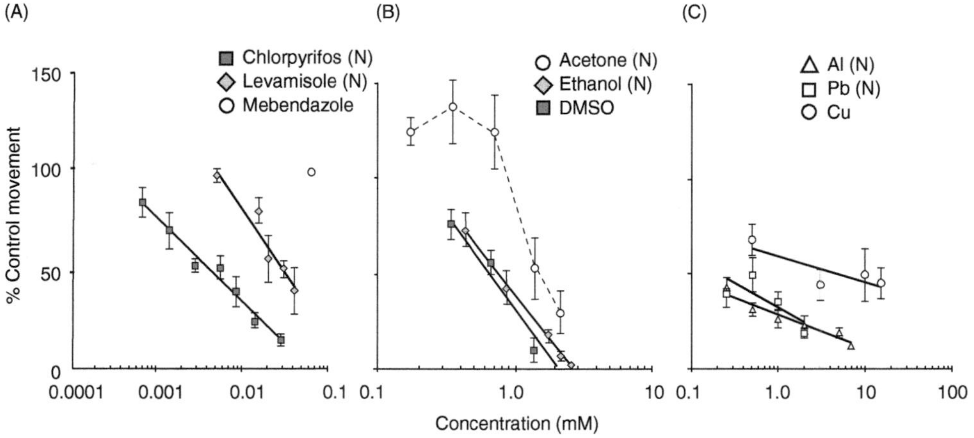

Fig. 13.4. Concentration–response relationship for compounds selected from three chemical classes: (A) organic pesticides; (B) organic solvents; (C) heavy metals. DMSO = dimethylsulfoxide; (N) = chemical with known neural toxicity. (Adapted with permission from Anderson *et al.*, 2004. Copyright SETAC, Pensacola, Florida, USA.)

(*Aphelenchoides parietinus*) were not able to take up beads ≥ 0.5 μm at all.

In sediments and soils, where nematodes occur, a large part of the contaminants may be bound to particles. Moving and feeding in the interstitial space, nematodes are exposed to contaminants via the dissolved and the particulate phase. As it is known that particle-bound contaminants are available to benthic organisms (Landrum *et al.*, 1990; Reinfelder and Fisher, 1991; Leppänen and Kukkonen, 1998; Lee *et al.*, 2000), dissolved and particle-bound contaminants might pose a risk for nematodes.

Studies with *C. elegans* in sediments and soils showed that the toxicity of pollutants, such as Cd, cationic surfactants and pyrene, related well to the concentrations in pore water measured for the respective chemicals (Höss *et al.*, 2001; Thomas *et al.*, 2009; Bielská *et al.*, 2017). Also, Boyd and Williams (2003) found two times higher LC_{50} values of Cu for three bacterial-feeding species, *C. elegans*, *P. pacificus* and *P. redivivus*, in a sandy soil (98% sand, 2% clay, 1.4% organic matter) compared to the respective LC_{50} in water. Similar results were found by Sochova *et al.* (2007) for the toxicity of six organic pollutants in natural soil and water. The ratios of $LC10_{sed}/LC10_{water}$ for mortality ranged here between 4 and 3000. Thus, the chemical properties as well as the characteristics of the sediment or soil are important for the bioavailability and toxicity for bacterivorous nematodes.

Although the toxicity and bioaccumulation of chemicals in *C. elegans* could be best explained by their dissolved concentrations in water (Höss *et al.*, 2011b; Spann *et al.*, 2015), bacterial-bound chemicals still considerably contribute to the toxicity for *C. elegans* (Fischer *et al.*, 2016).

13.8 Multi-species Toxicity Testing in Model Ecosystems

Although single-species toxicity testing is an important and reliable tool in environmental risk assessments of chemicals, extrapolation of results to possible effects in the field is challenging (Forbes and Calow, 2002). Here, micro- and mesocosms (small, well-defined representatives of ecosystems) might be an intermediate solution, as these test systems can be used as experimental units (model ecosystems) for a defined period of time. Micro- and mesocosm can be regarded as community-level or multi-species bioassays yielding insights into the biotic responses of natural assemblages. At the same time, they provide sufficient replication and precise control over relevant experimental variables (Van den Brink *et al.*, 2005).

Nematodes offer many advantages for use in micro- or mesocosm experiments. They reach high abundances and species richness in sediments and soils and can thus be tested in small laboratory set-ups (0.01–10 l of sediment or soil). Actually, effects of chemicals on nematode communities have already been assessed in microcosms-containing sediments (freshwater and marine) and soils (for an overview see Haegerbaeumer *et al.*, 2015 and Höss *et al.*, 2022a; more methodological details can be found in Chapter 15, this volume). For aquatic and soil microcosms it could be shown that native nematode communities respond more sensitively to chemicals, such as pharmaceuticals and pesticides, than single-species tests with nematodes (Brinke *et al.*, 2010; Höss *et al.*, 2022b).

13.9 Toxicity Testing of Contaminated Field Samples

Besides studies on the toxicity of known chemical substances, *C. elegans* has been used as a test organism to assess the toxicity of field samples, including very complex material, such as contaminated freshwater sediments, soils, wastewater and solid waste.

13.9.1 Case studies with freshwater sediments

Feiler *et al.* (2013) used a battery of sediment contact tests to assess the toxicity of various river sediments with a gradient of chemical contamination ranging from low to high. The test battery used test organisms from various trophic and organizational levels (bacteria, yeast, higher plants, nematodes, oligochaetes and fish embryos), representing various exposure routes of sediment- and pore-water-associated chemicals and different susceptibilities towards the chemicals. The different test systems provided non-redundant information which enabled the authors to reliably rank the river sediments according to their toxicity. Moreover, it was possible to relate sediment-associated effects with the toxic potential based on sediment quality guidelines provided by MacDonald *et al.* (2000). Within the test battery, *C. elegans* belonged to the most sensitive test species and their response could be well related to the toxic potential of the measured organic chemicals (maximal TEC-Q) (Fig. 13.5).

13.9.2 Case studies with wastewater and solid waste

Hitchcock *et al.* (1997) used *C. elegans* to assess the toxicity contribution of various industrial operations to the waste stream of a municipal wastewater treatment plant (WWTP). The water discharge from the WWTP had been found to be above acceptable levels and the source of this problem was believed to be one of the many industrial facilities that fed the plant. Three industrial facilities, fibreglass manufacturing, paper packaging and yarn-dyeing facilities, were identified as the most likely source of the problem. Over a ten-month time frame, 24-hour composite samples were taken monthly from each of these facilities, as well as a common point and discharge point from the WWTP. This approach allowed for the contribution of each industrial operation to be evaluated as well as the combined input from all sources and the treated discharge from the WWTP. Following sample collection from each site, the nematode was used to assess water quality, along with measurements of standard water quality parameters such as pH and the discharge volumes. Nematode mortality trends from the testing identified the sources (e.g. industrial facility) for the toxicity in the wastewater and provided the data to assist in controlling the releases. The

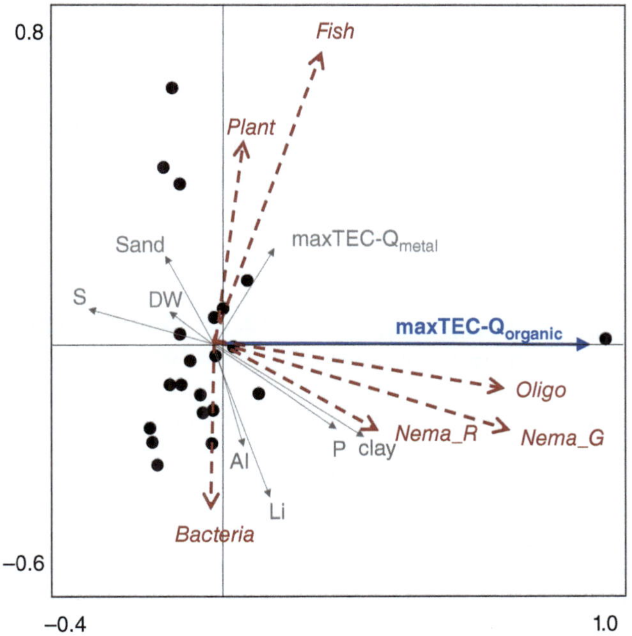

Fig. 13.5. Redundancy analysis (RDA) of sediment contact tests: multivariate analysis based on the performance of *Caenorhabditis elegans* (Nema_R = reproduction; Nema_G = growth) compared to that of routine test organisms in toxicity assessments of contaminated freshwater sediments (fish, plants, oligochaetes, bacteria). DW = sediment dry weight; P = total phosphorus; S = total sulfur; Al = aluminium; Li = lithium; maxTEC-$Q_{metal/organic}$ = maximal quotients of the measured concentrations of metals or organic pollutants divided by those defined in sediment quality guidelines (threshold effect concentrations) according to MacDonald *et al.* (2000). (Modified from Feiler *et al.*, 2013 with permission of John Wiley and Sons.)

study demonstrated the ease of use and low cost of this organism in ecotoxicological studies.

Höss and Römbke (2019) could demonstrate that *C. elegans* can also be used for evaluating the hazard of solid waste. Contaminated waste samples of various origins were mixed with standard soil (Lufa St.2.2) to achieve dilution series and tested for their toxicity on the nematodes' reproduction (ISO 10872:2020; ISO, 2020). Compared to other soil toxicity tests with earthworms, bacteria and plants, the nematodes provided additional, non-redundant information to classify the hazard of the waste samples.

13.9.3 Case studies with soils

In January of 2000, cyanide and heavy metals were accidently released from a mine waste lagoon in Romania, and this initial release was followed in March 2000 by two subsequent mine waste spills. These releases eventually reached the Tisza River that flows through Hungary. Concurrently with the Tisza River contamination, the area received heavy rainfall followed by severe flooding which potentially spread the contamination to soils adjacent to the river. The nematode *C. elegans* was used as a test organism to evaluate soil toxicity following this event and the study demonstrated the benefits of using this organism for soil toxicity studies (Black and Williams, 2001). Soil samples of 100 g from each field site were transported back to the laboratory in the USA and triplicate tests were performed with each sample. Following the nematode testing, enough of the soil sample remained for use in chemical and physical analyses. A similar approach with earthworms would have required kilogram samples from each site and added difficulty and expense in both transporting and the eventual disposal of the soil samples. Further, the lethality end point could be assessed following 24 hours of exposure compared to 14 days with the earthworms. The data were used to provide a preliminary assessment of the metal contamination of the river's floodplain.

C. elegans was used to assess the toxicity of soil from fields that were cultivated with transgenic maize (*Bt* maize; MON810). This maize variety is resistant to the European corn borer, because it contains genes for insecticidal δ-endotoxins from the bacterium *Bacillus thuringiensis* (*Bt*). The results showed that growth and reproduction of *C. elegans* were significantly lower in soil from fields with *Bt* maize compared to the isogenic maize and that both toxicity end points correlated negatively with the toxin (Cry1Ab protein) concentrations in soil (Höss *et al.*, 2008). However, a toxicity test with the pure Cry1Ab protein in aqueous medium revealed toxicity thresholds that were far above the detected soil concentrations (Höss *et al.*, 2008), indicating that the nematodes were not directly affected by the *Bt* toxin. This study demonstrated that *C. elegans* might be a suitable test organism for monitoring the risk of pest-resistant genetically modified plants, as the test is: (i) quick and cost effective; and (ii) can be carried out in aqueous medium (for testing the pure *Bt* toxin solution) and soils (for testing soils containing *Bt* toxin).

Other studies have found that nematodes can play a role in vectoring pathogenic strains of bacteria from the soil to edible plants. Gibbs *et al.* (2005) found that *Diploscapter* sp. strain LKC25 exposed to pathogenic bacteria in soil could ingest the bacteria and shed it to the surface of fruits and vegetables in contact with the soil. Anderson *et al.* (2006) found that *C. elegans* could have a similar role when exposed to pathogenic strains of *Escherichia coli* and *Salmonella* in the soil.

13.10 Standardization of Toxicity Tests with *C. elegans*

In environmental risk assessment there is a need for standardized tools, to produce reliable data that are able to bear up against legislation. Especially when testing complex matrices, such as sediments and soils, where test organisms are influenced by many different factors, readily standardized tests are required to differentiate effects of anthropogenic pollution from responses to natural sediment or soil properties, and thus to avoid false positive or negative results. For sediment and soil toxicity tests only a limited number of standards is available for invertebrates (Table 13.1). Among the standardized tests, *C. elegans* is the sole representative of the nematodes. However, there are established test protocols using *P. acuminatus* (Kammenga *et al.*,

1996) and *P. redivivus* (Sherry *et al.*, 1997) that can be considered as a suitable basis for future standards. In the USA, Williams and Dusenbery (1990a) developed a bioassay using *C. elegans* for testing aqueous substrates using mortality as the toxicity end point. In the following years, this test was adapted for testing soils (Donkin and Dusenbery, 1993; Freeman *et al.*, 2000). After confirming the reliability of the test system by testing reference substances (Freeman *et al.*, 1998; Peredney and Williams, 2000) and the tolerance to several abiotic factors (Khanna *et al.*, 1997), an ASTM standard was developed in 2001, and updated two times until 2022: the ASTM standard covers procedures for obtaining laboratory data to evaluate the adverse effects of chemicals associated with soil to nematodes from soil toxicity tests. The methods are designed to assess lethal or sublethal toxic effects on nematodes in short-term tests in terrestrial systems (ASTM, 2022).

Based on the work of Williams and his laboratory group, in Europe a toxicity test for aqueous medium and freshwater sediments was developed that used sublethal toxicity end points, such as growth and reproduction (Traunspurger *et al.*, 1997). The test protocol has been thoroughly standardized and ring-tested (Höss *et al.*, 1999, 2012) and transferred into an international standard (ISO 10872; ISO, 2010). This standard test proved to be suitable for testing freshwater sediments (Höss *et al.*, 2010; Tuikka *et al.*, 2011; Feiler *et al.*, 2013; Schertzinger *et al.*, 2019) and soils (Höss *et al.*, 2008, 2009, 2011a), as well as the toxicity of chemicals in artificial and natural sediments (Comber *et al.*, 2006, 2008; Höss *et al.*, 2001, 2020). After suggestions for improvements for testing soil samples (lower water content in soils: 80% of maximal water-holding capacity; Huguier *et al.*, 2013), the ISO 10872 has been revised to become the first ISO standard for 'water and soil quality' (ISO, 2020), addressing the need for ecosystem-overarching (aquatic and terrestrial) test systems.

13.11 Conclusions and Outlook

This chapter clearly shows that nematodes are suitable test organisms for ecotoxicity testing. Nematodes have significant ecological relevance for the use in sediment and soil toxicity testing. The easy culturing and short generation time of some nematode species, such as *C. elegans*, allow for quick and cost-effective tests. Moreover, toxicity tests can be carried out in a variety of test substrates, such as water (wastewater, pore water, eluates, toxicant solutions), sediment, soil or waste. Thereby, the sensitivity of nematodes (e.g. of *C. elegans*) to various pollutants is within the range of other benthic or soil invertebrates, in spite of the considerably lower exposure time. Innovative methods allow controlled exposure conditions (passive dosing) and high-throughput testing of chemicals, from the gene level to the population level. This exceptionally broad toolbox of nematode-based ecotoxicology supports their use in the challenging task of environmental risk assessment of chemicals contamination.

References

Alda Álvarez, O., Jager, T., Kooijman, S.A.L.M. and Kammenga, J.E. (2005) Responses to stress of *Caenorhabditis elegans* populations with different reproductive strategies. *Functional Ecology* 19, 656–664.

Alves, P.R.L. and Cardoso, E.J.B.N. (2016) Overview of the standard methods for soil ecotoxicology testing. In: Larramendy, M.L. and Soloneski, S. (eds) *Invertebrates – Experimental Models in Toxicity Screening*. IntechOpen, Rijeka, Croatia, pp. 35–56.

Anderson, G.L., Boyd, W.A. and Williams, P.L. (2001) Assessment of sublethal endpoints for toxicity testing with the nematode *Caenorhabditis elegans*. *Environmental Toxicology and Chemistry* 20, 833–838.

Anderson, G.L., Cole, R.D. and Williams, P.L. (2004) Assessing behavioral toxicity with *Caenorhabditis elegans*. *Environmental Toxicology and Chemistry* 23, 1235–1240.

Anderson, G.L., Kenney, S.J., Millner, P.D., Beuchat, L.R. and Williams, P.L. (2006) Shedding of foodborne pathogens by *Caenorhabditis elegans* in compost-amended and unamended soil. *Food Microbiology* 23, 146–153.

ASTM (American Society for Testing and Materials) (2022) *Standard Guide for Conducting Laboratory Soil Toxicity Tests with the Nematode Caenorhabditis elegans; ASTM E2172-01*. ASTM International, West Conshohocken, Pennsylvania.

Bezchlebová, J., Cernohlávková, J., Lana, J., Sochova, I., Kobeticova, K. and Hofman, J. (2007) Effects of toxaphene on soil organisms. *Ecotoxicology and Environmental Safety* 68, 326–334.
Bielská, L., Kah, M., Sigmund, G., Hofmann, T. and Höss, S. (2017) Bioavailability and toxicity of pyrene in soils upon biochar and compost addition. *Science of The Total Environment* 595, 132–140.
Black, M.C. and Williams, P.L. (2001) Preliminary assessment of metal toxicity in the Middle Tisza River (Hungary) flood plain. *Journal of Soils and Sediments* 1, 203–206.
Boroditsky, J.M. and Samoiloff, M.R. (1973) Effects of growth inhibitors on development of the reproductive system of the free-living nematode *Panagrellus redivivus* (Cephalobidae). *Canadian Journal of Zoology* 51, 483–492.
Boyd, W.A. and Williams, P.L. (2003) Comparison of the sensitivity of three nematode species to copper and their utility in aquatic and soil toxicity tests. *Environmental Toxicology and Chemistry* 22, 2768–2774.
Boyd, W.A., McBride, S.J., Rice, J.R., Snyder, D.W. and Freedman, J.H. (2010a) A high-throughput method for assessing chemical toxicity using a *Caenorhabditis elegans* reproduction assay. *Toxicology and Applied Pharmacology* 245, 153–159.
Boyd, W.A., Smith, M.V., Kissling, G.E. and Freedman, J.H. (2010b) Medium- and high-throughput screening of neurotoxicants using *C. elegans*. *Neurotoxicology and Teratology* 32, 68–73.
Brinke, M., Höss, S., Fink, G., Ternes, T.A., Heininger, P. and Traunspurger, W. (2010) Assessing effects of the pharmaceutical ivermectin on meiobenthic communities using freshwater microcosms. *Aquatic Toxicology* 99, 126–137.
Brinke, M., Heininger, P. and Traunspurger, W. (2013) Effects of a bioassay-derived ivermectin lowest observed effect concentration on life-cycle traits of the nematode *Caenorhabditis elegans*. *Ecotoxicology* 22, 148–155.
Cesnaitis, R., Sobanska, M.A., Versonnen, B., Sobanski, T., Bonnomet, V. *et al.* (2014) Analysis of the ecotoxicity data submitted within the framework of the REACH Regulation. Part 3. Experimental sediment toxicity assays. *Science of the Total Environment* 475, 116–122.
Chalfie, M., Tu, Y., Euskirchen, G., Ward, W.W. and Prasher, D.C. (1994) Green fluorescent protein as a marker gene expression. *Science* 263, 802–805.
Chen, H., Wang, C., Li, H., Ma, R., Yu, Z. *et al.* (2019) A review of toxicity induced by persistent organic pollutants (POPs) and endocrine-disrupting chemicals (EDCs) in the nematode *Caenorhabditis elegans*. *Journal of Environmental Management* 237, 519–525.
Comber, S.D.W., Conrad, A.U., Höss, S., Webb, S. and Marshall, S. (2006) Chronic toxicity of sediment-associated linear alkylbenzene sulphonates (LAS) to freshwater benthic organisms. *Environmental Pollution* 144, 661–668.
Comber, S.D.W., Rule, K.L., Conrad, A.U., Höss, S., Webb, S.F. and Marshall, S. (2008) Bioaccumulation and toxicity of a cationic surfactant (DODMAC) in sediment dwelling freshwater invertebrates. *Environmental Pollution* 153, 184–191.
De Ley, P. and Blaxter, M.L. (2002) Systematic position and phylogeny. In: Lee, D.L. (ed.) *The Biology of Nematodes*. Taylor & Francis, London, pp. 1–30.
Diepens, N.J., Arts, G.H.P., Brock, T.C.M., Smidt, H., Van den Brink, P.J. *et al.* (2014) Sediment toxicity testing of organic chemicals in the context of prospective risk assessment: a review. *Critical Reviews in Environmental Science and Technology* 44, 255–302.
Donkin, S.G. and Dusenbery, D.B. (1993) A soil toxicity test using the nematode *C. elegans* and an effective method of recovery. *Archives of Environmental Contamination and Toxicology* 25, 145–151.
Feiler, U., Höss, S., Ahlf, W., Gilberg, D., Hammers-Wirtz, M. *et al.* (2013) Sediment contact tests as a tool for the assessment of sediment quality in German waters. *Environmental Toxicology and Chemistry* 32, 144–155.
Félix, M.-A. and Braendle, C. (2010) The natural history of *Caenorhabditis elegans*. *Current Biology* 20, R965–R969.
Fischer, F., Böhm, L., Höss, S., Möhlenkamp, C., Claus, E. *et al.* (2016) Passive dosing in chronic toxicity tests with the nematode *Caenorhabditis elegans*. *Environmental Science & Technology* 50, 9708–9716.
Forbes, V.E. and Calow, P. (2002) Extrapolation in ecological risk assessment: balancing pragmatism precaution in chemical controls legislation. *Bioscience* 52, 249–257.
Freeman, M.N., Marse, T.J. and Williams, P.L. (1998) Establishment of quality assurance procedures for aquatic toxicity testing with the nematode *Caenorhabditis elegans.* In: Little, E.E., DeLonay, A.J. and Greenberg, B.M. (eds) *Environmental Toxicology and Risk Assessment: Seventh Volume, ASTM STP 1333*. American Society for Testing and Materials, West Conshohocken, Pennsylvania, pp. 45–59.
Freeman, M.N., Peredney, C.L. and Williams, P.L. (2000) A soil bioassay using the nematode *Caenorhabditis elegans.* In: Henshel, D.S., Black, M.C. and Harrass, M.C. (eds) *Environmental Toxicology and Risk Assessment: Standardization of Biomarkers for Endocrine Disruption and Environmental Assessment*. American Society for Testing and Materials, West Conshohocken, Pennsylvania, pp. 305–318.

Fueser, H., Mueller, M.-T., Weiss, L., Höss, S. and Traunspurger, W. (2019) Ingestion of microplastics by nematodes depends on feeding strategy and buccal cavity size. *Environmental Pollution* 255, 113227.

Gibbs, D.S., Anderson, G.L., Beuchat, L.R., Carta, L.K. and Williams, P.L. (2005) Potential role of *Diploscapter* sp. strain LKC25, a bacterivorous nematode from soil, as a vector of food-borne pathogenic bacteria to preharvest fruits and vegetables. *Applied and Environmental Microbiology* 71, 2433–2437.

Haegerbaeumer, A., Höss, S., Heininger, P. and Traunspurger, W. (2015) Experimental studies with nematodes in ecotoxicology: an overview. *Journal of Nematology* 47, 11–27.

Haegerbaeumer, A., Höss, S., Heininger, P. and Traunspurger, W. (2018) Is *Caenorhabditis elegans* representative of freshwater nematode species in toxicity testing? *Environmental Science and Pollution Research* 25, 2879–2888.

Haitzer, M., Burnison, B.K., Höss, S., Traunspurger, W. and Steinberg, C.E.W. (1999) Effects of quantity, quality and contact time of dissolved organic matter on bioconcentration of benzo[a]pyrene in the nematode *Caenorhabditis elegans*. *Environmental Toxicology and Chemistry* 18, 459–465.

Hitchcock, D.R., Black, M.C. and Williams, P.L. (1997) Investigations into using the nematode *Caenorhabditis elegans* for municipal and industrial wastewater toxicity testing. *Archives of Environmental Contamination and Toxicology* 33, 252–260.

Höss, S. (2021) Single- and multi-species toxicity testing with nematodes. In: Traunspurger, W. (ed.) *Ecology of Freshwater Nematodes*. CAB International, Wallingford, UK, pp. 298–322.

Höss, S. and Römbke, J. (2019) Effects of waste materials on *Caenorhabditis elegans* (Nematoda) using the ISO standard soil toxicity test. *Environmental Science and Pollution Research* 26, 26304–26312.

Höss, S. and Weltje, L. (2007) Endocrine disruption in nematodes: effects and mechanisms. *Ecotoxicology* 16, 15–28.

Höss, S., Haitzer, M., Traunspurger, W. and Steinberg, C.E.W. (1999) Growth and fertility of *Caenorhabditis elegans* (Nematoda) in unpolluted freshwater sediments – response to particle size distribution and organic content. *Environmental Toxicology and Chemistry* 18, 2921–2925.

Höss, S., Henschel, T., Haitzer, M., Traunspurger, W. and Steinberg, C.E.W. (2001) Toxicity of cadmium to *Caenorhabditis elegans* (Nematoda) in whole sediment and porewater – the ambiguous role of organic matter. *Environmental Toxicology and Chemistry* 20, 2794–2801.

Höss, S., Arndt, M., Baumgarte, S., Tebbe, C.C., Nguyen, H.T. and Jehle, J.A. (2008) Effects of transgenic corn and Cry1Ab protein on the nematode, *Caenorhabditis elegans*. *Ecotoxicology and Environmental Safety* 70, 334–340.

Höss, S., Jänsch, S., Junker, T., Moser, T. and Römbke, J. (2009) Assessing the toxicity of contaminated soils using the nematode *Caenorhabditis elegans* as test organism. *Ecotoxicology and Environmental Safety* 72, 1811–1818.

Höss, S., Ahlf, W., Fahnenstich, C., Gilberg, D., Hollert, H. *et al.* (2010) Variability of freshwater sediment contact tests in sediments with low-level anthropogenic contamination – determination of toxicity thresholds. *Environmental Pollution* 158, 2999–3010.

Höss, S., Nguyen, H.T., Menzel, R., Pagel-Wieder, S., Miethling-Graf, R. *et al.* (2011a) Assessing the risk posed to free-living soil nematodes by a genetically modified maize expressing the insecticidal Cry3Bb1 protein. *Science of The Total Environment* 409, 2674–2684.

Höss, S., Schlottmann, K. and Traunspurger, W. (2011b) Toxicity of ingested cadmium to the nematode *Caenorhabditis elegans*. *Environmental Science & Technology* 45, 10219–10225.

Höss, S., Ahlf, W., Bergtold, M., Bluebaum-Gronau, E., Brinke, M. *et al.* (2012) Interlaboratory comparison of a standardized toxicity test using the nematode *Caenorhabditis elegans* (ISO 10872). *Environmental Toxicology and Chemistry* 31, 1525–1535.

Höss, S., Reiff, N., Nguyen, H.T., Jehle, J.A., Hermes, H. and Traunspurger, W. (2014) Small-scale microcosms to detect chemical induced changes in soil nematode communities – effects of crystal proteins and *Bt*-maize plant material. *Science of The Total Environment* 472, 662–671.

Höss, S., Roessink, I., Brock, T.C.M. and Traunspurger, W. (2020) Response of a nematode community to the fungicide fludioxonil in sediments of outdoor freshwater microcosms compared to a single species toxicity test. *Science of The Total Environment* 710, 135627.

Höss, S., Gansfort, B. and Traunspurger, W. (2022a) Nematodes as ecological and environmental indicators. In: Glazer, I., Shapiro-Ilan, D. and Sternberg, P.W. (eds) *Nematodes as Model Organisms*. CAB International, Wallingford, UK, pp. 165–195.

Höss, S., Reiff, N., Asekunowo, J. and Helder, J. (2022b) Nematode community of a natural grassland responds sensitively to the broad-spectrum fungicide mancozeb in soil microcosms. *Environmental Toxicology and Chemistry* 41, 2420–2430.

Huguier, P., Manier, N., Méline, C., Bauda, P. and Pandard, P. (2013) Improvement of the *Caenorhabditis elegans* growth and reproduction test to assess the ecotoxicity of soils and complex matrices. *Environmental Toxicology and Chemistry* 32, 2100–2108.

Hunt, P.R. (2017) The *C. elegans* model in toxicity testing. *Journal of Applied Toxicology* 37, 50–59.

ISO (International Organization for Standardization) (2010) *Water quality – Determination of the toxic effect of sediment and soil samples on growth, fertility and reproduction of Caenorhabditis elegans (Nematoda); ISO 10872:2010*. ISO, Geneva, Switzerland.

ISO (International Organization for Standardization) (2020) *Water and soil quality – Determination of the toxic effect of sediment and soil samples on growth, fertility and reproduction of Caenorhabditis elegans (Nematoda); ISO 10872:2020*. ISO, Geneva, Switzerland.

Jackson, B.P., Williams, P.L., Lanzirotti, A. and Bertsch, P.M. (2005) Evidence for biogenic pyromorphite formation by the nematode *Caenorhabditis elegans*. *Environmental Science & Technology* 39, 5620–5625.

Kammenga, J.E., Van Gestel, C.A.M. and Bakker, J. (1994) Patterns of sensitivity to cadmium and pentachlorophenol among nematode species from different taxonomic and ecological groups. *Archives of Environmental Contamination and Toxicology* 27, 88–94.

Kammenga, J.E., Van Koert, P.H.G., Riksen, J.A.G., Korthals, G.W. and Bakker, J. (1996) A toxicity test in artificial soil based on the life history strategy of the nematode *Plectus acuminatus*. *Environmental Toxicology and Chemistry* 15, 722–725.

Khanna, N., Cressman, C.P. III, Tatara, C.P. and Williams, P.L. (1997) Tolerance of the nematode *Caenorhabditis elegans* to pH, salinity, and hardness in aquatic media. *Archives of Environmental Contamination and Toxicology* 32, 110–114.

Kim, D., Kim, H. and An, Y.-J. (2023) Species sensitivity distributions of micro- and nanoplastics in soil based on particle characteristics. *Journal of Hazardous Materials* 452, 131229.

Kwon, H., Roh, J.Y., Lim, D., Choi, J. and Kwon, J. (2011) Maintaining the constant exposure condition for an acute *Caenorhabditis elegans* mortality test using passive dosing. *Environmental Health and Toxicology* 26, 1–7.

Landrum, P.F., Robbins, J.A., Baudo, R., Giesy, J. and Muntau, H. (1990) Bioavailability of sediment associated contaminants to benthic invertebrates. In: Baudo, R., Giesy, J.P. and Muntau, H. (eds) *Sediments: Chemistry and Toxicity of In-Place Pollutants*. Lewis Publisher, Ann Arbor, Michigan, pp. 237–263.

Lee, B.G., Griscom, S.B., Lee, J.S., Choi, H.J., Koh, C.H. *et al.* (2000) Influences of dietary uptake and reactive sulfides on metal bioavailability from aquatic sediments. *Science* 287, 282–284.

Leppänen, M.T. and Kukkonen, J. (1998) Relative importance of ingested sediment and pore water as bioaccumulation routes for pyrene to oligochaete (*Lumbriculus variegatus*, Müller). *Environmental Science & Technology* 32, 1503–1508.

Leung, M.C.K., Williams, P.L., Benedetto, A., Au, C., Helmcke, K.J. *et al.* (2008) *Caenorhabditis elegans*: an emerging model in biomedical and environmental toxicology. *Toxicological Sciences* 106, 5–28.

Ma, H., Glenn, T.C., Jagoe, C.H., Jones, K.L. and Williams, P.L. (2009) A transgenic strain of the nematode *Caenorhabditis elegans* as a biomonitor for heavy metal contamination. *Environmental Toxicology and Chemistry* 28, 1311–1318.

MacDonald, D.D., Ingersoll, C.G. and Berger, T.A. (2000) Development and evaluation of consensus-based sediment quality guidelines for freshwater ecosystems. *Archives of Environmental Contamination and Toxicology* 39, 20–31.

Meyer, D. and Williams, P.L. (2014) Toxicity testing of neurotoxic pesticides in *Caenorhabditis elegans*. *Journal of Toxicology and Environmental Health, Part B: Critical Reviews* 17, 284–306.

Mayer, P., Wernsing, J., Tolls, J., de Maagd, P.G.-J. and Sijm, D.T.H.M. (1999) Establishing and controlling dissolved concentrations of hydrophobic organics by partitioning from a solid phase. *Environmental Science & Technology* 33, 2284–2290.

Miranda-Vizuete, A. and Veal, E.A. (2017) *Caenorhabditis elegans* as a model for understanding ROS function in physiology and disease. *Redox Biology* 11, 708–714.

Mueller, M.-T., Fueser, H., Trac, L.N., Mayer, P., Traunspurger, W. and Höss, S. (2020) Surface-related toxicity of polystyrene beads to nematodes and the role of food availability. *Environmental Science & Technology* 54, 1790–1798.

Peña Santiago, R. (2020) *Morphology and Bionomics of Dorylaims (Nematoda, Dorylaimida), Nematology*. Brill, Leiden, The Netherlands.

Peredney, C.L. and Williams, P.L. (2000) Utility of *Caenorhabditis elegans* for assessing heavy metal contamination in artificial soil. *Archives of Environmental Contamination and Toxicology* 39, 113–118.

Rand, G.M. (2003) *Fundamentals of Aquatic Toxicology: Effects, Environmental Fate, and Risk Assessment*, 2nd edn. Taylor & Francis, New York.

Reinfelder, J.R. and Fisher, N.S. (1991) The assimilation of elements ingested by marine copepods. *Science* 251, 794–798.

Ristau, K., Akgül, Y., Bartel, A.S., Fremming, J., Müller, M.-T. *et al.* (2015) Toxicity in relation to mode of action for the nematode *Caenorhabditis elegans*: acute to chronic ratios and quantitative structure–activity relationships. *Environmental Toxicology and Chemistry* 34, 2347–2353.

Saleem, S., Böhme, A. and Schüürmann, G. (2023) Baseline narcosis for the glass-vial 96-h growth inhibition of the nematode *C. elegans* and its use for identifying electrophilic and pro-electrophilic toxicity. *Environmental Science & Technology* 57, 1692–1700.

Sandhu, A., Badal, D., Sheokand, R., Tyagi, S. and Singh, V. (2021) Specific collagens maintain the cuticle permeability barrier in *Caenorhabditis elegans*. *Genetics* 217, iyaa047.

Schertzinger, G., Zimmermann, S. and Sures, B. (2019) Predicted sediment toxicity downstream of combined sewer overflows corresponds with effects measured in two sediment contact bioassays. *Environmental Pollution* 248, 782–791.

Sherry, J., Scott, B. and Dutka, B. (1997) Use of various acute, sublethal and early life stage tests to evaluate the toxicity of refinery effluents. *Environmental Toxicology and Chemistry* 16, 2249–2257.

Simpson, S.L., Campana, O. and Ho, K.T. (2016) Sediment toxicity testing. In: Blasco, J., Chapman, P.M., Campana, O. and Hampel, M. (eds) *Marine Ecotoxicology: Current Knowledge and Future Issues*. Academic Press, New York, pp. 199–237.

Smith, K.E.C., Dom, N., Blust, R. and Mayer, P. (2010) Controlling and maintaining exposure of hydrophobic organic compounds in aquatic toxicity tests by passive dosing. *Aquatic Toxicology* 98, 15–24.

Sochova, I., Hofman, J. and Holoubek, I. (2007) Effects of seven organic pollutants on soil nematode *Caenorhabditis elegans*. *Environment International* 33, 798–804.

Spann, N., Goedkoop, W. and Traunspurger, W. (2015) Phenanthrene bioaccumulation in the nematode *Caenorhabditis elegans*. *Environmental Science & Technology* 49, 1842–1850.

Tang, B., Williams, P.L., Xue, K.S., Wang, J.-S. and Tang, L. (2020) Detoxification mechanisms of nickel sulfate in nematode *Caenorhabditis elegans*. *Chemosphere* 260, 127627.

Tejeda-Benitez, L. and Olivero-Verbel, J. (2016) *Caenorhabditis elegans*, a biological model for research in toxicology. *Reviews of Environmental Contamination and Toxicology* 237, 1–35.

Thomas, P.C., Velthoven, K., Geurts, M. and Van Wijk, D. (2009) Bioavailability and detoxification of cationics: II. Relationship between toxicity and CEC of cationic surfactants on *Caenorhabditis elegans* (Nematoda) in artificial and natural substrates. *Chemosphere* 75, 310–318.

Trac, L.N., Schmidt, S.N. and Mayer, P. (2018) Headspace passive dosing of volatile hydrophobic chemicals – aquatic toxicity testing exactly at the saturation level. *Chemosphere* 211, 694–700.

Traunspurger, W., Haitzer, M., Höss, S., Beier, S., Ahlf, W. and Steinberg, C.E.W. (1997) Ecotoxicological assessment of aquatic sediments with *Caenorhabditis elegans* (Nematoda) – a method for testing in liquid medium and whole sediment samples. *Environmental Toxicology and Chemistry* 16, 245–250.

Tuikka, A., Schmitt, C., Höss, S., Bandow, N., Von der Ohe, P.C. *et al.* (2011) Toxicity assessment of sediments from three European river basins using a sediment contact test battery. *Ecotoxicology and Environmental Safety* 74, 123–131.

Van den Brink, P.J., Tarazona, J.V, Solomon, K.R., Knacker, T., Van den Brink, N.W. *et al.* (2005) The use of terrestrial and aquatic microcosms and mesocosms for the ecological risk assessment of veterinary medical products. *Environmental Toxicology and Chemistry* 24, 820–829.

Wei, J.Z., Hale, K., Carta, L., Platzer, E., Wong, C. *et al.* (2003) *Bacillus thuringiensis* crystal proteins that target nematodes. *Proceedings of the National Academy of Sciences USA* 100, 2760–2765.

Williams, P.L. and Dusenbery, D.B. (1990a) Aquatic toxicity testing using the nematode *Caenorhabditis elegans*. *Environmental Toxicology and Chemistry* 9, 1285–1290.

Williams, P.L. and Dusenbery, D.B. (1990b) A promising indicator of neurobehavioral toxicity using the nematode *Caenorhabditis elegans* and computer tracking. *Toxicology and Industrial Health* 6, 425–440.

Wilson, M. and Kakouli-Duarte, T. (2009) *Nematodes as Environmental Indicators*. CAB International, Wallingford, UK.

Wittkowski, P., Marx-Stoelting, P., Violet, N., Fetz, V., Schwarz, F. *et al.* (2019) *Caenorhabditis elegans* as a promising alternative model for environmental chemical mixture effect assessment – a comparative study. *Environmental Science & Technology* 53, 12725–12733.

Wu, T., Xu, H., Liang, X. and Tang, M. (2019) *Caenorhabditis elegans* as a complete model organism for biosafety assessments of nanoparticles. *Chemosphere* 221, 708–726.

Zhao, J., Lin, M., Wang, Z., Cao, X. and Xing, B. (2021) Engineered nanomaterials in the environment: are they safe? *Critical Reviews in Environmental Science & Technology* 51, 1443–1478.

14 *Caenorhabditis elegans* and Ecotoxicogenomics

Cristina Lagido[1]*, Chubin Zhang[2], Lilian Atira[3], Alejandra Vieyra-Ramirez[3], Thomais Kakouli-Duarte[3] and Stephen R. Stürzenbaum[2]*

[1]Department of Chemistry and Bioscience, University of Aalborg, Denmark; [2]Analytical, Environmental and Forensic Sciences Department, Faculty of Life Sciences & Medicine, King's College London, UK; [3]MENRG, enviroCORE, Department of Applied Science, Department of Applied Science, South East Technological University, Ireland

Abstract

The free-living, bacterial-feeding nematode *Caenorhabditis elegans* is a well-established multicellular eukaryotic model organism. This chapter presents an overview of how available genomic tools can be utilized to enhance ecotoxicological research. It first elucidates how biosensors are generated, describes their use in biomonitoring, and reviews historic and recent developments the topic of stress-response biology. Subsequently, it examines how the profiling of genome-wide changes in gene expression, following exposure to environmental chemical stressors, has been assessed by DNA microarrays and more recently RNA sequencing technologies. The chapter concludes by providing an outlook regarding *C. elegans*' potential as a powerful tool to advance ecotoxicogenomics.

14.1 Introduction

The free-living nematode *Caenorhabditis elegans* is one of the best-characterized metazoans pertaining its genetics, development, anatomy and behaviour. This nematode is also a model organism for ecotoxicogenomics studies due to its sensitivity to contaminants and its representative role in key ecological functional levels across different environmental niches (Queirós *et al.*, 2019; Calarco *et al.*, 2025). Its amenability for genetic modification has been exploited to generate transgenic biosensor strains that are sensitive and provide information on mechanistic function in the organism's response to a toxic insult. A large percentage of the *C. elegans* genome is evolutionarily conserved and many worm genes have counterparts in higher (vertebrate) organisms (Culetto and Sattelle, 2000). This supports the use of an invertebrate biosensor to study the xenobiotic-induced activation of specific genetic pathways within an environmental sample. Furthermore, this approach can be extended to examine genetic markers associated with human diseases.

A key challenge of the post-genomic era is to translate the vast amounts of information generated by large-scale sequencing projects into units that enhance the understanding of genome function and biological processes, both in general and in relation to environmental triggers and/or stressors. Over the past decades,

*Email: cmr@bio.aau.dk and stephen.sturzenbaum@kcl.ac.uk

DOI: 10.1079/9781800624221.0014

high-throughput technologies have emerged as powerful resources to study the complexity of biological interactions at the genomic, epigenomic and transcriptomic levels (Lee, 2023). A pioneering technology was the DNA microarray, which enabled the expression profiling of thousands of genes in a single experiment. This approach to transcriptomics was, more recently, replaced by RNA sequencing (RNA-Seq) which produces a more data-rich output, is less costly and is not reliant on an existing genome sequence (Wang *et al.*, 2009; Abbass *et al.*, 2021; Chen *et al.*, 2024). Both technologies provide the tools to deliver a greater understanding of whole-genome expression profiles in response to chemical stressors and, in turn, of ecological and toxicological modes of action. The terms 'ecogenomics' (Chapman, 2001) and 'ecotoxicogenomics' (Snape *et al.*, 2004), in their broadest sense, encompass not only transcript profiling (*transcriptomics*) but can also incorporate information linked to protein composition (*proteomics*) and the metabolic constituents (*metabolomics*) of a cell.

Transcription is the initial step in gene expression; thus, a transcriptional response can give an indication of cellular mechanisms that are affected by a pollutant and in consequence provide a sensitive starting point to assess, often acute, ecological and toxicological (i.e. ecotoxicological) responses. However, this assumes some basic principles (Oberemm *et al.*, 2005; Fedorenkova *et al.*, 2010), namely that: (i) all ecotoxicologically relevant effects are accompanied by a meaningful alteration in gene expression; and (ii) a specific change in expression pattern, induced by a chemical stressor, is similar for all chemicals with similar toxic modes of action. These notions form the basis for predictive ecotoxicology and are used to categorize/differentiate between the mode of action of single-chemical and/or mixture exposures. Several studies have generated large volumes of gene expression data, validated via modular classification systems (Burczynski *et al.*, 2000) and organized in dedicated databases, such as the Chemical Effects in Biological Systems (CEBS) (Waters *et al.*, 2003) and Comparative Toxicogenomics Database (CTD) (Davis *et al.*, 2023).

High-throughput gene expression technologies enhance, rather than replace, classical toxicological approaches which study parameters such as morphological changes, altered reproductive capacity or mortality (Hamadeh *et al.*, 2001; Menzel *et al.*, 2005a; Long *et al.*, 2023). The value of classical ecotoxicology remains undisputed; however, ecotoxicogenomics can add a powerful new dimension that unravels mechanistic processes, reveals novel modes of action and provides the opportunity to characterize the dynamic nature of biological systems. Ecotoxicogenomics aims to develop new predictive models to identify environmental or human health hazards and specific and rapid molecular biomarkers of exposure to natural as well as man-made chemical stressors.

Key aims of ecotoxicogenomics include:

- The characterization of the modes of action of known chemical environmental stressors.
- The identification of the modes of action of previously uncharacterized chemical stressors with the intention of correlating the observed molecular signatures with well-characterized pollutant-induced expression patterns.
- The assessment of stress-specific gene expression patterns as a biomarker of chemical exposure.
- The identification of chemical-specific and species-overlapping gene expression patterns.
- The characterization of complex mixtures of natural and anthropogenic chemicals.
- The examination of acute versus chronic effects, as well as low-concentration versus high-concentration exposures.

Throughout this chapter, the terms 'ecotoxicological' and 'ecotoxicology' comprise both man-made and natural chemical environmental stressors. It is also worth noting that the presence of natural endogenic and exogenic chemical stressors has been instrumental for, and driven the development of, stress defence systems, examples of which include the antioxidant or biotransformation systems, expression of stress proteins or metal-binding proteins. The purpose of ecotoxicogenomics is to provide an insight into the physiological status of organisms and decipher responses and interactions of organisms within the environment. Ultimately, the interaction(s) between organisms and their environments is analogous to the complex interactions of genes and proteins at the cellular level (Chapman, 2001).

14.2 Transgenic *C. elegans* Biosensors

Transgenic *C. elegans* biosensors offer the possibility of monitoring the effects of stressors at the whole organism level in a cost-effective manner. Transformation techniques are well established in *C. elegans*, allowing for the routine construction of transgenic animals carrying reporter genes (Sands *et al.*, 2018). Due to the worm's transparent nature, it is possible to visualize signals derived from β-galactosidase, fluorescent proteins and luciferase (Lagido, 2009). Standard cloning techniques introduced over two decades ago were used to assemble the constructs for the genetic transformation of *C. elegans* via their multi-copy microinjection into the gonad (Boulin *et al.*, 2006; Rieckher and Tavernarakis, 2017). This can result in heritable extrachromosomal arrays. However, the stable and chromosome-location-specific (targeted) integration by MosSCI (Mos1-mediated single-copy insertion) or CRISPR (clustered regularly interspaced short palindromic repeats)/CRISPR-associated (Cas) methods is now more common (Friedland *et al.*, 2013; Chen *et al.*, 2023a). As *C. elegans* can be cryopreserved, most strains are deposited (and sourceable) via the Caenorhabditis Genetics Center (https://cgc.umn.edu/, accessed 23 August 2025) or the National BioResource Project (https://nbrp.jp/en/resource/c-elegans-en/, accessed 23 August 2025).

14.2.1 Detection limit and sensitivity of biosensors

The usefulness of transgenic strains in biomonitoring is best evaluated by comparison with other physiological parameters commonly used as descriptors of toxicity in *C. elegans* (see Chapter 13, this volume). Table 14.1 provides an overview of the detection limits obtained with transgenic strains when compared to the lethal concentration where 50% of animals perish (LC_{50}). Transgenic strains are typically more sensitive and provide a faster stress detection than conventional LC_{50} tests. They respond to sublethal levels of stress which affect sensitive end points such as feeding, development and reproduction.

Later in this chapter, examples of transgenic biosensor strains are presented in relation to their response to various stressors (Section 14.4) and the testing of environmental samples (Section 14.5).

14.2.2 Dose–response of biosensors

To assess the bioavailability of toxicants, it is important to determine the linearity range of biosensor response to specific toxicants for each biosensor strain. Cadmium has been identified as a strong inducer of multiple biosensor strains, particularly those incorporating heat-shock (*hsp*) and metallothionein (*mtl*)-responsive promoters. For instance, the CB4027 (*hsp-70::lacZ*) strain exhibited a concentration-dependent response to cadmium in the 1–16 mg l^{-1} range at 32°C, whereas zinc, mercury, lindane, tin, tributyltin (TBT) and manganese were all weak inducers of this strain (Guven *et al.*, 1994). Similarly, the PC72 strain (*hsp16.1::lacZ*) showed a significant concentration-dependent induction of β-galactosidase in response to cadmium 0–15 mg l^{-1} in a laboratory exposure (Dennis *et al.*, 1997) and cadmium to up to 500 μg g^{-1} in spiked soil and in extracted soil water samples (Power and de Pomerai, 1999). Other strains that displayed dose–response relationships to cadmium include PC161 (*hsp16.1::lacZ::gfp*), JF.9 (*cdr-1::lacZ*) and JF2.1 (*mtl-2::lacZ*) (Cioci *et al.*, 2000; Liao *et al.*, 2002; David *et al.*, 2003). These strains are mostly of historic value, since the use of the green fluorescent protein (GFP) allows for *in vivo* assessment and greater versatility (Fig. 14.1).

Dose–response relationships were also reported with persistent organic pollutants (POPs) in various biosensor strains, such as the insecticide endosulfan and the endocrine disrupter nonylphenol (Chen *et al.*, 2019). Exposure to endosulfan (0.1 to 10 μM) significantly increased germ-cell apoptosis in a time- and dose-dependent manner, an effect that was dependent on *jnk-1*/JNK-*MAPK*, *sek-1*/MAP2K and *pmk-1*/p38-MAPK (Wang *et al.*, 2017). Moreover, bisphenol A (BPA) exposure resulted in a dose-dependent decrease in GFP expression in the strains SJ4005 (*hsp-4::gfp*), LD1171 (*gcs-1::gfp*) and CL2166 (*gst-4::gfp*), but an increased expression in TJ375 (*hsp-16.2::gfp*) (Tan *et al.*, 2015). The quantification of these concentration-dependent effects

Table 14.1. Minimum stress level that elicits a response in a transgenic strain (detection limit). Lethality data is expressed as lethal concentration where 50% of animals perish (LC_{50}) for comparison (wild-type strain unless otherwise stated).

Stress	Strain	Detection limit[a]	LC_{50}[b]
Cadmium ($CdCl_2 \cdot 2.5H_2O$)	JF2.1	0.1 μM (24 h, 20°C, mixed stages)[c]	8042μM[d]
	JF9	1 μM (24 h, 20°C, mixed stages)[g]	
	mtl-1::gfp mtl-2::gfp	2.5μM (24 h) or >20 μM (1 h)[h]	8390μM[e]
	CB4027	4.5 μM (7 h, 32°C, adults)[i]	
	PC71/PC72/PC73	25 μM (16 h, ?°C, L2/L3)[f]	
	PC161	43.8 μM (GFP expression) or 21.9 μM (*lacZ* expression) (16 h, 24°C, ? stage)[j]	185μM, PC71 strain[f]
	KC136	272.8 μM (5 h, 22°C, L3)[k]	
	KC334	54.6 μM (5 h, 22°C, L3)[l]	
	hsp-16.48::gfp	463.7μM (24 h, 20°C, young adults)[m]	
Zinc ($ZnSO_4 \cdot 7H_2O$)	JF2.1	10 μM (24 h)[c]	
	CB4027	30.6 μM (7 h, 32°C, adults)[i]	
$ZnCl_2$			3089.2μM;[d] 2169 μM[e]
Mercury ($HgCl_2$)	JF2.1	0.2 μM (24 h, 20°C, mixed stages)[c]	49.9μM[d]
	CB4027	2.5 μM (7 h, 32°C, adults)[i]	72μM (24 h);[e] approx. 47.9μM[f]
	PC71/PC72/PC73	3.7 μM (24 h, ?°C, L2/L3)[f]	
	KC136	15 μM (5 h, 22°C, L3)[k]	
Copper ($CuCl_2$)	PC71/PC72/PC73	74.4 μM (8 h, ?°C, L2/L3)[f]	100.4μM;[f] 346.2 μM;[d] 1420μM[e]
$CuSO_4 \cdot 5H_2O$	PE39	80 μM (2 h, 25°C, mixed stages)[n]	
	KC136	100 μM (5 h, 22°C, L3)[k]	100μM (48 h)[k]
Lead ($Pb(NO_3)_2$)	PC71/PC72/PC73	30.2 μM (16 h, ?°C, L2/L3)[f]	622.6μM;[d] 260 μM[e]
	PE39	50 μM (2 h, 25°C, mixed stages)[n]	
Arsenite ($NaAsO_2$)	PC71/PC72/PC73	50 μM (8 h, ?°C, L2/L3)[f]	2429.2[d];> 200 μM;[e] 170μM (48 h), PC72[f]
Silver ($AgNO_3$)	CB4027	1.4/4.6 μM (7 h, 32°C, adults)[i]	46.4μM[d]
Nickel ($NiSO_4 \cdot 6H_2O$)	JF2.1	10 μM (24 h, 20°C, mixed stages)[c]	$NiCl_2$: 49'682 μM;[d] 63'400μM[e]
Captan	PC72	10 mg l^{-1} (48 h, 15°C, L3)[o]	
Captafol	PC72	1 mg l^{-1} (48 h, 15°C, L3)[o]	
Folpet	PC72	10 mg l^{-1} (48 h, 15°C, L3)[o]	

[a]Nominal concentrations, exposure conditions in brackets; [b]wild-type strain and 24 h exposure, unless otherwise stated; [c]Cioci *et al*. (2000); [d]Williams and Dusenbery (1990); [e]Tatara *et al*. (1997); [f]Stringham and Candido (1994); [g]Liao *et al*. (2002); [h]Swain *et al*. (2004); [i]Guven *et al*. (1994); [j]David *et al*. (2003); [k]Chu and Chow (2002); [l]Chu *et al*. (2005); [m]Roh *et al*. (2006); [n]Lagido *et al*. (2001); [o]Jones *et al*. (1996).

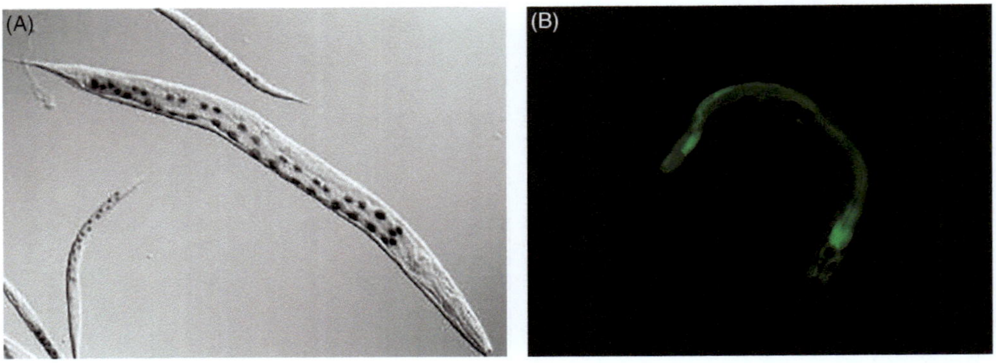

Fig. 14.1. (A) A *Caenorhabditis elegans* strain expressing the *mtl-2::lacZ* transgene displays β-galactosidase activity in intestinal cell nuclei, following treatment with 100 μM $CdCl_2$ for 24 h. (Reproduced from Cioci *et al.*, 2000 with permission from Allen Press Publishing Services and Jonathan Freedman.)
(B) *cyp-35A2::gfp* reporter strain displays intense green fluorescence when exposed to β-naphthoflavone 1 mg l^{-1} for 24 h and observed under fluorescence microscopy. (Photograph courtesy of Ralph Menzel.)

provides a better understanding of the causal relationship between a toxicant and its effects (Tsatsakis *et al.*, 2018).

14.2.3 Variability in response of biosensors

For consistent results in biomonitoring, standardized experimental conditions need to be established. To deal with the inherent variability of bioassays, sets of multiple replicates are recommended within each experiment (Dennis *et al.*, 1997; Lagido, 2009). One important biological aspect that influences the outcome is the developmental stage of the test population of nematodes, as different larval stages may vary in their stress response (Guven *et al.*, 1994; Stringham and Candido, 1994; Jones *et al.*, 1996; Almutairi *et al.*, 2024). However, mixed populations of worms including several developmental stages are easier to generate and many studies have opted for these (Freedman *et al.*, 1993; Mutwakil *et al.*, 1997; Lagido *et al.*, 2001), at the cost of increasing variability between experiments.

14.3 Biosensor Profiling by Transcriptomics

In transcriptomic experiments replications are, as with all 'omic' approaches, the key to reliable results (Menzel *et al.*, 2009a,b). Given that gene expression can be sensitive to environmental change, it is important that test organisms should be maintained under well-controlled laboratory conditions. In addition, it might be necessary to analyse several time points or developmental stages before reaching a conclusion that is statistically robust. Taking these points into account will avoid, or at least minimize, the interference and variability caused by physical parameters (e.g. temperature, light, medium structure) during development, or effects that are dependent on ploidy, gender, age or nutritional status. One of the most challenging issues is the identification of changes in gene expression that represent an adaptive response to external stimuli. These may have no ecotoxicological significance or increased risk and are distinct from early stages of disease progression (Pennie *et al.*, 2000). For example, some changes in gene expression might reflect a non-specific and fully reversible response to stress and thus have no biological consequence. Dose–response studies are a powerful tool to: (i) differentiate adaptive versus toxic responses; and (ii) establish ecotoxicogenomic thresholds that need to be exceeded prior to the initiation of the cascade of molecular responses leading to adverse effects.

Because of the importance of statistical analysis in transcriptomic studies (Nadon and Shoemaker, 2002; Allison *et al.*, 2006), at least two crucial points need to be emphasized. First, although a large change in gene expression may reflect a response of biological importance, a small change does not imply the opposite. A simple numerical listing of up- and downregulating is therefore not a sufficient evaluation of complex transcriptomic data. Second, the simultaneous analysis of a large number of genes will, by definition, include false positives (type I errors). Multiple test correction methods are however conservative, thus result in a high false negative (type II) error rate. Several approaches are available to address these problems, the most common being false discovery rate (FDR) analysis, introduced by Benjamini and Hochberg (1995).

The analysis of expression patterns and cluster analysis are a so-called unsupervised method, which does not use any kind of sample classification as input, but asks the simple question: 'Are there unexpected, but biologically meaningful patterns that exist in the data?' Clustering can identify similar expression patterns within a cohort of genes and may suggest a cascade of similar responses or a common regulatory circuit. Two of the most widely used unsupervised approaches are hierarchical clustering (Weinstein *et al.*, 1997; Eisen *et al.*, 1998) and *k*-means partitional clustering (Soukas *et al.*, 2000). A more detailed description of hierarchical algorithms can be found in Menzel *et al.* (2009a).

Transcriptomic experiments rely on functional annotations such as gene ontology (GO) to describe species-overlapping attributes of gene products. The use of functional annotation tools offers the possibility to study coordinated changes that control specific pathways or processes. Three GO terms are typically scrutinized: the 'Molecular Function' of a gene product; its role in multi-step 'Biological Processes'; and its localization to 'Cellular Components'. Each GO term is assigned to one of the three subdivisions and consists of a unique alphanumerical identifier, a common name and a definition. GO term analysis facilitates uniform queries across transcriptomic experiments and highlights trends that have previously been identified. Several online resources are available to assist with functional GO annotation. The Database for Annotation, Visualization and Integrated Discovery (DAVID), from the US National Institutes of Health, is open access and easy to use. It functionally groups transcripts using predefined levels of stringency. Further suitable tools are listed in Table 14.2.

A weakness of functional annotation and pathway mapping is that it relies on previously defined processes, evolutionary conserved pathways, and struggles to identify gene regulatory networks. Any putative targets identified from a transcriptomic experiment therefore require validation by Northern blotting or quantitative real-time PCR (RT-PCR, and in some cases at the protein level.

An important step of any transcriptomic analysis pertains to downstream data mining. This facilitates the detailed comparisons with related or even apparently unrelated research projects and may well substantiate knowledge and the mechanistic understanding of gene function. Several useful databases for data mining are presented in Table 14.2. The largest and most comprehensive of these repositories is the Gene Expression Omnibus (GEO) at the National Center for Biotechnology Information (NCBI).

The Microarray and Gene Expression Data (MGED) Society, now the Functional Genomics Data (FGED) Society, has developed guidelines and standards for the user community. Now widely accepted, MIAME (Minimum Information About a Microarray Experiment) and MINSEQE (Minimum Information About a Next-generation Sequencing Experiment) define the minimum quantity and quality of information that is required to interpret and verify results of transcriptomic studies (Brazma *et al.*, 2001; Brazma, 2009).

14.3.1 Biosensor profiling: microarrays

The first true whole-genome transcriptomics typically utilized DNA microarrays, which represent organized arrangements of multiple DNA probes fixed on an immobilized surface. Several thousand DNA spots (each representing a gene or DNA sequence), usually ranging from 60 to 100 μm in diameter, are spotted on to a single slide. Oligonucleotide arrays comprise short, immobilized DNA sequences (approximately 25–80 DNA bases) designed as complementary probes to specific genes or genome sequences. However, issues of signal intensity, the signal variability and cost are limiting factors of oligonucleotide arrays, and they were superseded by RNA-Seq technology. Today, microarrays can still serve a purpose when studying, in a routine

Table 14.2. Gene expression databases and software tools for large-scale transcriptomic analysis.

Name	Website;[a] https://	Description
Databases		
Gene Expression Omnibus (GEO)	www.ncbi.nlm.nih.gov/geo/	Searchable database of microarray experiments submitted to National Center for Biotechnology Information (NCBI)
ArrayExpress	www.ebi.ac.uk/biostudies/arrayexpress	Searchable database of microarray experiments submitted to European Bioinformatics Institute (EBI)
ArrayTrack	www.fda.gov/science-research/bioinformatics-tools/arraytracktm-hca-pca-standalone-package-powerful-data-exploring-tools	Provides an integrated solution for managing, analysing and interpreting microarray gene expression data associated with pharmaco- or toxicogenomics studies
Chemical Effects in Biological Systems (CEBS)	cebs.niehs.nih.gov/cebs/	House data from multiple gene expression platforms, protein expression, and changes in low-molecular-weight metabolite levels aligned by their detailed toxicological context
Comparative Toxicogenomics Database (CTD)	ctdbase.org/	Interactions between chemicals and genes/proteins in diverse organisms are curated from literature and integrated with diverse data (human diseases, references, sequences, vertebrate and invertebrate organisms, and the gene ontology) to facilitate environmental health research
Annotation/pathway tools		
Database for Annotation, Visualization and Integrated Discovery (DAVID)	davidbioinformatics.nih.gov/	A comprehensive set of functional annotation tools for investigators to understand biological meaning behind large lists of genes, developed by National Institute of Allergy and Infectious Diseases (NIAID)
GoMiner	discover.nci.nih.gov/gominer/ (retired, no longer available)	To identify the biological processes, functions and components represented in microarray gene lists
Selected data analysis software		
Gene Set Enrichment Analysis (GSEA)	www.gsea-msigdb.org/gsea/index.jsp	A computational method that determines which in a given set of genes show statistically significant differences between two biological states
Bioconductor	www.bioconductor.org/	Open source and open development software project for the analysis and comprehension of genomic data
GenePattern	genepattern.org/	For the analysis of molecular profiles, e.g. genome-wide microarray expression signatures by clustering

[a]All accessed 23 August 2025.

screening approach, a small number of well-defined and carefully selected genes of known function.

14.3.2 Biosensor profiling: RNA-Seq

Recent advances have increased the speed and reduced the cost of sequencing, which in turn has resulted in the development of very powerful RNA-Seq technology. RNA-Seq delivers the full sequencing of the entire transcriptome, to identify differentially modulated transcripts, splice variants and non-coding transcripts that can be of toxicological interest (Rao *et al.*, 2019; Rhiannon, 2020). Modern toxicogenomics now makes full-use of RNA-Seq technology (Rao *et al.*, 2019). In detail, RNA-Seq generates data by randomly sequencing complementary DNA (cDNA) fragments derived from RNA samples. The sequenced reads (typically 150 to 600 bp in length from 30–40 million per sample/replicate) are then aligned to the reference genome or transcriptome and classified as exonic, junction or poly(A) end-reads, and used to create an expression profile for each gene (Wang *et al.*, 2009). RNA-Seq has been developed and used because of its low background signal and the possibility to detect genes with low expression levels (Rhiannon, 2020). No a priori knowledge of the genome is needed, making RNA-Seq a valuable tool in ecotoxicogenomics when studying environmental sentinel non-model organisms.

14.4 Ecotoxicogenomics in *C. elegans*

The nematode *C. elegans* is not only known for its exquisite molecular genetic tools but also satisfies all typical criteria to function as an ecotoxicological test organism (see Chapter 13, this volume). In addition, *C. elegans* is a widespread model organism for environmental toxicity testing due to its ortholog and homolog genes for key mammalian functions (Karengera *et al.*, 2022). Moreover, due to its completely sequenced genome, *C. elegans* is ideally suited for ecotoxicogenomics. For example, several studies have demonstrated that specific genes can affect lifespan via the modulation of signal transduction, transcription and mitochondrial function (Lund *et al.*, 2002; Wang and Kim, 2003; Golden and Melov, 2004). Whole-transcriptome technology has facilitated the identification of many genes that exhibit altered expression following xenobiotic exposure and, therefore, are putative biosensors of a stress response. However, the significance and specificity of the stress-induced change in gene expression are defined by downstream biological effects (Feder and Walser, 2005). A suitable approach to characterize and validate a stress-responsive gene is by application of loss-of-function mutants or by means of gene silencing via RNA interference (RNAi) (Menzel *et al.*, 2005a, 2007; Cui *et al.*, 2007; Abbass *et al.*, 2021). Once the physiological role has been established, transgenic strains can be generated containing reporter genes under the control of the regulatory regions of stress-inducible genes.

14.4.1 Biosensors of the heat-shock response

Stress-induced protein denaturation elicits a universal cellular response, known as the heat-shock response, involving the rapid and coordinated induction of heat-shock proteins (HSPs) (Westerheide and Morimoto, 2005). There has been much interest in monitoring the expression of HSPs, particularly HSP70 and HSP16, as a proxy of environmental contamination. HSP70s act as molecular chaperones, promoting folding and minimizing protein denaturation and aggregation; less is known about HSP16 proteins, however, but they appear to be strictly stress-inducible. Table 14.3 summarizes relevant ecotoxicological studies using biosensors of the heat-shock response in *C. elegans*. For reviews of the heat-shock response, please refer to Westerheide and Morimoto (2005) and Akerfelt *et al.* (2010).

14.4.2 Biosensors of metal exposure

Cadmium is a persistent environmental toxicant that is associated with a variety of diseases. At low levels of exposure, the toxicological effects can be buffered by activation of intracellular defence and repair systems, the so-called stress response. Cui *et al.* (2007) exposed *C. elegans* to 100 μM Cd for 4 and 24 h and subsequent microarray experiments revealed 237 upregulated and 53 downregulated genes that changed significantly. These genes were clustered into early and late response genes. The former encompasses pathways which regulate the localization and transportation of different chemical species (in particular, metal ions) (Fig. 14.2). This suggests that the first response to cadmium intoxication is a transcriptional adjustment to maintain ion homeostasis and readjustment of perturbed energy supply. During the 24 h exposure period, metabolic and localization pathways were enriched within the up- and downregulated gene lists (Fig. 14.2). Cadmium exposure resulted in the overexpression of 14 cytochrome P450 (CYP) genes, six glutathione *S*-transferase (GST) genes and five UDP-glucuronosyltransferase (UGT) genes. In addition to these biotransformation enzymes, the expression of four ABC

Table 14.3. Transgenic strains that report on induction of the heat-shock response.

Strain	Promoter	Reporter gene/fusion	Expression of reporter	Stress	Medium
PC72 PC71 PC73	*hsp-16.1* *hsp-16.1* *hsp-16.48*	*lacZ* (plus SV40 NLS)/ translational	Nuclei of cells. Heat-shock, arsenite (AsO_2^-): generalized; other stresses: differential expression e.g. gut (Hg^{2+}, Cd^{2+}), anterior pharynx (Cu^{2+}, captan, captafol), pharynx (Cd^{2+}, folpet), pharynx and nerve ring (Hg^{2+}), posterior pharynx (Pb^{2+}), hypodermal cells (Zn^{2+})[a,b]	Heat, Cd^{2+}, Cu^{2+}, Hg^{2+}, Zn^{2+}, Pb^{2+}, AsO_2^-, fungicides (captan, captafol, folpet, mancozeb) herbicide (paraquat);[a,b,c] nematicide (pyrantel), insecticide (Deltamethrin);[c] metal-polluted river water;[d] metal mixtures (Cd^{2+}, Cu^{2+}, and Cd^{2+} + Zn^{2+}).[e] Non-inducers: phthalimide, tetrahydrophthalimide, *N*-phthaloyl-L-glutamic acid.[b] Poor inducers: fungicides (maneb, carbendazim), herbicide (Carbetamex), insecticide (dimethoate);[c] Mn^{2+}[f]	Aqueous: distilled water;[a,d] buffered K medium;[b] K medium;[c,f] extracted soil water;[c,e] polluted river water[d] Soil (Lufa 2.2)[e]
PC161	*hsp-16.1*	*lacZ::gfp* (plus SV40 NLS)/translational	Heat-shock: generalized; Cd^{2+}: pharynx[g]	Heat (30°C), Cd^{2+}[g]	Aqueous: distilled water[g]
CB4027	*hsp70*	*lacZ*/transcriptional	Pharynx to general expression depending on stress	Heat (32°C), Cd^{2+}, Hg^{2+}, Zn^{2+}, Ag^+, Sn^{2+}, Mn^{2+}, tributyltin (TBT), lindane;[h] thiabendazole, ivermectin;[c] metal-polluted river water;[d] Cd^{2+} + Ca^{2+} and Cd^{2+} + Zn^{2+}[i]	Aqueous: K medium;[h,i] distilled water, polluted river water[d]
	hsp-16.2 *hsp-16.48*	*gfp*/translational	Generalized	Heat, EtOH, NaN_3, Cd^{2+};[j,k] weak induction by di(2-ethylhexyl)phthalate (DEHP).[l] Non-inducers:[k] Pb^{2+}, $Cr_2O_7^{2-}$, AsO_2^-	Aqueous: M9;[j] K medium;[k] DEHP-contaminated soils[l]
KC136[m]	*hsp-16.2*	*gfp*/transcriptional	Pharynx and intestine	Cd^{2+}, Hg^{2+}, metal mixtures (Cd^{2+}+ Cu^{2+}, Cd^{2+}+ Cu^{2+}+ $Cr_2O_7^{2-}$, Cd^{2+}+ Cu^{2+}+ Hg^{2+}). Non-inducer: Zn^{2+}	Aqueous: K medium
KC334[n]	*hsp-16.2*	*gfp*/transcriptional	Pharynx and intestine (Cd^{2+}); pharynx (extracted field sample)	Cd^{2+}; field sample extracted from river sediments, diluted with K medium	Aqueous: K medium
CL2070[o] CL2071[o]	*hsp-16.2*	*gfp*/transcriptional	Pharynx	Heat (35°C), juglone, paraquat, plumbagin	Nematode growth medium (NGM) plates; aqueous: liquid survival medium

[a]Stringham and Candido (1994); [b]Jones *et al.* (1996); [c]Guven *et al.* (1999); [d]Mutwakil *et al.* (1997); [e]Power and de Pomerai (1999); [f]Dennis *et al.* (1997); [g]David *et al.* (2003); [h]Guven *et al.* (1994); [i]Guven *et al.* (1995); [j]Hong *et al.* (2004); [k]Roh *et al.* (2006); [l]Roh *et al.* (2007); [m]Chu and Chow (2002); [n]strain genetic background: *daf-16(m26); unc-75(e950)*, Chu *et al.* (2005); [o]Link *et al.* (1999).

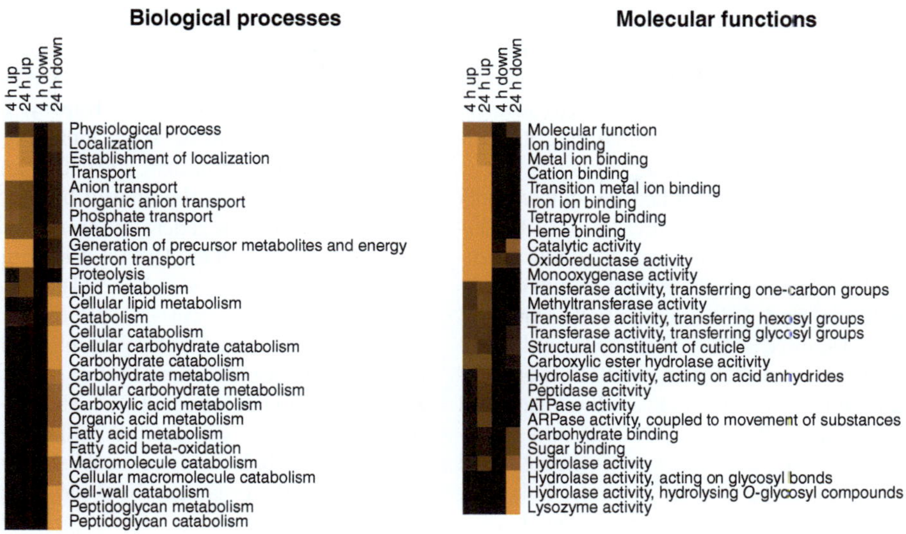

Fig. 14.2. Biological processes and molecular functions enriched with cadmium-responsive genes. Only gene ontology (GO) terms with $P < 0.05$ and ≥4-fold changed genes in at least one of four conditions (up- or downregulated after 4 or 24 h of cadmium exposure) are displayed. Brightness (orange) translates into significance and enrichment of the pathway. (From Cui *et al.*, 2007. CC BY 2.0.)

transporters, *pgp-1*, *pgp-8*, *pgp-9* and *mrp-3*, was induced by cadmium, and *pmp-5* was repressed. Moreover, proteolysis was enriched significantly, possibly leading to an accumulation of damaged proteins. Pathways that were overpopulated with downregulated genes included fatty acid metabolism, cellular lipid metabolism and cell-wall catabolism, indicating that multiple cellular functions may be disrupted by cadmium toxicity. Of the 53 downregulated genes, several have documented RNAi phenotypes (http://www.wormbase.org, accessed 23 August 2025), including embryonic lethality, slow growth, larval growth arrest and sterility. To study the biological role of upregulated genes, their expression was inhibited when *mtl-2* expression was silenced by RNAi. In the presence of cadmium, 50 of the 92 genes tested resulted in reduced growth. Although several of the genes have previously been reported to drive cadmium detoxification, the majority are uncharacterized targets involved in metal toxicity. The discovery of novel genes involved in the resistance to cadmium toxicity provided valuable information in understanding the biological function of the transcriptional changes caused by cadmium.

Metallothioneins (MTs) are small (~ 60 amino acid residues), cysteine-rich proteins that are ubiquitous among eukaryote organisms (Kagi, 1991). They complex metals, reducing their bioavailability. MTs are not only induced by physiological concentrations of essential metals such as zinc and copper, but are also strongly induced by non-essential metals, e.g. cadmium and mercury, suggesting that they have a dual role in metal homeostasis and detoxification (Palmiter, 1998). Additionally, MTs respond to heat shock (Freedman *et al.*, 1993).

C. elegans has two metallothionein genes, *mtl-1* and *mtl-2*, encoding polypeptides that are 75 and 63 amino acids long. These two MTs differ not only in length but also their metal-binding capacity, metal affinity and expression. Utilizing single-copy, integrated, transgenic metallothionein transgenes combined with isoform-specific quantitative PCR (qPCR) uncovered that, in the absence of metal overload, *mtl-1* is expressed at very low levels, in contrast *mtl-2* is constitutively expressed at high levels in the gut cells. Upon metal overload (or likely an oxidative stress challenge) both MTs are induced in the gut. This has suggested that *mtl-1* is involved in the detoxification of non-essential metals and *mtl-2* maintains the homeostatic control of essential metals but also contributes to detoxification (Essig *et al.*, 2024). The importance of *C. elegans* metallothionein genes for the protection against cadmium has been demonstrated experimentally: reduction in brood size and lifespan caused by cadmium was magnified when the *mtl-1* or *mtl-2* genes were knocked down by RNAi or when these genes were deleted (Swain *et al.*, 2004; Hughes and Sturzenbaum, 2007; Almutairi *et al.*, 2024; Essig *et al.*, 2024).

A recent study identified 26 differentially expressed circular RNAs (DEcircRNAs), 143 differentially expressed long non-coding RNAs (DElncRNAs), 69 differentially expressed microRNAs (DEmiRNAs) and 6209 differentially expressed mRNAs (DEmRNAs) in cadmium-exposed *C. elegans*, where the circRNAs and lncRNAs were suggested to influence reproductive, neurological and ageing processes (Qu *et al.*, 2022). Besides metallothioneins (*mtl-1* and *mtl-2*), an uncharacterized hyperresponsive gene was recently linked to stage-specific detoxification against metal stress and/or possibly oxidative stress (Almutairi *et al.*, 2024). These findings highlight that stage-specific effects are of paramount importance when assessing the impact of environmental toxicants.

C. elegans has been reported as a highly sensitive organism used in ecotoxicological studies, and in particular an effective bioindicator of metal pollution, as well as in the nano-remediation of metal-polluted soils. Some of the strains that have been used in this regard are shown in Table 14.4. Interestingly, the percentage of differentially expressed genes of *C. elegans* exposed to metal-polluted soil subjected to a nano-remediation treatment for 120 days decreased with exposure time (Fajardo *et al.*, 2020). Specifically, *clec*-genes involved in innate immune response, as well as *asp-*, *cpr-* and *nas*-genes involved in proteolysis, metal-responsive genes and the oxidative-stress-related genes (*gst*) seem to play key roles (Fajardo *et al.*, 2020). Genome sequencing and genome mapping technologies have allowed the study of genes in *C. elegans* involved in metal detoxification and their response to toxic environments (Caito *et al.*, 2012).

14.4.3 Biosensors of energy balance perturbation

An alternative approach to biomonitoring relies on the assessment of stress-induced metabolic perturbation (Lagido *et al.*, 2008, 2009, 2015; Gonzalez-Hunt *et al.*, 2021), extended to include a dynamic energy budget modelling approach (Wren *et al.*, 2011). In healthy cells, ATP consumption and generation are tightly coupled, and cells strive to maintain stable ratios of ATP:ADP and of ATP:AMP (Hardie and Hawley, 2001). However, many forms of stress, including heat, metabolic poisoning, anoxia, starvation and oxidative stress, perturb the energy balance in a variety of organisms, generally lowering cellular ATP levels (Findley *et al.*, 1983; Gasbarrini *et al.*, 1992; Houthoofd *et al.*, 2002; Tiwari *et al.*, 2002; Apfeld *et al.*, 2004). This triggers the activation of compensatory mechanisms to restore the energy balance (Corton *et al.*, 1994; Zhang and Haldenwang, 2005). The type and severity of stress experienced will determine whether such balance is re-established and at what cost to the organism. In *C. elegans*, transgenic strains that express firefly luciferase are available and under controlled experimental conditions produce light in proportion to their cellular ATP levels, acting as an *in vivo* sensor (Lagido *et al.*, 2008; Luz *et al.*, 2016).

Transgenic strains reporting on the perturbation of energy have been utilized as rapid quantitative tools to report on metabolic impacts from environmental pollution. McLaggan *et al.* (2012) reported the impacts of sublethal levels of environmental pollutants found in sewage sludge using the firefly-luciferase-expressing PE254 and PE255 strains. A short-term exposure (up to 8 h) of the worms resulted in a concentration-dependent decrease in bioluminescence, suggesting a decrease in ATP levels of the organism and a response to the stress. Interestingly, a longer exposure (22–24 h) of the worms to the pollutants resulted in the recovery of the bioluminescence, possibly reflecting an increase in ATP levels of the organism. The probable explanation for the recovery was attributed to the increased resilience of the worms to pollution. This is not surprising, as *C. elegans* is known to have many detoxification mechanisms, such as short-chain dehydrogenases and CYPs for example, which aid in the buffering of xenobiotic stress (Hartman *et al.*, 2021).

Recent studies have demonstrated that exposure to nanoplastics significantly influences energy-related metabolic pathways in *C. elegans*, particularly affecting the tricarboxylic acid (TCA) cycle and lipid metabolism (Long *et al.*, 2023). The ingestion of plastic particles disrupts the absorption of essential energy sources,

Table 14.4. Transgenic strains that report on stress responses other than the heat-shock response. All of these strains are transcriptional reporters.

Strain	Promoter	Reporter gene	Expression of reporter	Stress	Medium
	mtl-1[a]	*lacZ* (plus SV40 NLS)	Intestinal cells of larvae; markedly reduced in adults. Un-induced controls: constitutive expression in pharynx	Cd^{2+}	Aqueous: S medium
JF2.1[a,b]	*mtl-2*	*lacZ* (plus SV40 NLS)	Nuclei of intestinal cells of post-embryonic stages	Cd^{2+}, Hg^{2+}, Ni^{2+} and Zn^{2+}	Aqueous: S medium;[a] S basal pH 6.0 + cholesterol (exposure to Cd^{2+}, Hg^{2+});[b] K medium (Ni^{2+}, Zn^{2+})[b]
JF9[c]	*cdr-1*	*lacZ* (plus SV40 NLS)	Nuclei of intestinal cells	Cd^{2+} non-inducers: Pb^{2+}, Hg^{2+}, Cu^{2+}, Zn^{2+}, paraquat, juglone, heat shock (33°C)	Aqueous: K medium
	mtl-1[d] *mtl-2*[d]	*gfp*	As described for *mtl-1::lacZ*[a] and *mtl-2::lacZ*[a]	Cd^{2+} poor inducers: Cu^{2+} and Zn^{2+}	Nematode growth medium (NGM) agar with *Escherichia coli*
	CYP35A2[e]	*gfp*	Whole intestine. Un-induced controls: anterior and posterior intestine	PCB52	Aqueous: S complete with *E. coli*
	cyp-35A3[f] *dhs-23*[f] *cyp-14A3*[f]	*gfp*	Intestinal cells (and for *cyp-14A3* also hypodermis and head neurons). Un-induced controls: vestigial	PCB52	NGM agar with *E. coli*
PE39[g]	*let-858*	*luc* (no peroxisome targeting sequence)	Constitutive in most cells	3,5-Dichlorophenol (3.5-DCP), Pb^{2+}, Cu^{2+}, heat. Response: perturbation of energy balance	Aqueous: ddH_2O
	mtl-2[h]	*gfp*	Intestinal cells	Cd^{2+}, Hg^{2+}, Cu^{2+}, Zn^{2+}	K medium

[a]Freedman *et al.* (1993); [b]Cioci *et al.* (2000); [c]Liao *et al.* (2002); [d]Swain *et al.* (2004); [e]Menzel *et al.* (2001); [f]Menzel *et al.* (2007); [g]Lagido *et al.* (2001); [h]Ma *et al.* (2009).

ultimately leading to the dysregulation of metabolic homeostasis. In addition to nanoplastics, exposure to the herbicide atrazine has been linked to metabolic perturbations in *C. elegans*. Notably, atrazine affects metabolic pathways involved in glycolysis, pyrimidine metabolism and glycerophospholipid metabolism, with alterations observed in key metabolites associated with amino acid, lipid and energy metabolism (Yin *et al.*, 2020). Similarly, the antibacterial agent triclosan, commonly found in household and personal care products, has been reported to disrupt TCA cycle intermediates, as well as carbohydrate, amino acid and polyamine metabolism. Furthermore, phenotypic analyses have indicated that triclosan exposure affects locomotion, reproduction and reactive oxygen species (ROS) levels, triggering stress responses in transgenic *C. elegans* strains (Kim *et al.*, 2019).

14.4.4 Biosensors of steroid hormone response

In order to assist in the identification of possible endocrine-disrupting chemicals (EDCs), Custodia *et al.* (2001) studied the global gene expression patterns of *C. elegans* to three natural steroid hormones: testosterone, 17β-oestradiol and progesterone. The nematodes were exposed for 3–4 days to the hormones and all experiments were performed in triplicate using the Stanford microarray facilities and the Stanford Microarray Database (SMD).

The study confirmed that *C. elegans* is responsive to the three vertebrate steroid hormones, but the overall pattern of gene expression differed among the three treatments. While a serial dilution (10 μmol, 100 nmol and 1 nmol) of progesterone did not result in concentration-dependent responses, 17β-oestradiol (10 μmol) and progesterone (100 nmol) induced the most dramatic effects. 17β-Oestradiol and progesterone induced expression of 1496 and 1077 genes, respectively. Moreover, 349 genes were found to be downregulated by 17β-oestradiol, and 939 by progesterone (the cut-off was set to 2.6-fold up-/downregulation). In particular, genes involved in xenobiotic metabolism and responses (e.g. GSTs, CYPs, MT and several HSP70 proteins) were shown to be upregulated. Transcription of several vitellogenin genes was increased by 17β-oestradiol but decreased by progesterone. Although the expression patterns could be partially confirmed by Western blot experiments, the overall results remained preliminary, but at the very least substantiated that *C. elegans* is a useful model for screening EDC.

Besides microarrays studies, RNA-Seq has also been incorporated into ecotoxicological studies of *C. elegans* on the effect of steroid compounds that result from the microbial degradation of bile acids present in agricultural manure (Mortazavi *et al.*, 2008; Mendelski *et al.*, 2019). It was found that these steroids possibly influence the nematode NHR-69 testosterone receptor, which in turn affects worm reproductive processes; these effects were confirmed by behavioural assays. In addition, the study concluded that the steroids affected the expression of lectins or galectin antimicrobials, implicating a negative effect on the innate immune response of *C. elegans* (Mendelski *et al.*, 2019).

Nuclear hormone receptors (NHRs) are involved in the response to xenobiotics. They have been implicated in response to progesterone, oestradiol and cholesterol, with of 25 out of 284 *C. elegans* NHR genes showing altered expression (Novillo *et al.*, 2005). NHR-based biosensors may find an application in the detection of EDCs in the environment, as supported by recent studies. A dual-labelled reporter strain (*cyp-35A2::mCherry; unc-47::gfp*) suggested that the nuclear hormone receptor *nhr-49* may serve as a key regulator in the detoxification of the environmental carcinogen benzo[a]pyrene (BaP) (Chen *et al.*, 2023b). These findings indicate the potential for *nhr-49* to be developed as a biomarker for BaP exposure assessment. Furthermore, RNAi studies targeting nuclear hormone receptor genes (*daf-12*, *nhr-14*, *nhr-47* and *nhr-12*) revealed that they play a critical role in mediating the transgenerational toxicity of nanoplastics at environmentally relevant concentrations (Liu *et al.*, 2024).

14.4.5 Biosensors of ethanol exposure

C. elegans responds to acute ethanol exposure in a similar way to humans, mice and fruit flies (Kwon *et al.*, 2004). Seven independent experiments were performed in which nematodes were exposed to ethanol for different time periods, namely four sets of 6 h exposure, two sets of 15 min exposure, and one set of 30 min exposure. DNA microarray experiments yielded 230 genes, which were deemed to be significantly affected by ethanol. Fifty randomly selected genes were confirmed by Northern blot analysis. While the ethanol response of some of these genes was changed significantly at the early time points (indicative of genes that may be involved in the ethanol-stress response), most changes were observed at 6 h, representing the physiological consequence of the ethanol exposure. The early response genes included many heat-shock protein genes, confirming that the concentration used (7% ethanol) acts as a strong stressor to the nematodes. By analysing the promoter regions of the early response genes, a regulatory element (TCTGCGTCTCT) was identified that was necessary for the expression of a subset of ethanol-response genes. All ethanol-responsive genes were plotted on to gene expression maps (Kim *et al.*, 2001) which highlighted that many downregulated genes were found to be clustered on mountain 8, corresponding to an enrichment of genes expressed in the intestine. The descriptions of the corresponding genes imply that intestinal proteins play a role in the response to ethanol stress and may contribute to a reduced metabolic function in the nematodes. Microarrays have also been used to evaluate the effects of time exposure to ethanol in *C. elegans*. Less than 2 h of exposure to ethanol induced the expression of enzymes related to the metabolism of ethanol and retinol (Sterken *et al.*, 2021).

Lower levels of ethanol (0.017 M) increased longevity of *C. elegans* under controlled conditions (food-deprived, stress-resistant L1 larval stage) (Patananan *et al.*, 2015). RNA-Seq analysis revealed the differential expression of 649 genes, many of them associated with ethanol metabolism and triglycerides oxidation into glucose and acetyl-CoA (Patananan *et al.*, 2015). Of the 230 genes affected by higher ethanol concentrations (Kwon *et al.*, 2004), only 29 were also expressed at lower ethanol levels, five of which were down- or upregulated in similar ways (Patananan *et al.*, 2015). These differences in gene expression suggest that different pathways become active, depending on the developmental stage and the concentration of ethanol nematodes are exposed to.

Although the precise molecular genetic basis of ethanol exposure remained elusive, many of the genes identified had human homologs and their expression profile may serve as putative leads in the context of ethanol exposure or the state of alcohol-related diseases.

14.4.6 Biosensors of humic substances

Besides investigating the impact of xenobiotics on organisms it is also of interest to determine the influence of naturally occurring substances that may chemically stress organisms. Humic materials are complex organic molecules constituting the most abundant source of natural organic matter (NOM) in freshwater and soil ecosystems (Steinberg and Münster, 1985; Thurman, 1985). Over the past years they have been shown to interfere with biological systems, for example via the induction of biotransformation enzymes, the inhibition of photosynthetic oxygen release (in freshwater plants), the production of internal oxidative stress or through the feminization of fish and amphibians (Steinberg, 2003; Steinberg *et al.*, 2006). In *C. elegans*, most humic materials tested were found to stimulate reproduction (Höss *et al.*, 2001). Moreover, it has been shown that humic substances act as powerful chemoattractant in *C. elegans* (Menzel *et al.*, 2005b).

The humic-like substance HS1500 had the strongest effect, marked by an upregulation of 554 genes, compared to 240 in NOM Fuchskuhle. Moreover, 685 genes were found to be downregulated by HS1500, and 350 by NOM Fuchskuhle. Considering that both sources of humic materials were able to attract the nematode, it is remarkable that 18 genes, coding for putative chemoreceptors, were found to be differentially regulated. Furthermore, a limited number of enzymes involved in biotransformation were found to be differentially expressed, namely CYPs, GSTs, UGTs as well as acyl- and glycosyltransferases. Humic materials are likely to produce oxidative stress caused by H_2O_2 and other ROS (Pflugmacher *et al.*, 2006), which in turn may generate oxidative stress in the nematode. Interestingly, one of the most responsive genes induced was F32A5.2, a putative peroxidase. Similarly, the nuclear hormone receptor SDC-1 was found to be significantly overexpressed in response to both humic materials; *sdc-1* encodes a zinc-finger protein of the C_2H_2-type that affects egg laying and vulval development. Humic materials seemingly act as a messenger that impacts on essential bodily functions and may have the potential to act as an environmental signal. This study on humic material, as natural chemical stressors and simultaneous attractants to *C. elegans*, led to a subsequent paper which tried to provide an answer to the question: 'Why does an organism actively seek a stressful environment?' Although the exposed organisms have to spend energy to rid themselves of the chemicals they are exposed to, they also have the potential to expand their lifespan (Steinberg *et al.*, 2007) – the ecological consequence remains obscure.

14.4.7 Biosensors of polychlorinated biphenyl exposure

Polychlorinated biphenyls (PCBs) are ubiquitous organic chemicals that pose a global environmental health problem. Menzel *et al.* (2007) performed whole-genome DNA microarray experiments to elucidate the response of *C. elegans* exposed to PCB52, an *ortho*-substituted 2,2′,5,5′-tetrachlorobiphenyl. Overall, the DNA microarray experiment identified 1158 upregulated and 560 downregulated genes, which significantly responded to PCB52 treatment. The results from the GO screen resulted in clearly arranged Venn diagrams (Fig. 14.3), where the upregulated genes encompassed the processes of oxidative catabolism. In addition, the esterase/lipase/thioesterase class, lipid binding, and chemoreceptors of the serpentine receptor class epsilon (SRE) were modulated. Furthermore, the presence of general toxicity was evidenced by the induction of several small *hsp-20* genes and caspases. The increased oxidoreductase activity suggests that the nematode metabolizes PCB52 and the upregulation of the lipid-modifying enzymes implies that PCB52 disrupts the lipid homeostasis in *C. elegans*. The vast majority of downregulated genes were members of general physiological processes such as cellular metabolism, binding, organelle biogenesis and cell integrity.

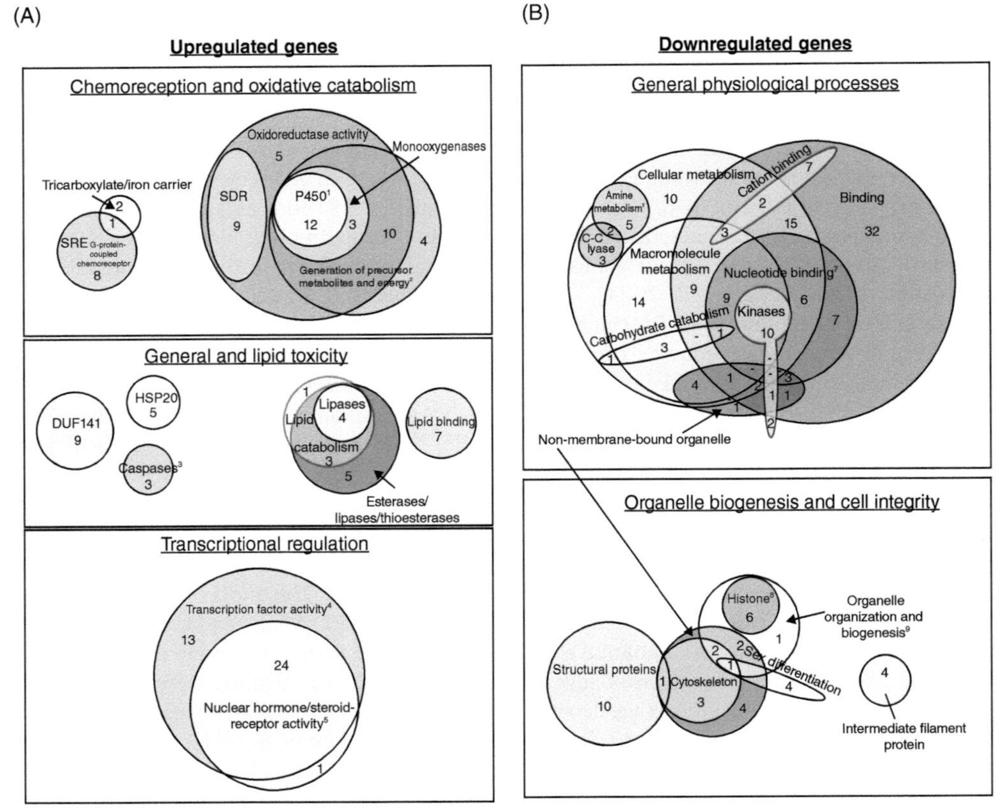

Fig. 14.3. Venn diagrams illustrating the overlap between over-represented gene classes of significantly PCB52-induced (A) and -repressed (B) genes. Included are only gene class members displaying changes in expression of at least twofold ($P < 0.05$). The descriptors of the superscript numbers can be obtained from the original paper. (Reproduced from Menzel *et al.*, 2007 with permission from Elsevier.)

The knockdown of 30 genes by RNAi was used to determine the consequences in reproductive fitness of the presence and absence of PCB52. The screen revealed hypersensitivity to PCB52 upon the RNAi of *dhs-2* and *dhs-23* as well as hyposensitivity to PCB52 when *cyp-14A3* or *cyp-35A1-5* was silenced. Figure 14.4 summarizes these results, including the validation by RT-PCR. Localization experiments by means of promoter–*gfp* fusions indicated that all genes investigated were predominantly expressed in the intestine of *C. elegans*. Four of the five selected *cyp-35A1-5* cytochromes P450 were shown to be involved in fat storage, with PCB52 exposure increasing the fat content in wild-type worms as indicated by Nile Red staining. CYP-14A3 is a potential candidate that may hydroxylate PCB52. In summary, the paper by Menzel *et al.* (2007) provided strong evidence of the molecular mechanisms underlying the toxicity of non-coplanar PCBs and verified that the activation of lipid metabolism and increase in lipid storage are likely factors that drive the toxic effect of PCB52. Moreover, it serves as a positive example of how experiments can succeed in revealing toxic modes of action on the basis of transcriptomic data.

In a further study, which analysed the genome-wide gene expression of *C. elegans* in response to 24 h exposure to 30 μM PCB1254 in the L4 larval stage, Karengera *et al.* (2022) identified 321 differentially expressed genes, many of which related to metabolic processes and catalytic activity. During that study, 17 CYP genes were found to be related to the mammalian CYP2 (*C. elegans cyp-14A, cyp-33B, cyp-33C, cyp-34A, cyp-35A, cyp-35C* subfamilies) and CYP3 (*C. elegans cyp-13A* subfamily). The study suggested that *C. elegans* is able to metabolize PCB1254 via worm orthologs of the human or rodent CYP enzymes (CYB2B, 2C and 3A subfamilies).

14.4.8 Biosensors of phthalate exposure

Di(2-ethylhexyl)phthalate (DEHP) is added as a softener to polyvinyl chloride (PVC), a core component of cables, floor tiles, garden hoses, containers, footwear and clothing. Although DEHP has been shown to be present in the environment, little is known about its ecotoxicological properties. Roh *et al.* (2007) investigated toxicity of DEHP to *C. elegans*, assessing changes in mortality, growth, reproduction and stress-related gene expression. The authors suggested that the exposure to DEHP resulted in the strong induction of the cytochrome P450 *cyp-35A2*, the UDP-glucuronosyltransferase *ugt-21*, the ABC transporters *pgp-1* and the vitellogenin *vit-1*. This occurred concomitantly with the deterioration of the physiological state, which suggests an increase in expression of those genes is likely to be a reaction to toxicity, rather than a homeostatic response.

DEHP was also shown to exhibit a concentration-dependent effect in *C. elegans* exposed for 72 h (Li *et al.*, 2021). At higher concentrations (10 mg l^{-1}), DEHP inhibited the expression of key autophagy-related genes (*atg-18, atg-7, bec-1, lgg-1* and *unc-51*), whereas lower concentrations (0.1 and 1 mg l^{-1}) promoted their expression. These findings suggest a dose-dependent modulation of autophagic pathways in response to DEHP exposure. Chronic exposure to environmentally relevant levels of DEHP has been implicated in the decline of innate immune function, potentially leading to immunosenescence in *C. elegans*. This immunosenescence is characterized by the downregulation of the antimicrobial gene *lys-7* and the increased nuclear translocation of HLH-30, an ortholog of mammalian transcription factor EB (TFEB) (Yen *et al.*, 2023). Furthermore, DEHP exposure upregulates *riok-1*, a homolog of human RIO kinase, which has been associated with DEHP-induced HLH-30/TFEB translocation.

14.4.9 Biosensors of nanoparticle exposure

In recent years, nanomaterials (1–100 nm in diameter) have been identified as a new class of chemical stressor. Nanomaterials are widely used as components in industrial products; for example, gold nanoparticles (NPs) are used in bioimaging, biosensing, facial creams, and for therapeutic purposes.

Hu *et al.* (2018) performed DNA microarray assays to study transcriptional, physiological and behavioural effects in *C. elegans* exposed to gold NPs coated with or without 11-mercaptoundecanoic acid. Ten genes were affected due to the exposure,

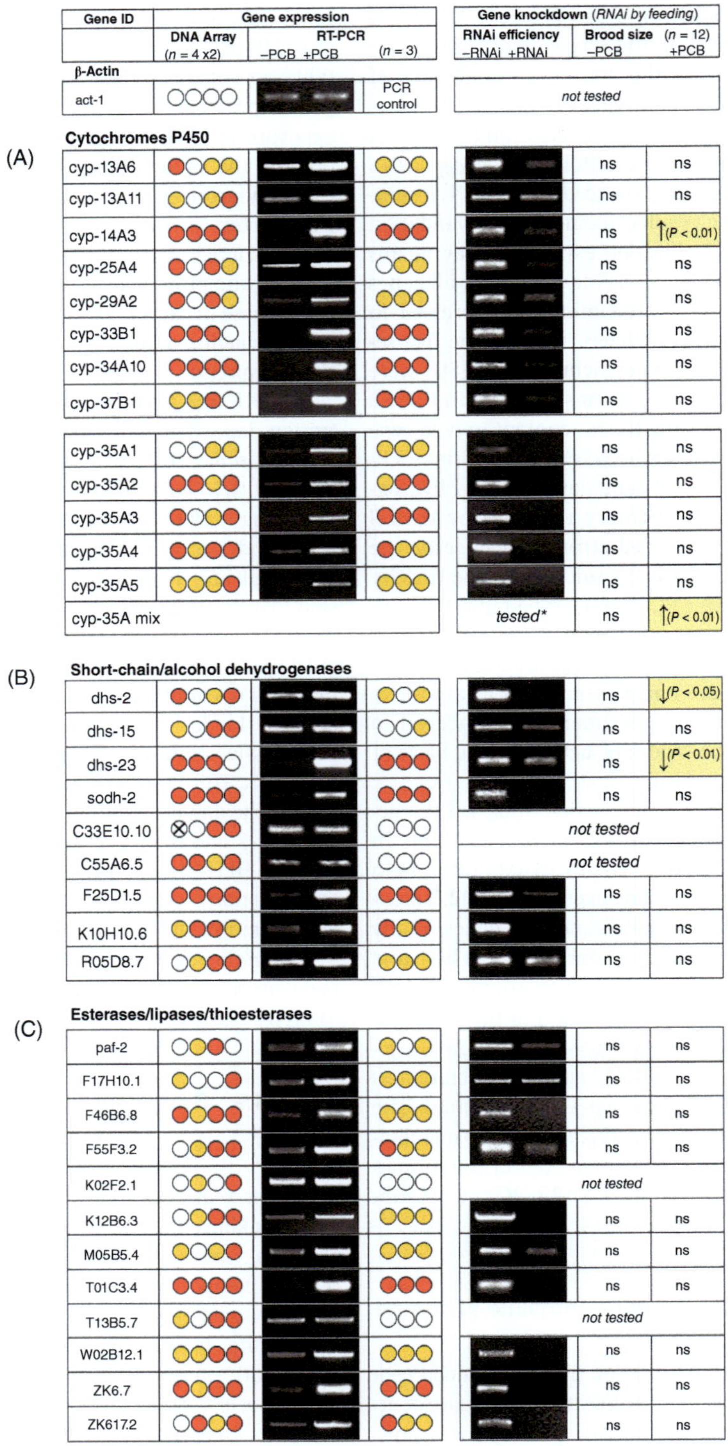

Fig. 14.4. Summarized data of PCB52-induced gene expression, RNA interference (RNAi)-coupled reproduction assay and RNAi efficiency: (A) cytochrome P450 (CYP) genes; (B) short-chain dehydrogenases/reductases (SDR) genes; (C) esterase/lipase/thioesterase genes. ●, > 4-fold increase; ●, 2–4-fold increase; ○, < 2-fold increase; ⊗, no results. RNAi efficiency was determined only in PCB52-treated nematodes (ns = not significantly different). (Reproduced from Menzel *et al.*, 2007 with permission from Elsevier.)

including *Y65B4BR.1* and *clec-174*, both involved in cellular defence processes. Genes linked to lipid storage, namely *ZK593.3* and *fil-1*, also changed at the transcriptional level, as did *cut-3*, involved in body morphogenesis. The suppression of *acdh-1* expression was linked to a reduced lifespan, and the *mtl-1* gene, involved in metal detoxification and nematode growth and fertility, was also affected (Hunt, 2016).

The application of RNA-Seq technology identified transcriptomic hallmarks of exposure to silver and SiO_2 NPs which were correlated with phenotypic traits such as locomotion, reproduction and growth in *C. elegans* (Viau *et al.*, 2020). The authors identified 20 GO biological processes that participate in the regulation of locomotion (transcription factor *che-1* and MiRP K-channel accessory subunit *mps-1*), reproduction (caveolin *cav-1*) and cell growth (cyclic nucleotide-gated cation channel *tax-4* and protein *let-756*).

RNA-Seq also revealed 56 downregulated and 89 upregulated transcripts, when *C. elegans* were challenged with polystyrene NPs 100 µg l^{-1} for 72 h. GO analysis identified pathways that are linked to the synthesis and deposition of collagen, cuticle development, and the moulting cycle (Shang *et al.*, 2021). Analogous results were obtained in *C. elegans* exposed to suldifized silver NPs (Starnes *et al.*, 2016).

The increased use of polystyrene has raised concerns regarding the bioavailability and potential toxicity of polystyrene nanoparticles (PS-NPs) in environmental organisms. For example, *dpy-5* and *rol-6* were shown to be involved in the regulation of induced neurodevelopmental toxicity and oxidative damage in response to PS-NPs exposure (Shang *et al.*, 2021). Epigenetic responses serve as key indicators of exposure to pollutants, with *C. elegans* displaying alterations in histone methylation in response to PS-NPs. The histone methyltransferase MET-2, responsible for H3K9 methylation, has been identified as a protective factor against nanoparticle-induced toxicity. Prolonged exposure to environmentally relevant concentrations of PS-NPs (1–100 µg l^{-1}) results in a decrease in *met-2* expression, increasing the sensitivity towards toxic outcomes (Wang *et al.*, 2021), a notion which was substantiated by RNAi knockdown of *met-2* in intestinal and germline cells. PS-NPs may also be able to trigger transgenerational effects. Exposure to PS-NPs results in induction of germline fibroblast growth factor (FGF) genes *egl-17* and *egl-15*, suggesting that toxicity can be inherited (Hua *et al.*, 2023). This underlines how *C. elegans* can be used to assess long-term ecological risks posed by environmental pollutants. PS-NPs, combined with high glucose levels (10–100 µg l^{-1}), were shown to reduce lifespan and impair movement, regulated by key insulin-signalling components *daf-2*, *age-1*, *akt-1*, *akt-2*, *daf-16* and *sod-3* (Zhuang *et al.*, 2024). The nematode's ability to respond to metabolic stress highlights its value in studying environmental interactions with metabolic disorders. Furthermore, *C. elegans* demonstrates its utility in neurotoxicity assessment through altered gene expression patterns in response to co-exposure to PS-NPs and herbicide metabolites such as 2-hydroxyatrazine. Changes in neurodegeneration-associated genes (*mec-4*, *crt-1*, *itr-1*, *asp-3* and *asp-4*) and stress-response-related neuronal signalling molecules (*mpk-1*, *jnk-1*, *glb-10* and *daf-7*) indicate its suitability for detecting neurotoxic effects of environmental pollutants (Wang *et al.*, 2023).

14.5 *C. elegans* Biosensors and Environmental Samples

Early research on transgenic *C. elegans* as environmental monitors often focused on single toxicants or simple mixtures. Yet environmental samples are often complex mixtures that are poorly characterized in terms of their individual components. This has triggered the interest in studying environmentally relevant exposures, including wastewater, sewage sludge, industrial processes, mining sites, polluted sediments (Menzel *et al.*, 2009a), or the risk assessment of exposure to genetically modified maize (Höss *et al.*, 2011).

Abbas *et al.* (2018) used the transgenic strain carrying *cyp-35A3::gfp* as a tool to understand the environmental impacts of surface water and wastewater from conventional and advanced treatment technologies, such as ozonation and granular activated carbon (GAC). In one of the conventionally treated effluents, the authors reported a significant induction of GFP which was further enhanced after the ozonation treatment, even though the targeted micropollutants reduced in concentration after the treatment. This indicated that unmonitored xenobiotic compounds might have been responsible for the gene expression. Alternatively, the sensor may respond to transformation products resulting from the ozonation procedure.

In England, water samples from the Cannon River basin waters were assessed for using heat-shock-responsive *C. elegans* strains (Mutwakil *et al.*, 1997). The strength of the induction of β-galactosidase was a good indication of the relative toxicity of the samples as inferred from metal analysis and macroinvertebrate data. The least polluted sample induced a clear response even though it contained a mixture of metals at low concentrations that, in isolation, would not elicit a response. This showed a synergistic effect of mixtures of toxicants, also observed in other studies (Dennis *et al.*, 1997; Power and de Pomerai, 1999; Chu and Chow, 2002). Biosensor strains may therefore be more sensitive monitors of toxicity in field conditions than predictions based on single stressors in the laboratory. In China, interstitial water from sediment samples from the metal-polluted rivers Shing Mun and Xihe were tested with *C. elegans* biosensor and wild-type strains (Chu *et al.*, 2005; Kang *et al.*, 2023). Exposure resulted in pharyngeal GFP expression and measurable effects on survival, growth and reproduction, providing a good indication of the toxicity of the river sediments.

Toxicity of contaminated soils revealed a complex picture, with bioavailability of metals depending on the local conditions. Higher concentrations of cadmium were required to elicit a response in soil compared to aqueous media (Guven *et al.*, 1994; Power and de Pomerai, 1999). This is likely due to its binding to organic matter and consequent reduction in bioavailability. Interestingly, cadmium and copper co-exposure increased the response of PC72 worms relative to cadmium alone, possibly because copper displaces cadmium from soil binding sites, resulting in its increased availability in soil pore water. A more complex soil matrix also affected the availability of the fungicide mancozeb, but the fraction entering the soil water compartment was sufficient to induce a mild stress response in PC72 (Guven *et al.*, 1999).

Reaching a comprehensive understanding of all stress responses induced by the components in the samples using nematode tests presents a challenge. This is mainly because using a single or multiple strains may only present a single stress pathway in the sample, representing a one-dimensional toxicological assessment. Considering the complexity of environmental samples, using multiple strains could also fail to identify effects of combined toxicants on gene expression. Mathematical modelling by De Pomerai *et al.* (2010) and empirical studies by Anbalagan *et al.* (2012) demonstrated that certain metal combinations exhibit additive or inhibitory effects on gene induction, highlighting the complexity of stress responses that single-strain assays mail fail to capture. By limiting experiments to a single strain there is a risk that biochemical complexities are not accounted for.

The use of multiple transgenic *C. elegans* to test multiple stress pathways in environmental samples is now being adopted. Tejeda-Benitez *et al.* (2016) applied GFP transgenic strains to evaluate the toxicity profile of Magdalena River sediments in Colombia. The transgenic strains *hsp-6::gfp*, *hsp-16.2::gfp*, *hsp-70::gfp*, *sod-1::gfp*, *sod-4::gfp*, *gpx-6::gfp*, *mtl-2::gfp*, *mtl-1::gfp*, *gst-1::gfp* and *cyp-34A9::gfp* were screened to identify changes in their gene expression. After sediment exposure, the *mtl-2* and *sod-4* genes were the most highly expressed, suggesting together with other toxicity end points that metal contamination was the most pronounced toxicant. Subsequently, cadmium and lead levels were found to be the main toxicants associated with the sediment toxicity of the river. More interestingly, samples from sites with heavy industrial activities, particularly from soil refining and petrochemical processes, had the highest expression of these genes tested; providing an indication that these sites were the most probable sources of pollution to the Magdalena River. Using multiple transgenic strains, as exemplified here, provides a high-throughput, real-time and low-cost means to screen for toxicant effects across multiple stress pathways (Anbalagan *et al.*, 2012). Turner *et al.* (2013) also utilized multiple transgenic *C. elegans* strains to test the mechanisms of toxicity associated with water and sediments from a mountaintop removal/valley fill coal-mining-affected watershed in West Virginia, USA. This concept can be extrapolated to other environmental pollutants, such as pesticides and organic inputs from agricultural soils and mining sites (Anbalagan *et al.*, 2013).

Last but not least, a particularly valuable application of biosensor strains is to interrogate environmental samples to determine whether they activate mechanisms implicated in human disease. Oxidative stress is a major component of disease initiation and progression, including diabetes, atherosclerosis, cancer and ageing. Transgenic *C. elegans* have therefore been utilized as bioindicators to assess environmental contaminants and consequences of oxidative stress. The RB1072 (*sod-2(ok1030) I*) strain has been developed as a sensitive indicator for metal-induced oxidative stress in both aqueous and soil environments (Fajardo *et al.*, 2022). Similarly, the VP596 (*gst-4p::gfp::nls III*) strain has been employed to screen environmentally relevant chemicals and to identify novel oxidative-stress-inducing compounds with potential cellular toxicity (Huynh and Wu, 2022).

The above examples highlight the value of transgenic strains in environmental monitoring and toxicology. The identification of novel stress-responsive genes by intricate transcriptomic studies, as well as RNAi screens, and current genetics approaches, will lead to the construction of novel biosensor strains so that tools for environmental monitoring are continually evolving.

14.6 Perspectives, Challenges and Outlook

The application of DNA microarrays and more recently RNA-Seq are now well-established methodologies to study ecotoxicogenomics in *C. elegans*. As reviewed in previous sections, interest in the use of biosensors and ecotoxicogenomics in real-life scenarios is increasing. More effort in validating transgenic biosensor strains is needed for their use in routine monitoring of the environment. Controlled baseline studies will help define which 'normal' natural environment conditions trigger the biosensor strains. The distinction between response to natural variation or to hazardous conditions may be possible based on two criteria: (i) the magnitude of response in individual worms; and (ii) the proportion of worms that show a response. A high level of sensitivity is an advantage in monitoring, especially for toxicants that have the potential to bioaccumulate along the food chain. Situations in which direct applicability of transgenic strains is envisaged include the monitoring of potential sources of environmental contamination, such as discharges from industrial wastewater, sewage treatment plants and agricultural runoff, and the routine screening of drinking water supplies and of extracted water from soils (Donkin and Dusenbery, 1993).

The development of improved technologies that enable the automation of assays and response quantification could further ensure cost-efficiency during toxicity testing. Machines such as the COPAS (Complex Object Parametric Analyser and Sorter) BIOSORT dispense predefined numbers of *C. elegans*, at specific developmental stages, into a microplate format (Pulak, 2006). Among other measurements, it can determine growth rates and distribution of developmental stages. In addition, it can perform the functions of a fluorometer, thus reporting on the level of reporter gene expression (e.g. GFP). It can also be used in conjunction with a luminometer for luciferase-based assays. Moreover, microfluidic chips allow for miniaturization, enhanced throughput and experiments with living worms (Levine and Lee, 2020).

Many transgenic strains are available to the *C. elegans* community, as GFP fusions are routinely generated to describe the expression pattern of a gene (McKay *et al.*, 2003; http://elegans.bcgsc.bc.ca, accessed 23 August 2025). While detailed information about these transgenic strains is available via WormBase (https://www.wormbase.org, accessed 23 August 2025), their potential as biosensors tends to be underused. As mentioned above, further work is needed to: (i) validate the use of existing strains; but also to (ii) generate new sensors for stress-responsive genetic elements which may serve as novel biomarkers of exposure.

The creation of new transgenic *C. elegans* strains, specifically designed to act as biosensors, will connect different levels of biological organization: not only will transgenic *C. elegans* define whole-organismal responses as representatives of an ecologically important phylum, but also unravel mechanistic details at a cellular and molecular level (Koga *et al.*, 2000; Henderson and Johnson, 2001; Lindblom *et al.*, 2001; An and Blackwell, 2003; An *et al.*, 2005; Essers *et al.*,

2005; Inoue *et al.*, 2005; Oh *et al.*, 2005; Dupuy *et al.*, 2007).

This will, in turn, facilitate the identification and classification of gene responses to drugs and environmental pollutants which will inform/ guide future risk assessment and biomarker research. For biomarkers to be useful, more work needs to focus on the link between molecular genetic data and their contribution to changes in life-cycle end points that are of ecotoxicological importance. Ideally, the integration of 'omic' data with traditional ecotoxicological parameters will identify mechanistically based biomarkers and elucidate mechanistic networks that can be used to establish thresholds of toxicity and predict exposure levels of a contaminant or complex mixture: in short, a predictive quantitative biomarker (Boverhof and Zacharewski, 2006). Adverse effects, however, can rarely be attributed to an individual event. Most responses to a pollutant will involve complex interactions between genes, proteins and metabolites. At present, only ecotoxicogenomics provides the key to determine (simultaneously) the broad molecular status of an organism experiencing toxicological effects. Ecotoxicogenomics is therefore not limited to identifying novel candidate genes as biomarkers but also aims to pinpoint stressor-specific expression signatures and reveal stress-modulated molecular pathways, to contribute to the establishment of environmental safety assessments.

Acknowledgements

We thank Professor Christian Steinberg and Dr Ralph Menzel for their invaluable contribution to the original chapter that was part of the first edition of this book, *Nematodes as Environmental Indicators*.

References

Abbas, A., Valek, L., Schneider, I., Bollmann, A., Knopp, G. *et al.* (2018) Ecotoxicological impacts of surface water and wastewater from conventional and advanced treatment technologies on brood size, larval length, and cytochrome P450 (35A3) expression in *Caenorhabditis elegans*. *Environmental Science and Pollution Research* 25(14), 13868–13880. DOI: 10.1007/s11356-018-1605-2.

Abbass, M., Chen, Y., Arlt, V.M. and Stürzenbaum, S.R. (2021) Benzo[a]pyrene and *Caenorhabditis elegans*: defining the genotoxic potential in an organism lacking the classical CYP1A1 pathway. *Archives of Toxicology* 95(3), 1055–1069.

Akerfelt, M., Morimoto, R.I. and Sistonen, L. (2010) Heat shock factors: integrators of cell stress, development and lifespan. *Nature Reviews Molecular Cell Biology* 11(8), 545–555. DOI: 10.1038/nrm2938.

Allison, D.B., Cui, X., Page, G.P. and Sabripour, M. (2006) Microarray data analysis: from disarray to consolidation and consensus. *Nature Reviews Genetics* 7, 55–65.

Almutairi, N., Khan, N., Harrison-Smith, A., Arlt, V.M. and Stürzenbaum, S.R. (2024) Stage-specific exposure of *Caenorhabditis elegans* to cadmium identifies unique transcriptomic response cascades and an uncharacterized cadmium responsive transcript. *Metallomics* 16(5), mfae016. DOI: 10.1093/mtomcs/mfae016.

An, J.H. and Blackwell, T.K. (2003) SKN-1 links *C. elegans* mesendodermal specification to a conserved oxidative stress response. *Genes and Development* 17, 1882–1893.

An, J.H., Vranas, K., Lucke, M., Inoue, H., Hisamoto, N. *et al.* (2005) Regulation of the *Caenorhabditis elegans* oxidative stress defense protein SKN-1 by glycogen synthase kinase-3. *Proceedings of the National Academy of Sciences USA* 102, 16275–16280.

Anbalagan, C., Lafayette, I., Antoniou-Kourounioti, M., Haque, M., King, J. *et al.* (2012) Transgenic nematodes as biosensors for metal stress in soil pore water samples. *Ecotoxicology* 21(2), 439–455. DOI: 10.1007/s10646-011-0804-0.

Anbalagan, C., Lafayette, I., Antoniou-Kourounioti, M., Gutierrez, C., Martin, J.R. *et al.* (2013) Use of transgenic GFP reporter strains of the nematode *Caenorhabditis elegans* to investigate the patterns of stress responses induced by pesticides and by organic extracts from agricultural soils. *Ecotoxicology* 22(1), 72–85. DOI: 10.1007/s10646-012-1004-2.

Apfeld, J., O'Connor, G., McDonagh, T., DiStefano, P.S. and Curtis, R. (2004) The AMP-activated protein kinase AAK-2 links energy levels and insulin-like signals to lifespan in *C. elegans*. *Genes and Development* 18, 3004–3009.

Benjamini, Y. and Hochberg, Y. (1995) Controlling the false discovery rate: a practical and powerful approach to multiple testing. *Journal of the Royal Statistical Society: Series B (Statistical Methodology)* 57, 289–300.

Boulin, T., Etchberger, J.F., Obert, O. and Hughes, H. (2006) Reporter gene fusions. In: WormBook (ed.) The *C. elegans* Research Community. Wormbook. DOI: 10.1895/worm-book.1.106.1. Available at: http://www.wormbook.org (accessed 12 February 2025).

Boverhof, D.R. and Zacharewski, T.R. (2006) Toxicogenomics in risk assessment: applications and needs. *Toxicological Sciences* 89, 352–360.

Brazma, A. (2009) Minimum information about a microarray experiment (MIAME)- successes, failures, challenges. *The Scientific World Journal* 29(9), 420–423.

Brazma, A., Hingamp, P., Quackenbush, J., Sherlock, G., Spellman, P. *et al.* (2001) Minimum information about a microarray experiment (MIAME) – towards standards for microarray data. *Nature Genetics* 29, 365–371.

Burczynski, M.E., McMillian, M., Ciervo, J., Li, L., Parker, J.B. *et al.* (2000) Toxicogenomics-based discrimination of toxic mechanism in HepG2 human hepatoma cells. *Toxicological Sciences* 58, 399–415.

Caito, S., Fretham, S., Martinez-Finley, E., Chakraborty, S., Avila, D. *et al.* (2012) Genome-wide analyses of metal responsive genes in *Caenorhabditis elegans*. *Frontiers in Genetics* 3, 52.

Calarco, J.A., Taylor, S.R. and Miller, D.M. (2025) Detecting gene expression in *Caenorhabditis elegans*. *Genetics* 229(1), 1–108.

Chapman, R.W. (2001) EcoGenomics – a consilience for comparative immunology? *Developmental and Comparative Immunology* 25, 549–551.

Chen, H., Wang, C., Li, H., Ma, R., Yu, Z. *et al.* (2019) A review of toxicity induced by persistent organic pollutants (POPs) and endocrine-disrupting chemicals (EDCs) in the nematode *Caenorhabditis elegans*. *Journal of Environmental Management* 237, 519–525. DOI: 10.1016/j.jenvman.2019.02.102.

Chen, Y., Arlt, V.M. and Stürzenbaum S.R. (2023a) MosSCi mediated exogenous gene expression is modulated by genomic positioning. *Biotechnology Journal* 18(9), e2300062.

Chen, Y., Abbass, M., Brock, T., Hobbs, G., Ciufo, L.A. *et al.* (2023b) Environmental carcinogen benzo[a]pyrene alters neutral lipid storage via a *cyp-35A2* mediated pathway in *Caenorhabditis elegans*. *Environmental Pollution* 339, 122731. DOI: 10.1016/j.envpol.2023.122731.

Chen, Y., Jiang, Y., Sarvanantharajah, N., Apirakkan, O., Yang, M. *et al.* (2024) Genome-modified *Caenorhabditis elegans* expressing the human cytochrome P450 (CYP1A1 and CYP1A2) pathway: an experimental model for environmental carcinogenesis and pharmacological research. *Environment International* 194, 109187.

Chu, K.W. and Chow, K.L. (2002) Synergistic toxicity of multiple heavy metals is revealed by a biological assay using a nematode and its transgenic derivative. *Aquatic Toxicology* 61, 53–64.

Chu, K.W., Chan, S.K. and Chow, K.L. (2005) Improvement of heavy metal stress and toxicity assays by coupling a transgenic reporter in a mutant nematode strain. *Aquatic Toxicology* 74, 320–332.

Cioci, L.K., Qiu, L. and Freedman, J.H. (2000) Transgenic strains of the nematode *Caenorhabditis elegans* as biomonitors of metal contamination. *Environmental Toxicology and Chemistry* 19, 2122–2129.

Corton, J.M., Gillespie, J.G. and Hardie, D.G. (1994) Role of the AMP-activated protein kinase in the cellular stress response. *Current Biology* 4, 315–324.

Cui, Y., McBride, S.J., Boyd, W.A., Alper, S. and Freedman, J.H. (2007) Toxicogenomic analysis of *Caenorhabditis elegans* reveals novel genes and pathways involved in the resistance to cadmium toxicity. *Genome Biology* 8, R122.

Culetto, E. and Sattelle, D.B. (2000) A role for *Caenorhabditis elegans* in understanding the function and interactions of human disease genes. *Human Molecular Genetics* 9, 869–877.

Custodia, N., Won, S.J., Novillo, A., Wieland, M., Li, C. and Callard, I.P. (2001) *Caenorhabditis elegans* as an environmental monitor using DNA microarray analysis. *Annals of the New York Academy of Sciences* 948, 32–42.

David, H.E., Dawe, A.S., de Pomerai, D.I., Jones, D., Candido, E.P. and Daniells, C. (2003) Construction and evaluation of a transgenic *hsp16-GFP-lacZ Caenorhabditis elegans* strain for environmental monitoring. *Environmental Toxicology and Chemistry* 22, 111–118.

Davis, A.P., Wiegers, T.C., Johnson, R.J., Sciaky, D., Wiegers, J. and Mattingly, C.J. (2023) Comparative Toxicogenomics Database (CTD): update 2023. *Nucleic Acids Research* 51(D1), D1257–D1262. DOI: 10.1093/nar/gkac833.

Dennis, J.L., Mutwakil, M.H.A.Z., Lowe, K.L. and de Pomerai, D.I. (1997) Effects of metal ions in combination with a non-ionic surfactant on stress responses in a transgenic nematode. *Aquatic Toxicology* 40, 37–50.

De Pomerai, D., Anbalagan, C., Lafayette, I., Rajagopalan, D., Loose, M. *et al.* (2010) High-throughput analysis of multiple stress pathways using GFP reporters in *C. elegans*. *WIT Transactions on Ecology and the Environment* 132, 177–187. DOI: 10.2495/ETOX100171.

Donkin, S.G. and Dusenbery, D.B. (1993) A soil toxicity test using the nematode *Caenorhabditis elegans* and an effective method of recovery. *Archives of Environmental Contamination and Toxicology* 25, 145–151.

Dupuy, D., Bertin, N., Hidalgo, C.A., Venkatesan, K., Tu, D. *et al.* (2007) Genome-scale analysis of *in vivo* spatiotemporal promoter activity in *Caenorhabditis elegans. Nature Biotechnology* 25, 663–668.

Eisen, M.B., Spellman, P.T., Brown, P.O. and Botstein, D. (1998) Cluster analysis and display of genome-wide expression patterns. *Proceedings of the National Academy of Sciences USA* 95, 14863–4868.

Essers, M.A., de Vries-Smits, L.M., Barker, N., Polderman, P.E., Burgering, B.M. and Korswagen, H.C. (2005) Functional interaction between beta-catenin and FOXO in oxidative stress signaling. *Science* 308, 1181–1184.

Essig, Y.J., Leszczyszyn, O.I., Almutairi, N., Harrison-Smith, A., Blease, A. *et al.* (2024) Juggling cadmium detoxification and zinc homeostasis: a division of labour between the two *C. elegans* metallothioneins. *Chemosphere* 350, 141021. DOI: 10.1016/j.chemosphere.2023.141021.

Fajardo, C., Martín, M., Nande, M., Botías, P., García-Cantalejo, J. *et al.* (2020) Ecotoxicogenomic analysis of stress-induced on *Caenorhabditis elegans* in heavy metal contaminated soil after nZVI treatment. *Chemosphere* 254, 126909.

Fajardo, C., Martín, C., Garrido, E., Sánchez-Fortún, S., Nande, M. *et al.* (2022) Copper and chromium toxicity is mediated by oxidative stress in *Caenorhabditis elegans*: the use of nanoparticles as an immobilization strategy. *Environmental Toxicology and Pharmacology* 92, 103846. DOI: 10.1016/j.etap.2022.103846.

Feder, M.E. and Walser, J.C. (2005) The biological implications of transcriptomics in elucidating stress and stress responses. *Journal of Evolutionary Biology* 18, 901–910.

Fedorenkova, A., Vonk, J.A., Lenders H.J.R., Ouborg, N.J., Breure, A.M. and Hendriks, A.J. (2010) Ecotoxicogenomics: bridging the gap between genes and populations. *Environmental Science & Technology* 44(11), 4328–4333.

Findley, R.C., Gillies, R.J. and Schulman, R.G. (1983) *In vivo* phosphorus-31 nuclear magnetic resonance reveals lowered ATP during heat shock of *Tetrahymena. Science* 219, 1223–1225.

Freedman, J.H., Slice, L.W., Dixon, D., Fire, A. and Rubin, C.S. (1993) The novel metallothionein genes of *Caenorhabditis elegans*. Structural organization and inducible, cell-specific expression. *Journal of Biological Chemistry* 268, 2554–2564.

Friedland, A.E., Tzur, Y.B., Esvelt, K.M., Colaiácovo, M.P., Church, G.M. and Calarco, J.A. (2013) Heritable genome editing in *C. elegans* via a CRISPR-Cas9 system. *Nature Methods* 10(8), 741–743. DOI: 10.1038/nmeth.2532.

Gasbarrini, A., Borle, A.B., Farghali, H., Bender, C., Francavilla, A. and Van Thiel, D. (1992) Effect of anoxia on intracellular ATP, Na^{+}_{i}, Ca^{2+}_{i}, Mg^{2+}_{i}, and cytotoxicity in rat hepatocytes. *Journal of Biological Chemistry* 267, 6654–6663.

Golden, T.R. and Melov, S. (2004) Microarray analysis of gene expression with age in individual nematodes. *Aging Cell* 3, 111–124.

Gonzalez-Hunt, C.P., Luz, A.L., Ryde, I.T., Turner, E.A., Ilkayeva, O.R. *et al.* (2021) Multiple metabolic changes mediate the response of *Caenorhabditis elegans* to the complex I inhibitor rotenone. *Toxicology* 447, 152630. DOI: 10.1016/j.tox.2020.152630.

Guven, K., Duce, J.A. and de Pomerai, D.I. (1994) Evaluation of a stress-inducible transgenic nematode strain for rapid aquatic toxicity testing. *Aquatic Toxicology* 29, 119–137.

Guven, K., Duce, J.A. and de Pomerai, D.I. (1995) Calcium moderation of cadmium stress explored using a stress-inducible transgenic strain of *Caenorhabditis elegans. Comparative Biochemistry and Physiology* 110C, 61–70.

Guven, K., Power, R.S., Avramides, S., Allender, R. and de Pomerai, D.I. (1999) The toxicity of dithiocarbamate fungicides to soil nematodes, assessed using a stress-inducible transgenic strain of *Caenorhabditis elegans*. *Journal of Biochemical and Molecular Toxicology* 13, 324–333.

Hamadeh, H.K., Bushel, P., Paules, R. and Afshari, C.A. (2001) Discovery in toxicology: mediation by gene expression array technology. *Journal of Biochemical and Molecular Toxicology* 15, 231–242.

Hardie, D.G. and Hawley, S.A. (2001) AMP-activated protein kinase: the energy charge hypothesis revisited. *BioEssays* 23, 1112–1119.

Hartman, J.H., Widmayer, S.J., Bergemann, C.M., King, D.E., Morton, K.S. *et al.* (2021) Xenobiotic metabolism and transport in *Caenorhabditis elegans*. *Journal of Toxicology and Environmental Health, Part B: Critical Reviews* 24(2), 51–94. DOI: 10.1080/10937404.2021.1884921.

Henderson, S.T. and Johnson, T.E. (2001) *daf-16* integrates developmental and environmental inputs to mediate aging in the nematode *Caenorhabditis elegans*. *Current Biology* 11, 1975–1980.

Hong, M., Kwon, J.Y., Shim, J. and Lee, J. (2004) Differential hypoxia response of *hsp-16* genes in the nematode. *Journal of Molecular Biology* 344, 369–381.

Höss, S., Bergtold, M., Haitzer, M., Traunspurger, W. and Steinberg, C.E.W. (2001) Refractory dissolved organic matter can influence the reproduction of *Caenorhabditis elegans* (Nematoda). *Freshwater Biology* 46, 1–10.

Höss, S., Nguyen, H.T., Menzel, R., Pagel-Wieder, S., Miethling-Graf, R. *et al.* (2011) Assessing the risk posed to free-living soil nematodes by a genetically modified maize expressing the insecticidal Cry3Bb1 protein. *Science of The Total Environment* 409(13), 2674–2684.

Houthoofd, K., Braeckman, B.P., Lenaerts, I., Brys, K., De Vreese, A. *et al.* (2002) No reduction of metabolic rate in food restricted *Caenorhabditis elegans*. *Experimental Gerontology* 37, 1359–1369.

Hu, C.C., Wu, G.H., Lai, S.F., Shanmugam, M.M., Hwu, Y. *et al.* (2018) Toxic effects of size-tunable gold nanoparticles on *Caenorhabditis elegans* development and gene regulation. *Scientific Reports* 8, 15245.

Hua, X., Cao, C., Zhang, L. and Wang, D. (2023) Activation of FGF signal in germline mediates transgenerational toxicity of polystyrene nanoparticles at predicted environmental concentrations in *Caenorhabditis elegans*. *Journal of Hazardous Materials* 451, 131174. DOI: 10.1016/j.jhazmat.2023.131174.

Hughes, S. and Sturzenbaum, S.R. (2007) Single and double metallothionein knockout in the nematode *C. elegans* reveals cadmium dependent and independent toxic effects on life history traits. *Environmental Pollution* 145, 395–400.

Hunt, P.R. (2017) The *C. elegans* model in toxicity testing. *Journal of Applied Toxicology* 37(1), 50–59.

Huynh, D. and Wu, C.-W. (2022) High-throughput identification of oxidative stress-inducing environmental chemicals in the *C. elegans* system. *Free Radical Biology and Medicine* 191, 59–65. DOI: 10.1016/j.freeradbiomed.2022.08.032.

Inoue, H., Hisamoto, N., An, J.H., Oliveira, R.P., Nishida, E. *et al.* (2005) The *C. elegans* p38 MAPK pathway regulates nuclear localization of the transcription factor SKN-1 in oxidative stress response. *Genes and Development* 19, 2278–2283.

Jones, D., Stringham, E.G., Babich, S.L. and Candido, E.P. (1996) Transgenic strains of the nematode *C. elegans* in biomonitoring and toxicology: effects of captan and related compounds on the stress response. *Toxicology* 109, 119–127.

Kagi, J.H. (1991) Overview of metallothionein. *Methods in Enzymology* 205, 613–626.

Kang, Y., Zheng, S., Wan, T., Wang, L., Yang, Q. and Zhang, J. (2023) Nematode as a biomonitoring model for evaluating ecological risks of heavy metals in sediments from an urban river. *Ecological Indicators* 147, 110013. DOI: 10.1016/j.ecolind.2023.110013.

Karengera, A., Sterken, M.G., Kammenga, J.E., Riksen, J.A., Dinkla, I.J. and Murk, A.J. (2022) Differential expression of genes in *C. elegans* reveals transcriptional responses to indirect-acting xenobiotic compounds and insensitivity to 2,3,7,8-tetrachlorodibenzodioxin. *Ecotoxicology and Environmental Safety* 233, 113344.

Kim, H.M., Long, N.P., Yoon, S.J., Nguyen, H.T. and Kwon, S.W. (2019) Metabolomics and phenotype assessment reveal cellular toxicity of triclosan in *Caenorhabditis elegans*. *Chemosphere* 236, 124306. DOI: 10.1016/j.chemosphere.2019.07.037.

Kim, S.K., Lund, J., Kiraly, M., Duke, K., Jiang, M. *et al.* (2001) A gene expression map for *Caenorhabditis elegans*. *Science* 293, 2087–2092.

Koga, M., Zwaal, R., Guan, K.L., Avery, L. and Ohshima, Y. (2000) A *Caenorhabditis elegans* MAP kinase kinase, MEK-1, is involved in stress responses. *EMBO Journal* 19, 5148–5156.

Kwon, J.Y., Hong, M., Choi, M.S., Kang, S., Duke, K. *et al.* (2004) Ethanol-response genes and their regulation analyzed by a microarray and comparative genomic approach in the nematode *Caenorhabditis elegans*. *Genomics* 83, 600–614.

Lagido, C. (2009) Transgenic *Caenorhabditis elegans* as biosensors. In: Wilson, M.J. and Kakouli-Duarte, T. (eds) *Nematodes as Environmental Indicators*. CAB International, Wallingford, UK, pp. 225–252.

Lagido, C., Pettitt, J., Porter, A.J., Paton, G.I. and Glover, L.A. (2001) Development and application of bioluminescent *Caenorhabditis elegans* as multicellular eukaryotic biosensors. *FEBS Letters* 493, 36–39.

Lagido, C., Pettitt, J., Flett, A. and Glover, L.A. (2008) Bridging the phenotypic gap: real-time assessment of mitochondrial function and metabolism of the nematode *Caenorhabditis elegans*. *BMC Physiology* 8, 7. DOI: 10.1186/1472-6793-8-7.

Lagido, C., McLaggan, D., Flett, A., Pettitt, J. and Glover, A.L. (2009) Rapid sublethal toxicity assessment using bioluminescent *Caenorhabditis elegans*, a novel whole-animal metabolic biosensor. *Toxicological Sciences* 109(1), 88–95. DOI: 10.1093/toxsci/kfp058.

Lagido, C., McLaggan, D. and Glover, L.A. (2015) A screenable *in vivo* assay for mitochondrial modulators using transgenic bioluminescent *Caenorhabditis elegans*. *Journal of Visualized Experiments* 104, e53083. DOI: 10.3791/53083.

Lee, J.Y. (2023) The principles and applications of high-throughput sequencing technologies. *Development & Reproduction* 27(1), 9.

Levine, E. and Lee, K.S. (2020) Microfluidic approaches for *Caenorhabditis elegans* research. *Animal Cells and Systems* 24(6), 311–320. DOI: 10.1080/19768354.2020.1837951.

Li, J., Qu, M., Wang, M., Yue, Y., Chen, Z. *et al.* (2021) Reproductive toxicity and underlying mechanisms of di(2-ethylhexyl) phthalate in nematode *Caenorhabditis elegans*. *Journal of Environmental Sciences* 105, 1–10. DOI: 10.1016/j.jes.2020.12.016.

Liao, V.H., Dong, J. and Freedman, J.H. (2002) Molecular characterization of a novel, cadmium-inducible gene from the nematode *Caenorhabditis elegans*. A new gene that contributes to the resistance to cadmium toxicity. *Journal of Biological Chemistry* 277, 42049–42059.

Lindblom, T.H., Pierce, G.J. and Sluder, A.E. (2001) A *C. elegans* orphan nuclear receptor contributes to xenobiotic resistance. *Current Biology* 11, 864–868.

Link, C.D., Cypser, J.R., Johnson, C.J. and Johnson, T.E. (1999) Direct observation of stress response in *Caenorhabditis elegans* using a reporter transgene. *Cell Stress and Chaperones* 4, 235–242.

Liu, Z., Wang, Y., Bian, Q. and Wang, D. (2024) Transgenerational response of germline nuclear hormone receptor genes to nanoplastics at predicted environmental doses in *Caenorhabditis elegans*. *Toxics* 12(6), 420. DOI: 10.3390/toxics12060420.

Long, N.P., Kang, J.S. and Kim, H.M. (2023) *Caenorhabditis elegans*: a model organism in the toxicity assessment of environmental pollutants. *Environmental Science and Pollution Research* 30, 39273–39287.

Lund, J., Tedesco, P., Duke, K., Wang, J., Kim, S.K. and Johnson, T.E. (2002) Transcriptional profile of aging in *C. elegans*. *Current Biology* 12, 1566–1573.

Luz, A.L., Lagido, C., Hirschey, M.D. and Meyer, J.N. (2016) *In vivo* determination of mitochondrial function using luciferase-expressing *Caenorhabditis elegans*: contribution of oxidative phosphorylation, glycolysis, and fatty acid oxidation to toxicant-induced dysfunction. *Current Protocols in Toxicology* 69, 25.8.1–25.8.2. DOI: 10.1002/cptx.10.

Ma, H., Glenn, T.C., Jagoe, C.H., Jones, K.L. and Williams, P.L. (2009) A transgenic strain of the nematode *Caenorhabditis elegans* as a biomonitor for heavy metal contamination. *Environmental Toxicology and Chemistry* 28(6), 1311–1318. DOI: 10.1897/08-496.1.

McKay, S.J., Johnsen, R., Khattra, J., Asano, J., Baillie, D.L. *et al.* (2003) Gene expression profiling of cells, tissues, and developmental stages of the nematode *C. elegans*. *Cold Spring Harbor Symposium on Quantitative Biology* 68, 159–169.

McLaggan, D., Amezaga, M.R., Petra, E., Frost, A., Duff, E.I. *et al.* (2012) Impact of sublethal levels of environmental pollutants found in sewage sludge on a novel *Caenorhabditis elegans* model biosensor. *PLoS ONE* 7(10), e46503. DOI: 10.1371/journal.pone.0046503.

Mendelski, M.N., Dölling, R., Feller, F.M., Hoffmann, D., Ramos Fangmeier, L. *et al.* (2019) Steroids originating from bacterial bile acid degradation affect *Caenorhabditis elegans* and indicate potential risks for the fauna of manured soils. *Scientific Reports* 9(1), 11120.

Menzel, R., Bogaert, T. and Achazi, R. (2001) A systematic gene expression screen of *Caenorhabditis elegans* cytochrome P450 genes reveals CYP35 as strongly xenobiotic inducible. *Archives of Biochemistry and Biophysics* 395, 158–168.

Menzel, R., Rodel, M., Kulas, J. and Steinberg, C.E. (2005a) CYP35: xenobiotically induced gene expression in the nematode *Caenorhabditis elegans*. *Archives of Biochemistry and Biophysics* 438, 93–102.

Menzel, R., Stürzenbaum, S., Bärenwaldt, A., Kulas, J. and Steinberg, C.E.W. (2005b) Humic material induces behavioral and global transcriptional responses in the nematode *Caenorhabditis elegans*. *Environmental Science and Technology* 39, 8324–8332.

Menzel, R., Yeo, H.L., Rienau, S., Li, S., Steinberg, C.E. and Sturzenbaum, S.R. (2007) Cytochrome P450s and short-chain dehydrogenases mediate the toxicogenomic response of PCB52 in the nematode *Caenorhabditis elegans*. *Journal of Molecular Biology* 370, 1–13.

Menzel, R., Swain, S.C., Hoess, S., Claus, E., Menzel, S. *et al.* (2009a) Gene expression profiling to characterize sediment toxicity – a pilot study using *Caenorhabditis elegans* whole genome microarrays. *BMC Genomics* 10, 160.

Menzel, R., Steinberg, C.E.W. and Stürzenbaum, S.R. (2009b) *Caenorhabditis elegans* and chemical environmental stressors: from DNA microarrays and data analysis to genomic data of true ecotoxicological relevance. In: Wilson, M.J. and Kakouli-Duarte, T. (eds) *Nematodes as Environmental Indicators*. CAB International, Wallingford, UK, pp. 253–274.

Mortazavi, A., Williams, B.A., McCue, K., Schaeffer, L. and Wold, B. (2008) Mapping and quantifying mammalian transcriptomes by RNA-Seq. *Nature Methods* 5(7), 621–628.

Mutwakil, M.H., Reader, J.P., Holdich, D.M., Smithurst, P.R., Candido, E.P.M. *et al.* (1997) Use of stress-inducible transgenic nematodes as biomarkers of heavy metal pollution in water samples from an English river system. *Archives of Environmental Contamination and Toxicology* 32, 146–153.

Nadon, R. and Shoemaker, J. (2002) Statistical issues with microarrays: processing and analysis. *Trends in Genetics* 18, 265–271.

Novillo, A., Won, S.J., Li, C. and Callard, I.P. (2005) Changes in nuclear receptor and vitellogenin gene expression in response to steroids and heavy metal in *Caenorhabditis elegans*. *Integrative and Comparative Biology* 45, 61–71.

Oberemm, A., Onyon, L. and Gundert-Remy, U. (2005) How can toxicogenomics inform risk assessment? *Toxicology and Applied Pharmacology* 207, S592–S598.

Oh, S.W., Mukhopadhyay, A., Svrzikapa, N., Jiang, F., Davis, R.J. and Tissenbaum, H.A. (2005) JNK regulates lifespan in *Caenorhabditis elegans* by modulating nuclear translocation of forkhead transcription factor/DAF-16. *Proceedings of the National Academy of Sciences USA* 102, 4494–4499.

Palmiter, R.D. (1998) The elusive function of metallothioneins. *Proceedings of the National Academy of Sciences USA* 95, 8428–8430.

Patananan, A.N., Budenholzer, L.M., Eskin, A., Torres, E.R. and Clarke, S.G. (2015) Ethanol-induced differential gene expression and acetyl-CoA metabolism in a longevity model of the nematode *Caenorhabditis elegans*. *Experimental Gerontology* 61, 20–30.

Pennie, W.D., Tugwood, J.D., Oliver, G.J. and Kimber, I. (2000) The principles and practice of toxicogenomics: applications and opportunities. *Toxicological Sciences* 54, 277–283.

Pflugmacher, S., Pietsch, C., Rieger, W. and Steinberg, C.E.W. (2006) Influence of structural moieties of dissolved organic matter on the photosynthetic oxygen production of aquatic plants. *Science of The Total Environment* 357, 169–175.

Power, R.S. and de Pomerai, D.I. (1999) Effect of single and paired metal inputs in soil on a stress-inducible transgenic nematode. *Archives of Environmental Contamination and Toxicology* 37, 503–511.

Pulak, R. (2006) Techniques for analysis, sorting, and dispensing of *C. elegans* on the COPAS™ flow-sorting system. In: Strange, K. (ed.) *C. elegans. Methods in Molecular Biology*, vol. 351. Humana Press, Totowa, New Jersey, pp. 275–286. DOI: 10.1385/1-59745-151-7:275.

Qu, Z., Guo, P., Zheng, S., Yu, Z., Liu, L. *et al.* (2022) Cadmium is detrimental to *Caenorhabditis elegans* via a network involving circRNA, lncRNA and phosphorylated protein (preprint). *bioRxiv*, posted 16 April. DOI: 10.1101/2022.04.16.486470.

Queirós, L., Pereira, J.L., Gonçalves, F.J.M., Pacheco, M., Aschner, M. and Pereira, P. (2019) *Caenorhabditis elegans* as a tool for environmental risk assessment: emerging and promising applications for a 'nobelized worm'. *Critical Reviews in Toxicology* 49(5), 411–429. DOI: 10.1080/10408444.2019.1626801.

Rao, M.S., Van Vleet, T.R., Ciurlionis, R., Buck, W.R., Mittelstadt, S.W. *et al.* (2019) Comparison of RNA-Seq and microarray gene expression platforms for the toxicogenomic evaluation of liver from short-term rat toxicity studies. *Frontiers in Genetics* 9, 636.

Rhiannon, D. (2020) The promise of toxicogenomics for genetic toxicology: past, present and future. *Mutagenesis* 35(2), 153–159.

Rieckher, M. and Tavernarakis, N. (2017) Generation of *Caenorhabditis elegans* transgenic animals by DNA microinjection. *Bio-protocol* 7(19), e2565. DOI: 10.21769/BioProtoc.2565.

Roh, J.Y., Lee, J. and Choi, J. (2006) Assessment of stress-related gene expression in the heavy metal-exposed nematode *Caenorhabditis elegans*: a potential biomarker for metal-induced toxicity monitoring and environmental risk assessment. *Environmental Toxicology and Chemistry* 25, 2946–2956.

Roh, J.Y., Jung, I.H., Lee, J.Y. and Choi, J. (2007) Toxic effects of di(2-ethylhexyl)phthalate on mortality, growth, reproduction and stress-related gene expression in the soil nematode *Caenorhabditis elegans*. *Toxicology* 237, 126–133.

Sands, B., Burnaevskiy, N., Yun, S.R., Crane, M.M., Kaeberlein, M. and Mendenhall, A. (2018) A toolkit for DNA assembly, genome engineering and multicolor imaging for *C. elegans*. *Translational Medicine of Aging* 2(1), 1–10.

Shang, Y., Wang, S., Jin, Y., Xue, W., Zhong, Y. *et al.* (2021) Polystyrene nanoparticles induced neurodevelopmental toxicity in *Caenorhabditis elegans* through regulation of *dpy-5* and *rol-6*. *Ecotoxicology and Environmental Safety* 222, 112523.

Snape, J.R., Maund, S.J., Pickford, D.B. and Hutchinson, T.H. (2004) Ecotoxicogenomics: the challenge of integrating genomics into aquatic and terrestrial ecotoxicology. *Aquatic Toxicology* 67, 143–154.

Soukas, A., Cohen, P., Socci, N.D. and Friedman, J.M. (2000) Leptin-specific patterns of gene expression in white adipose tissue. *Genes and Development* 14, 963–980.

Starnes, D.L., Lichtenberg, S.S., Unrine, J.M., Starnes, C.P., Oostveen, E.K. *et al.* (2016) Distinct transcriptomic responses of *Caenorhabditis elegans* to pristine and sulfidized silver nanoparticles. *Environmental Pollution* 213, 314–321.

Steinberg, C.E.W. (2003) *Ecology of Humic Substances in Freshwaters – Determinants from Geochemistry to Ecological Niches*. Springer, Heidelberg, Germany.

Steinberg, C.E.W. and Münster, U. (1985) Geochemistry and ecological role of humic substances in lakewater. In: Aiken, G.R., McKnight, D.M., Wershaw, R.L. and MacCarthy, P. (eds) *Humic Substances in Soil, Sediment and Water: Geochemistry, Isolation and Characterization*. Wiley, New York, pp. 105–146.

Steinberg, C.E.W., Kamara, S., Peuthert, A., Prokhotskaya, V.Y., Grigutyte, R. *et al.* (2006) From individuals to ecosystems: dissolved humic substances as ecological driving forces in unproductive freshwater lakes. *Freshwater Biology* 51, 1189–1210.

Steinberg, C.E.W., Saul, N., Pietsch, K., Meinelt, T., Rienau, S. and Menzel, R. (2007) Dissolved humic substances facilitate fish life in extreme aquatic environments and have the potential to extend lifespan of *Caenorhabditis elegans*. *Annals of Environmental Science* 1, 81–90.

Sterken, M.G., van Wijk, M.H., Quamme, E.C., Riksen, J.A.G., Carnell, L. *et al.* (2021) Transcriptional analysis of the response of *C. elegans* to ethanol exposure. *Scientific Reports* 11(1), 10993.

Stringham, E.G. and Candido, E.P. (1994) Transgenic *hsp16-lacZ* strains of the soil nematode *Caenorhabditis elegans* as biological monitors of environmental stress. *Environmental Toxicology and Chemistry* 13, 1211–1220.

Swain, S.C., Keusekotten, K., Baumeister, R. and Sturzenbaum, S.R. (2004) *C. elegans* metallothioneins: new insights into the phenotypic effects of cadmium toxicosis. *Journal of Molecular Biology* 341, 951–959.

Tan, L., Wang, S., Wang, Y., He, M. and Liu, D. (2015) Bisphenol A exposure accelerated the aging process in the nematode *Caenorhabditis elegans*. *Toxicology Letters* 235(2), 75–83. DOI: 10.1016/j.toxlet.2015.03.010.

Tatara, C.P., Newman, M.C., McCloskey, J.T. and Williams, P.L. (1997) Predicting relative metal toxicity with ion characteristics: *Caenorhabditis elegans* LC_{50}. *Aquatic Toxicology* 39, 279–290.

Tejeda-Benitez, L., Flegal, R., Odigie, K. and Olivero-Verbel, J. (2016) Pollution by metals and toxicity assessment using *Caenorhabditis elegans* in sediments from the Magdalena River, Colombia. *Environmental Pollution* 212, 238–250. DOI: 10.1016/j.envpol.2016.01.057.

Thurman, E.M. (1985) Aquatic humic substances. In: *Organic Geochemistry of Natural Waters*. Nijhoff-Junk, Dordrecht, The Netherlands, pp. 273–361.

Tiwari, B.S., Belenghi, B. and Levine, A. (2002) Oxidative stress increased respiration and generation of reactive oxygen species, resulting in ATP depletion, opening of mitochondrial permeability transition, and programmed cell death. *Plant Physiology* 128, 1271–1281.

Tsatsakis, A.M., Vassilopoulou, L., Kovatsi, L., Tsitsimpikou, C., Karamanou, M. *et al.* (2018) The dose response principle from philosophy to modern toxicology: the impact of ancient philosophy and medicine in modern toxicology science. *Toxicology Reports* 5, 1107–1113.

Turner, E.A., Kroeger, G.L., Arnold, M.C., Thornton, B.L., Di Giulio, R.T. and Meyer, J.N. (2013) Assessing different mechanisms of toxicity in mountaintop removal/valley fill coal mining-affected watershed samples using *Caenorhabditis elegans*. *PLoS ONE* 8(9), e75329. DOI: 10.1371/journal.pone.0075329.

Viau, C., Haçariz, O., Karimian, F. and Xia, J. (2020) Comprehensive phenotyping and transcriptome profiling to study nanotoxicity in *C. elegans*. *PeerJ* 8, e8684.

Wang, J. and Kim, S. (2003) Global analysis of dauer gene expression in *Caenorhabditis elegans*. *Development* 130, 1621–1634.

Wang, J., Du, H., Nie, Y., Wang, Y., Dai, H. *et al.* (2017) Mitochondria and MAPK cascades modulate endosulfan-induced germline apoptosis in *Caenorhabditis elegans*. *Toxicology Research* 6(4), 412–419. DOI: 10.1039/C7TX00046D.

Wang, S., Zhang, R. and Wang, D. (2021) Induction of protective response to polystyrene nanoparticles associated with methylation regulation in *Caenorhabditis elegans*. *Chemosphere* 271, 129589. DOI: 10.1016/j.chemosphere.2021.129589.

Wang, Y., Yuan, X., Zhou, R., Bu, Y. and Wang, D. (2023) Combinational exposure to hydroxyatrazine increases neurotoxicity of polystyrene nanoparticles on *Caenorhabditis elegans*. *Science of The Total Environment* 880, 163283. DOI: 10.1016/j.scitotenv.2023.163283.

Wang, Z., Gerstein, M. and Snyder, M. (2009) RNA-Seq: a revolutionary tool for transcriptomics. *Nature Reviews Genetics* 10(1), 57–63.

Waters, M., Boorman, G., Bushel, P., Cunningham, M., Irwin, R. *et al.* (2003) Systems toxicology and the Chemical Effects in Biological Systems (CEBS) knowledge base. *EHP Toxicogenomics* 111, 15–28.

Weinstein, J.N., Myers, T.G., O'Connor, P.M., Friend, S.H., Fornace, A.J. Jr. *et al.* (1997) An information-intensive approach to the molecular pharmacology of cancer. *Science* 275, 343–349.

Westerheide, S.D. and Morimoto, R.I. (2005) Heat shock response modulators as therapeutic tools for diseases of protein conformation. *Journal of Biological Chemistry* 280, 33097–33100.

Williams, P.L. and Dusenbery, D.B. (1990) A promising indicator of neurobehavioral toxicity using the nematode *Caenorhabditis elegans* and computer tracking. *Toxicology and Industrial Health* 6, 425–440.

Wren, J.F., Kille, P., Spurgeon, D.J., Swain, S.C., Stürzenbaum, S.R. and Jager, T. (2011) Application of physiologically based modelling and transcriptomics to probe the systems toxicology of aldicarb for *Caenorhabditis elegans* (Maupas 1900). *Ecotoxicology* 20(2), 397–408.

Yen, P.-L., Yang, C.-R., Huang, M.-L., Lin, T.-A. and Liao, V.H.-C. (2023) Chronic exposure to di(2-ethylhexyl) phthalate (DEHP) weakens innate immunity and leads to immunosenescence in *C. elegans*. *Environmental Toxicology and Pharmacology* 98, 104071. DOI: 10.1016/j.etap.2023.104071.

Yin, J., Hong, X., Ma, L., Liu, R. and Bu, Y. (2020) Non-targeted metabolomic profiling of atrazine in *Caenorhabditis elegans* using UHPLC-QE Orbitrap/MS. *Ecotoxicology and Environmental Safety* 206, 111170. DOI: 10.1016/j.ecoenv.2020.111170.

Zhang, S. and Haldenwang, W.G. (2005) Contributions of ATP, GTP, and redox state to nutritional stress activation of the *Bacillus subtilis* σB transcription factor. *Journal of Bacteriology* 187, 7554–7560.

Zhuang, Z., Liu, T., Liu, Z. and Wang, D. (2024) Polystyrene nanoparticles strengthen high glucose toxicity associated with alteration in insulin signaling pathway in *C. elegans*. *Ecotoxicology and Environmental Safety* 272, 116056. DOI: 10.1016/j.ecoenv.2024.116056.

15 Protocols for Nematode Sampling, Extraction, Morphological and Molecular Identification, Index Calculation and Toxicity Testing

Ron G.M. de Goede[1]*, Gerhard du Preez[2], Stefan Geisen[3], Sebastian Höss[4] and Thomais Kakouli-Duarte[5]

[1]Soil Biology Group, Wageningen University and Research, The Netherlands; [2]Unit for Environmental Sciences and Management, North-West University, Potchefstroom, South Africa; [3]Laboratory of Nematology, Wageningen University and Research, The Netherlands; [4]Ecossa, Starnberg, Germany; [5]MENRG, enviroCORE, Department of Applied Science, South East Technological University, Ireland

Abstract

This chapter presents guidelines for choosing the most appropriate methodology for conducting research on communities of free-living nematodes. Each of the chapter's six sections begins with a flow chart that guides the researcher through the most common research options in the context of basic and applied research on nematodes as environmental indicators. The accompanying texts provide assistance in selecting the most appropriate research methods. For details on the research methods, the reader is referred to chapters elsewhere in this volume.

15.1 Nematode Sampling

15.1.1 First things first: research questions and hypotheses

Nematode sampling might seem to be a straightforward activity, but this is somewhat misleading. The flow chart in Fig. 15.1 presents a large number of considerations to be made on site selection, sampling methodology and sample treatment.

Ultimately, all these decisions depend on the goal of the research project. The aim may be to investigate basic ecological research questions *in vivo*, assess environmental quality issues for a third party, or collect soil for *in vitro* research. Formulation of research questions and preferably setting hypotheses are essential. Especially the formulation of hypotheses will be helpful to design the most optimal sampling strategy. Hypotheses should be as specific as possible, avoiding formulations such as 'effects are expected', 'communities will differ', etc. Better, is to specify the effects by naming the parameters related to the effects and by making use of words such as 'will be larger', 'smaller', 'increase', 'decrease' and if possible quantify the effect sizes. Such hypotheses will make very specific what data should be generated to test the hypotheses and

*Email: ron.degoede@gmail.com

 Nematodes as Environmental Indicators: From Theory to Practice, 2nd Edition (T. Kakouli-Duarte *et al.* eds)

DOI: 10.1079/9781800624221.0015

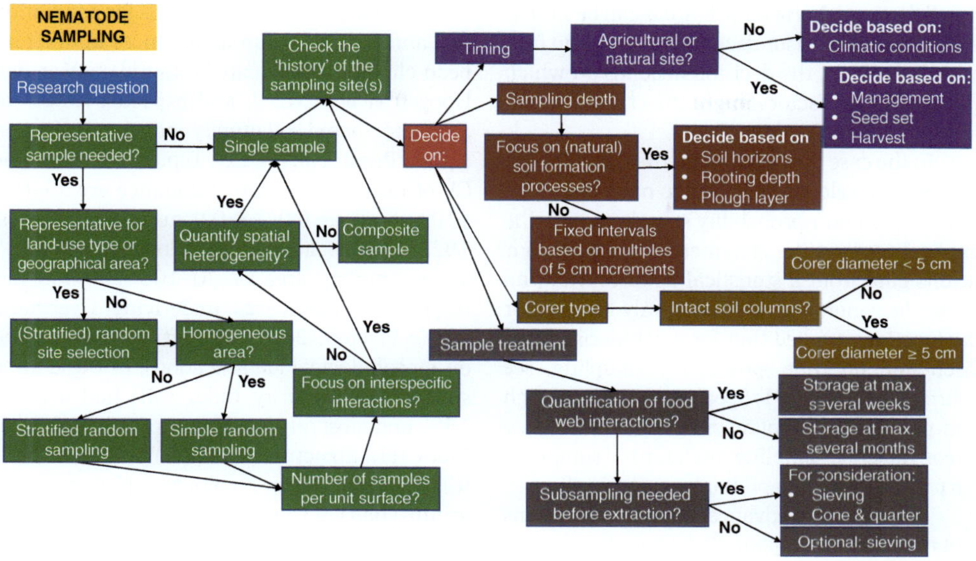

Fig. 15.1. Flow chart structuring the various steps in the nematode sampling process. Blue: defining goals. Green: defining general sampling strategy. Red: decisions on the timing of sample collection (purple), depth of sampling (dark brown), selection of corer (light brown) and procedures after collecting the sample (grey). (Author's own image.)

to answer the research questions. In the case of research activities related to environmental quality assessment (see above), for example, it can be difficult to formulate such specific hypotheses to the research questions. Then, a valid strategy to determine the most appropriate sampling plan can start by creating a general research plan, such as a Gantt chart, to which details are added in a second step, starting at the end of the Gantt chart, i.e. what needs to be reported, and then filling in the research plan from back to front (Knotters, 2008). If you follow the above recommendations, you will increase the chances of your project being a success.

15.1.2 Site and sample location selection

When collecting samples, one needs to know how many samples of what size to take where, when and how, in order to make statements with predetermined accuracy. This involves e.g. site selection, the spatial distribution of samples within a site, composite or single samples, timing, sampling depth, corer type, number of cores and sample treatment (Fig. 15.1). In general, the aim is to collect samples that are representative of the area under investigation, i.e. probability sampling. From a statistical point of view, this requires a random selection of sampling sites from a larger population of potentially relevant sites, such that all potentially suitable sites have an equal chance of being selected. The procedure is known as simple random sampling. In the special case of a field experiment with a set of different treatments, simply all plots are sampled, following the same guidelines as for sampling larger landscape units. Here the only question could be why the experimental field is located where it is. For other cases, before actually carrying out the field research, the site(s) and preferably also the locations within each site should be selected based on random selection procedures (for some practical examples, see Stevens, 1994). It is advisable to designate reserve points that could replace selected points that are found not to meet the predetermined criteria during sampling. In the case of strong heterogeneity within a survey site, simple random sampling is not sufficient. In this case, stratified random sampling can be applied, where the survey site is divided into two or more fairly homogeneous sub-areas. Each sub-area is sampled proportionally to its size. Selection criteria for sub-areas can be similar to the criteria used for deciding on blocks in field experiments, i.e. the decision depends on which (management) factors might affect your nematological 'end points'.

In the case that you loosen the rules around the random selection of sample points, you are dealing with non-probability sampling, with the implication that the statements based on your results can, from a statistical point of view, no longer be generalized and may only be valid for the locations studied (but for a comparison and discussion on (non-)probability sampling, see Schreuder *et al.*, 2001; Speak *et al.*, 2018). Such non-probability sampling procedures are known as convenience sampling, preferential sampling, purposive sampling and haphazard sampling.

To be aware in advance of notable situations that the selected sampling site(s) have faced in the past, it is strongly recommended to use open-source online tools that provide timelines of satellite images or consult landowners.

15.1.3 Number of samples and number of cores per sample

When representative samples of an area are needed, several spatially separated probes are often collected and, if information on the spatial distribution of nematodes is not needed, combined into a composite sample. The number of probes (or composite samples) per site is seldom based on knowledge of the heterogeneity in the spatial distribution of the free-living nematodes and is at best related to the size of the investigated area (e.g. fixed number of probes per hectare).

Formulas exist that can help predict the number of probes needed to estimate, for example, the mean nematode population density in a composite sample with predetermined accuracy. These formulas require information on the coefficient of variation (CV) and/or the aggregation value (*k*, lower values indicate stronger aggregation) of the population (Ekschmitt, 1998; Been and Schomaker, 2022). The aggregation value *k* can be calculated as the mean^2/(sd^2 – mean) (where SD is the standard deviation) and is sample size-specific. This requires that there is some preliminary expectation of the size of the mean and variation.

Estimates of expected means for population size and feeding group abundances in different pedo-climatic zones can be found in Van den Hoogen *et al.* (2020), and expected ranges of nematode-based soil quality indices are given in e.g. Du Preez *et al.* (2022). Typical values for the CV of mean population abundance are 35–60–110% (*k*-values: 8.0, 2.8, 0.9; number of samples: 192, 100, 80; volume of a sample: 42, 44, 251 cm^3) for a sampling area of 30–100–180,000 m^2, respectively (databases underlying Overmars, 1985; Viketoft, 2013; Ikoyi *et al.*, 2021; based on morphological identification). In these three studies, the Maturity Index (MI) had a CV < 15%, and most other indicators (i.e. Enrichment Index (EI), Structure Index (SI), Channel Index (CI), feeding groups, colonizer–persister (*c-p*) groups) had CV values ranging from 30 to 50%. In the case of low scores (i.e. <5%), in general, these indicators had CV values >50%. CV values for individual nematode taxa typically varied between 50 and 100%, with values >100% for rare taxa (i.e. proportion < 1–5%).

Assuming a normal distribution of the sample mean (which is not unreasonable due to the Central Limit Theorem; Quinn and Keough, 2002), the sample size can be roughly estimated based on predefined accuracy standards. For instance, we might require that the probability that the sample mean differs more than 10% from the population mean is less than or equal to 15%. This boils down to a required sample size of at least $(t_{0.95} \times \mathrm{CV}/0.15)^2$, where CV is the coefficient of variation (i.e. the ratio of its standard deviation and its mean) and $t_{0.95}$ is the critical value of a Gaussian distribution (Quinn and Keough, 2002). The formula implies that, given the quality standards proposed above, the number of probes needed for a composite sample to estimate the true population mean in a field with a nematode population with a CV of 0.35 is 15 probes. On the other hand, if not 15 but say 50 probes had been collected, the probability of a deviation greater than 10% would have been 8% instead of 15%.

15.1.4 How and when to collect and process the samples

Finally, decisions should be made on the timing of the sampling, sampling depth, type of sampling device and sample treatment (Fig. 15.1). Timing can be based on climatic conditions, vegetation development or (agricultural) land management, although nematodes can be successfully sampled and extracted from soil year-round. Sampling depth can be related to management factors such as tillage depth or can be tailored to the occurrence of natural soil horizons or the rooting depth of plants. The diameter of the auger should not be e.g. <20 mm, as smaller diameters physically damage part of the nematode population, resulting in an underestimation of nematode numbers (Seinhorst, 1988). In the case of a gouge auger, it is important that the upper part of the sample is not lost during sampling, something that can easily happen in grasslands. Storing samples until nematode extraction should be as short as possible. In the case that it is impossible to extract the entire sample, subsampling is necessary and should preferably be standardized and based on an official procedure, such as the cone and quarter subsampling method. Sieving should be done carefully, as it may damage the nematodes. ISO (2022) can be consulted for suggestions on standardizing a sampling protocol.

15.2 Nematode Extraction

Nematodes can be extracted from soil, sediments and plant material using a multitude of different methods. The global database of terrestrial nematodes initiated by Van den Hoogen *et al.* (2020) indicates that Baermann funnel extraction *sensu lato* is the most widely used method, followed by extraction methods using an elutriator, centrifuge or decanting, and sieving, respectively. Therefore, this section focuses mainly on a discussion of these extraction methodologies.

15.2.1 Which method to select?

Often the selection of an extraction technique will depend on the equipment available in a laboratory and less on the special aims of the study. If there is room for selecting between different extraction methods, selection criteria can be based on sample size, number of samples, extraction efficiency, time available and extraction of immotile nematodes (Fig. 15.2). Irrespective of which method is selected, the extraction

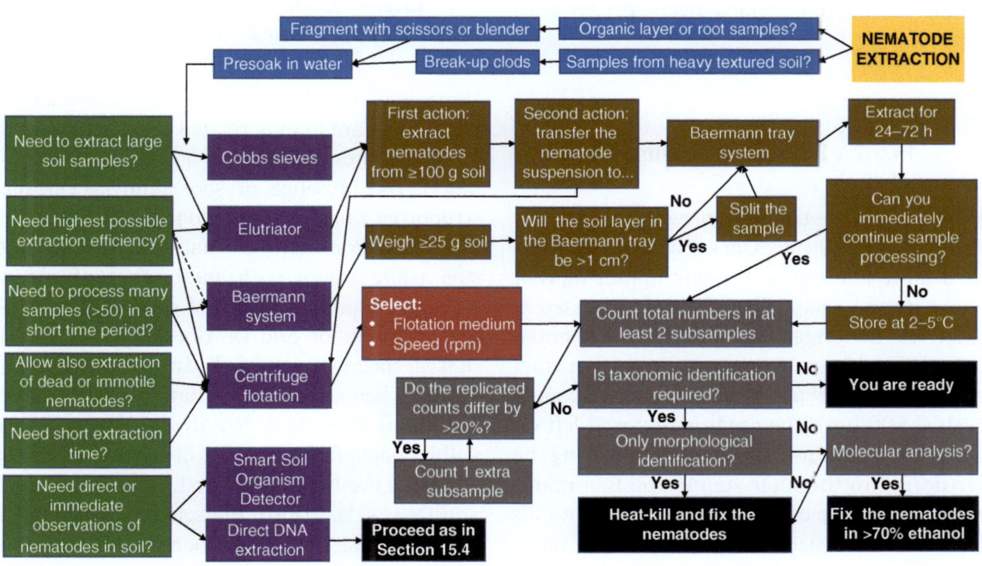

Fig. 15.2. Flow chart structuring the various nematode extraction tools. Dotted lines indicate weak relationships. Blue: sample preparation. Green: information to determine the extraction method. Purple: extraction methods. Brown: elutriator–Baermann extraction methods. Red: centrifugation methods. Grey: enumerating total abundance. Black: towards next steps in the workflow. (Author's own image.)

process will make use of either Baermann tray separation, centrifugation, or direct *in situ* observation. Cobbs sieving and (Oostenbrink or Seinhorst) elutriator extraction can precede Baermann funnel extraction and centrifugation. The advantage of Cobbs sieving and elutriator extraction is that large sample sizes can be used and that they provide a first clean-up of the samples by separating relatively heavy mineral soil particles from nematodes and lighter soil components such as clay particles and organic matter. This cleaner nematode suspension should be further processed by centrifugation or can be transferred to a Baermann extraction system, i.e. a wire basket with a paper filter. As the extraction efficiency of the Baermann extraction system is negatively affected by the thickness of the soil or debris layer on top of the paper filter, it strongly benefits from the clean-up by Cobbs sieving or elutriation. Care should always be taken to ensure that the soil or debris layer thickness on the paper filter is less than 10 mm, but preferably no thicker than 1–2 mm (Whitehead and Hemming, 1965; Cesarz *et al.*, 2019). If no use can be made of Cobbs sieving or elutriator extraction, the diameter of the Baermann wire baskets and the size of the soil sample should be matched, and if necessary the sample should be divided over several wire baskets.

15.2.2 Sample pretreatment

Depending on the soil type, pretreatment of the sample might be necessary. Samples containing a large amount of organic matter or root material can be soaked in water or processed in a blender (no longer than 10 s) or fragmented with scissors. In extraction methods that start with sieving the fresh sample over a coarse mesh (≥800 µm), the organic material left on this sieve can directly or after grinding be transferred to the filter step in the Baermann extraction. Samples from clayey or loamy soils should be crumbled before entering the Baermann extraction and also soaked in water before entering the Cobbs sieving or elutriator extraction. In general, extraction from sandy soils is relatively easy and requires no special pretreatment.

15.2.3 Extraction time

The time needed to obtain the final clean nematode suspension strongly depends on the extraction method selected. When using centrifugation, the final nematode suspension can be available within 1 h. Machines such as the Smart Soil Organism Detector can process samples even faster and produce total counts, morphometric information, and even provide taxonomic information (Filgueiras *et al.*, 2023). They do not physically separate nematodes from soil but distinguish nematodes from dirt based on high-resolution imaging and multispectral sensing. In contrast, the total extraction time for methods that make use of a paper filter step will be >24 h. Extraction time can vary between 24 and 72 h. Longer extraction times allow for reproduction and the hatching of eggs and the migration of endoparasitic nematodes from roots. If it is decided to extend the extraction time to >72 h, it is recommended to collect and further process the extracted nematodes by time cohort to be able to distinguish the active soil nematode community and the newly produced offspring.

15.2.4 Losses during extraction and extraction efficiency

Losses of nematodes during the various extraction procedures are mainly related to morphological characteristics such as body length, mass, annulation, the presence of setae and specific lip structures, tail glands, and to motility.

Extraction methods that include a sieving step using sieves with mesh sizes of about ≥45 µm are prone to losses of small-sized nematodes. Verschoor and De Goede (2000) found that all specimens with a body length ≤100 µm are lost when using a 45 µm sieve, and 50% of the specimens with a body length of 200 µm. Losses will be negligible when the nematode length is ≥10× the mesh size (Seinhorst, 1955; Verschoor and De Goede, 2000). Ptatscheck *et al.* (2020) found for freshwater nematodes that 23% of the organisms were not retained by a 41 µm sieve, and still 10% passed the 20 µm sieve, with significant consequences on the results for species number, age structure and evenness. They propose to use a maximum mesh size of 20 µm.

Annulation, setae, lip structures, tail glands and motility are serious constraints during the extraction over a paper filter. Motility can hamper the extraction of slow-moving species but also moulting specimens. In combination with Oostenbrink elutriator extraction, the net loss of nematodes on the paper filter was 32–50% (Verschoor and De Goede, 2000), i.e. a situation with a debris layer of maximum 1–2 mm. An extrapolation based on the findings of Cesarz *et al.* (2019) suggests that losses from a soil layer of 10 mm could be as high as 75%.

Finally, care should be taken in the selection of the flotation solution for the centrifuge-flotation extraction as this (e.g. sugar, but not Ludox) can cause damage to the cuticula due to osmotic forces in specific taxonomical groups (Van Bezooijen, 2006).

15.2.5 Standardization

All presented extraction methods have their advantages, disadvantages and complexities. Therefore, these methods should be standardized per study or project to allow comparisons across spatial and temporal scales. Also, it might be worthwhile testing methods (especially if used for the first time) before applying them for research purposes.

15.2.6 Further processing of the extracted nematode suspension

Counting total numbers of nematodes is preferably done within 2 weeks after extraction (T. Sálànki, SBL, WUR, The Netherlands, 2024, personal communication) when they are still alive and easy to detect under the microscope. If possible, count the entire solution or else a random subsample containing at least 100 individuals. If taxonomic identification is necessary, kill and fix the nematodes immediately after counting the total numbers. Heat fixation ensures that the nematodes will stretch, which facilitates identification. The selection of the fixation solution needs to take account of your own safety (toxic compounds) and requirements related to the further processing of the nematodes (microscopic identification and/or molecular analyses).

15.2.7 Detailed information on the extraction of nematodes

Descriptions of equipment and procedures can be found in Van Bezooijen (2006), Hallmann and Viaene (2013), Cesarz *et al.* (2019), Perry *et al.* (2020), Schenk and Traunspurger (2021), ISO (2022) and Filgueiras *et al.* (2023).

15.3 Morphological Nematode Identification

Tools for morphological identification of nematodes are presented in the flow chart of Fig. 15.3.

15.3.1 Nematode killing and fixing

For nematode community analyses it is necessary to identify the specimens to their family or at least to their feeding group (see Section 15.5). Accurate identification of whole nematode communities is possible only after killing and fixing the nematodes. After extraction and counting, for microscopic identification nematodes are usually killed by the addition of 4% (v/v) formaldehyde in combination with a heat shock (if also molecular taxonomic analyses are required ethanol should be used). If stored in airtight vials, they can be stored for months to years.

15.3.2 Slide preparation and microscope selection

There can be at least three situations in which it might not be necessary to transfer the nematodes to slides that can be used for microscopic identification: (i) determination of the basic food web status (see Section 15.5) using a dissection microscope; (ii) taxonomic identification using an inverted compound microscope; and (iii) direct observation in soil (see Section 15.2; Filgueiras *et al.*, 2023). In general, using a dissecting microscope at low magnification (40×) is not recommended, as not all nematodes can be reliably identified to their corresponding feeding category (e.g. distinguishing the fungal-feeding *Filenchus* species from other plant-feeding

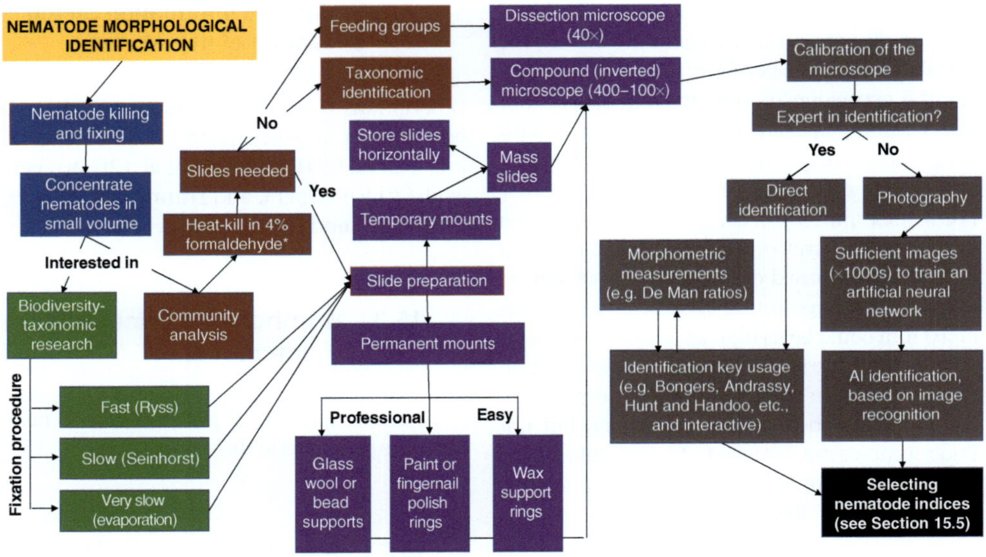

Fig. 15.3. Flow chart structuring the various morphological nematode identification tools. Blue: processing before slide preparation. Green: fixation for taxonomic research. Brown: fixation for community analysis. Purple: slide preparation and handling. Grey: microscopy and identification. Black: next step in workflow. *Can affect human health and should be used in a fume chamber. AI = artificial intelligence. (Author's own image.)

Tylenchidae (see Section 15.5) or assigning dorylaimids into feeding groups). An inverted compound microscope (magnification ≥400×) can be used to examine whole nematode communities, especially if one is familiar with the recognition of taxa occurring in a particular environment. In most other cases, it is necessary to mount the nematodes on slides.

Slides needed for taxonomic research require high standards of fixation and preservation of the mounted nematodes (for a description of methods, see Seinhorst, 1959; Van Bezooijen, 2006; Ryss, 2017), while for analyses of nematode communities more simple methods suffice. Most practical for nematode community analyses is the use of mass slides that usually contain 100–200 nematodes. After concentrating the nematode solution with fixed nematodes to a volume of approximately 1 ml and homogenization, a droplet of suspension can be pipped (use a micropipette) in the middle of a glass slide (e.g. 76 mm × 50 mm) fitted with a paraffin square (e.g. 45 mm × 45 mm). After the placing of a coverslip (e.g. 45 mm × 45 mm), a hot plate (65°C; prevent boiling the nematodes) can be used to melt the paraffin to encapsule the nematode suspension (Van Bezooijen, 2006). The mass slide can be sealed with nail varnish to prevent leakage or drying out and can be used for nematode identification for several weeks to months. It is essential to store mass slides horizontally to prevent nematodes from drifting away.

15.3.3 Identification

Nematode identification can be a time-consuming activity and therefore a personally adjusted chair and desk are of great importance. Also, careful adjustment of the microscope is necessary for optimal image quality. Consult the manual of the microscope to adjust the size and alignment of the diaphragm and condenser, the height of the stage and the diopter of the oculars. Turn up the brightness of the illumination to maximum. When using 76 mm × 50 mm mass slides you need a microscope with a large mount clamp to hold the glass slide.

For nematode community analyses usually the first 100–150 individuals you come across are identified to the appropriate taxonomic level by systemically scrolling through the mass slide and carefully studying each specimen encountered. Good-quality specimens that cannot be identified at the correct taxonomic level must also be included in the count. Work efficiency can be optimized by identifying the nematodes with a maximum magnification of 400× (without using oil immersion) first. The position of specimens that need further study (e.g. identification to species) can be noted down by using the *x*- and *y*-coordinates of the microscope stage. After finishing the first round of identification, a second round can then be used to improve or check the identification of the specimens that were noted down in round one. Now oil immersion and a 1000× magnification can be used.

It is important to identify enough nematodes to obtain sufficient data for adequate determination of taxonomic nematode community structure and nematode-based indices (see Section 15.5). It is general practice to identify at least 100–150 nematodes (Geisen *et al.*, 2018), but it should be realized that the calculation of some indices is based on only a fraction of this total number. For example, the calculation of the MI is based only on the non-plant-feeding nematodes. Simulations ($n = 10$ simulations for nine nematode community samples with true MI values ranging from 1.2 to 3.2) showed that when at least 60–80 specimens were identified, none of the samples deviated >0.15 MI units from the true MI value, which was the case for smaller subsamples (R. de Goede, personal observation). The mean absolute difference from the true MI value was 0.06 MI units, and this appeared to be independent of the total number of nematodes in the extracted solution. Note that indices such as the EI, SI and CI (see Section 15.5) use even a smaller fraction of the total nematode community than the MI. For an acceptable estimate of such indices, we suggest as a rule of thumb to identify at least 100 nematodes to determine the nematode community structure and then continue identifying non-plant feeders until their number exceeds 80 individuals. If it is decided not to include dauer larvae in the index calculations, then they should also not be included when counting additional nematodes.

The identification can be based on identification keys such as published in Bongers (1988) (in Dutch), Andrassy (1984, 2005, 2007) and online keys (e.g. Nemys: https://www.nemys.ugent.be/, accessed 24 August 2025). The website Nemaplex (http://nemaplex.ucdavis.edu/, accessed 24 August 2025) provides a wealth of information on nematode taxonomy, biology and ecology as well as links to web-based resources. See Chapter 17, this volume, for suggestions of courses on nematode taxonomic and ecological education.

For the identification it might be necessary to measure the dimensions of the specimens. These include the total body length and maximum body width, but also the De Man indices of which a, b, b′ c, c′ and V are most relevant. The indices stand for body length/greatest body diameter (a), body length/oesophagus length excluding elongated oesophageal glands (b), body length/oesophagus length including elongated oesophageal glands if present (b′), body length/tail length (c), tail length/tail diameter at anus or cloaca (c′), and distance of the vulva from the anterior/body length × 100 (V), respectively (see the Nemaplex website for a more complete overview).

15.3.4 Machine learning identification

Artificial intelligence (AI), such as machine or deep learning, opens a new era in the wider scientific discipline, and nematology is not excluded. Availability of automated and high-throughput systems for morphological nematode identification and quantification, especially in complex backgrounds as environmental samples can be, is on the ecologist's wish list. These systems rely on trained computer algorithms like, for example, convolutional neural networks (CNNs). The CNN processes multidimensional, grid-like forms of data, such as images and video (LeCun *et al.*, 2015; Goodfellow *et al.*, 2016). Moreover, the accuracy of these types of networks has been increasing steadily (Tan and Le, 2019; Abade *et al.*, 2022; Shabrina *et al.*, 2023).

Akintayo *et al.* (2018) have trained a deep learning architecture, initially developed for rare object identification in clutter-filled images, to identify eggs of the soybean cyst nematode, *Heterodera glycines*. Moreover, a new CNN model, named NemaNet (Abade *et al.*, 2022), was created and used to identify the five main nematode species damaging to soyabean crops (*Helicotylenchus dihystera*, *H. glycines*, *Meloidogyne incognita*, *Pratylenchus brachyurus* and *Rotylenchulus reniformis*) with higher than 95% accuracy. These are developments that point towards the potential of these technologies to identify different nematode species and stages in environmental samples. Hakim *et al.* (2018) combined the capabilities of various image-processing software packages to develop a platform called WorMachine to process simultaneously and in full automation phenotypic features in *Caenorhabditis elegans*. WorMachine is capable of distinguishing between life stages and a range of intermediate phenotypes in *C. elegans* mutants. It also specified the level of masculinization by measuring tails, gonad width, body length and area, and brightness of head and tail. Another model, Xception, was able to differentiate nematodes from the same trophic group and from the same genus, by comparing juvenile sizes and adult diagnostic morphological features, even between individuals of the same genus (Uhlemann *et al.*, 2020; also see Filgueiras *et al.*, 2023). This is considered an achievement based on miniscule differences between species.

These technologies not only generate possibilities for nematode taxonomic identification, but also for further development of morphological trait-based (body dimensions and appendages) analyses of nematode communities. Until now, most community studies have been based on generic nematode traits, such as feeding guilds or *c-p* scale assignment, rather than morphological traits observed and measured on individuals present under the microscope, with the studies of Mulder and Vonk (2011) and Zhao *et al.* (2019) being exceptions. Therefore, the De Man indices (see above) and many other morphological characteristics can be recruited for this purpose.

Machine learning methods have the potential to allow any user to employ them, thus circumventing the current shortage of trained nematode taxonomists. They also allow for standardization, which avoids human bias, reduce processing time, and overall show a great promise for the advancement of environmental nematology. However, before the development of universal and widely accepted and utilized AI nematode identification tools, there are some important aspects that need to be carefully considered. Currently, various CNNs are available (Uhlemann *et al.*, 2020; Abade *et al.*, 2022). Joint training of these models can be efficient in accelerating their perfecting. Standards are needed for image selection and quality, and it should be avoided that a certain level of minor inconsistencies manifest in repeated image generation when training the algorithm. For this, a team of experts can be used, rather than a single user, capturing a large number of images (anything from 500 to several thousand) for each taxon and stage to be identified and annotated, minimizing subjectivity. Each image must be of a different individual nematode and not merely a recapturing image from the same individual, ensuring that natural morphological variations and nuances of the taxon are captured to allow the algorithm to be trained accordingly. This might be a hurdle with taxa which cannot be cultured or isolated in large numbers from the environment.

Starting guides towards nematode image generation can be the works of Eisenback (2012) and the University of California, Riverside website (https://tinyurl.com/ywkea2ja, accessed 24 August 2025) based on the work by De Ley and Bert (2002). However, as technology is developing fast, any effort based on these studies must ensure adjustment to newer versions of microscopes, cameras and software.

15.4 Molecular Identification of Nematodes

15.4.1 The importance of molecular methods

Biodiversity, including that of nematodes, is increasingly studied using molecular methods. There are several reasons for this change in focus. The main advantage of molecular methods is their often ease-of-use, as many are highly standardized and based on simple protocols, such as for DNA extraction and (q)PCR. As a result, studies have a reduced (but not complete) dependence on expertise, as researchers do not need to identify organisms morphologically, but rather apply fairly simple methods and protocols to obtain definitive data such as abundances or taxon lists. Especially in high-income countries, the reduced workload and hands-on time (molecular methods often run automatically for hours or days after the set-up) translate to a reduction of total costs as person-costs are commonly the main part of the budget (Geisen *et al.*, 2018). Finally, some molecular methods including quantitative PCR (qPCR), long-read amplicon sequencing or mito-/metagenomics allow taxon identification often to the species or even strain level (Derycke *et al.*, 2012; Gendron *et al.*, 2023). Therefore, molecular methods are promising tools to study nematode communities. We subsequently list some of the most commonly used molecular tools to study nematodes and list likely future developments. Furthermore, we highlight prerequisites to study soil nematodes with molecular tools and list limitations. We end with recommendations and guides to study nematodes in soil with molecular tools to help researchers in choosing and applying the most suitable methods of choice (Fig. 15.4).

15.4.2 Overview of molecular methods

To better understand the potential usefulness of molecular methods, we here provide an overview of molecular methods that are suitable to study nematodes communities. But first, a definition is needed of what molecular methods are. In principle, molecular methods focus on some molecules in organisms. In microbial ecology, molecular methods often focus on nucleic acids (DNA or RNA) that differentiate those from biochemical methods (Geisen *et al.*, 2019), a definition we apply here as well. The currently most frequently applied molecular methods in microbial ecology and also on nematodes are qPCR and short-read amplicon sequencing (Geisen *et al.*, 2018; Nisa *et al.*, 2022). Both of these approaches, and almost all other molecular tools with the exception of 'omics' tools (e.g. mitogenomics and metagenomics, introduced later), depend on primer-based amplification of barcoding regions that are subsequently studied. For qPCR, primers are designed to target a given taxon (e.g. a family, genus or species, but also broader targets are possible) and amplify a specific sequence of DNA of this taxon. The amount of amplicon produced is subsequently measured and used to provide a measure of abundance for the targeted taxon (Vervoort *et al.*, 2012). Amplicon sequencing, mostly using Illumina platforms, is arguably the most used tool to study microbes and is increasingly being applied to nematode communities (Geisen *et al.*, 2018). For amplicon sequencing also, primers are designed to target specific taxa, such as the group of nematodes. The sequence of the produced amplicons is determined e.g. by Illumina-based techniques and taxonomically assigned with the help of a reference database. The advantage of these methods is that products of any kind of amplified sequence (in the thousands per sample) can be assigned taxonomically. This technique

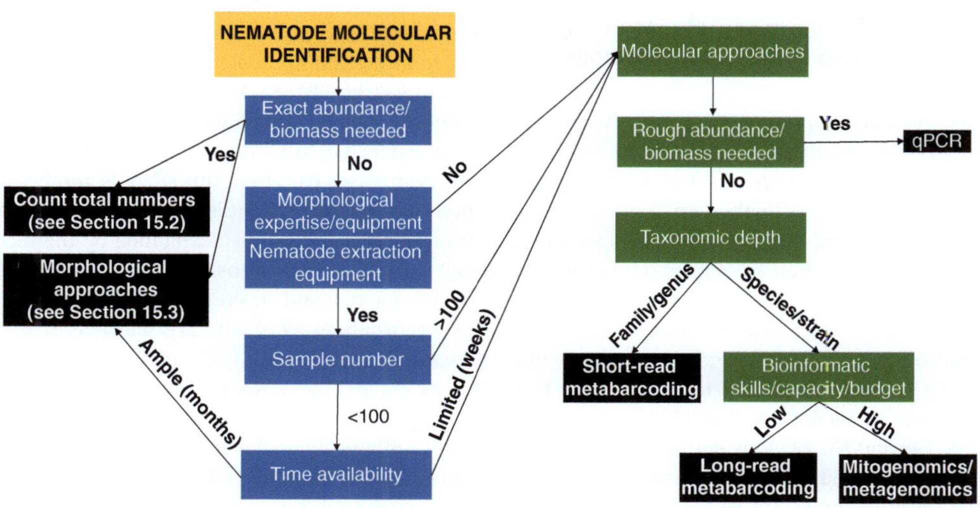

Fig. 15.4. Flow chart structuring the choice to use either morphological or various molecular nematode identification tools. Blue: information to determine which method to choose. Green: information to determine which molecular method to select. Black: final method to apply. (Author's own image.)

allows for the analysis of complex (nematode) communities. Upcoming approaches already applied for microbes are long-read sequencing using PacBio or Nanopore platforms, which have the same principle as Illumina-based methods but can provide higher taxonomic resolution due to the longer nucleotide regions that are sequenced (Tedersoo and Anslan, 2019). The first long-read studies on nematodes are now appearing (Van Himbeeck *et al.*, 2024). Obsolete primer-based approaches are denaturing gradient gel electrophoresis (DGGE) or terminal restriction fragment length polymorphism (T-RFLP) because of their limited taxonomic resolution. In addition to these primer-based approaches are omics tools including metagenomics, metatranscriptomics and mitogenomics. These approaches sequence genomes, transcriptomes or mitochondrial genomes of entire communities without any primers. Especially for extracted nematode communities there are huge possibilities for metagenomics approaches, as complete genomes can be assembled, a common approach when studying bacteria (Bowers *et al.*, 2017). A cost-effective alternative that avoids sequencing the many genomes of nematodes, which are much more complex than those of bacteria, is mitogenomics, in which mitochondria from nematodes are enriched and then fully sequenced (Gendron *et al.*, 2023).

15.4.3 Prerequisites, shortcomings and guides to molecular methods

There are certain prerequisites to apply molecular methods. Sampling needs to be tailored to nematodes (see Section 15.1) and a larger soil sample subjected to further analyses might be needed compared to the <1 g soil extraction standard that is often used for studying microorganisms (Wiesel *et al.*, 2015; but for potential improvements after sample preprocessing, see Oliverio *et al.*, 2018). Currently, the most reliable way to study nematodes is to isolate them first from soils (see Section 15.2) and then extract their DNA. Beyond the obvious equipment to extract DNA or RNA and perform PCRs/qPCRs/sequencing (or outsource some of those), the analyses can be more difficult. qPCR primers need to be designed to be specific to the target taxa and standardization of the assay must take place to accurately estimate taxon abundance or biomass (see Section 15.2). For all sequencing-based approaches, reference databases are crucial to assign resulting sequences to. The most commonly used primer-based approaches target the 18S rRNA gene or more rarely the 28S rRNA gene. While the National Center for Biotechnology Information's (NCBI) GenBank contains an enormous amount of these sequences and also more, not all are correct or correctly classified. Platforms like Silva (for 18S and 28S rRNA gene sequences) and PR2 (for 18S rRNA gene sequences) contain curated sequences that allow distinguishing nematodes and other eukaryotes (Balvočiūtė and Huson, 2017; Sun *et al.*, 2022). With these or custom-made databases, a given sequence can be taxonomically assigned and thereby other information like feeding mode obtained. For upcoming approaches like mitogenomics or metagenomics, the issues detailed above, particularly shortages in reference databases, are even more prevalent and need to be addressed to increase the value of these omics-based approaches. Also, experts on nematodes should be involved as only then can shortcomings and actual nematode ecology reliably be interpreted.

There are several limitations of molecular methods that have to be acknowledged before application and interpretation. Generally, data obtained with molecular methods cannot directly be compared to data obtained with morphological tools, as method-specific biases are inherent to any given approach (Geisen *et al.*, 2018; Griffiths *et al.*, 2018). For example, differences even exist for the same molecular tool applied to DNA extracted from nematodes isolated from soil compared to when the DNA is analysed directly from soil (Griffiths *et al.*, 2018). These discrepancies can only be explained by the fact that nematode extractions, underlying nearly all morphological and molecular nematode community analyses from soil, are selective. Issues associated mainly with molecular methods are, for example, that absolute abundance data is usually not obtained. Reasons underlying differences among molecular methods are manyfold and include primer biases, length variability, and differences in DNA extraction efficiency from nematode extracts (Sun *et al.*, 2022). This fact does not need to be detrimental to a particular study as treatments within a study are comparable and ecological patterns are similarly recovered, but the realism of the observed composition of the nematode community needs to be critically scrutinized. Also, comparisons between studies using different methods become difficult (Geisen *et al.*, 2018), an issue also prevalent for different methods applied in morphological studies.

Some recommendations and guidelines to potentially choose and optimize a given molecular method to study nematode communities are summarized in Fig. 15.4. The first and most essential is to know the method, what information it provides, and what its strengths and weaknesses are. This knowledge helps in choosing the method that matches the goal (or objective/aim) of a study (see Section 15.1). Sometimes know-how and availability/funding may limit the choice to be considered, but it is essential that any available method is not just blindly applied because it may not match the goals. Also importantly, interpretation benefits greatly from expertise in nematology, as only then can information on differences between different nematode taxa be reliably interpreted and potential errors in automated assignments identified.

Regarding sample processing, we propose the following steps.

1. Isolate nematodes from soil as described in Section 15.2 and concentrate to a volume of 1–2 ml.
2. Extract DNA immediately (e.g. commercial protocols such as for soils or cells will work); if not possible, store the concentrated nematode suspension in >70% (v/v) ethanol (replace ethanol with water/buffer before further use).
3. Perform subsequent molecular analyses methods on the DNA samples, depending on the research question and available analytical techniques (Fig. 15.4):

a. qPCR for determining the abundance/biomass (Schenk *et al.*, 2019) of target taxa with primers such as in Vervoort *et al.* (2012). This is especially useful for plant feeders and specific target species, not for complex communities.

b. Short-read (mostly Illumina-based) amplicon sequencing for community profiling that enables detecting rare species with identification, on average, to the genus level. Primers targeting variable regions in the 18S rRNA gene are recommended because of the highest coverage of species in databases (Sun *et al.*, 2022). With increasing references in databases, barcodes in the 28S rRNA gene could be preferred because of the increased taxonomic resolution (Schenk *et al.*, 2019).

c. Long-read (PacBio/Nanopore) amplicon sequencing for community profiling with possibilities for identification to species, sometimes to beyond the species, if the internal transcribed spacer (ITS) region is included in addition to the 18S/28S rRNA gene region (Derycke *et al.*, 2012). This approach is not yet well established and the availability of databases covering non-18S rRNA gene regions is limited.

d. Mitogenomic/metagenomic sequencing provides the deepest taxonomic insights (Gendron *et al.*, 2023), even with functional gene information for metagenomics. However, currently these methods are expensive and annotation to known species is limited by the virtual absence of references in databases.

e. Final analyses on nematode communities, including bioinformatics, depend on the molecular approach adopted. Costs can determine the method as costs increase from a to d. Subsequent data analyses are comparable to morphology-based approaches (see Section 15.5) and depend on the research question (see Section 15.1). However, some data analysis methods including the calculation of nematode indices (Du Preez *et al.*, 2022) are not directly applicable to molecular data and at least not comparable to results obtained with morphological approaches. Such analyses require careful, expert interpretations, as method calibrations are usually still needed (Griffiths *et al.*, 2018).

In summary, molecular tools can be very useful, but a guided choice of methodologies is needed as well as careful interpretation of the data.

15.5 Selecting Appropriate Indices for the Ecological Interpretation of Nematode Community Data

Nematodes are best known as plant parasites that cause substantial damage and economic loss in crop production systems (Mesa-Valle *et al.*, 2020). Nevertheless, since the 1960s and 1970s an increased interest in the ecological function of nematodes has spurred the development of approaches using nematode community data and ecological indices as bioindicators of environmental conditions (Ferris and Bongers, 2009). To find a way through the possibilities of applying nematode data in ecological functioning studies, this section explores a decision flow chart (Fig. 15.5) based on the research question (see Section 15.1), taxonomic abilities (see Sections 15.3 and 15.4), available time and the studied habitat (i.e. soil versus fresh water). Three main branches were identified, and each branch is rooted in a research question related to (i) community heterogeneity, (ii) basic food web status and (iii) ecosystem status, respectively. Community heterogeneity refers to the variation in taxa composition and abundance across a community and helps us understand the complexity, structure and health of an ecosystem. This is because biodiversity is central to the functioning of soil ecosystems and the delivery of services (Wall *et al.*, 2015). Basic food web status, in turn, deals with how well each trophic (feeding) group is represented and serves as a rudimentary assessment of energy flow within the studied system. Healthier soil ecosystems are generally characterized by balanced trophic levels (Ferris *et al.*, 2001). Finally, ecosystem status refers to the health and functioning of a soil ecosystem and its ability to perform key processes that contribute to the delivery of ecosystem functions, including nutrient cycling, carbon and climate regulation, water regulation and purification, and pest and disease regulation (Creamer *et al.*, 2022).

15.5.1 Community heterogeneity

In the decision flow chart, key considerations are listed relating to the taxonomic expertise and time availability for counting and identifying nematode taxa in community heterogeneity assessments. Due to the high variability in ecophysiological characteristics (i.e. behaviour, activity and size) presented by different nematodes species (Ferris and Tuomisto, 2015; Du Preez *et al.*, 2022), it is not advised to evaluate community heterogeneity at family or higher taxonomic levels. After generating nematode community

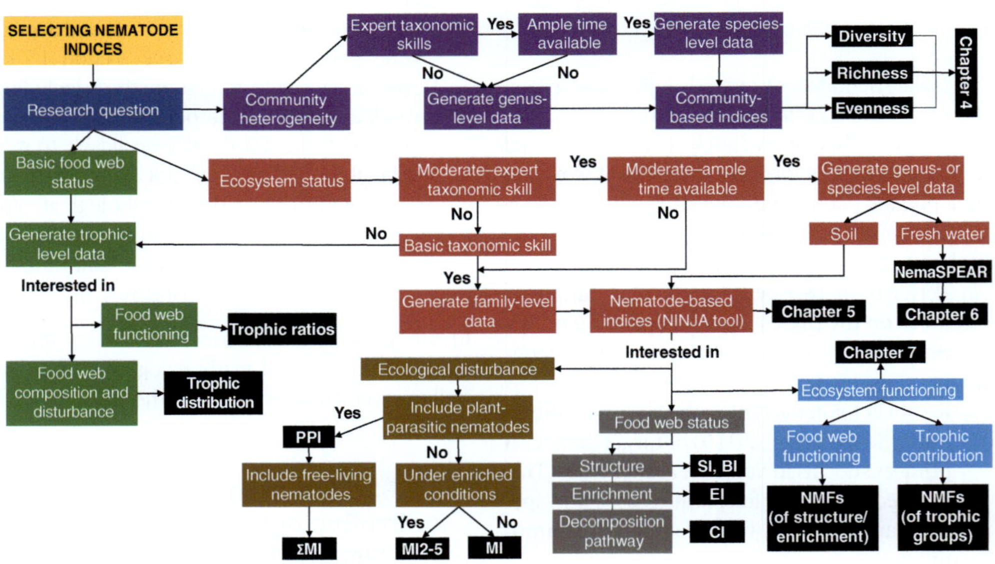

Fig. 15.5. Flow chart structuring the various nematode-based indices of soil ecosystem status. Depending on the research goals (dark blue), information is presented to determine which method to choose for basic food web analysis (green), ecosystem status (red), community heterogeneity (purple), ecological disturbance (brown), food web status (grey) and ecosystem functioning (light blue). Black: final method to apply. PPI = Plant-Parasitic Index; ΣMI = Sigma Maturity Index; MI2-5 = Maturity Index 2-5; MI = Maturity Index; NINJA; Nematode Indicator Joint Analysis; SI = Structure Index; BI = Basal Index; EI = Enrichment Index; CI = Channel Index; NemaSPEAR = Nematode SPEcies At Risk index; NMF = nematode metabolic footprint. (Author's own image.)

data at species or genus level, multiple general community indices can be calculated, evaluating heterogeneity by considering taxa diversity, richness and evenness. Detailed information on the calculation and application of these indices is provided in Chapter 4 of this volume.

15.5.2 Basic food web status

The second branch of the decision flow chart evaluates basic food web status, which requires only nematode abundances per trophic group. Users must select between an interest in food web functioning (e.g. decomposition) and food web composition and disturbance. Food web functioning can be evaluated using abundance-based trophic ratios, for example the fungivores (F) to bacterivores (B) ratio (Sohlenius and Boström, 1984) that later evolved into the Nematode Channel Ratio (NCR = B/(B + F) (Yeates, 2003). Food web composition, in turn, is simply the relative abundance of each trophic group. Even though the NBI toolbox (see next paragraph) is arguably more powerful in the assessment of food web status, recent publications have solely relied on trophic-level data to evaluate trophic structure (Kamath *et al.*, 2022), ecological disturbance (Kou *et al.*, 2020) and energy flow (Wan *et al.*, 2022; Zhu *et al.*, 2023) in soil food webs.

15.5.3 Ecosystem status

The third branch in the decision flow chart evaluates the ecosystem status branch. It is more elaborate, yet still simple to follow. It allows the user to select a subset of indices that form part of a powerful ecological toolset developed to assess soil ecosystem status (Bongers, 1990; Ferris *et al.*, 2001; Ferris, 2010). These indices are collectively referred to as nematode-based indices (NBIs) (Du Preez *et al.*, 2022). Similarly, the NemaSPEAR (Nematode SPEcies At Risk) index was developed and refined to evaluate the chemically induced changes in freshwater habitats using nematode community data (Höss *et al.*, 2017; Schenk *et al.*, 2022). As with the previous branch, the user must first consider their ability to identify nematodes. With ‘moderate–expert taxonomic skill’ and ‘moderate–ample time available’ a user will be able to generate genus-level community data, which is the taxonomic resolution required for the NemaSPEAR index. For soil habitats, users will already be able to calculate NBIs with only ‘basic taxonomic skills’ for family-level nematode identification. Also, family-level data are usually generated with the lack of moderate–ample time for counting and identification. However, it should be noted that genus-level data is highly recommended as some genera do not conform to the life-history classification assigned at family level (Du Preez *et al.*, 2022). An example of this would the fungal-feeding genus *Filenchus*, which belongs to the epidermal-cell and root-hair feeder family Tylenchidae (Okada *et al.*, 2002). Furthermore, for convenience and to minimize human error, users are advised to use the Nematode Indicator Joint Analysis (NINJA) online tool to calculate NBIs (Sieriebriennikov *et al.*, 2014).

The NBI toolset enables users to produce detailed data on ecological disturbance, food web status and ecosystem functioning, all of which fall under the ‘ecosystem status’ branch. Ecological disturbance refers to any physical, chemical or biological interference from natural or anthropogenic origins that disrupts the soil environment and alters the structure and function of the soil ecosystem (Neher *et al.*, 2005). This is reflected in the nematode community and can be assessed and monitored using the Maturity Index family where the choice of index is largely dependent on nutrient enrichment levels and the inclusion of plant-parasitic nematodes (Sánchez-Moreno and Ferris, 2018; Du Preez *et al.*, 2022). The soil food web, in turn, represents the complex interactions among soil organisms. Using NBIs, the food web status can be assessed considering its structure (i.e. connectance between trophic levels), availability of nutrients (i.e. enrichment) and the predominant decomposition pathway of organic material (i.e. bacterial versus fungal dominated) (Ferris *et al.*, 2001). Different NBIs are used to evaluate each of these states. Finally, ecosystem functioning represents the latest addition to the NBI toolbox and evaluates the magnitude of processes fulfilled by the soil ecosystem that contribute to ecosystem functioning and service delivery (Ferris, 2010). Using nematode metabolic footprints (NMFs), the ecosystem functioning of the food web can be evaluated in its entirety, or per trophic group. For more detailed information on the selection, calculation and application of

NBIs, please see Chapter 5 of this volume. Additional useful resources include Sánchez-Moreno and Ferris (2018) and Du Preez *et al.* (2022).

15.6 Toxicity Testing with Nematodes

For nematodes a large variety of standardized and non-standardized methods for evaluating the toxicity of chemicals in various matrices exist (see also Chapters 13 and 14, this volume). In this chapter, we give an overview of currently used methods for laboratory bioassays using nematodes as test organisms. Based on a flow chart (Fig. 15.6), the reader is instructed how to select the appropriate method for specific research or regulatory purposes. So far, the only toxicity test systems that are available as officially acknowledged international standards are based on single-species tests. Besides, a plethora of single-species and multi-species test is available for testing fundamental scientific research questions related to toxicity testing.

15.6.1 Single-species toxicity testing

Currently, for first-tier risk assessment for regulatory purposes (i.e. registration of chemicals), single-species tests are required to determine effect thresholds (no observed effect concentration (NOEC), 10% effect concentration (EC_{10}), 50% effect concentration (EC_{50})) that can be used for deriving risk quotients based on predicted environmental concentrations (e.g. toxicity–exposure ratio, TER). Moreover, such single-species tests are suitable also for studying toxicity mechanisms or toxicokinetics and toxicodynamics (TKTD). While for regulatory purposes standardized test systems are required, scientific questions can be studied using non-standardized tests, if reliable test protocols are available.

Standardized test systems

Standards for toxicity test protocols only exist for the nematode species *Caenorhabditis elegans* (see also Chapter 13, this volume):

- ISO 10872:2020 (ISO, 2020), with the European Standard (EN ISO 10872:2021) and the respective German version (DIN EN ISO 10872:2021; DIN, 2021). See Table 15.1 for details.
- ASTM E2172-22 (ASTM, 2022). See Table 15.1 for details.

Non-standardized test systems

Non-standardized toxicity test systems exist for a variety of nematode species (reviewed in

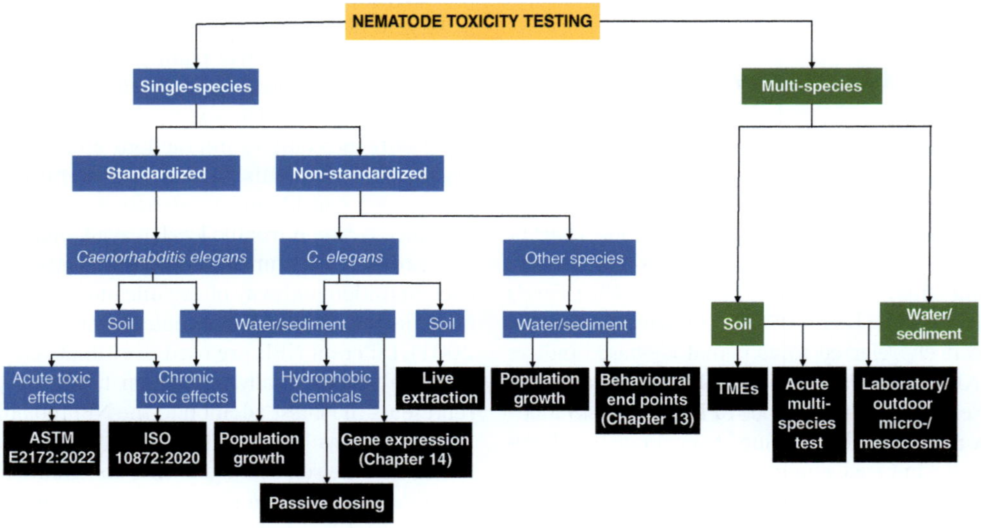

Fig. 15.6. Flow chart structuring the various nematode-based methods in toxicity testing. Information to determine which method to choose for single-species toxicity tests (blue) or multi-species toxicity tests (green). Black: final method to apply. TME = terrestrial model ecosystem. (Author's own image.)

Table 15.1. Standards for toxicity testing with nematodes. (Author's own table.)

Test conditions	ISO 10872:2020	ASTM E2172:2022
Test substrate	Water, sediment, soil	Soil
Exposure	Chronic	Acute
Quantity	0.5 ml water; 0.5 g fresh sediment; 0.5 g dry soil	2.33 g soil
Test containers	12-well (water, sediment) or 24-well (soil) multidishes	Petri dishes (35 mm × 10 mm)
Water content	Sediment: min. 40% (w/w) water content; soil: 80% (w/w) max. water-holding capacity (WHC_{max})	1.5 ml/2.33 g soil
Food	Bacteria: *Escherichia coli* (OP50)	–
Test duration	96 h	24 h
Temperature	20°C	20°C
Nematode separation from soil/sediment	Density flotation and centrifugation technique using a silica suspension (Ludox®)	
Toxicity end point	Growth, fertility, reproduction	Survival
Validity criteria	80–120% recovery of test organisms; ≥80% fertility; ≥50 offspring/test organism; ≤20% males	>80% recovery of test organisms

Haegerbaeumer *et al.*, 2015; Höss *et al.*, 2022a; see also Chapter 13, this volume). Here only a few examples are presented to highlight the broad range of application in nematode-based toxicity testing. The examples focus on population growth, passive dosing and alternative nematode extraction techniques. Examples on gene expression and behavioural end points are presented in Chapters 13 and 14 of this volume, respectively.

POPULATION GROWTH. Studying effects of chemicals on the population growth of single species in more elaborated test systems, using detailed life-cycle parameters and time series instead of single time points, provides additional information on the dynamic interactions of the toxicant with the test organism (TKTD modelling). Several methodological approaches have been published (e.g. Alda Álvarez *et al.*, 2005; Brinke *et al.*, 2013; Mueller *et al.*, 2020). Alda Álavarez *et al.* (2005) tested the effects of cadmium on *C. elegans* using 12-well multidishes and 6 cm Petri dishes filled with agar. Single, synchronized individuals (2–4 h after hatching) were incubated at 15°C and population growth was determined using dynamic energy budget modelling (DEBTox; Jager, 2020). Brinke *et al.* (2013) studied ivermectin effects on *C. elegans* using the hanging drop method (Muschiol *et al.*, 2009). Synchronized, single individuals were exposed to the chemical at 20°C in 12 µl drops of a Gerlite®-based medium (CNGG). Mueller *et al.* (2020) tested the effects of polystyrene beads on *C. elegans*, *Acrobeloides nanus* and *Plectus acuminatus*. They used a stable population distribution (i.e. 20 juveniles and 20 adults) in glass vials containing, similar to Brinke *et al.* (2013), CNGG. In all three studies, nematodes were fed with bacteria (*Escherichia coli*).

PASSIVE DOSING. For substances which are scarcely soluble in water, usually solvents are used for adding the chemical to the test medium. This so-called 'active spiking' method has several drawbacks: (i) with the solvent, another potentially toxic substance is added to the test system; and (ii) by binding to any solid phases in the test system (walls, organic debris, etc.) or by evaporation, hydrophobic and/or volatile chemicals are deprived of the water, so that aimed dissolved concentrations are decreasing during the experiment. Therefore, passive dosing (PD) systems have been developed, where: (i) no solvent is required; (ii) the chemical losses are compensated by partitioning from a reservoir loaded with the hydrophobic chemical (e.g. silicone materials, oils; Mayer *et al.*, 1999; Smith *et al.*, 2010; Trac *et al.*, 2018). This innovative dosing technique has already been applied for *C. elegans* when testing the effects of benzyl butyl phthalate (BBP) (Kwon *et al.*, 2011), different polycyclic

Table 15.2. Model ecosystems for multi-species toxicity testing with nematodes in soils and fresh water. (Author's own table.)

Model ecosystem	Type	Cosm size	Analysed substrate	Tested chemicals	Community end points	References
Soil						
Acute MS toxicity test	Laboratory	No information	50 g soil	Cu	Survival	Bongers *et al.* (2001)
Small-scale microcosms	Laboratory	6 cm^2; 50 cm^3 (50 ml plastic centrifugation tubes)	30g fresh soil	*Bt*-toxins, Zn, fluoranthene, fludioxonil, mancozeb, nano-enabled tebuconazole	Ab, PRC, MI, SI, EI, FT	Haegerbaeumer *et al.* (2019); Höss *et al.* (2014, 2021, 2022b); Eghbalinejad *et al.* (2024)
Microcosms	Laboratory	Equivalents to 3 kg dry soil in plastic bag	100 g soil	Metals	Ab, MI, FT	Korthals *et al.* (1996)
TME	Outdoor	240 cm^2, 0.96 m^3; soil cores in HDPE tubes	35 cm^3 fresh soil	Carbendazim	Ab, Div, MI, FT	Moser *et al.* (2004)
Field mesocosms	Outdoor	0.5 m^2/plot; 0.25 m^3; plots separated by steel plates	120 g fresh soil	Zn	Ab, PRC; Div; MI	Smit *et al.* (2002)
Fresh water						
Acute MS toxicity test	Laboratory	4 ml water in 20 ml glass dishes	4 ml water	Metals, PAHs	Survival	Haegerbaeumer *et al.* (2018a)
Microcosms	Laboratory	0.5 l sediment in 1.7 l glass bottles	100 ml sediment	E.g. ivermectin, iron oxide colloids, metals, PAHs	Ab, PRC, Div, MI, NemaSPEAR	Brinke *et al.* (2010); Höss *et al.* (2015); Haegerbaeumer *et al.* (2018b)
Microcosms	Outdoor	80 ml sediment (in HDPE pots placed in larger cosms)	80 ml sediment	Fludioxonil	Ab, PRC, Div, NemaSPEAR, FT	Höss *et al.* (2020)
Microcosms	Outdoor	50 l sediment in steel containers	10 ml sediment	Nonylphenol	Ab, PRC, Div	Höss *et al.* (2004)

MS = multi-species; TME = terrestrial model ecosystem; HDPE = high-density polyethylene; *Bt*, *Bacillus thuringiensis*; PAH, polycyclic aromatic hydrocarbon; PRC = principal response curves (multivariate analysis); Ab = total abundance; MI = Maturity Index; SI = Structure Index; EI = Enrichment Index; FT = feeding type composition; Div = taxonomic diversity; NemaSPEAR = Nematode SPEcies At Risk index.

aromatic hydrocarbons (PAHs) (Fischer *et al.*, 2016) and styrene (Mueller *et al.*, 2020), using polydimethylsiloxan tubing, a silicone O-ring and triglyceride oil as PD device, respectively. The time allowed for equilibrium partitioning of the tested chemical between the PD device and water before adding test organisms varied from 2 to 24 h and survival and reproduction were selected as toxicity end points.

ALTERNATIVE NEMATODE EXTRACTION TECHNIQUE. Both standards for toxicity testing with *C. elegans* (ASTM E2172:2022; ISO 10872:2020) instruct to use the flotation/centrifugation (FC) technique using colloidal silica for separating the test organisms (with offspring) from the soil matrix. An alternative method, using the migration of living nematodes from the test soil towards a bacterial lawn on agar plates (MLN), was suggested by Kim *et al.* (2015). Essential differences between both methods relate to the inclusion of dead specimens and the possibility to store samples (possible for FC, not for MLN), the precision of the test duration (FC: ± 1 min, independent of sample amount; MLN: ± several hours, depending on sample amount) and the required equipment (FC: centrifuge and Ludox TM50®; MLN: Petri dish, agar and bacterial lawn). Furthermore, for the FC technique there is no limitation on the sample amount, whereas this is the case for the MLN method.

15.6.2 Multi-species toxicity testing

For higher- and intermediate-tier testing within the regulation process for the authorization of chemicals, multi-species test systems (MSTSs), such as micro- and mesocosms (model ecosystems), are necessary tools (e.g. Van Gestel *et al.*, 2021). Table 15.2 shows several MSTS approaches for investigating the effects of chemicals on freshwater and soil nematode communities (reviewed in Haegerbaeumer *et al.*, 2015). Besides, also easy-to-use experimental multi-species set-ups exist for testing acute toxicity of chemicals (Kammenga *et al.*, 1994; Bongers *et al.*, 2001; Haegerbaeumer *et al.*, 2018a).

References

Abade, A., Faria Porto, L., Ferreira, P.A. and de Barros Vidal, F. (2022) NemaNet: a convolutional neural network model for identification of soybean nematodes. *Biosystems Engineering* 213, 39–62. DOI: 10.1016/j.biosystemseng.2021.11.016.

Akintayo, A., Tylka, G.L., Singh, A.K., Ganapathysubramanian, B. Singh, A. and Sarkar, S. (2018) A deep learning framework to discern and count microscopic nematode eggs. *Scientific Reports* 8, 9145.

Alda Álvarez, O., Jager, T., Kooijman, S.A.L.M. and Kammenga, J.E. (2005) Responses to stress of *Caenorhabditis elegans* populations with different reproductive strategies. *Functional Ecology* 19, 656–664. DOI: 10.1111/j.1365-2435.2005.01012.x

Andrassy, I. (1984) *Klassa Nematoda (Ordnungen Monhysterida, Desmoscolecida, Araeolaimida, Chromadorida, Rhabditida)*. Gustav Fischer Verlag, Stuttgart, Germany.

Andrassy, I. (2005) *Free-Living Nematodes of Hungary (Nematoda Errantia)*, Vol. I. Hungarian Natural History Museum, Budapest.

Andrassy, I. (2007) *Free-Living Nematodes of Hungary (Nematoda Errantia)*, Vol. II. Hungarian Natural History Museum, Budapest.

ASTM (American Society for Testing and Materials) (2022) *Standard Guide for Conducting Laboratory Soil Toxicity Tests with the Nematode Caenorhabditis elegans; ASTM E2172-01*. ASTM International, West Conshohocken, Pennsylvania.

Balvočiūtė, M. and Huson, D.H. (2017) SILVA, RDP, Greengenes, NCBI and OTT – how do these taxonomies compare? *BMC Genomics* 18, 114.

Been, T.H. and Schomaker, C.H. (2022) Distribution patterns and sampling. In: Perry, R.N. and Moens, M. (eds) *Plant Nematology*. CAB International, Wallingford, UK, pp. 302–326.

Bongers, T. (1988) *De nematoden van Nederland*. KNNV Bibliotheekuitgave No. 46. KNNV Uitgeverij, Utrecht, The Netherlands.

Bongers, T. (1990) The maturity index: an ecological measure of environmental disturbance based on nematode species composition. *Oecologia* 83(1), 14–19. DOI: 10.1007/BF00324627.

Bongers, T., Ilieva-Makulec, K. and Ekschmitt, K. (2001) Acute sensitivity of nematode taxa to $CuSO_4$ and relationships with feeding-type and life-history classification. *Environmental Toxicology and Chemistry* 20, 1511–1516.

Bowers, R.M., Kyrpides, N.C., Stepanauskas, R., Harmon-Smith, M., Doud, D. *et al.* (2017) Minimum information about a single amplified genome (MISAG) and a metagenome-assembled genome (MIMAG) of bacteria and archaea. *Nature Biotechnology* 35, 725–731.

Brinke, M., Höss, S., Fink, G., Ternes, T.A., Heininger, P. and Traunspurger, W. (2010) Assessing effects of the pharmaceutical ivermectin on meiobenthic communities using freshwater microcosms. *Aquatic Toxicology* 99, 126–137.

Brinke, M., Heininger, P. and Traunspurger, W. (2013) Effects of a bioassay-derived ivermectin lowest observed effect concentration on life-cycle traits of the nematode *Caenorhabditis elegans*. *Ecotoxicology* 22, 148–155.

Cesarz, S, Schulz, A.E., Beugnon, R. and Eisenhauer, N. (2019) Testing soil nematode extraction efficiency using different variations of the Baermann-funnel method. *Soil Organisms* 91, 61–72.

Creamer, R.E., Barel, J.M., Bongiorno, G.R. and Zwetsloot, M.J. (2022) The life of soils: integrating the who and how of multifunctionality. *Soil Biology and Biochemistry* 166, 108561. DOI: 10.1016/j.soilbio.2022.108561.

De Ley, P. and Bert, W. (2002) Video capture and editing as a tool for the storage, distribution and illustration of morphological characters of nematodes. *Journal of Nematology* 34, 296–302.

Derycke, S., Sheibani Tezerji, R., Rigaux, A. and Moens, T. (2012) Investigating the ecology and evolution of cryptic marine nematode species through quantitative real-time PCR of the ribosomal ITS region. *Molecular Ecology Resources* 12, 607–619.

DIN (Deutsches Institut für Normung) (2021) *Wasserbeschaffenheit und Bodenbeschaffenheit – Bestimmung der toxischen Wirkung von Sediment- und Bodenproben auf Wachstum, Fertilität und Reproduktion von Caenorhabditis elegans (Nematoda); DIN ISO 10872:2012-10*. DIN, Berlin.

Du Preez, G., Daneel, M., De Goede, R., Du Toit, M.J., Ferris, H. *et al.* (2022) Nematode-based indices in soil ecology: application, utility, and future directions. *Soil Biology and Biochemistry* 169, 108640. DOI: 10.1016/j.soilbio.2022.108640.

Eghbalinejad, M., Hofman, J., Kotouček, J., Grillo, R., Bílková, Z.H. *et al.* (2024) Nano-enabled pesticides: a comprehensive toxicity assessment of tebuconazole nanoformulations with nematodes at single species and community level. *Environmental Science Europe* 36, 51. DOI: 10.1186/s12302-024-00879-9.

Eisenback, J.D. (2012) A technique for making high-resolution megapixel mosaic photomicrographs of nematodes. *Journal of Nematology* 44(3), 260–263.

Ekschmitt, K. (1998) Population assessment of soil fauna: general criteria for the planning of sampling schemes. *Applied Soil Ecology* 9, 439–445.

Ferris, H. (2010) Form and function: metabolic footprints of nematodes in the soil food web. *European Journal of Soil Biology* 46(2), 97–104. DOI: 10.1016/j.ejsobi.2010.01.003.

Ferris, H. and Bongers, T. (2009) Indices developed specifically for analysis of nematode assemblages. In: Wilson, M.J. and Kakouli-Duarte, T. (eds) *Nematodes as Environmental Indicators*. CAB International, Wallingford, UK, pp. 124–145. DOI: 10.1079/9781845933852.0124.

Ferris, H. and Tuomisto, H. (2015) Unearthing the role of biological diversity in soil health. *Soil Biology and Biochemistry* 85, 101–109. DOI: 10.1016/j.soilbio.2015.02.037.

Ferris, H., Bongers, T. and De Goede, R.G.M. (2001) A framework for soil food web diagnostics: extension of the nematode faunal analysis concept. *Applied Soil Ecology* 18(1), 13–29. DOI: 10.1016/S0929-1393(01)00152-4.

Filgueiras, C.C., Yongwoon, K., Wickings, K.G., El Borai, F., Duncan, L.W. and Willett, D.S. (2023) The Smart Soil Organism Detector: an instrument and machine learning pipeline for soil species identification. *Biosensors and Bioelectronics* 221, 114417. DOI: 10.1016/j.bios.2022.114417.

Fischer, F., Böhm, L., Höss, S., Möhlenkamp, C., Claus, E. *et al.* (2016) Passive dosing in chronic toxicity tests with the nematode *Caenorhabditis elegans*. *Environmental Science & Technology* 50, 9708–9716.

Geisen, S., Snoek, L.B., ten Hooven, F.C., Duyts, H., Kostenko, O. *et al.* (2018) Integrating quantitative morphological and qualitative molecular methods to analyse soil nematode community responses to plant range expansion. *Methods in Ecology and Evolution* 9, 1366–1378.

Geisen, S., Briones, M.J.I., Gan, H., Behan-Pelletier, V.M., Friman, V.-P. *et al.* (2019) A methodological framework to embrace soil biodiversity. *Soil Biology and Biochemistry* 136, 107536.

Gendron, E.M., Sevigny, J.L., Byiringiro, I., Thomas, W.K., Powers, T.O. and Porazinska, D.L. (2023) Nematode mitochondrial metagenomics: a new tool for biodiversity analysis. *Molecular Ecology Resources* 23, 975–989.

Goodfellow, I., Bengio, Y. and Courville, A. (2016) *Deep Learning*. MIT Press, Cambridge, Massachusetts.

Griffiths, B.S., de Groot, G.A., Laros, I., Stone, D. and Geisen, S. (2018) The need for standardisation: exemplified by a description of the diversity, community structure and ecological indices of soil nematodes. *Ecological Indicators* 87, 43–46.

Haegerbaeumer, A., Höss, S., Heininger, P. and Traunspurger, W. (2015) Experimental studies with nematodes in ecotoxicology: an overview. *Journal of Nematology* 47, 11–27.

Haegerbaeumer, A., Höss, S., Heininger, P. and Traunspurger, W. (2018a) Is *Caenorhabditis elegans* representative of freshwater nematode species in toxicity testing? *Environmental Science and Pollution Research* 25, 2879–2888.

Haegerbaeumer, A., Höss, S., Heininger, P. and Traunspurger, W. (2018b) Response of nematode communities to metals and PAHs in freshwater microcosms. *Ecotoxicology and Environmental Safety* 148, 244–253.

Haegerbaeumer, A., Raschke, R., Reiff, N., Traunspurger, W. and Höss, S. (2019) Comparing the effects of fludioxonil on non-target soil invertebrates using ecotoxicological methods from single-species bioassays to model ecosystems. *Ecotoxicology and Environmental Safety* 183, 109596.

Hakim, A., Mor, Y., Toker, I.A., Levine, A., Neuhof, M. *et al.* (2018) WorMachine: machine learning-based phenotypic analysis tools for worms. *BMC Biology* 16, 8.

Hallmann, J. and Viaene, N. (2013) PM 7/119 (1) Nematode extraction. *Bulletin OEPP/EPPO Bulletin* 43, 471–495. DOI: 10.1111/epp.12077.

Höss, S., Traunspurger, W., Severin, G.F., Jüttner, I., Pfister, G. and Schramm, K.-W.W. (2004) Influence of 4-nonylphenol on the structure of nematode communities in freshwater microcosms. *Environmental Toxicology and Chemistry* 23, 1268–1275.

Höss, S., Reiff, N., Nguyen, H.T., Jehle, J.A., Hermes, H. and Traunspurger, W. (2014) Small-scale microcosms to detect chemical induced changes in soil nematode communities – effects of crystal proteins and *Bt*-maize plant material. *Science of The Total Environment* 472, 662–671.

Höss, S., Frank-Fahle, B., Lüders, T. and Traunspurger, W. (2015) Response of bacteria and meiofauna to iron oxide colloids in sediments of freshwater microcosms. *Environmental Toxicology and Chemistry* 34, 2660–2669.

Höss, S., Heininger, P., Claus, E., Mohlenkamp, C., Brinke, M. and Traunspurger, W. (2017) Validating the NemaSPEAR[%]-index for assessing sediment quality regarding chemical-induced effects on benthic communities in rivers. *Ecological Indicators* 73, 52–60. DOI: 10.1016/j.ecolind.2016.09.022.

Höss, S., Roessink, I., Brock, T.C.M. and Traunspurger, W. (2020) Response of a nematode community to the fungicide fludioxonil in sediments of outdoor freshwater microcosms compared to a single species toxicity test. *Science of The Total Environment* 710, 135627.

Höss, S., Reiff, N., Traunspurger, W. and Helder, J.A. (2021) On the balance between practical relevance and standardization – testing the effects of zinc and pyrene on native nematode communities in soil microcosms. *Science of The Total Environment* 788, 147742.

Höss, S., Gansfort, B. and Traunspurger, W. (2022a) Nematodes as ecological and environmental indicators. In: Glazer, I., Shapiro-Ilan, D. and Sternberg, P.W. (eds) *Nematodes as Model Organisms*. CAB International, Wallingford, UK, pp. 165–196.

Höss, S., Reiff, N., Asekunowo, J. and Helder, J. (2022b) Nematode community of a natural grassland responds sensitively to the broad-spectrum fungicide mancozeb in soil microcosms. *Environmental Toxicology and Chemistry* 41, 2420–2430.

Hunt, D.J. and Handoo, Z.A. (2009) Taxonomy, identification and principal species. In: Perry, R.N., Moens, M. and Starr, J.L. (eds) *Root-Knot Nematodes*. CAB International, Wallingford, UK, pp. 55–97. DOI: 10.1079/9781845934927.0055.

Ikoyi, I., Heuvelink, G.B.M. and De Goede R.G.M. (2021) Geostatistical modelling and mapping of nematode-based soil ecological quality indices in a polluted nature reserve. *Pedosphere* 31, 670–682.

ISO (International Organization for Standardization) (2020) Water and soil quality – Determination of the toxic effect of sediment and soil samples on growth, fertility and reproduction of *Caenorhabditis elegans* (Nematoda); ISO 10872:2020. Available at: https://www.iso.org/standard/71352.html (accessed 25 August 2025).

ISO (International Organization for Standardization) (2022) Soil quality – Sampling of soil invertebrates. Part 4: Sampling, extraction and identification of soil-inhabiting nematodes; ISO 23611-4:2022. Available at: https://www.iso.org/standard/77794.html (accessed 25 August 2025).

Jager, T. (2020) Revisiting simplified DEBtox models for analysing ecotoxicity data. *Ecological Modelling* 416, 108904.

Kamath, D., Barreto, C. and Lindo, Z. (2022) Nematode contributions to the soil food web trophic structure of two contrasting boreal peatlands in Canada. *Pedobiologia* 93–94, 150809. DOI: 10.1016/j.pedobi.2022.150809.

Kammenga, J.E., Van Gestel, C.A.M. and Bakker, J. (1994) Patterns of sensitivity to cadmium and pentachlorophenol among nematode species from different taxonomic and ecological groups. *Archives of Environmental Contamination and Toxicology* 27, 88–94.

Kim, S.W., Moon, J. and An, Y.J. (2015) A highly efficient nonchemical method for isolating live nematodes (*Caenorhabditis elegans*) from soil during toxicity assays. *Environmental Toxicology and Chemistry* 34, 208–213.

Knotters, M. (2008) *Een blik op monitoring van de natuurlijke leefomgeving (A Look at Natural Habitat Monitoring)*. WOT Nature & Environment (WOt-study 6). Wageningen University and Research, Wageningen, The Netherlands.

Korthals, G.W., Van de Ende, A., Van Megen, H., Lexmond, T.M., Kammenga, J.E. and Bongers, T. (1996) Short-term effects of cadmium, copper, nickel, and zinc on soil nematodes from different feeding and life-history strategy groups. *Applied Soil Ecology* 4, 107–117.

Kou, X., Zhang, X., Bai, W., Cai, Q., Wu, Z. *et al.* (2020) Exploring N fertilizer reduction and organic material addition practices: an examination of their alleviating effect on the nematode food web in cropland. *Land Degradation & Development* 31(18), 2952–2961. DOI: 10.1002/ldr.3685.

Kwon, H., Roh, J.Y., Lim, D., Choi, J. and Kwon, J. (2011) Maintaining the constant exposure condition for an acute *Caenorhabditis elegans* mortality test using passive dosing. *Environmental Health and Toxicology* 26, 1–7.

LeCun, Y., Bengio, Y. and Hinton, G. (2015) Deep learning. *Nature* 521(7553), 436–444.

Mayer, P., Wernsing, J., Tolls, J., de Maagd, P.G.-J. and Sijm, D.T.H.M. (1999) Establishing and controlling dissolved concentrations of hydrophobic organics by partitioning from a solid phase. *Environmental Science & Technology* 33, 2284–2290.

Mesa-Valle, C.M., Garrido-Cardenas, J.A., Cebrian-Carmona, J., Talavera, M. and Manzano-Agugliaro, F. (2020) Global research on plant nematodes. *Agronomy* 10(8), 1148. DOI: 10.3390/agronomy10081148.

Moser, T., Schallnass, H.J., Jones, S., Van Gestel, C.A.M., Koolhaas, J.E. *et al.* (2004) Ring-testing and field-validation of a terrestrial model ecosystem (TME) – an instrument for testing potentially harmful substances: effects of carbendazim on nematodes. *Ecotoxicology* 13, 61–74.

Mueller, M.-T., Fueser, H., Trac, L.N., Mayer, P., Traunspurger, W. and Höss, S. (2020) Surface-related toxicity of polystyrene beads to nematodes and the role of food availability. *Environmental Science & Technology* 54, 1790–1798.

Mulder, C. and Vonk, J.A. (2011) Nematode traits and environmental constraints in 200 soil systems: scaling within the 60–6000 μm body size range. *Ecology* 92, 2004.

Muschiol, D., Schroeder, F. and Traunspurger, W. (2009) Life cycle and population growth rate of *Caenorhabditis elegans* studied by a new method. *BMC Ecology* 9, 14.

Neher, D.A., Wu, J., Barbercheck, M.E. and Anas, O. (2005) Ecosystem type affects interpretation of soil nematode community measures. *Applied Soil Ecology* 30(1), 47–64. DOI: 10.1016/j.apsoil.2005.01.002.

Nisa, R.U., Tantray, A.Y. and Shah, A.A. (2022) Shift from morphological to recent advanced molecular approaches for the identification of nematodes. *Genomics* 114, 110295. DOI: 10.1016/j.ygeno.2022.110295.

Okada, H., Tsukiboshi, T. and Kadota, I. (2002) Mycetophagy in *Filenchus misellus* (Andrássy, 1958) Lownsbery & Lownsbery, 1985 (Nematoda: Tylenchidae), with notes on its morphology. *Nematology* 4(7), 795–801. DOI: 10.1163/156854102760402586.

Oliverio, A.M., Gan, H., Wickings, K. and Fierer, N. (2018) A DNA metabarcoding approach to characterize soil arthropod communities. *Soil Biology and Biochemistry* 125, 37–43.

Overmars, H. (1985) *Horizontale en verticale verdeling van nematoden met het oog op hun bemonstering (Horizontal and Vertical Distribution of Nematodes for the Purpose of Their Sampling)*. Student Report. Wageningen University and Research, Wageningen, The Netherlands.

Perry, R.N., Hunt, D.J. and Subbotin, S.A. (2020) *Techniques for Work with Plant and Soil Nematodes*. CAB International, Wallingford, UK.

Ptatscheck, C., Gehner, S. and Traunspurger, W. (2020) Should we redefine meiofaunal organisms? The impact of mesh size on collection of meiofauna with special regard to nematodes. *Aquatic Ecology* 54, 1135–1143.

Quinn, G.P. and Keough, M.J. (2002) *Experimental Design and Data Analysis for Biologists*. Cambridge University Press, Cambridge.

Ryss, A.Y. (2017) A simple express technique to process nematodes for collection slide mounts. *Journal of Nematology* 49, 27–32. DOI: 10.21307/jofnem-2017-043.

Sánchez-Moreno, S. and Ferris, H. (2018) Nematode ecology and soil health. In: Sikora, R., Coyne, D., Hallmann, J. and Timper, P. (eds) *Plant Parasitic Nematodes in Subtropical and Tropical Agriculture*. CAB International, Wallingford, UK, pp. 62–86.

Schenk, J. and Traunspurger, W. (2021) Sampling and processing of freshwater nematodes with emphasis on molecular methods. In: Traunspurger, W. (ed.) *Ecology of Freshwater Nematodes*. CAB International, Wallingford, UK, pp. 31–57.

Schenk, J., Geisen, S., Kleinboelting, N. and Traunspurger, W. (2019) Metabarcoding data allow for reliable biomass estimates in the most abundant animals on earth. *Metabarcoding and Metagenomics* 3, e46704.

Schenk, J., Höss, S., Kleinbölting, N. and Traunspurger, W. (2022) Suitability of molecular taxonomy for assessing polluted sediments using the NemaSPEAR [%] index. *Ecological Indicators* 137, 108761.

Schreuder, H.T., Gregoire, T.G. and Weyer, J.P. (2001) For what applications can probability and non-probability sampling be used? *Environmental Monitoring and Assessment* 66, 281–291.

Seinhorst, J.W. (1955) A simple method for the separation of eelworms from soil. *Tijdschrift over Plantenziekten* 61, 188–190.

Seinhorst, J.W. (1959) A rapid method for the transfer of nematodes from fixative to anhydrous glycerine. *Nematologica* 4, 67–69.

Seinhorst, J.W. (1988) The estimation of densities of nematode populations in soils and plants. *Växtskyddsrapporter* 51, 107.

Shabrina, N.H., Lika, R.A. and Indarti, S. (2023) Deep learning models for automatic identification of plant-parasitic nematode. *Artificial Intelligence in Agriculture* 7, 1–12.

Sieriebriennikov, B., Ferris, H. and De Goede, R.G.M. (2014) NINJA: an automated calculation system for nematode-based biological monitoring. *European Journal of Soil Biology* 61, 90–93. DOI: 10.1016/j.ejsobi.2014.02.004.

Smit, C.E., Schouten, A.J., Van den Brink, P.J., Van Esbroek, M.L.P. and Posthuma, L. (2002) Effects of zinc contamination on a natural nematode community in outdoor soil mesocosms. *Archives of Environmental Contamination and Toxicology* 42, 205–216.

Smith, K.E.C., Dom, N., Blust, R. and Mayer, P. (2010) Controlling and maintaining exposure of hydrophobic organic compounds in aquatic toxicity tests by passive dosing. *Aquatic Toxicology* 98, 15–24.

Sohlenius, B. and Boström, S. (1984) Colonization, population development and metabolic activity of nematodes in buried barley straw. *Pedobiologia* 27, 67–78.

Speak, A., Escobar, F.J., Russo, A. and Zerbe, S. (2018) Comparing convenience and probability sampling for urban ecology applications. *Journal of Applied Ecology* 55, 2332–2342. DOI: 10.1111/1365-2664.13167.

Stevens, D.L. (1994) Implementation of a national monitoring program. *Journal of Environmental Management* 42, 1–29.

Sun, Y., Du, X., Li, Y., Han, X., Fang, S. *et al.* (2022) Database and primer selections affect nematode community composition under different vegetations of Changbai Mountain. *Soil Ecology Letters* 5, 142–150.

Tan, M. and Le, Q.V. (2019) EfficientNet: rethinking model scaling for convolutional neural networks. In: *Proceedings of the 36th International Conference of Machine Learning, 9-15 June 2019, Long Beach, California, USA*. PMLR 97. Available at: https://proceedings.mlr.press/v97/ (accessed 25 August 2025).

Tedersoo, L. and Anslan, S. (2019) Towards PacBio-based pan-eukaryote metabarcoding using full-length ITS sequences. *Environmental Microbiology Reports* 11, 659–668.

Trac, L.N., Schmidt, S.N. and Mayer, P. (2018) Headspace passive dosing of volatile hydrophobic chemicals – aquatic toxicity testing exactly at the saturation level. *Chemosphere* 211, 694–700.

Uhlemann, J., Cawley, O. and Kakouli-Duarte, T. (2020) Nematode identification using artificial neural networks. In: *Proceedings of 1st International Conference on Deep Learning Theory and Applications (DeLTa 2020)*. SciTePress – Science and Technology Publications, Setúbal, Portugal, pp. 13–22.

Van Bezooijen, J. (2006) *Methods and Techniques for Nematology*. Wageningen University, Wageningen, The Netherlands.

Van den Hoogen, J., Geisen, S., Wall, D.H., Wardle, D.A., Traunspurger, W. *et al.* (2020) A global database of soil nematode abundance and functional group composition. *Scientific Data* 7, 103. DOI: 10.1038/s41597-020-0437-3.

Van Gestel, C.A.M., Mommer, L., Montanarella, L., Pieper, S., Coulson, M. *et al.* (2021) Soil biodiversity: state-of-the-art and possible implementation in chemical risk assessment. *Integrated Environmental Assessment and Management* 17, 541–551.

Van Himbeeck, R., Cazzaniga, S., Van den Elsen, S., Vrielink, J.O., Aslan, S.K. *et al.* (2024) A full length SSU rRNA-based workflow for high-resolution monitoring of nematode communities reveals direct and indirect responses to plant-based manipulations. *Soil Biology and Biochemistry* 189, 109263. DOI: 10.1016/j.soilbio.2023.109263.

Verschoor, B.C. and De Goede, R.G.M. (2000) The nematode extraction efficiency of the Oostenbrink elutriator–cottonwool filter method with special reference to nematode body size and life strategy. *Nematology* 2, 325–342.

Vervoort, M.T.W., Vonk, J.A., Mooijman, P.J.W., van den Elsen, S.J.J., van Megen, H.H.B. *et al.* (2012) SSU ribosomal DNA-based monitoring of nematode assemblages reveals distinct seasonal fluctuations within evolutionary heterogeneous feeding guilds. *PLoS One* 7, e47555.

Viketoft, M. (2013) Determinants of small-scale spatial patterns: importance of space, plants and abiotics for soil nematodes. *Soil Biology and Biochemistry* 62, 92–98.

Wall, D.H., Nielsen, U.N. and Six, J. (2015) Soil biodiversity and human health. *Nature* 528(7580), 69–76. DOI: 10.1038/nature15744.

Wan, B., Hu, Z., Liu, T., Yang, Q., Li, D. *et al.* (2022) Organic amendments increase the flow uniformity of energy across nematode food webs. *Soil Biology and Biochemistry* 170, 108695. DOI: 10.1016/j.soilbio.2022.108695.

Whitehead, A.G. and Hemming, J.R. (1965) A comparison of some quantitative methods of extracting small vermiform nematodes from soil. *Annals of Applied Biology* 55, 25–38. DOI: 10.1111/j.1744-7348.1965.tb07864.x.

Wiesel, L., Daniell, T.J., King, D. and Neilson, R. (2015) Determination of the optimal soil sample size to accurately characterise nematode communities in soil. *Soil Biology and Biochemistry* 80, 89–91.

Yeates, G.W. (2003) Nematodes as soil indicators: functional and biodiversity aspects. *Biology and Fertility of Soils* 37(4), 199–210. DOI: 10.1007/s00374-003-0586-5.

Zhao, J., Xiao, J., Zhang, W., Fu, Z., Zhang, M. *et al.* (2019) A method for estimating nematode body lengths for use in the calculation of biomass via a simplified formula. *Soil Biology and Biochemistry* 134, 36–41. DOI: 10.1016/j.soilbio.2019.03.021.

Zhu, B., Wan, B., Liu, T., Zhang, C., Cheng, L. *et al.* (2023) Biochar enhances multifunctionality by increasing the uniformity of energy flow through a soil nematode food web. *Soil Biology and Biochemistry* 183, 109056. DOI: 10.1016/j.soilbio.2023.109056.

16 Environmental Policy and Nematodes

Thomais Kakouli-Duarte[1]*, Lilian Atira[1], Luana Da Costa Monteiro[2], Anique Godjo[1], Candice Jansen van Rensburg[3], Anna Karpinska[1], Anusha Pulavarty[1], Federica Semprucci[4], Alejandra Vieyra-Ramírez[1] and Maria Viketoft[5]

[1]MENRG, enviroCORE, Department of Applied Science, South East Technological University, Ireland; [2]Marine Biology Research Group, Biology Department, University of Ghent, Belgium; [3]Department Zoology and Entomology, University of the Free State, Bloemfontein, South Africa; [4]Department of Biomolecular Sciences (DiSB), University of Urbino, Italy; [5]Department of Ecology, Swedish University of Agricultural Sciences, Uppsala, Sweden

Abstract

Integrating nematode-based tools in environmental policy could be parallelized to a 'holy grail' in achieving a wider acceptance of this useful phylum and its utilization in environmental biomonitoring and in various nature conservation and restoration efforts. This chapter reviews environmental legislation and policies in various locations around the world and in relation to various environmental challenges and habitats. Unfortunately, the review confirms the lack of this integration but highlights its benefits for the environment and calls for action.

16.1 Introduction

More than 16 years have elapsed between the current and the previous (Wilson and Kakouli-Duarte, 2009) editions of this book. This period of time demonstrated even more the usefulness of nematodes as environmental indicators and indeed works in this volume more than attest to this. In addition, a recent study by Du Preez *et al.* (2022) thoroughly reviewed past and latest literature to investigate the extent and the range of applicability of nematodes and their communities in environmental biomonitoring. The authors discovered the universality, at this stage, and the versatility of the applications of members of the phylum Nematoda as environmental bioindicators. At the same time, all authors and reviewers who worked to produce the present volume have, one way or another, worked with nematodes in addressing local and global environmental issues. What became apparent in the preparation of this book is that in the past 16 years since the first edition, numerous environmental legislations have been proposed and implemented at national, European Union (EU) and up to United Nations (UN) levels. As our planet, its lifeforms and our environment in general face numerous challenges, many of anthropogenic origin, legislators have acted accordingly for environmental protection. However, from the current editors' professional experience and perspective, it was concluded that this important but vastly underrated phylum that could be an ally in addressing challenges and help mitigate their negative impacts,

*Email: thomae.kakouli@setu.ie

DOI: 10.1079/9781800624221.0016

the phylum Nematoda, has mostly been neglected to be implemented in these policies. This was also realized in the case of standardization of toxicological assessments in soils and sediments (see Chapter 13, this volume). It was therefore decided to include the present chapter in this volume to provide the opportunity to review environmental policies and legislations in various locations around the world, in relation to terrestrial and aquatic ecosystems and in addressing various environmental challenges. The aim was to discover the relationship between environmental policy and nematodes. The authors are all experts in various areas of environmental nematology, from diverse locations around the world and with work experience in terrestrial and aquatic ecosystems. This chapter therefore reviews EU and global policy legislated to address specific environmental issues, and it is this format and structuring that has been applied in this text. After these reviews the chapter concludes with emphasis on the advantages of nematodes as environmental indicators and calls for action.

16.2 Environmental Policies on Soil and Water Pollution

Pollution is the direct or indirect introduction, as a result of human activity, of substances, vibrations, heat or noise into air, water or land, which may: (i) be harmful to human health or the quality of the environment; (ii) result in damage to material property; or (iii) impair or (iv) interfere with amenities and other legitimate uses of the environment (Directive 2010/75/EU, Article 3[2]; EU Monitor, 2010).

Worldwide, pollution has deleterious effects on human, animal and environmental health. Children, people with certain medical conditions and the elderly are the most vulnerable groups who suffer from multiple mental and physical pollution-related illnesses and premature death (European Commission, 2021a). The EU has adopted several policies to protect and improve the quality of water and soil in its countries. Recently the EU Zero Pollution Action Plan was adopted on 12 May 2021 to provide a healthy planet for all. This policy focuses on preventing contamination in air, water and soil and is in synergy with the targets of the European Green Deal (EGD). The Zero Pollution Action Plan can greatly contribute to the UN 2030 Agenda for Sustainable Development, especially Sustainable Development Goals (SDGs) 3, 6, 11, 12, 14 and 15 (European Commission, 2021a). In aquatic habitats, the European Water Framework Directive (WFD), currently in its third cycle 2022–2027, is considered an important piece of environmental legislation aiming to improve water quality around European countries. The Nitrate Directive (91/676/EEC) (EEC, 1991a) is an integral part of the WFD and requires EU Member States to monitor the quality of waters, especially in the areas where polluted water is draining or at risk of pollution due to agricultural activities. In addition to these legislations, the most recent policy on soil protection is the EU Soil Strategy for 2030. Its vision is that by 2050 all EU soil ecosystems will be in healthy condition and thus more resilient. A long-term objective among others is that by 2050 soil pollution should be reduced to a level no longer considered harmful to human health and natural ecosystems, and respect the boundaries our planet can cope with, thus creating a toxic-free environment (European Commission, 2021b).

In Africa, at least four international conventions addressing soil pollution have been established. These include the Basel Convention, the Rotterdam Convention, the Minamata Convention and the Stockholm Convention. To meet the aims of these conventions, relevant actions are required from the participating countries to protect humans and the environment from anthropogenic emissions, unregulated and irresponsible trade and use of hazardous and toxic chemicals, usually referred to as persistent organic pollutants (POPs), and uncontrolled transboundary movements of hazardous wastes (Tindwa and Singh, 2023). In addition to these regulations, the Bamako Convention which came into effect in 1998 is a treaty signed by 29 African countries that prohibits the import into Africa of any hazardous (including radioactive) waste and controls its transboundary movement within the continent (Wordsworth, 1993; Matemilola and Fadeyi, 2020; UNEP, 2020). In sub-Saharan Africa, apart from the international initiatives stated above, there is a general lack of national systems for pollution monitoring and assessment (Tindwa and Singh, 2023), with few exceptions.

In Nigeria, the extent of oil spills is monitored through the National Oil Spill Detection and Response Agency (NOSDRA) (Nigerian Oil Spill Monitor, 2023).

In South Africa, the reporting system focuses on activities in breach of the Integrated Coastal Management Act, 2008 (Act No. 24 of 2008) and the Waste Act, 2008 (Act No. 59 of 2008), and lists soil and water pollution in the National Norms and Standards for the Remediation of Contaminated Land and Soil Quality (Department of Environmental Affairs, 2020).

New measures for the reduction or elimination of hazardous wastes have been adopted in countries such as Côte d'Ivoire (decrees Order No. 0120/MINEDD/DGE of 15 May 2019 and Order No. 091/MINEDD/DGE of 30 April 2019), Rwanda (Rwanda Resource Efficient and Cleaner Production Centre and stakeholders), Republic of Congo (circular No. 0613/MEFDDE/CAB/DGE of 24 April 2017) and the Republic of Benin. In Benin, the government and the Environmental Agency have proposed the Anti-pollution Strategy, even though additional concrete actions are urgently needed to correctly perceive, monitor and mitigate environmental problems in the country (World Bank, 2010; UNEP, 2025).

Similar legislations have been adopted in other parts of the world towards soil and water protection from pollution. For instance, in India, the Water Act 1974, the Water (Prevention and Control of Pollution) Cess Act 1977 and the Environment Protection Act 1986 have been established. In the Water Act 1974, wastewater discharge standards were developed and recommended by the Central Pollution Control Board (CPCB), which monitors both point sources and non-point sources (Jadeja *et al.*, 2022). In 2016, the CPCB proposed the Zero Liquid Discharge Strategy to eliminate liquid discharge and promote recycling and reuse of wastewater by the fertilizer and distilleries sectors (Álvarez *et al.*, 2017; Vimalkumar *et al.*, 2018; Jadeja *et al.*, 2022). In the USA, additional standards have been provided for water, air and soil in the Environment Protection Act (1986) and the Environmental Protection Rules (1989), proposed by the Environmental Protection Agency (EPA) for gas emission and effluent discharge by industries (US EPA, 1990a). Similar policies, such as the Action Plan on Prevention and Control of Soil Pollution, have recently been adopted in China (Li *et al.*, 2019). The policy emphasizes that soil management should focus on pollution prevention, and a national survey of soil pollution should be conducted by the State Council at least once in each 10-year period (Li *et al.*, 2019).

In the USA, the control of pollution in all habitats is regulated by the EPA and other federal agencies by several policies. The Clean Water Act, last amended in 1987, is the main instrument used to regulate discharge pollutant materials to waters (Davies and Mazurek, 1998; Fesenmyer *et al.*, 2021). The Pollution Prevention Act (1990), the 2010–2014 Pollution Prevention Program Strategic Plan and the National Environmental Policy Act, last amended in 1977, help reducing or eliminating at the source hazardous substances and their release into the environment. This includes nitrogen pollution due to agricultural practices, which is becoming a high environmental concern around the world (US EPA, 1990b; Kanter *et al.*, 2020).

Despite the numerous policies and frameworks tailored towards addressing water and soil pollution challenges, most do not acknowledge the role of nematodes in the environment in their monitoring plans. Nematodes are excellent bioindicators for monitoring pollution including heavy metals, pesticides and other contaminants (see Chapters 6, 8, 9 and 13, this volume). In a few cases, nematodes are integrated in risk assessment schemes in countries which have policy on pollution-contaminated areas, with for example trigger values (Intervention Values) for remediation (M. Rutgers, The Netherlands, 2025, personal communication). However, a more concerted and widespread approach, to incorporate the assessment and monitoring of nematodes as indicators of soil health and water quality within environmental policy development frameworks, would be highly beneficial. For instance, in efforts to restore degraded sites through soil remediation, nematodes can be utilized as tools for identifying sites that require rehabilitation. In this regard, before issuing site closure certificates, proponents could be required to provide a full nematode community assessment report accompanied with a monitoring plan on the soil quality. In relation to addressing water pollution, monitoring systems can include periodic assessments of both aquatic and terrestrial nematodes for improving or maintaining good ecological status of waterbodies.

16.3 Environmental Policies in Agriculture

Agricultural soils sustain human society with production of food, fibres and energy, by supporting primary production and decomposition processes, regulating nutrient, carbon and water cycles, and providing habitat for soil organisms. However, some management practices in agricultural soils, such as use of mineral fertilizers, intensive tillage and application of pesticides, have a negative environmental impact and there is a great need for a sustainable agricultural intensification that is able to meet the future challenges of feeding a growing population, while reducing this environmental impact. Launched in 1962, the Common Agricultural Policy (CAP) supports European farmers and ensures Europe's food security. The current CAP specifically emphasizes the environment and climate, and there are stronger links to legislation in these areas. The adopted EGD includes a set of policy initiatives with the overarching aim of making Europe climate neutral by 2050 (European Commission, 2019); soils are a key factor for these initiatives (Montanarella and Panagos, 2021). The Farm to Fork Strategy is at the heart of the EGD aiming to make food systems fair, healthy and environmentally friendly, and this strategy is also central to achieving the UN's SDGs (European Commission, 2020). The specific targets of the Farm to Fork Strategy for 2030 include reducing the use and risk of chemical pesticides by 50%, cutting nutrient losses by 50%, lowering fertilizer use by 20% and increasing the share of agricultural land under organic farming to at least 25%.

Increasing agricultural productivity to meet the global food demands, while maintaining ecosystem functions, requires policies and frameworks that both conserve ecosystems and promote food production. However, it is a challenge achieving both conservation and increased food production, as many agricultural policies that evaluate farm inputs mostly focus on the nutrient composition of the products and not the associated impacts to the environment. Understanding both the impacts associated with agricultural inputs and the crops' nutrient composition is important in achieving sustainable food production. These management policies could greatly benefit from the application of nematodes as environmental bioindicators. For instance, legislation regulating use of sustainable bio-based fertilizers could adopt nematodes and their communities, to assess potential risks associated with the use of these fertilizers before their application by farmers (Karpinska *et al.*, 2021; Ryan *et al.*, 2022; Saju *et al.*, 2023).

Two important environmental topics related to agricultural systems, besides biodiversity loss (see Section 16.5), are carbon storage and greenhouse gas (GHG) emissions. Soil is the largest terrestrial carbon pool and renders high potential in carbon sequestration by storing biomass-based carbon in soil organic matter (SOM) with long stability (Hou, 2021). A small increase in soil carbon stocks could mitigate significant amounts of global GHG emissions, and calculations show that net soil carbon sequestration (SCS) could offset 4% of average annual global GHG emissions over the rest of the century (Henderson *et al.*, 2022). In addition, many global policies and initiatives require monitoring of soil carbon stocks, including the UN Convention to Combat Desertification, the UN Framework Convention on Climate Change (UNFCCC), the UN SDGs and the Protocol for the Assessment of Sustainable Soil Management put forward in 2020 by the Food and Agriculture Organization of the United Nations' (FAO) Intergovernmental Technical Panel on Soils (ITPS) (Faber *et al.*, 2022).

At the same time, as agricultural soils can serve as a carbon sink, they can also be a significant source of the three most important greenhouse gases: carbon dioxide (CO_2), methane (CH_4) and nitrous oxide (N_2O) (Oertel *et al.*, 2016). Reducing the emissions of GHGs is an important part of the Paris Agreement, which is the legally binding international treaty on climate change (UNFCCC, 2016). To aid with climate change mitigation, the Intergovernmental Panel on Climate Change (IPCC) provides regular assessments of the scientific basis of climate change, its impacts and future risks, and options for adaptation and mitigation. Within the IPCC, there is a Task Force on National Greenhouse Gas Inventories with the main objective to develop and refine a methodology for the calculation and reporting of national GHG emissions and removals (IPCC, 2006, 2019). On the European level, the European Parliament has adopted the European Climate Law (EU, 2021), which, like the Farm to Fork Strategy, is part of the EGD.

The Climate Law raises the EU's 2030 emissions reduction target from 40% to at least 55% through a package of legislation known as Fit for 55 in 2030, with the ultimate goal of climate neutrality by 2050; this implies achieving net zero GHG emissions for EU countries as a whole.

Agricultural soils are potential reserves for carbon storage if correct management practices are aimed at either reducing the amount of GHG released into the atmosphere or maintaining and enhancing soil organic carbon stocks. Soil organisms, through complex interactions, influence soil carbon stocks by taking part in decomposition, soil aggregation and carbon cycling. Nematodes are components of the complex soil food web and interact with soil microbial communities to impact SCS and the amount of soil organic carbon (Shi *et al.*, 2023). International frameworks such as UNFCCC and the UN Convention to Combat Desertification can use data on changes in nematode composition and diversity in their soil carbon assessment protocols to monitor sustainable agricultural management practices aimed at increasing carbon stocks. Terrestrial nematodes can also be used as a monitoring tool to infer the condition of agricultural land, in response to climate stress. In this case, climate change frameworks and policies aimed at enhancing climate change resilience of agricultural land can utilize soil nematodes as bioindicators to assess how soils respond to stress from global warming (see Chapter 11, this volume).

16.4 Environmental Policies on Pesticide Application

Increasing demand for food and fibre production globally has challenged scientists, growers and policy makers to introduce sustainable practices to avoid crop losses. Some of these practices include integrated pest management (IPM) and integrated nematode management (INM) (Sikora *et al.*, 2021). Every country has been working towards imposing various policies and legislations on pesticide application. The Government of India has adopted IPM as prime principle for plant protection since 1985. IPM-related operations are carried out by 31 Central Integrated Pest Management Centres (CIPMCs), which have been established in 28 States and one Union Territory in India. The Sub Mission on Plant Protection and Plant Quarantine (SMPPQ) programme was introduced by the Indian Ministry of Agriculture and Farmers' Welfare and the Department of Agriculture, Co-operation and Farmers' Welfare (DAC and FW) as part of the 13th Five-Year Plan EFC Memo (Acharya Balkrishna *et al.*, 2022). The National Mission on Agricultural Extension and Technology (NMAET) was formed, under which SMPPQ was introduced from 2014 (Government of India, 2014).

The Strengthening and Modernization of Pest Management Approach (SMPMA), in India, is a key element of SMPPQ that is incorporated to improve crop cultivation practices. For the production of biocontrol agents and biopesticides, there are currently 352 biocontrol laboratories operating in India. These laboratories were established by numerous organizations, including the central government, state governments, the Indian Council for Agricultural Research (ICAR), the Department of Biotechnology (DBT), various non-governmental organizations (NGOs) and private business owners.

In European Member States strict laws in relation to application of pesticides and their residues are enforced by the European Food Safety Authority (EFSA). Such regulations include Regulation (EC) No 1107/20093, referred to as 'the PPP (plant protection products) Regulation', and Regulation (EC) No 396/20054, referred to as 'the MRL (maximum residue levels) Regulation'; these regulations are in essence a risk management framework. The UN 2030 Agenda for Sustainable Development can be achieved to a certain extent because of the PPP and MRL Regulations. Since 2013, the current CAP has included steps to encourage the proper, sustainable use of PPPs as an environmental cross-compliance obligation. Farmers who fail to abide by the PPP usage guidelines may see their direct payments reduced. The Commission gave the sustainable use of PPPs high priority in the proposal for a new CAP (2018), through the objectives outlined in the proposal and through the suggested indicators (European Commission, 2018).

Isgren and Andersson (2021) have reported an environmental justice perspective on smallholder pesticide use in sub-Saharan Africa. Despite having some of the lowest reported pesticide application rates on the continent (Loha *et al.*, 2018), Uganda has recently received repeated warnings from the EU concerning chemical residues on agricultural imports (*The Independent*,

2019). CropLife Africa Middle East created the Spray Service Providers (SSP) initiative to promote the use of high-quality pesticides and guarantee proper application. After completing specialized training from the relevant industry association, SSP offer their services to farmers (Isgren and Anderson, 2021). In Paya subcounty, eastern Uganda, it was found that farmers with sufficient finances occasionally hired additional informal spraying workers (e.g. neighbours). The World Health Organization (WHO) has issued special guidelines on public health pesticide management policies for African regions (WHO, 2011). The Department of Agriculture, Forestry and Fisheries (DAFF) have adopted pesticide management policy for South Africa (Joemat-Pettersson, 2010). A few policies and regulations issued by various countries in relation to pesticide application have been briefed in Table 16.1.

In delivering their mandate, government agencies and regulatory authorities require manufacturers of PPPs to provide data on the behaviour and fate of their products in the environment for approval, before they are introduced to the market. While a wide range of organisms have been included for PPP toxicity testing, soil nematodes have not been included despite strong evidence of their capability for ecotoxicological studies and environmental health assessment. As elucidated in this work (see Chapter 13, this volume), *Caenorhabditis elegans* is highly sensitive to chemical exposure and can be utilized as a tool for toxicity tests associated with pesticide use in agriculture. Moreover, beneficial steinernematids have been documented as useful indicators to evaluate the potential toxicity of certain PPPs (Pulavarty *et al.*, 2020). Government agencies and regulatory authorities could require PPP manufacturers to provide toxicity test reports using nematodes under protocols that have been standardized for toxicity testing. It is therefore recommended that, alongside other organisms, nematodes should also be used to investigate toxicity effects of pesticides in agricultural soils.

16.5 Environmental Policy and Biodiversity Loss

Biodiversity is an essential component for achieving a healthy planet and human well-being (Cardinale *et al.*, 2012; SCBD, 2015). But the rate of biodiversity loss is continuing relentlessly, with species extinctions increasing at an alarming rate (Tittensor *et al.*, 2014; Steffen *et al.*, 2015). Human activities causing pollution, along with habitat destruction, overexploitation, and foreign plant and animal invasions are resulting in the ever-increasing loss of the Earth's biological wealth (Department of Environmental Affairs and Tourism, 1997; Millennium Ecosystem Assessment, 2005). It is estimated that by 2050 the cost to the global economy from biodiversity loss and loss of ecosystem functions will be €14 trillion (Ten Brink *et al.*, 2008).

The loss of biodiversity is a multifaceted issue, with biodiversity conservation relying heavily on strategies that include a variety of policy approaches (Mangalagiu *et al.*, 2019). These may include regulatory command and control, economic incentives, promoting innovation, building capacity and setting goals. The 1992 Convention on Biological Diversity (CBD), the Intergovernmental Science-Policy Platform on Biodiversity and Ecosystem Services (IPBES) and Protected Areas are some of the major international policies and governance responses that have been adopted by many countries (Mangalagiu *et al.*, 2019).

Bagnoli *et al.* (2008) noted that other multilateral environmental agreements were also in place to help protect biodiversity, for example the Convention on International Trade in Endangered Species (CITES), the Convention on Wetlands (Ramsar Convention), the World Heritage Convention, and the Convention on the Conservation of European Wildlife and Natural Habitats. These agreements all attempt to ensure a coordinated process for addressing biodiversity loss. The implementation of these policies is generally applied at national level that addresses the sources of impacts on biodiversity (Bagnoli *et al.*, 2008).

In 2011, the United Nations Environment Programme (UNEP) and the Secretariat of the Convention on Biological Diversity (SCBD) established the 20 Aichi Targets as part of the Strategic Plan for Biodiversity 2011–2020 (SCBD, 2011; UNEP, 2011). Their aim was to offer a framework in which the world and individual countries could address the growing rates of biodiversity loss and ways to decelerate it. By 2022, at the 15th meeting of the Conference of the

Table 16.1. Regulations and policies implemented by various countries in relation to pesticide application. (Author's own table.)

Policy/Regulation	Country	Description of the policy	Reference
Agriculture raisonée [good agricultural practice]	France	Measures to reduce the risks of using pesticides	Brouwer and Van Berkum, (1996)
UK National Action Plan for the Sustainable Use of Pesticides (Plant Protection Products)	UK	Pesticide use across the UK for the next five years to reduce associated risks	DEFRA (2013)
European Union (Withdrawal) Act 2018 with amendments, into UK law		The Department for Environment, Food and Rural Affairs (DEFRA) and the Health & Safety Executive (HSE) authorize pesticide products, regulate pesticide use and monitor occupational health impacts and pesticide residues in food	Kennard and Vagnoni, (2021)
Pesticides Regulation 2017	New South Wales (NSW), Australia	The Environment Protection Authority (EPA), Commonwealth and NSW government agencies and other stakeholder organizations manage pesticide use in NSW to minimize environmental harm	NSW EPA (2017)
State Council Decree 677	China	Release of the 'Regulations on the Management of Pesticides', for registration, production, distribution and use of pesticides	USDA (2017) (prepared in cooperation with CropLife China)
The Federal Insecticide, Fungicide, and Rodenticide Act (FIFRA) 7 U.S.C. §136 et seq. (1996)	USA	Federal regulation of pesticide distribution, sale and use All pesticides distributed or sold in the USA must be registered (licensed) by Environmental Protection Agency (EPA) The Office of Pesticide Programs (OPP) regulates use and limits for pesticide residues in food	US EPA (1996)
Bill 6.670/2017, PNARA (National Pesticide Reduction Policy)	Latin America	Preventing the use of 2,4-D-based herbicides (PL 21.273/2015 and PL 262/2014, in Brazil) Banning the aerial spraying of pesticides (PL 21.314/2015, and PL 263/2014, in Brazil)	Burity *et al*. (2020)
Eurasian Economic Union (EAEU) Decision No. 149	Eurasia Economic Union (prepared in Moscow, Russia)	Maximum residue limits (MRLs) for pesticides and 573 chemicals in the human body, soil, reservoir water, working air, open air, and in agricultural raw materials and food products	Kennard and Vagnoni (2021)

Parties (COP) of the CBD in Montreal, none of the countries had met these targets. The Kunming–Montreal Global Biodiversity Framework was adopted at this meeting after a four-year consultation process. This framework is aimed at recognizing the urgent need for policy action on global, national and regional scales, to achieve sustainable development (SCBD, 2022). It is hoped that by adopting this framework, biodiversity loss will be reduced and/or reversed, so as to allow the recovery of all ecosystems and the Convention's vision of Living in Harmony with Nature by 2050 will be achieved.

Clear commitments in terms of policies have already been made by countries to prevent biodiversity loss. These include the UK Environmental Bill (Environment Act 2021) which aims to halt species decline by 2030 (HM Government, 2021). They also include the EU Biodiversity Strategy 2030 (EU, 2020a), which aims to put Europe's biodiversity on a path to recovery by 2030 and contains specific actions and commitments. South Africa has a strong policy and legislative framework for the conservation, management and sustainable use of biodiversity, which includes 13 policies and frameworks. Some of these policies and frameworks include the White Paper on the Conservation and Sustainable use of Biodiversity (1997), the National Environmental Management Act, (Act 107 of 1998), the National Environmental Management and Biodiversity Act (Act 10 of 2004), and the second National Biodiversity Strategy and Action Plan 2015–2025 (Government of South Africa, 2015).

An important global reservoir of biodiversity is soil, which, according to Decaëns *et al.* (2006), hosts at least a quarter of all living organisms on the planet. Tibbet *et al.* (2020) noted that one-third of the world's land has been degraded with the agricultural sector contributing to two-thirds of this degradation.

The ubiquitous distribution of nematode species across different ecosystems makes them suitable indicators (see Chapters 6 and 7, this volume) to monitor biodiversity loss and restoration efforts on a global scale and over various natural and anthropogenic environments. Analyses of nematode communities and assemblages and nematode-based indices represent practical and standardized approaches (see Chapters 4 and 5, this volume) that can be incorporated into the measurable goals in the agendas of international biodiversity agreements. These tools can also be utilized in the evaluation of various biodiversity policy interventions. A partnership between nematologists, policy makers and other stakeholders could help to address the current gaps between biodiversity loss, restoration and monitoring efforts.

16.5.1 Species sensitivity distribution models and nematodes

Species can differ in their sensitivity to specific chemicals (Rizzi *et al.*, 2021), influenced by factors such as behaviour, physiology, morphology and life-history traits (De Zwart, 2002). This variability in sensitivity has led to the development of the species sensitivity distribution (SSD) approach, which uses statistical distributions – either parametric or non-parametric – to represent the sensitivity of multiple species (Van Straalen and Denneman, 1989; Rizzi *et al.*, 2021) to certain toxicants in the environment. SSDs assume that distinct species exhibit different degrees of sensitivity to contaminants, and can provide a statistical estimate of the probability that species will be affected by certain chemicals or chemical mixtures and/or predict harmful effect concentrations (Del Signore *et al.*, 2016; Posthuma *et al.*, 2019; Fox *et al.*, 2021).

The SSD approach is valuable in precautionary decision making, supporting the development of environmental quality criteria, benchmarks for screening assessments and ecological risk estimations (Stephan, 2002). It is widely used in global ecological risk assessments. In the USA, SSDs serve as screening benchmarks for contaminants in water under the National Ambient Water Quality Criteria (Staples *et al.*, 2004; Schmitt-Jansen and Altenburger, 2005). In the EU, SSDs are incorporated into several regulations, such as the Registration, Evaluation, Authorisation, and Restriction of Chemicals (REACH), PPP registrations and the WFD (EC, 2000, 2009). SSD models are used for water quality assessment and the formulation of water quality guidelines (Wheeler *et al.*, 2002), including the principles of the WFD (EC, 2000; Posthuma *et al.*, 2019).

Accurate models require data on the toxicity of different contaminants, including chemical mixtures. However, this represents a limitation, as information is lacking for many compounds and their mixtures (Posthuma *et al.*, 2019). Additionally, standard model organisms (e.g. the nematode *C. elegans* or the zebrafish *Danio rerio*) may not accurately reflect the reality of the status of the natural free-living communities (Ferraz *et al.*, 2022). Smit *et al.* (2002) also noted that a drawback of using SSDs is that most tests are laboratory-based and focus on the effects of toxicants on a single species, with results therefore not reflecting impacts on soil communities in the field. The main goal of environmental policy is in fact to protect natural populations, communities or whole ecosystems from adverse effects (Smit *et al.*, 2002). Therefore, development of methods that allow for predicting and assessing true field effects of toxic compounds on soil ecosystems becomes more important for risk assessment at polluted sites (Smit *et al.*, 2002).

By gathering data on species' responses to toxicants in an environment, SSDs may help policy makers set safe pollutant thresholds and regulatory limits to protect biodiversity. Nematodes, in particular, are important in this respect due to their widespread presence in soil, water and sediment, their sensitivity to contaminants and their essential role in ecosystem processes (Chauvin *et al.*, 2020). As bioindicators in ecotoxicology, nematodes provide a broader understanding of pollutant effects, extending beyond higher trophic levels (Hägerbäumer *et al.*, 2015). Their inclusion in SSDs is particularly valuable for assessing soil health and sustainability, as they contribute significantly to nutrient cycling and soil structure (Chauvin *et al.*, 2020). Incorporating nematodes into SSD models can thus offer key insights into how nematodes respond to environmental stresses, thus possibly enhancing SSDs which will aid environmental policies in protecting biodiversity.

So far, few studies have incorporated nematode species distributions into SSD models. For instance, a broad range of nematode taxa was studied for their acute sensitivity to cupric sulfate ($CuSO_4$) to determine whether changes in nematode community structure may be used as a useful bioindication method for heavy-metal contamination (Bongers *et al.*, 2001). In addition, Chen *et al.* (2018) investigated the toxicity of nanomaterials, while Wang *et al.* (2022) and Kim *et al.* (2023) examined the toxicity of micro- and nanoplastics. Moreover, nematode SSD analysis facilitated pesticide risk assessments. Specifically, Hennig *et al.* (2023) focused on nanopesticides, showing that *C. elegans* was highly sensitive to nano-encapsulated atrazine compared to *Enchytraeus crypticus* and *Pseudokirchneriella subcapitata*. Nematodes were also found more sensitive to the fungicide carbendazim and to copper toxicity than arthropods (Frampton *et al.*, 2006), when applying this approach.

There is no doubt that additional studies are needed to provide robust and reliable data on nematode species' sensitivity to different contaminants, perhaps including other nematode model organisms such as, for example, *Litoditis marina* (Semprucci *et al.*, 2025). However, this alone is not sufficient to fully investigate the long-term responses of ecosystems and their components to perturbations.

Indeed, model organisms, by definition, possess characteristics such as a high number of descendants, a short generation time and a high tolerance to laboratory-controlled conditions. In other words, many of these models are r-strategist species. As a result, the response of the other side of the 'coin' – namely, the response of more sensitive species (or K-strategist species) – remains completely unknown.

Therefore, supporting an ecological quality assessment solely with SSD models is limiting, particularly because they do not consider cumulative effects or food chain interactions, which generally occur in complex ecosystems and are crucial for estimating ecosystem functional loss (Hägerbäumer *et al.*, 2015; Posthuma *et al.*, 2019; Ferraz *et al.*, 2022; Schratzberger *et al.*, 2023). This means that the lack of official protocols and indices, that include free-living nematode communities as a level of organization in biomonitoring, within most legal framework directives remains not only unresolved but also a significant challenge.

16.5.2 Soil biodiversity assessment

Nematode diversity and abundance were identified as one of the soil biodiversity descriptors in the proposal for a European Soil Monitoring Law (European Commission, 2023). The directive aims to set up a soil monitoring framework for all soils throughout the EU to achieve healthy soils by

2050. The European Commission also supports the development of the Land Use and Coverage Area frame Survey Soil (LUCAS Soil), as the largest harmonized open-access data set of topsoil properties available for the EU at the global scale (Orgiazzi *et al.*, 2018; Köninger *et al.*, 2023). In 2018, the LUCAS Soil survey included soil biodiversity analysis based on DNA metabarcoding (Orgiazzi *et al.*, 2015) for the first time, and the biodiversity assessment continued in the following round of sampling that took place in 2022. The high-throughput sequencing method of nematode community analysis was performed (Sapkota and Nicolaisen, 2015), but the survey results for 2022 were not available for public review during the preparation of this chapter.

Furthermore, there have been projects and initiatives that strongly focused on soil biodiversity assessment and working towards environmental policy advisory. Three such projects, called SoildiverAgro (H2020 Project ID 817819; https://cordis.europa.eu/project/id/817819, accessed 8 September 2025), SOILGUARD (H2020 Project ID 101000371; https://cordis.europa.eu/project/id/101000371, accessed 8 September 2025) and EXCALIBUR (H2020 Project ID 817946; https://cordis.europa.eu/project/id/817946, accessed 8 September 2025), had secured their funding from the EU's Horizon 2020 research and innovation programme until May 2024, May 2025 and November 2024, respectively. The SoildiverAgro listed several project objectives, including the identification of the main agricultural activities threatening soil biodiversity and expanding knowledge on the genetic and functional diversity of soil macro- and microorganisms. Their activities will contribute to the ongoing development of protocols for soil biodiversity assessment and monitoring purposes, under different production systems and management practices across Europe. In the course of its biodiversity evaluation, the project included analysis of nematode diversity and community characterization via molecular approaches (Hendrickx, 1995; Waeyenberge *et al.*, 2019). The SOILGUARD project promoted sustainable soil management, to unleash soil biodiversity potential and increase environmental, social and economic well-being at EU and international levels. SOILGUARD aimed to create an analytical framework with the potential to become the global standard for future soil biodiversity assessment, as well as an evidence-based conservation recommendation towards soil policies. The EXCALIBUR project aimed to improve the overall knowledge on soil biodiversity dynamics to enhance the efficacy and application of biocontrol and biofertilization practices in horticultural farming. By exploiting the multifunctional potential of below-ground biodiversity, the project planned to provide a soil-biodiversity-driven management strategy for farmers. The project's biological survey encompassed analysis of nematode diversity and community composition via morphological approaches (Goodey, 1963; Bongers, 1990; Yeates *et al.*, 1993; Bongers, 1999; Ferris *et al.*, 2001; Sieriebriennikov *et al.*, 2014).

The Soil Biodiversity Observation Network (Soil BON) (https://geobon.org/bons/thematic-bon/soil-bon/, accessed 12 September 2025) was established in 2018, working closely with the Global Soil Biodiversity Initiative (GSBI) (https://www.globalsoilbiodiversity.org/, accessed 12 September 2025) and other global and regional partners (Guerra *et al.*, 2021a). It aims to create a foundation for the establishment of a soil biodiversity monitoring system globally (Guerra *et al.*, 2021b). The Soil BON initiative operates under the Group on Earth Observations Biodiversity Observation Network (GEOBON) (https://geobon.org/, accessed 12 September 2025). It is working towards developing a global biodiversity observation network, expanding soil-specific Essential Biodiversity Variables (EBVs) and supporting environmental policies (CBD, SDGs, Paris Agreement). Soil nematode total abundance, as well as taxonomic and functional diversity, were recognized as essential biodiversity variables and as part of a holistic representation of the soil system, its biodiversity and functionality (Guerra *et al.*, 2021a). The Soil BON Foodweb (https://soilbonfoodweb.org/, accessed 12 September 2025) is a voluntary initiative established in 2021 as an extension of Soil BON. The aim of the initiative is to expand the scope of the Soil BON by linking soil biodiversity to ecosystem functioning via a soil food web perspective (Potapov *et al.*, 2022). The biological survey outcome, including nematode analysis details, can be uploaded to a publicly available EUdaphobase (https://www.eudaphobase.eu/, accessed 12 September 2025) data warehouse, designated for soil biodiversity data storage and evaluation in Europe.

In recent years, the different approaches to address the analyses of nematode community assemblages (i.e. DNA metabarcoding, morphological analyses) and the application of nematode indices have allowed for their adoption in projects, initiatives, monitoring frameworks and laws aimed at assessing and monitoring soil biodiversity. Their main focus is to allow for the development of strategies for farmers and soil policy makers to halt soil biodiversity loss. The continuous development of nematodes as biomonitoring indicators can help these efforts by linking soil biodiversity to: (i) specific ecosystem functions (see Chapter 7, this volume); and (ii) the expansion of transdisciplinary networks with different stakeholders to address soil biodiversity loss from several perspectives.

16.6 Environmental Policies in Aquatic Environments

Aquatic environments are subjected to threats from multiple human activities; where species and habitats have been assessed, cumulative impacts that impair the overall ecosystem resilience are revealed. Many directives, especially in the EU, prioritize the need for the assessment of ecological quality status (EQS), the achievement or maintenance of good environmental status of ground- and surface waters, as well as the establishment of environmental targets and associated indicators to assure a high resolution of the EQS (Table 16.2). Despite the urgent need for an improved understanding of ecosystem health status and functioning, for making accurate decisions and plan managements and/or conservation strategies, only macro- and mega-bioindicators are considered (Table 16.2) (Semprucci *et al.*, 2015; Schratzberger *et al.*, 2023). None of the European directives prioritizes aquatic nematodes and there is no single ongoing, nor planned, monitoring strategy of a national agency based on, or prioritizing, marine nematode diversity in their standard protocols (Danovaro *et al.*, 2021). Yet monitoring early changes, by establishing indicators or sentinels at the base of food webs and ecosystem functions, such as nematodes, rather than among its end members, might allow more efficient monitoring and therefore timely conservation responses (Ingels *et al.*, 2021). The erroneous idea that only large, tasty or dangerous organisms are important is not the only reason that impairs the adoption of marine nematodes in relevant environmental policies. A secondary factor is related to their difficult and time-consuming taxonomic identification, combined with the enormous biodiversity of the phylum Nematoda, especially in marine habitats (Semprucci *et al.*, 2022). Certainly, the implementation of environmental DNA (eDNA) metabarcoding is a challenge for the future, but further efforts are necessary to integrate nematodes into international and national environmental policies.

Free-living marine and freshwater nematodes have different characteristics that make them ideal indicators of EQS (see Chapters 1, 2 and 3, this volume). Therefore, nematodes are a powerful tool to achieve the goals of the different directives mentioned in Table 16.2. For instance, to assess the quality of drinking water (Drinking Water Directive (DWD)), nematode species known as sensitive to pollutants are suitable test sentinels and bioindicators of water quality (see Chapter 6, this volume). Similarly, toxicity testing with nematodes is a rapid and relevant alternative to other standard toxicity tests and can assess the quality of waters (see Chapter 13, this volume). This approach can also be applied for assessing the quality of bathing or drinking waters, in accordance with the relevant directives (i.e. Bathing Water Directive and DWD; Table 16.2).

Besides, community assessment of nematode assemblages has been proven to be an efficient tool to investigate environmental quality and provide support for risk assessment policies, coastal management and pollution assessment (see Chapters 6 and 13, this volume). Therefore, the investigation of the composition and distribution of marine and freshwater nematode communities is a strong tool for achieving the goals of different environmental policies such as the Marine Strategy Framework Directive and the Oceans Act.

The WFD (EC, 2000) aims at achieving good quality status of groundwaters and surface waters, by focusing on their biological, chemical, physical and quantitative status. Although the focus is on surface waters, the directive recognizes the importance of benthic assessment for the quality of waterbodies as a whole. Benthic invertebrate fauna is already listed as one of

Table 16.2. Overview of European water-related policies. (Author's own table.)

Policy/Regulation	Country	Description of the policy	Reference
Floods Directive (2007/60/EC)	Europe	This legislation establishes a framework on the assessment and management of flood risk, aiming at the reduction of the adverse consequences of floods for human health, the environment, cultural heritage and economic activities	These Regulations may be cited as the Water Environment (Floods Directive) Regulations (Northern Ireland) 2009 and have been in operation since 14 December 2009 (HM Government, 2009)
Drinking Water Directive (DWD) (98/83/EC) – 2020/2184	Europe	Protects human health from the adverse effects of any contamination of water intended for human consumption by ensuring that it is wholesome and clean	EU (2020b)
Urban Water Directive (91/271/EEC)	Europe	Protects the water environment from the adverse effects of discharges of urban wastewater and from certain industrial discharges	EEC (1991b)
Bathing Water Directive (06/7/EC)	Europe	Concerns the management of bathing water quality and repeals Directive 76/160/EEC	EC (2006)
Water Framework Directive (WFD) (2000/60/EC)	Europe	Aims to achieve good qualitative and quantitative status of all waterbodies (including marine waters up to 1 nautical mile from shore) by 2015	EC (2000)
Nitrates Directive (91/676/EC)	Europe	Intends to improve water quality by protecting water (rivers, lakes, marine) against pollution caused by nitrates from agricultural sources	EEC (1991a)
Marine Strategy Framework Directive (2008/56/EC)	Europe	Requires Member States to achieve or maintain good environmental status (GES) in the marine environment by the year 2020 at the latest. It defines 11 high-level descriptors of GES including 'Maintaining the biological diversity of marine habitats and species' and 'Limiting contaminants to the marine environment to levels which do not cause pollution'	EC (2008)
Habitats Directive (92/43/EEC)	Europe	Adopted after the Birds Directive and requires all member states to establish a strict protection regime for species listed in Annex IV, both inside and outside Natura 2000 sites	EEC (1992)
Water Reuse (EU Regulation 2020/741)	Europe	Aims to alleviate climate change-related water scarcity by fostering water reuse in agricultural irrigation. To further promote and facilitate its implementation, new rules were implemented by June 2023	EU (2020c)
National Water Act 1998	South Africa	Intends to provide for fundamental reform of the law relating to water resources in terms of protection, use, development, conservation, management and control	South African Government (1998)
Oceans Act 1997	Canada, USA	Promotes integrated, collaborative and inclusive decision making for oceans-related activities, the sustainable use of ocean and coastal resources, and the protection of biodiversity and ecosystem health, including pollution prevention	US Congress (1997)

the main 'biological quality elements' to be assessed. Such specification is an opportunity to integrate and promote aquatic nematodes in environmental policies. But there is still a need for a bigger effort from environmental nematologists to achieve this goal and bring awareness to policy makers, as well as to society in general, about the importance of marine and freshwater nematode species and assemblages and why they should no longer be ignored in environmental policies. Related to that is also the opportunity for marine nematologists to engage in the UN Decade for Ocean Science for Sustainable Development 2021–2030, which is connected to the UN SDG 14 'Life below water' which aims to 'conserve and sustainably use the world's oceans, seas and marine resources' (UN, 2015).

Coordinated efforts among nematologists, environmental scientists and policy makers can offer an opportunity to develop integral indicators to monitor water quality and pollution and, at the same time, aquatic biodiversity loss and restoration efforts. The utilization of nematodes and their communities as environmental indicators across diverse ecosystems (marine, aquatic, terrestrial) can provide valuable insights on food web structures and ecosystem health. It can also serve as a common ground for assessing these diverse ecosystems and the ecosystem services they provide.

16.7 Conclusion

Nematodes constitute the only phylum whose members have been thoroughly studied, developed and applied as environmental indicators at gene, cellular, organism and community levels. To date, there is no other phylum to offer this service at all these levels of hierarchical order. There have been several reviews which have identified the best features that certain organisms need to have in order to be utilized as effective bioindicators. Among other characteristics, ideal bioindicator organisms should be:

- Abundant in their ecosystems and easily manipulated.
- Easily sampled and categorized.
- Cost-effective and easy to rear in the laboratory.
- Representative of their habitat, whether this is terrestrial or aquatic.
- Known to display well-defined responses to environmental challenges.

In light of the above criteria, no other group of organisms offers as much potential as nematodes for this purpose. Nematodes can be used to study gene expression in relation to environmental challenges, they can be used in laboratory ecotoxicity tests, or their assemblages can be studied *in situ* to monitor the environmental impact of a wide range of challenges.

Although the new EU Soil Strategy for 2030 (https://eur-lex.europa.eu/legal-content/EN/TXT/?uri=CELEX%3A52021DC0699, accessed 10 September 2025) has recently been established, and the WFD (https://environment.ec.europa.eu/topics/water/water-framework-directive_en, accessed 10 September 2025) has been in implementation for years, neither of these policies includes the usage of nematodes for environmental biomonitoring. This is not only unfortunate but also may pose certain threats to various environmental management efforts, as these cannot achieve their full potential due to this deficiency. It is therefore imperative that the international community of environmental nematologists mobilize, as sole scientists and/or also collectively, as circumstances dictate, and take action to inform policy makers about the value and the necessity of adopting and integrating nematodes and relevant methodology in environmental monitoring, restoration and management efforts.

References

Acharya Balkrishna, A.K., Arya, V., Sharma, G., Srivastava, D., Sharma, A. and Khan, N. (2022) Situational analysis of government initiatives for the welfare of farmers in India: impact and futuristic insights. *Indian Journal of Ecology* 49(3), 1023–1035.

Álvarez, J.A., Ávila, C., Otter, P., Kilian, R., Istenič, D. *et al.* (2017) Constructed wetlands and solar-driven disinfection technologies for sustainable wastewater treatment and reclamation in rural India: SWINGS project. *Water Science and Technology* 76(6), 1474–1489. DOI: 10.2166/wst.2017.329.

Bagnoli, P., Goeschl, T. and Kovacs, E. (2008) *People and Biodiversity Policies: Impacts, Issues and Strategies for Policy Action*. OECD Publishing, Paris. DOI: 10.1787/9789264034341-en.

Bongers, T. (1990) The maturity index: an ecological measure of environmental disturbance based on nematode species composition. *Oecologia* 83, 14–19.

Bongers, T. (1999) The Maturity Index, the evolution of nematode life history traits, adaptive radiation and cp-scaling. *Plant and Soil* 212, 13–22.

Bongers, T., Ilieva-Makulec, K. and Ekschmitt, K. (2001) Acute sensitivity of nematode taxa to $CuSO_4$ and relationships with feeding-type and life-history classification. *Environmental Toxicology and Chemistry* 20(7), 1511–1516. DOI: 10.1002/etc.5620200714.

Brouwer, F.M. and Van Berkum, S. (1996) *CAP and Environment in the European Union: Analysis of the Effects of the CAP on the Environment and Assessment of Eexisting Environmental Conditions in Policy*. Wageningen Press, Wageningen, The Netherlands, p. 171.

Burity, V.T.A., Melgarejo, L., Gonzáles, J.C.M., Prates, L.A. and Rocha, N.C. (2020) *Pesticides in Latin America: Violations Against the Right to Adequate Food and Nutrition. 2020 Regional Report*. FIAN Brasil – Organization for the Human Right to Adequate Food and Nutrition, Brasília.

Cardinale, B.J., Duffy, J.E., Gonzalez, A., Hooper D.U., Perrings, C. *et al.* (2012) Biodiversity loss and its impact on humanity. *Nature* 486, 59–67. DOI: 10.1038/nature11148.

Chauvin, C., Trambolho, M., Hedde, M., Makowski, D., Cérémonie, H. *et al.* (2020) Soil nematodes as indicators of heavy metal pollution: a meta-analysis. *Open Journal of Science* 10, 579–601. DOI: 10.4236/ojss.2020.1012028.

Chen, M., Zhou, S., Zhu, Y, Sun, Y, Zeng, G. *et al.* (2018) Toxicity of carbon nanomaterials to plants, animals and microbes: recent progress from 2015–present. *Chemosphere* 206, 255–264.

Danovaro, R., Fanelli, E., Aguzzi, J., Billett, D., Carugati, L. *et al.* (2021) Reply to: Ecological variables for deep-ocean monitoring must include microbiota and meiofauna for effective conservation. *Nature Ecology & Evolution* 5, 30–31. DOI: 10.1038/s41559-020-01337-4.

Davies, J.C. and Mazurek, J. (1998) *Pollution Control in the United States: Evaluating the System.* Resources for the Future, New York, p. 319. DOI: 10.4324/9781315060705.

Decaëns, T, Jiménez, J.J., Gioia, C., Measey, G.J. and Lavelle, P. (2006) The values of soil animals for conservation biology. *European Journal of Soil Biology* 42, S23–S38. DOI: 10.1016/j.ejsobi.2006.07.001.

DEFRA (Department for Environment, Food and Rural Affairs) (2013) UK National Action Plan for the Sustainable Use of Pesticides (Plant Protection Products). Available at: https://www.gov.uk/government/publications/pesticides-uk-national-action-plan/ (accessed 14 November 2024).

Del Signore, A., Hendriks, A.J., Lenders, H.J., Leuven, R.S. and Breure, A.M. (2016) Development and application of the SSD approach in scientific case studies for ecological risk assessment. *Environmental Toxicology and Chemistry* 35, 2149–2161. DOI: 10.1002/etc.3474.

Department of Environmental Affairs (2020) *Annual Report 2019/20*. Department of Environmental Affairs, Pretoria. Available at: https://www.gov.za/sites/default/files/gcis_document/202104/department-environmental-affairs-annual-report-2019-20_0.pdf (accessed 10 March 2023).

Department of Environmental Affairs and Tourism (1997) White Paper on the Conservation and Sustainable Use of South Africa's Biological Diversity. *Government Gazette*, 28 July 1997. Available at: https://faolex.fao.org/docs/pdf/saf211449.pdf (accessed 10 April 2023).

De Zwart, D. (2002) Observed regularities in SSDs for aquatic species. In: Posthuma, L., Suter, G.W. and Traas, T.P. (eds) *Species Sensitivity Distributions in Ecotoxicology*. CRC Press, Boca Raton, Florida, pp. 133–154.

Du Preez, G., Daneel, M., De Goede, R., Du Toit, M.J., Ferris, H. *et al.* (2022) Nematode-based indices in soil ecology: application, utility and future directions. *Soil Biology and Biochemistry* 169, 108–640. DOI: 10.1016/j.soilbio.2022.108640.

EC (European Community) (2000) Directive 2000/60/EC of the European Parliament and of the Council of 23 October 2000 establishing a framework for Community action in the field of water policy. Available at: https://eur-lex.europa.eu/eli/dir/2000/60/oj/eng (accessed 16 March 2025).

EC (European Community) (2006) Directive 2006/7/EC of the European Parliament and of the Council of 15 February 2006 concerning the management of bathing water quality and repealing Directive 76/160/EEC. Available at: https://eur-lex.europa.eu/eli/dir/2006/7/oj/eng (accessed 16 March 2025).

EC (European Community) (2008) Directive 2008/56/EC of the European Parliament and of the Council of 17 June 2008 establishing a framework for community action in the field of marine environmental policy (Marine Strategy Framework Directive). Available at: https://eur-lex.europa.eu/eli/dir/2008/56/oj/eng (accessed 16 March 2025).

EC (European Community) (2009) Regulation (EC) No 1107/2009 of the European Parliament and of the Council of 21 October 2009 concerning the placing of plant protection products on the market and repealing Council Directives 79/117/EEC and 91/414/EEC. Available at: https://eur-lex.europa.eu/eli/reg/2009/1107/oj/eng (accessed 16 March 2025).

EEC (European Economic Community) (1991a) Council Directive 91/676/EEC of 12 December 1991 concerning the protection of waters against pollution caused by nitrates from agricultural sources. Available at: https://eur-lex.europa.eu/eli/dir/1991/676/oj/eng (accessed 16 March 2025).

EEC (European Economic Community) (1991b) Council Directive 91/271/EEC of 21 May 1991 concerning urban waste-water treatment. Available at: https://eur-lex.europa.eu/eli/dir/1991/271/oj/eng (accessed 16 March 2025).

EEC (European Economic Community) (1992) Council Directive 92/43/EEC of 21 May 1992 on the conservation of natural habitats and of wild fauna and flora. Available at: https://eur-lex.europa.eu/eli/dir/1992/43/oj/eng (accessed 16 March 2025).

European Commission (2018) Proposal for a Regulation of the European Parliament and of the Council establishing rules on support for strategic plans to be drawn up by Member States under the Common agricultural policy (CAP Strategic Plans) and financed by the European Agricultural Guarantee Fund (EAGF) and by the European Agricultural Fund for Rural Development (EAFRD) and repealing Regulation (EU) No 1305/2013 of the European Parliament and of the Council and Regulation (EU) No 1307/2013 of the European Parliament and of the Council. COM (2018), 392. Available at: https://eur-lex.europa.eu/legal-content/EN/TXT/PDF/?uri=CELEX:52018PC0392 (accessed 16 March 2025).

European Commission (2019) The European Green Deal. COM (2019) 640 final. Available at: https://commission.europa.eu/strategy-and-policy/priorities-2019-2024/european-green-deal_en (accessed 16 March 2025).

European Commission (2020) A Farm to Fork Strategy for a fair, healthy and environmentally friendly food system. COM (2020) 381 final. Available at: https://www.europarl.europa.eu/factsheets/en/sheet/293547/the-farm-to-fork-strategy (accessed 16 March 2025).

European Commission (2021a) Pathway to a Healthy Planet for All. Available at: https://eur-lex.europa.eu/legalcontent/EN/TXT/?uri=CELEX%3A52021DC0400&qid=1623311742827 (accessed 18 April 2023).

European Commission (2021b) EU Soil Strategy for 2030. Available at: https://environment.ec.europa.eu/publications/eu-soil-strategy-2030_en (accessed 8 April 2023).

European Commission (2023) Proposal for a Directive of the European Parliament and of the Council on Soil Monitoring and Resilience (Soil Monitoring Law). Available at: https://eur-lex.europa.eu/legal-content/EN/TXT/?uri=CELEX:52023PC0416 (accessed 26 July 2024).

EU (European Union) (2020a) EU Biodiversity Strategy for 2030: Bringing Nature Back into Our Lives. Available at: https://op.europa.eu/en/publication-detail/-/publication/31e4609f-b91e-11eb-8aca-01aa75ed71a1 (accessed 16 March 2025).

EU (European Union) (2020b) Directive (EU) 2020/2184 of the European Parliament and of the Council of 16 December 2020 on the quality of water intended for human consumption (recast). Available at: https://eur-lex.europa.eu/eli/dir/2020/2184/oj/eng (accessed 16 March 2025).

EU (European Union) (2020c) Regulation (EU) 2020/741 of the European Parliament and of the Council of 25 May 2020 on minimum requirements for water reuse. Available at: https://eur-lex.europa.eu/eli/reg/2020/741/oj/eng (accessed 16 March 2025).

EU (European Union) (2021) Regulation (EU) 2021/1119 of The European Parliament and of The Council of 30 June 2021 establishing the framework for achieving climate neutrality and amending Regulations (EC) No 401/2009 and (EU) 2018/1999 ('European Climate Law'). Available at: https://op.europa.eu/en/publication-detail/-/publication/365a2e8e-e04f-11eb-895a-01aa75ed71a1/language-en (accessed 16 March 2025).

EU Monitor (2010) Directive 2010/75/EU of the European Parliament and of the Council of 24 November 2010 on industrial emissions (integrated pollution prevention and control) (recast). Available at: https://eur-lex.europa.eu/eli/dir/2010/75/oj/eng (accessed 12 November 2024).

Faber, J.H., Cousin, I., Meurer, K.H.E., Hendriks, C.M.J., Bispo, A. *et al.* (2022) Stocktaking for Agricultural Soil Quality and Ecosystem Services Indicators and their Reference Values (SIREN). EJP SOIL Internal Project SIREN Deliverable 2 Report. Available at: https://edepot.wur.nl/582329 (12 September 2025).

Ferraz, A.F., Pastorinho, M.R., Palmeira-de-Oliveira, A. and Sousa, A.C.A. (2022) Ecotoxicity of plant extracts and essential oils: a review. *Environmental Pollution* 292(Part B), 118319. DOI: 10.1016/j.envpol.2021.118319.

Ferris, H., Bongers, T. and Goede, R. (2001) A framework for soil food web diagnostics: extension of the nematode faunal analysis concept. *Applied Soil Ecology* 18, 13–29.

Fesenmyer, K.A., Wenger, S.J., Leigh, D.S. and Neville, H.M. (2021) Large portion of USA streams lose protection with new interpretation of Clean Water Act. *Freshwater Science* 40(1), 252–258. DOI: 10.1086/713084.

Fox, D.R., Van Dam, R.A., Fisher, R., Batley, G.E., Tillmans, A.R. *et al.* (2021) Recent developments in species sensitivity distribution modeling. *Environmental Toxicology and Chemistry* 40(2), 293–308. DOI: 10.1002/etc.4925.

Frampton, G.K., Jänsch, S., Janeck J.S., Römbke, J. and Van den Brink, P.J. (2006) Effects of pesticides on soil invertebrates in laboratory studies: a review and analysis using species sensitivity distributions. *Environmental Toxicology and Chemistry* 25(9), 2480–2489.

Goodey, J.B. (1963) *Laboratory Methods for Work with Plant and Soil Nematodes*. Technical Bulletin No. 2. Ministry of Agriculture, Fisheries and Food, London.

Government of India (2014) Integrated Pest Management. Available at: https://pib.gov.in/newsite/PrintRelease.aspx?relid=110364 (accessed 12 November 2024).

Government of South Africa (2015) *South Africa's National Biodiversity Strategy and Action Plan 2015–2025*. Department of Environmental Affairs, Pretoria. Available at: https://www.dffe.gov.za/sites/default/files/docs/publications/SAsnationalbiodiversity_strategyandactionplan2015_2025.pdf (accessed 26 August 2023).

Guerra, C.A., Bardgett, R.D., Caon, L., Crowther, T.W., Delgado-Baquerizo, M. *et al.* (2021a) Tracking, targeting, and conserving soil biodiversity. *Science* 371(6526), 239–241. DOI: 10.1126/science.abd7926.

Guerra, C.A., Wall, D. and Eisenhauer, N. (2021b) Unearthing soil ecological observations: see supporting information as supplementary material. *Soil Organisms* 93(2), 79–81. DOI: 10.25674/so93iss2id164.

Hägerbäumer, A., Höss, S., Heininger, P. and Traunspurger, W. (2015) Experimental studies with nematodes in ecotoxicology: an overview. *Journal of Nematology* 47, 11–27.

Henderson, B., Lankoski, J., Flynn, E., Sykes, A., Payen, F. and MacLeod, M. (2022) *Soil Carbon Sequestration by Agriculture – Policy Options*. OECD Food, Agriculture and Fisheries Paper No. 174. OECD Publishing, Paris.

Hendrickx, G. (1995) An automatic apparatus for extracting free-living nematode stages from soil. *Nematologica* 41, 308.

Hennig, T.B., Ogliari Bandeira, F., Costa Puerari, R., Fernandes Fraceto, L. and Gerson Matias, W. (2023) A systematic review of the toxic effects of a nanopesticide on non-target organisms: estimation of protective concentrations using a species sensitivity distribution (SSD) approach – the case of atrazine. *Science of The Total Environment* 871, 162094. DOI: 10.1016/j.scitotenv.2023.162094.

HM Government (2009) The Water Environment (Floods Directive) Regulations (Northern Ireland) 2009. Available at: https://www.legislation.gov.uk/nisr/2009/376/contents/made (accessed 16 March 2025).

HM Government (2021) Environment Act 2021. Available at: https://www.legislation.gov.uk/ukpga/2021/30/contents (accessed 14 November 2024).

Hou, D. (2021) Expediting climate-smart soils management. *Soil Use and Management* 38, 1–6.

The Independent (2019) EU threatens to slap fresh ban on Uganda's agro exports to Europe. Available at https://www.independent.co.ug/eu-threatens-to-slap-fresh-ban-on-ugandas-agro-exports-to-europe/ (accessed 8 March 2025).

Ingels, J., Vanreusel, A., Pape, E., Pasotti, F., Macheriotou, L. *et al.* (2021) Ecological variables for deep-ocean monitoring must include microbiota and meiofauna for effective conservation. *Nature Ecology & Evolution* 5, 27–29. DOI: 10.1038/s41559-020-01335-6.

IPCC (Intergovernmental Panel on Climate Change) (2006) *2006 IPCC Guidelines for National Greenhouse Gas Inventories. Prepared by the National Greenhouse Gas Inventories Programme* [Eggleston, H.S., Buendia, L., Miwa, K., Ngara, T. and Tanabe, K. (eds)]. Institute for Global Environmental Strategies, Hayama, Japan.

IPCC (Intergovernmental Panel on Climate Change) (2019) *2019 Refinement to the 2006 IPCC Guidelines for National Greenhouse Gas Inventories* [Calvo Buendia, E., Tanabe, K., Kranjc, A., Baasansuren, J., Fukuda, M. *et al.* (eds)]. IPCC, Geneva, Switzerland.

Isgren, E. and Andersson, E. (2021) An environmental justice perspective on smallholder pesticide use in sub-Saharan Africa. *The Journal of Environment & Development* 30(1), 68–97.

Jadeja, N.B., Banerji, T., Kapley, A. and Kumar, R. (2022) Water pollution in India – current scenario. *Water Security* 16, 100–119. DOI: 10.1016/j.wasec.2022.100119.

Joemat-Pettersson, T. (2010) Adoption of pesticide management policy for South Africa. Minister of Agriculture Forestry and Fisheries acting under Section 7bis of the Fertilizers, Farm Feeds, Agricultural Remedies and Stock Remedies Act, 1947 (Act No. 36 of 1947) hereby publish the Pesticide Management Policy for South Africa. Available at: https://www.gov.za/sites/default/files/gcis_document/201409/338991120.pdf (accessed 16 March 2025).

Kanter, D.R., Bartolini, F., Kugelberg, S., Leip, A., Oenema, O. and Uwizeye, A. (2020) Nitrogen pollution policy beyond the farm. *Nature Food* 1(1), 27–32.

Karpinska, A., Ryan, D., Germaine, K., Dowling, D., Forrestal, P. and Kakouli-Duarte, T. (2021) Soil microbial and nematode community response to the field application of recycled bio-based fertilisers in Irish grassland. *Sustainability* 13(22), 12342. DOI: 10.3390/su132212342.

Kennard, N. and Vagnoni, C. (2021) Pesticides and health. POSTbrief 43, UK Parliament POST. Accessible at: https://post.parliament.uk/research-briefings/post-pb-0043/ (accessed 8 March 2025).

Kim, D., Kim, H. and An, Y.J. (2023) Species sensitivity distributions of micro- and nanoplastics in soil based on particle characteristics. *Journal of Hazardous Materials* 452, 131229. DOI: 10.1016/j.jhazmat.2023.131229.

Köninger, J., Ballabio, C., Panagos, P., Jones, A., Schmid, M.W. *et al.* (2023) Ecosystem type drives soil eukaryotic diversity and composition in Europe. *Global Change Biology* 29, 5706–5719. DOI: 10.1111/gcb.16871.

Li, T., Liu, Y., Lin, S., Liu, Y. and Xie, Y. (2019) Soil pollution management in China: a brief introduction. *Sustainability* 11(3), 556. DOI: 10.3390/su11030556.

Loha, K.M., Lamoree, M., Weiss, J.M. and de Boer, J. (2018) Import, disposal, and health impacts of pesticides in the East Africa Rift (EAR) zone: a review on management and policy analysis. *Crop Protection* 112, 322–331.

Mangalagiu, D., Mukherjee, N., Pascal, D.A., Davies, J., Gonçalves LR. *et al.* (2019) Biodiversity policy. In: Ekins, P., Gupta, J. and Boileau, P. (eds) *Global Environmental Outlook 6: Healthy Planet, Healthy People*. Cambridge University Press, Cambridge, pp. 140–173. DOI: 10.1017/9781108627146.

Matemilola, S. and Fadeyi, O. (2020) Bamako Convention. In: Idowu, S., Schmidpeter, R., Capaldi, N., Zu, L., Del Baldo, M. and Abreu, R. (eds) *Encyclopaedia of Sustainable Management*. Springer, Cham, Switzerland, pp. 215–219.

Millennium Ecosystem Assessment (2005) *Ecosystems and Human Well-Being*. Island Press, Washington, DC. Available at: https://www.millenniumassessment.org/documents/document.356.aspx.pdf (accessed 10 April 2023).

Montanarella, L. and Panagos, P. (2021) The relevance of sustainable soil management within the European Green Deal. *Land Use Policy* 100, 104950.

Nigerian Oil Spill Monitor (2023) Visualising Oil Spill Data from NOSDRA 2023. Available at: https://nosdra.oilspillmonitor.ng (accessed 10 April 2023).

NSW EPA (New South Wales Environment Protection Authority) (2017) Regulating pesticides in NSW. Available at: https://www.epa.nsw.gov.au/your-environment/pesticides/pesticides-nsw-overview/regulating-pesticides-nsw (accessed 14 November 2024).

Oertel, C., Matschullat, J., Zurba, K., Zimmermann, F. and Erasmi, S. (2016) Greenhouse gas emissions from soils – a review. *Chemie der Erde* 76, 327–352.

Orgiazzi, A., Dunbar, M.B., Panagos, P., de Groot, A.G. and Lemanceau, P. (2015) Soil biodiversity and DNA barcodes: opportunities and challenges. *Soil Biology and Biochemistry* 80, 244–250. DOI: 10.1016/j.soilbio.2014.10.014.

Orgiazzi, A., Ballabio, C., Panagos, P., Jones, A. and Fernández-Ugalde, O. (2018) LUCAS Soil, the largest expandable soil dataset for Europe: A review. *European Journal of Soil Science* 69, 140–153. DOI: 10.1111/ejss.12499.

Posthuma, L., van Gils, J., Zijp, M.C., van de Meent, D. and de Zwart, D. (2019) Species sensitivity distributions for use in environmental protection, assessment, and management of aquatic ecosystems for 12386 chemicals. *Environmental Toxicology and Chemistry* 38, 905–917. DOI: 10.1002/etc.4373.

Potapov, A.M., Sun, X., Barnes, A.D., Briones, M.J., Brown, G.G. *et al.* (2022) Global monitoring of soil animal communities using a common methodology. *Soil Organisms* 94(1), 55–68. DOI: 10.25674/so94iss1id178.

Pulavarty, A., Horgan, K. and Kakouli-Duarte, T. (2020) Effect of an Alltech soil health product on entomopathogenic nematodes, root-knot nematodes and on the growth of tomato plants in the greenhouse. *Journal of Nematology* 52(1), e2020-14. DOI: 10.21307/JOFNEM-2020-014.

Rizzi, C., Villa, S., Cuzzeri, A.S. and Finizio, A. (2021) Use of species sensitivity distribution approach to derive ecological threshold of toxicological concern (eco-TTC) for pesticides. *International Journal of Environmental Research and Public Health* 18(22), 12078. DOI: 10.3390/ijerph182212078.

Ryan, D., Karpinska, A., Forrestal, P.J., Ashekuzzaman, S.M., Kakouli-Duarte, T. *et al.* (2022) The impact of bio-based fertilizer integration into conventional grassland fertilization programmes on soil bacterial, fungal, and nematode communities. *Frontiers in Sustainable Food Systems* 6, 832841. DOI: 10.3389/fsufs.2022.832841.

Saju, A., Van De Sande, T., Ryan, D., Karpinska, A., Sigurnjak, I. *et al.* (2023) Exploring the short-term in-field performance of Recovered Nitrogen from Manure (RENURE) materials to substitute synthetic nitrogen fertilisers. *Cleaner and Circular Bioeconomy* 5, 100043. DOI: 10.1016/j.clcb.2023.100043.

Sapkota, R. and Nicolaisen, M. (2015) High-throughput sequencing of nematode communities from total soil DNA extractions. *BMC Ecology* 15, 3. DOI: 10.1186/s12898-014-0034-4.

SCBD (Secretariat of the Convention on Biological Diversity) (2011) *Year in Review 2011: The Convention on Biological Diversity*. SCBD, Montreal, Canada. Available at: https://www.cbd.int/doc/reports/cbd-report-2011-en.pdf (accessed 16 March 2025).

SCBD (Secretariat of the Convention on Biological Diversity) (2015) *Synthetic Biology. Part I: Potential Impacts of Synthetic Biology on Biological Diversity. Part II: Gaps and Overlaps with the Provisions of the Convention and Other Agreements*. CBD Technical Series No. 82. SCBD, Montreal, Canada. Available at: https://www.cbd.int/ts/cbd-ts-82-en.pdf (accessed 10 April 2023).

SCBD (Secretariat of the Convention on Biological Diversity) (2022) Decision adopted by the Conference of the Parties to the Convention on Biological Diversity: 15/4. Kunming–Montreal Global Biodiversity Framework. Available at: https://www.cbd.int/doc/decisions/cop-15/cop-15-dec-04-en.pdf (accessed 10 April 2023).

Schmitt-Jansen, M. and Altenburger, R. (2005) Predicting and observing responses of algal communities to photosystem II–herbicide exposure using pollution-induced community tolerance and species-sensitivity distributions. *Environmental Toxicology and Chemistry* 24, 304–312. DOI: 10.1897/03-647.1.

Schratzberger, M., Danovaro, R., Ingels, J., Montagna, P., Rohal, M. *et al.* (2023) Hidden players – meiofauna mediate ecosystem effects of anthropogenic disturbances. In: Giere, O. and Schratzberger, M. (eds) *New Horizons in Meiobenthos Research – Profiles, Patterns and Potentials.* Springer, Heidelberg, Germany, pp. 175–255. DOI: 10.1007/978-3-031-21622-0_7.

Semprucci, F., Losi, V. and Moreno, M. (2015) A review of Italian research on free-living marine nematodes and the future perspectives on their use as Ecological Indicators (EcoInds). *Mediterranean Marine Science* 16, 352–365.

Semprucci, F., Grassi, E. and Balsamo, M. (2022) Simple is the best: an alternative method for the analysis of free-living nematode assemblage structure. *Water* 14, 1114.

Semprucci, F., Grassi, E., Cocozza di Montanara, A., Sandulli, R. and Baldrighi, E. (2025) Emerging marine nematodes as model organisms: which species for which question? *Diversity* 17, 59. DOI:10.3390/d17010059.

Shi, G., Luan, L., Zhu, G., Zeng, Z., Zheng, J. *et al.* (2023) Interaction between nematodes and bacteria enhances soil carbon sequestration under organic material amendments. *Frontiers in Microbiology* 14, 1155088. DOI: 10.3389/fmicb.2023.1155088.

Sieriebriennikov, B., Ferris, H. and Goede, R. (2014) NINJA: an automated calculation system for nematode-based biological monitoring. *European Journal of Soil Biology* 61, 90–93. DOI: 10.1016/j.ejsobi.2014.02.004.

Sikora, R.A., Desaeger, J. and Molendijk, L.P.G. (eds) (2021) Integrated nematode management and crop health: future challenges and opportunities. In: *Integrated Nematode Management: State-of-the-Art and Visions for the Future*. CAB International, Wallingford, UK, pp. 3–10. DOI: 10.1079/9781789247541.0000.

Smit, C.E., Schouten, A.J., Van den Brink, P.J., van Esbroek, M.L.P. and Posthuma, L. (2002) Effects of zinc contamination on a natural nematode community in outdoor soil mesocosms. *Archives of Environmental Contamination and Toxicology* 42, 2005–2216. DOI: 10.1007/s00244-001-0029-y.

Staples, C., Mihaich, E., Carbone, J., Woodburn, K. and Klecka, G. (2004) A weight of evidence analysis of the chronic ecotoxicity of nonylphenol ethoxylates, nonylphenol ether carboxylates, and nonylphenol. *Human and Ecological Risk Assessment* 10, 999–1017. DOI: 10.1080/10807030490887122.

South African Government (1998) National Water Act 36 of 1998. Available at: https://www.gov.za/documents/national-water-act (accessed 16 March 2025).

Steffen, W., Richardson, K., Rockström, J., Cornell, S.E., Fetzer, I. *et al.* (2015) Planetary boundaries: guiding human development on a changing planet. *Science* 347(6223), 1259855. DOI:10.1126/science.1259855.

Stephan, C.E. (2002) Use of species sensitivity distributions in the derivation of water quality criteria for aquatic life by the US Environmental Protection Agency. In: Posthuma, L., Suter, G.W. and Traas, T.P. (eds) *Species Sensitivity Distributions in Ecotoxicology*. CRC Press, Boca Raton, Florida, pp. 211–220.

Ten Brink, P., Chiabai, A., Rayment, M., Braeuer, I., Peralta, N. *et al.* (2008) The cost of policy inaction – in monetary terms. In: Braat, L. and ten Brink, P. (eds) *The Cost of Policy Inaction: The Cost of Not Meeting the 2010 Biodiversity Targets*. Alterra Report No. 1718. Alterra, Wageningen, Netherlands. Available at: edepot.wur.nl/152014 (accessed 10 April 2023).

Tibbet, M., Fraser, T.D. and Duddigan. S. (2020) Identifying potential threats to soil biodiversity. *PeerJ* 12(8), e9271. DOI: 10.7717/peerj.9271.

Tindwa, H.J. and Singh, B.R. (2023) Soil pollution and agriculture in sub-Saharan Africa: state of the knowledge and remediation technologies. *Frontiers in Soil Science* 2, 1101944. DOI: 10.3389/fsoil.2022.1101944.

Tittensor, D.P., Walpole, M., Hill, S.L.L., Boyce, D.G., Britten, G.L. *et al.* (2014) A mid-term analysis of progress toward international biodiversity targets. *Science* 346(6206), 241–244. DOI: 10.1126/science.1257484.

UN (United Nations) (2015) Transforming Our World: The 2030 Agenda for Sustainable Development. Available at: https://sdgs.un.org/2030agenda (accessed 5 May 2023).

UNEP (United Nations Environment Programme) (2011) UNEP 2011 Annual Report. Available at: https://www.unep.org/resources/annual-report/unep-2011-annual-report (accessed 16 March 2025).

UNEP (United Nations Environment Programme) (2020) Report of the Secretariat on the implementation of the Bamako Convention (UNEP/BC/COP.3/14), third meeting Brazzaville. (12–14 February 2020). Available at: https://wedocs.unep.org/bitstream/handle/20.500.11822/31479/ReportofSecretariat.pdf?sequence=1&isAllowed=y (accessed 12 September 2025).

UNEP (United Nations Environment Programme) (2025) Benin. Available at: https://www.unep.org/benin (accessed 28 February 2025).

UNFCCC (UN Framework Convention on Climate Change) (2016) The Paris Agreement – Publication. Available at: https://unfccc.int/documents/184656 (accessed 8 March 2025).

US Congress (1997) S.1213 – Oceans Act of 1997. Available at: https://www.congress.gov/bill/105th-congress/senate-bill/1213/text/rfh (accessed 16 March 2025).

USDA (US Department of Agriculture) (2017) China – Peoples Republic of China Releases Regulations on Pesticide Management. GAIN Report No. CH17030. Available at: https://apps.fas.usda.gov/newgainapi/api/report/downloadreportbyfilename?filename=China%20Released%20the%20Regulations%20on%20the%20Management%20of%20Pesticides_Beijing_China%20-%20Peoples%20Republic%20of_4-19-2017.pdf (accessed 16 March 2025).

US EPA (Environmental Protection Agency) (1990a) Pollution Prevention Act of 1990. Available at: https://www.epa.gov/p2/pollution-prevention-act-1990 (accessed 25 February 2025).

US EPA (Environmental Protection Agency) (1990b) Pollution Prevention Law and Policies. Available at: https://www.epa.gov/p2/pollution-prevention-law-and-policies (accessed 12 November 2024).

US EPA (Environmental Protection Agency) (1996) Summary of the Federal Insecticide, Fungicide, and Rodenticide Act. Available at: https://www.epa.gov/laws-regulations/summary-federal-insecticide-fungicide-and-rodenticide-act (accessed 14 November 2024).

Van Straalen, N.M. and Denneman, C.A.J. (1989) Ecotoxicological evaluation of soil quality criteria. *Ecotoxicology and Environmental Safety* 18, 241–251. DOI: 10.1016/0147-6513(89)90018-3.

Vimalkumar, K., Arun, E., Krishna-Kumar, S., Poopal, R.K., Nikhil, N.P. *et al.* (2018) Occurrence of triclocarban and benzotriazole ultraviolet stabilizers in water, sediment, and fish from Indian rivers. *Science of The Total Environment* 625, 1351–1360. DOI: 10.1016/j.scitotenv.2018.01.042.

Waeyenberge, L., de Sutter, N., Viaene, N. and Haegeman, A. (2019) New insights into nematode DNA-metabarcoding as revealed by the characterization of artificial and spiked nematode communities. *Diversity* 11(4), 52. DOI: 10.3390/d11040052.

Wang, W.M., Chen, C.Y., Lu, T.H. and Liao, C.M. (2022) Soil-dwelling species-based biomarker as a sensitivity-risk measure of terrestrial ecosystems response to microplastics: a dose–response modelling approach. *Science of The Total Environment* 833, 155178. DOI: 10.1016/j.scitotenv.2022.155178.

Wheeler, J.R., Grist, E.P.M., Leung, K.M.Y., Morritt, D. and Crane, M. (2002) Species sensitivity distributions: data and model choice. *Marine Pollution Bulletin* 45, 192–202. DOI: 10.1016/S0025-326X(01)00327-7.

WHO (World Health Organization) (2011) Guidelines on public health pesticide management policy for the WHO African Region. WHO/HTM/NTD/WHOPES/2011.2. Available at: https://www.who.int/publications/i/item/9789241501231 (accessed 16 March 2025).

Wilson, M.J. and Kakouli-Duarte, T. (2009) *Nematodes as Environmental Indicators*, 1st edn. CABI International, Wallingford, UK.

Wordsworth, F.J. (1993) The evolution of the Bamako Convention: an African perspective. *Colorado Journal of International Environmental Law and Policy* 4, 324–345.

World Bank (2010) Republic of Benin, Country Environmental Analysis: Benin – Analyse Environnementale Pays. Available at: https://openknowledge.worldbank.org/entities/publication/e24e425d-7991-5c31-a6a7-6e4c810e92c7 (accessed 26 August 2023).

Yeates, G.W., Bongers, T., de Goede, R.G.M., Freckman, D.W. and Georgieva, S.S. (1993) Feeding habits in soil nematode families and genera – an outline for soil ecologists. *Journal of Nematology* 25, 315–331.

17 Education and Environmental Nematology

Hendrika Fourie[1]*, Inge Dehennin[2], Ron G.M. de Goede[3], Laura Cortada[2], Gerard Korthals[4], Deborah A. Neher[5], Mike Hodda[6] and Wim Bert[2]

[1]Unit for Environmental Sciences and Management, North-West University, Potchefstroom, South Africa; [2]Nematology Research Unit, Faculty of Sciences, Ghent University, Belgium; [3]Soil Biology Group, Wageningen University and Research, The Netherlands; [4]Wageningen Plant Research, Wageningen University and Research, The Netherlands; [5]Department of Agriculture, Landscape and Environment, University of Vermont, Vermont, USA; [6]National Research Collections Australia, CSIRO, Canberra, Australia

Abstract

Training and education in biomonitoring require a strong understanding of nematode taxonomy, biology and ecology, and therefore considerable background and depth. To achieve effective training, there have been various educational programmes employing diverse approaches, including comprehensive long-term courses, units on biomonitoring within more general nematology programmes, modular courses, and short, intensive courses. Despite these efforts to improve content and delivery, the training opportunities remain geographically fragmented and often lack long-term support. To address these challenges, additional strategies are suggested to integrate environmental nematology into both formal and non-formal education curricula. Many strategies for improving training and education in biomonitoring particularly, and nematology more broadly, require increased funding, meaning that researchers and educators must effectively communicate the ecological significance of nematodes and the potential benefits of biomonitoring.

17.1 Introduction

Nematodes' ubiquity, abundance, diversity, rapid reproduction and involvement in numerous ecological processes make them ideal candidates for biomonitoring and as bioindicators. However, this complex field with diverse traits also complicates training and education in nematology. Current nematology training is predominantly centred on economically important plant parasites and their interactions with plants, with less focus on other facets of the discipline, such as biomonitoring – a situation that is far from ideal (Cortada *et al.*, 2019). Expanding education and training in other areas of nematology is crucial, not only for their intrinsic value but also to facilitate the broader use of nematodes as bioindicators. However, these efforts are currently limited, fragmented and often inconsistently scheduled. This chapter consolidates information on existing

*Email: driekie.fourie@nwu.ac.za

 Nematodes as Environmental Indicators: From Theory to Practice, 2nd Edition (T. Kakouli-Duarte *et al.* eds)

DOI: 10.1079/9781800624221.0017

and potential training initiatives related to biomonitoring with nematodes, while also proposing strategies to overcome the significant challenges faced.

Because there are few specific training courses or degree specializations in biomonitoring using nematodes, a substantial part of the training and education in this field of study is conducted as part of larger or more general nematology training. Training and education in nematology often require a wider approach than only teaching the correct and consistent identification of economically important plant-parasitic nematodes (PPNs) and animal-parasitic nematodes (APNs). It is necessary to interpret the effects of such species in the wider context of other nematodes, viz. free-living nematodes, and organisms in specific substrates. This is because the composition of nematode communities can vary greatly and significantly influence many ecosystem parameters and vice versa. Therefore, nematodes are particularly valuable for biomonitoring due to their high abundance, diversity (especially in non-polluted environments) and short life cycles. However, education and training in free-living nematodes, including biomonitoring, are often incorporated only as a smaller component of training focused on economically important groups like PPNs.

Free-living nematodes numerically dominate most habitats; they constitute overall approximately 73% of the total nematode community in terrestrial soil ecosystems compared to 27% for PPNs (van den Hoogen *et al.*, 2020), with an even larger proportion in marine and freshwater aquatic sediments (Hodda, 2022a). In all types of habitats, free-living nematodes are major components of food webs and perform critical ecological functions such as decomposition, carbon and nutrient cycling, sediment agitation, parasitism and predation on animals, plants and microbes. These are all ecosystem properties of concern in biomonitoring (Lawton *et al.*, 1995; Bloemers *et al.*, 1997; Hodda *et al.*, 1997; Hodda, 2022b). Free-living nematodes are a major component of most schemes for biomonitoring or using nematode indicators for broader ecosystem properties, such as resilience, evolutionary hotspots, and energy and nutrient transfers (Platt *et al.*, 1984; Hodda and Nicholas, 1986; Lambshead, 1986; Austen and McEvoy, 1997; Bongers and Ferris, 1999; Hodda *et al.*, 2009; Balsamo *et al.*, 2012; Traunspurger, 2021). However, free-living nematodes are under-represented in species descriptions, research and education (Neher, 2010; Sikora *et al.*, 2021; Hodda, 2022c). This means that even courses specifically focused on biomonitoring with nematodes require a lot of general nematology background. Therefore, it is desirable to include free-living nematodes from terrestrial, freshwater and marine habitats as part of more popular general nematology training.

With an emphasis on the diversity of nematodes and the broader soil ecosystem, this chapter outlines the basic and specialized methods developed and used for nematode biomonitoring (constituting smaller or elective parts of general nematology courses). These approaches enable the translation of nematode community data into relationships with soil or sediment properties and functions. This is a very active area of more specialized research for which education can greatly assist in interpreting the plethora of new findings. The focus of this chapter is to highlight the extent to which free-living or beneficial nematodes, and the use of nematodes as bioindicators are included in educational materials, academic curricula and courses. Because these topics are often part of broader training or courses, all types of training are considered.

In agronomy and biology education programmes, the field of nematology is often overlooked, in contrast to virology, mycology and entomology. This is a global phenomenon and is unjustified considering their abundances and the significant roles that nematodes play in soils and sediments. Consequently, there are generally very few courses specifically focused on nematology education and training. Nevertheless, there have been several training and educational programmes attempting to rectify the deficiency in all areas of nematology: general, economic parasites and environmental. Online courses and training programmes have been part of this and are also reviewed in this chapter.

Current or recent education and training initiatives in nematology are nevertheless global in scope, and thus reflective of the widespread distribution of our subject nematodes (Table 17.1).

Table 17.1. Ongoing and completed (online) training courses on free-living nematodes and their use as bioindicators. (Author's own table.)

Institution and main contact persons	Course/identification name	Focus and tools	Target audience	Ongoing?	Source(s)
Europe					
Ghent University (UGent), Belgium: Nematology Research Unit Dr Wim Bert, Inge Dehennin, Dr Ron de Goede	Soil Biodiversity and Nematodes as Bioindicators of Soil Health is a course unit of the International Master of Science in Agro- and Environmental Nematology (IMaNema)	Sampling, extraction, identification of terrestrial nematodes, functional groups, food web analysis and calculation of relevant indices	Postgraduate students	Yes	https://imanema.ugent.be/; https://studiekiezer.ugent.be/international-master-of-science-in-agro-and-environmental-nematology-en; https://studiekiezer.ugent.be/studiefiche/en/C003954/2023 (all accessed 27 August 2025)
	Identification of Aquatic Nematodes	Identification of marine and freshwater nematodes	Anyone interested	In 2018, upon request	https://imanema.ugent.be/aquatic-nema-course/ (accessed 27 August 2025)
Wageningen University and Research (WUR), The Netherlands: Laboratory of Nematology, and Biointeractions and Plant Health Dr Gerard Korthals, Dr S. Brandt	Identification of terrestrial and freshwater nematodes for ecologists/environmentalists	Basic identification skills, sampling, various extraction techniques, functional groups, food web analysis and calculation of relevant indices	Postgraduate students, consultants, extensionists and other interested persons	Yes	https://www.wur.nl/en/research-results/chair-groups/plant-sciences/laboratory-of-nematology/education/course-identification-of-terrestrial-and-freshwater-nematodes-for-ecologists-environmentalists.htm (accessed 27 August 2025)
WUR, The Netherlands: Soil Biology Group Dr Giulia Bongiorno, Dr Ron de Goede, Tamás Salánki	Soil Biology Lab Skills Course for Assessing Soil Functions	Five-day course to define which measurements of soil biology are relevant for which soil functions. The training is hybrid approach including lectures, laboratory and field sessions, of which one day is focused on nematodes	Postgraduate students, consultants, extensionists and other interested persons	Yes	https://biosisplatform.eu/services/training-and-workshops/soil-biology-lab-skills-course-assessing-soil-functions (accessed 27 August 2025)

USA					
University of Vermont: Plant and Soil Science Department Dr Deborah Neher	Soil Ecology	Lectures and hands-on laboratory sessions. Use of simple key to functional groups and computing indices including Maturity Index (MI), Structure Index (SI) and Enrichment Index (EI)	Graduate students	Yes	https://www.uvm.edu/cals/ale/profile/deborah-neher (accessed 10 September 2025)
University of Florida: Entomology and Nematology Dr Dorota Porazinska	Nematode Morphology and Anatomy Nematode Systematics and Molecular Phylogeny	Basic skills in taxonomy, classification, and skills in environmental nematology	Graduate students	Yes	Online: NEM6101, https://entnemdept.ufl.edu/academics/syllabi/NEM_6101_syllabus.pdf (accessed 10 September 2025) In person: NEM6101L, https://entnemdept.ufl.edu/academics/syllabi/NEM6101L.pdf (accessed 10 September 2025) Online: NEM6102, https://entnemdept.ufl.edu/academics/syllabi/NEM_6102.pdf (accessed 10 September 2025). Also ad hoc online nematode metabarcoding
University of California, Davis: College of Agricultural and Environmental Sciences Dr Shahid Siddique	Introduction to Nematology	Relationship of nematodes to the human environment. Classification, morphology, ecology, distribution and importance of nematodes occurring in water and soil as parasites of plants and animals	Undergraduate students	Yes	https://catalog.ucdavis.edu/departments-programs-degrees/entomology-nematology/nematology-minor/#requirementstext (accessed 27 August 2025)
University of California, Riverside: Department of Nematology Prof. Ole Becker	Soil Ecology	Examination of soil biota and relationships with plants and the soil environment. Importance of different groups examined from rhizosphere to ecosystem	Undergraduate students	Yes	https://nematology.ucr.edu/academics/courses (accessed 27 August 2025)
	Biology of Nematodes	Morphology, physiology, development, genetics, behaviour and ecology of nematodes from parasitic and free-living habitats	Undergraduate students	Yes	

Continued

Table 17.1. Continued.

Institution and main contact persons	Course/identification name	Focus and tools	Target audience	Ongoing?	Source(s)
Africa					
Institute of Tropical Nematology (IITA) and International Centre of Insect Physiology and Ecology (ICIPE): NemAfrica team in collaboration with IMaNema, UGent Dr Danny Coyne (IITA, Kenya), Dr Solveig Haukeland (ICIPE, Kenya), Dr Laura Cortada (UGent)	Basic Crash Course in Nematology	Basic identification skills, sampling and various extraction techniques	Now mainly extension officers, but previously undergraduate and postgraduate students, consultants and other interested persons	Yes, since 2018	https://imanema.ugent.be/basic-crash-course-nematology-kenya-bccn/; https://studiekiezer.ugent.be/international-master-of-science-in-agro-and-environmental-nematology-en (both accessed 27 August 2025)
University of the Free State (UFS), South Africa: Natural and Agricultural Sciences Dr Candice Janse van Rensburg (Dr Mieke Daneel and Prof. Gerhard du Preez, North-West University)	Short learning programmes in: Free-living nematode identification for ecologists and taxonomists; and Applied plant nematology	Free-living nematode systematics, trophic group identification, identification of nematode genera and aspects of nematode ecology. Basic identification skills, sampling, extraction techniques. Utilization of nematode-based indices (NBIs) and Nematode Indicator Joint Analysis (NINJA) online tool	Anyone with a secondary qualification as well as graduate and postgraduate students	To be continued at UFS	https://www.ufs.ac.za/kpa/programmes/free-living-nematode-identification-for-ecologists-and-taxonomists; and https://www.ufs.ac.za/kpa/programmes/applied-plant-nematology (both accessed 12 September 2025)
Asia and Oceania					
CSIRO, National Research Collections Australia Dr Mike Hodda, Dr Kerrie Davies (University of Adelaide)	Nematode Identification and Techniques	Basic identification skills, sampling, various extraction techniques, nematode ecology, including trophic groups, indicators and soil health	Diagnosticians, postgraduate students, consultants, extension officers, plant breeders	Yes, approx. biennial since 1999, next course 2025	mike.hodda@csiro.au

Dr Mike Hodda, Dr Dan Huston	Nematode Identification and Techniques	As above plus molecular techniques	Diagnosticians, extension officers, regulators	Yes, periodic since 2011	
Biological Crop Protection Pty Ltd. (private) Dr Graham Stirling	Master classes in nematology, soil biology and soil health at various venues in Australia and New Zealand	Basic identification, pests particularly of turf, soil health, biological control	Technical staff, agricultural consultants	2022, 2023, not ongoing	graham.stirling@biolcrop.com.au
Japan ASEAN Integration Fund Dr Hideaki Iwahori (Ryukoku University, Japan)	Diagnostics of Plant-Parasitic Nematodes	Diagnostics of plant-parasitic nematodes	Agriculture Department staff	2020, under review	
Fudan University, Shanghai, China	Soil Nematology, Taxonomy and Ecology	Terrestrial nematode identification and ecology	Undergraduate students	2016 and 2019	

17.2 Milestones in the Education of Environmental Nematology

Education on the environmental aspects of nematology, especially soil nematology, is recent and formally published information regarding education in environmental nematology before 2006 is scarce (Neher and Darby, 2006). We suspect that education about beneficial terrestrial nematodes began shortly after the discovery of the bacterivorous vinegar eel nematode *Turbatrix aceti* (Müller, 1783) Peters, 1927, by Petrus Borellus in 1656 (Peters, 1927). Other significant milestones linking nematodes to their environment were in marine environments, with the work of Wieser (1953) on feeding types and Gerlach (1953) on the species-salinity curve. In the 1970s, several nematologists expanded on this foundation, exploring various habitats and using nematodes as environmental bioindicators in terrestrial, freshwater and marine systems. For example, *Panagrellus redivivus* has been used as a bioindicator to detect toxin concentrations of any of 400 individual chemicals that affect development, survival and fecundity (Neher, 2001). Another significant milestone was the development of axenic culture techniques for nematodes in the 1950s and 1960s by Ellsworth C. Dougherty. These techniques were later applied to another rhabditid bacterivore, *Caenorhabditis elegans* (Maupas, 1900) Osche, 1952, which ultimately led to it becoming 'one of the best-understood animals in the world' (Riddle *et al.*, 1997). The ability to culture specific lines with no other organisms in totally defined media allowed even more detailed insights into the effects of various chemicals on organisms generally and use as an indicator species for the likely effects of chemicals on other organisms. *C. elegans* became the model organism for assessing the toxicity of chemicals in aqueous media, particularly fresh waters (Höss *et al.*, 2011, 2023). It was the first multicellular animal whose entire genome was mapped and is of high value in molecular and developmental biology, as well as in research for human diseases. Tietjen and Lee (1975) succeeded in the axenic culture of the marine nematode *Pellioditis marina* (Bastian, 1965) Andrássy, 1983, but its use in biomonitoring or toxicity testing has not been as widespread as that of the previously mentioned terrestrial nematodes, perhaps because this species is from a taxonomic group that is otherwise very uncommon in marine habitats and there is considerable cryptic genetic and biological diversity within what is now recognized as a species complex (Moens and Vincx, 1998; Derycke *et al.*, 2012).

In the 1970s, it became popular to use the nematode:copepod ratio for monitoring marine systems conditions at population level (Rafaelli and Mason, 1981; Neher, 2001). These measures were widely used by researchers and professionals but quickly became refined by Warwick (1981) and later replaced by the *k*-dominance curve developed by Lambshead *et al.* (1983). The nematode:copepod ratio proved to be practical for general ecological surveys, particularly for those who could not identify nematodes or copepods and were focused solely on abundances, but it was not without issues (Coull *et al.*, 1981; Hodda and Nicholas, 1986).

17.3 From Research to Education in Environmental Nematology

Efforts to educate students and scientists about using nematodes as bioindicators of ecosystem health were limited and fragmented until awareness grew of the importance and roles of free-living nematodes from the 1970s. In marine ecological studies, considerable attention was given to trophic diversity, particularly using the trophic index in nematodes, and its relationship to environmental parameters, ecosystem functioning and species richness (Heip *et al.*, 1985). The *k*-dominance curve (Lambshead *et al.*, 1983) was also employed as a tool in pollution studies to assess these relationships. Our understanding of environmental relationships of marine nematodes was reviewed recently by Warwick (2019).

During this early period, before the introduction of the Maturity Index (MI) (Bongers, 1990), research in terrestrial environments was mainly focused on nematode abundance and diversity (Yeates, 1970; Freckman, 1988; Wasilewska, 1979, 1995a,b, 1997). For all these methods, training in efficient extraction and basic recognition of nematodes was required and was provided for marine nematodes in general handbooks (e.g. McIntyre, 1971). More advanced training in identification of trophic types or morphospecies (for *k*-dominance curves in marine habitats) had to be learnt on the job from taxonomic texts (e.g. Platt and Warwick, 1983, 1988; Warwick *et al.*, 1998).

In 1987, Gregor Yeates was invited to present a series of eight lectures on resource utilization by nematodes for MSc students at Wageningen University with the aim 'to stimulate thought and to provoke discussion about the relation of nematode populations and ecosystem processes' (Yeates, 1987). Subsequent contributions to research in environmental nematology in terrestrial habitats included: the MI (Bongers 1990), initially focusing only on free-living terrestrial nematodes, but later expanded to other trophic groups and modified (De Goede *et al.*, 1993; Yeates, 1994; Bongers *et al.*, 1995; Korthals *et al.*, 1996; Bongers and Ferris, 1999); and the Enrichment Index (EI) and Structure Index (SI) (Ferris *et al.*, 2001; Ferris, 2010). These discoveries led to intensified educational efforts aimed at expanding the nematode toolset for studying soil ecosystem status. Linking nematodes to ecosystem function and status added relevance necessary to garner interest by policy makers and land managers (Neher, 2010). Significant strides were taken to educate individuals about this specific aspect of nematology, resulting in an enhanced understanding among both scientists and non-scientists.

Following on the contributions mentioned above, Tom Bongers has since 1998 run training sessions for ecologists on the identification of terrestrial and freshwater nematodes at Wageningen University and Research (WUR) (Table 17.1). This immersive 2-week training course is still held annually.

At Ghent University (UGent), a nematology educational programme was established in 1992, including a course on 'Ecology and Population Dynamics of Free-Living Nematodes' (focused on marine nematodes) as part of a 1-year curriculum. By 2003, the programme offered two majors: 'Nematology Applied to Agro-Ecosystems' and 'Nematology Applied to Natural Ecosystems'. When the 2-year Postgraduate International Nematology Course (PINC) launched in 2007, students could also choose the 'Biodiversity/Systematics' track. However, most opted for the 'Agro-Ecosystems' module, receiving limited exposure to beneficial nematodes aside from entomopathogenic nematodes. To address this, since 2018, the succeeding International Master of Science in Agro- and Environmental Nematology (IMaNema) has no longer offered separate modules. Although the programme remains largely agriculture-oriented, nearly all students now take the 6-credit course 'Soil Biodiversity and Nematodes as Bioindicators of Soil Health'.

In 1999, on the other side of the world, Mike Hodda, Kerrie Davies and colleagues from The University of Adelaide began offering training on nematological identification and techniques for students and scientists from Australia, Asia and the Pacific. The intensive 1-week course for up to 15 participants has run biennially since. While initially focused on PPNs, it quickly became clear that there was a need to include free-living nematodes and their relationships with environmental conditions, and other nematodes. Hence, identification of free-living nematodes, allocation to trophic groups, and biomonitoring or soil health were included. The course has been held at various universities in Australia and South-East Asia to reduce travel barriers (Table 17.1). Despite offering university credits, none of the hundreds of participants have used this option.

Other notable examples illustrating the progress of education in nematodes and their use as bioindicators include a workshop at the Second International Congress of Nematology (Veldhoven, The Netherlands, 1990) which produced a resource for soil ecologists to consistently assign nematodes to life-history and trophic groups, respectively (Yeates *et al.*, 1993). The paper represented the best professional agreement about ecological assignments to provide consistency among researchers and studies (Yeates *et al.*, 1993). A recent paper expanded this to include all trophic groups, including vertebrate parasites and marine nematodes (Hodda, 2022b). In 2004, the Society of Nematologists hosted a workshop led by D.A. Neher, T. Bongers and H. Ferris on the 'Computation of nematode community indices'. The workshop's unique approach enables researchers and students to assess soil health by classifying soils using nematodes as bioindicators. Index computations and interpretation were later published as a pair of book chapters (Neher and Darby, 2006, 2009). In 2018, a 2-week identification course on freshwater and marine nematodes was held at

UGent (Belgium), bringing together an international group of 16 participants. The course covered basic nematology techniques, including sampling, taxonomy and ecology, with the first week focused on marine nematodes (N. Smol and T.N. Bezerra) and the second week on freshwater nematodes (A. Zullini).

17.4 Where Are We Now?

As interest has grown in nematodes as bioindicators, soil and sediment ecology, and free-living nematodes, so has the focus on the subject in education. Major progress has been driven by positive developments such as the release of the first edition of this book, the advancement of the United Nation's Sustainable Development Goals (https://sdgs.un.org/goals, accessed 27 August 2025), the Global Soil Biology Initiative (https://www.globalsoilbiodiversity.org/, accessed 27 August 2025), renewed interest in conservation and regeneration agriculture (Kassam *et al.*, 2021), and recognition of the value of diverse soil food webs for food production (http://www.fao.org/conservation-agriculture/en/, accessed 27 August 2025). Assessing negative changes, such as soil degradation in agricultural fields, industrial development areas and natural vegetation, is important too. The exploration and development of new and different bioindicators to monitor the health of terrestrial and aquatic ecosystems have also contributed to the progress and increased interest in nematode biomonitoring. The increased scientific recognition of the diversity, roles and functions of beneficial nematodes (Du Preez *et al.*, 2022) has led to greater public awareness and appreciation. This has also expanded the demand for training to include students and various other stakeholders. Connected with this has been the continuation of intensive, small-group, informal, and in-house training as discussed in the following sections.

17.4.1 Existing nematology education

This section presents a survey of the two main approaches to nematology education:

1. Formal educational programmes.
2. Dedicated, hands-on short courses or non-formal education.

Although most courses are typically conducted face-to-face, some are also offered online in hybrid formats, combining contact and online components (see section on *Dedicated hands-on short courses* below and Table 17.1). However, contact courses with extensive practical work remain the primary and preferred approach. This is particularly relevant when working with free-living nematodes. Despite the development of molecular tools and libraries, visual identification with a microscope is still necessary for identifying most free-living nematodes because few have been sequenced, and sequencing effort lags well behind that of economic nematode groups such as plant parasites.

Formal educational programmes

Accredited, comprehensive nematology education programmes at higher education institutions (HEIs) vary considerably and encompass many topics and concepts apart from bioindicators. As discussed above, these foundational courses necessarily focus on a deeper understanding of the nematode morphology, biology and identification. Consequently, they require experienced nematologists as trainers and much specialized equipment, are limited in availability, and primarily target MSc and PhD levels. Comprehensive coverage of nematode identification, for example, requires experts in many different groups of nematodes, notwithstanding handbooks which cover identification of substantial portions of the phylum (e.g. Goodey, 1963; Andrássy, 1976, 1984; or the more recent Schmidt-Rhaesa, 2013) and the online identification resources listed in Table 17.2.

To the authors' best knowledge, there are no formal, stand-alone courses dedicated to Nematodes as Bioindicators besides those offered at the IMaNema programme from UGent in Belgium and at WUR, The Netherlands. Only at UGent is the Nematodes as Bioindicators course integral to the formal teaching programmes, with a dedicated, 2-year curriculum including lectures and practical sessions. Currently, all students must take a short course on 'Soil biodiversity and nematodes as bioindicators for soil health'. Additionally, there is an option to write a master's thesis on nematode-based indices (NBIs); but from 1999 to 2022, only 6.5% of students accepted this opportunity. The IMaNema programme provides a strong and unique

Table 17.2. Online tools used to understand and analyse data about nematode-based indices (NBIs), faunal analysis, and related parameters to be used as bioindicators of ecosystem health. (Author's own table.)

Inventor/manager and institution	Tool name	Discipline(s)	Focus	Target audience	Source(s)
Howard Ferris (University of California, Davis, USA)	Nemaplex	Nematology	A comprehensive, virtual encyclopaedia about nematodes. Includes a detailed section dedicated to nematodes as environmental indicators. Provides essential information for calculating ecological indices, including NBIs. Forms the starting point for the Nematode Indicator Joint Analysis (NINJA) online tool	Anyone with experience in nematology. Need to have and mastered basic diagnostic and research skills, e.g. nematode counting, picking, identifying. Experience in microscope use	http://nemaplex.ucdavis.edu/Uppermnus/topmnu.htm (accessed 27 August 2025)
Armen C. Tarjan (University of Florida, Lake Alfred) Robert P. Esser (Florida Department of Agriculture, Gainesville) Shih L. Chang (Environmental Protection Agency, Cincinnati, Ohio)	Interactive Diagnostic Key to Plant Parasitic, Free-living and Predaceous Nematodes	Nematology	Exploring the diversity of life on earth as reflected by the nematode world	Students and interested scientists who have a background in nematology	https://nematode.unl.edu/key/nemakey.htm (accessed 27 August 2025)
Bogdan Sieriebriennikov (originally Ghent University, Belgium) Sara Sánchez-Moreno (CSIC/INIA, Spain) Howard Ferris (University of California, Davis, USA) Ron de Goede & Carmen Vazquez-Martin (Wageningen University and Research, The Netherlands)	NINJA: an automated calculation system for nematode-based biological monitoring	Nematology	Development of an R code, compiled in html and deployed across the web, to perform the faunal analysis calculations. NINJA has a user-friendly interface and is freely available. It permits flexibility to override coded default feeding habits of individual nematode taxa. An input table with taxonomic inventory data of nematodes recorded in soil samples only is required. An output is provided within a few seconds indicating a summary of NBI data, faunal analysis, trophic group composition, metabolic footprint data and compost footprint data. There is also a check for correct scientific names. NINJA uses Nemaplex as its database	Nematologists, students and interested scientists (mostly ecologists) with a background in nematology	https://shiny.wur.nl/ninja/ (accessed 27 August 2025)

Continued

Table 17.2. Continued.

Inventor/manager and institution	Tool name	Discipline(s)	Focus	Target audience	Source(s)
Mike Hodda (CSIRO, National Research Collections, Australia)	Universal key to nematode feeding habits	Nematology	Polychotomous and pictorial keys to nematode trophic groups for terrestrial, freshwater and marine habitats, comparison of schemes proposed in the past and for different habitats	Undergraduate and postgraduate students, soil and sediment biologists, researchers	https://www.mapress.com/zt/article/view/zootaxa.5114.1.3 (accessed 27 August 2025)
	Alphabetical list of valid genera and biotope	Nematology	List of valid genera, recently described species, classification and biotope they are found in	Researchers, molecular ecologists	https://www.mapress.com/zt/article/view/zootaxa.5114.1.1 (accessed 25 August 2025)
Deborah Neher (University of Vermont, USA)	Simple key	Soil Biology	Simple key to functional groups that can be used by novices sufficient to compute indices including Maturity Index (MI), Structure Index (SI) and Enrichment Index (EI)	Undergraduate students	Neher (2023)

foundation in nematology and will be operating under a project-based approach until 2027.

In other formal education across universities, nematology courses are presented mostly as stand-alone topics within broader programmes or integrated into other course units such as agriculture, entomology, biology or ecology. Specific elements of NBIs are covered in various contact and online training sessions for soil ecology and/or PPNs at the graduate level at universities and/or at other institutions. Various facets of nematodes as bioindicators of environmental health are currently presented in courses at many universities (in addition to those listed in Table 17.1: University of Michigan, USA; Federal University of Pernambuco, Brazil; Iligan Institute of Technology, Philippines; and University of Western Sydney, Australia). While these programmes may involve formal classes, they often consist of small-group training within laboratory settings for students or guests from other institutes and are frequently associated with research projects.

A detailed survey in 2021 (I. Dehennin, 2021, unpublished results) illustrates the level to which training in nematodes as bioindicators is embedded in university courses. Of 33 HEIs surveyed in sub-Saharan Africa from nine different countries, NBIs were taught at eight out of 33 universities. At the BSc level, 36 course units with a nematology component existed, but only ten included training on nematodes as bioindicators. At the MSc level, only six out of the 35 courses included nematology (Fig. 17.1).

Dedicated hands-on short courses

Specialized nematology courses and workshops have emerged which include a focus on free-living nematodes and their use as environmental indicators. This addresses the lack of formal, accredited nematology programmes and the emphasis on PPNs only. These short courses and workshops offer opportunities to develop specific expertise on this topic but still require substantial background knowledge to understand the complexities of nematodes as environmental indicators. This background may be obtained from courses in general nematology or PPN, or within the short course itself. Training in general or for PPNs and free-living nematodes often intersect, and the crossover between PPNs and free-living nematodes is crucial for mastering nematode morphology, identification and biology.

The diverse statistical approaches to analyse nematode community data are included in many short courses and workshops but can also be found online (e.g. the online NINJA (Nematode

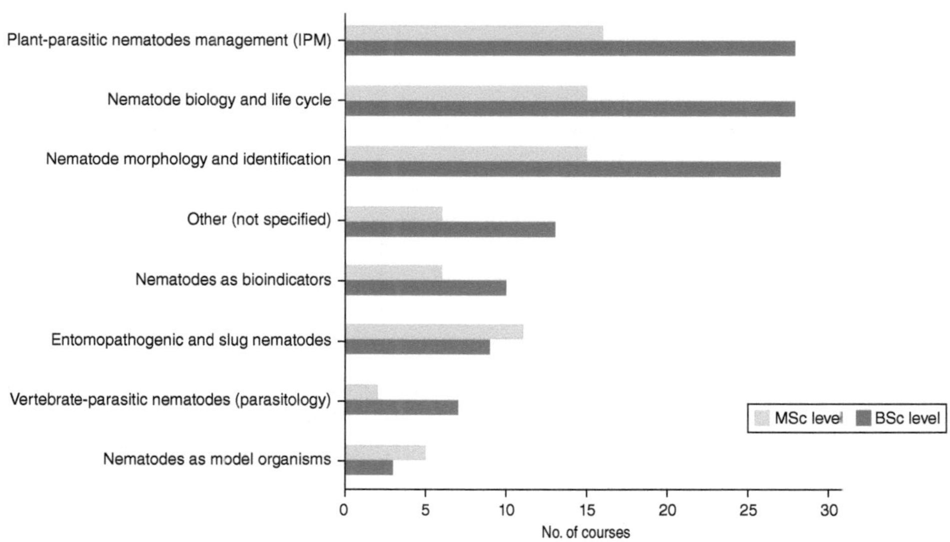

Fig. 17.1. Representation of nematology subjects in existing BSc and MSc programmes in Agriculture/Biology in sub-Saharan Africa. IPM = integrated pest management. (Compiled from I. Dehennin, 2021, unpublished results.) (Author's own image.)

Indicator Joint Analysis) application; https://www.wur.nl/en/product/ninja-nematode-indicator-joint-analysis.htm, accessed 27 August 2025) (Sieriebriennikov *et al.*, 2014). Understanding automatically calculated indices and dealing with raw data from high-throughput, whole-genome or barcode molecular analyses is crucial for accurate interpretation of data.

17.4.2 Online platforms as tools of education

Many online nematology resources, although not necessarily intended as educational resources per se, may still have a role in educating about nematode bioindicators. Online resources can help address the wide geographic dispersion of students, expand access to such students and promote consistency in curricula. Through online platforms, students can access various resources like recorded lectures, interactive tutorials and virtual labs for self-paced learning. These teaching methods are being integrated into some form of formal and informal education (e.g. the course presented at The University of Florida; Table 17.1). Practical experience of fieldwork techniques, microscopy and molecular techniques is still widely regarded as necessary for effectively using nematodes as environmental indicators. These practical skills, however, cannot be taught online.

A potential solution involves combining universally accessible online courses with concise, practical modules tailored for students who have completed online learning. Such hybrid courses can serve as a partial alternative for those with limited ability to engage in hands-on training and can also alleviate financial constraints for organizations and participants in longer formal courses. Examples include the Nematology Digital Learning Platform (Ufora and https://www.youtube.com/@imanemaugent9505, accessed 27 August 2025; UGent, Belgium) for remote practice of nematode identification and analysis, as well as virtual microscopy offering high-resolution images and multifocal video clips from Video Capturing and Editing (De Ley and Bert, 2002). Unfortunately, current initiatives in this area are scattered and focus primarily on PPNs.

There are several online tools available to support and encourage independent learning (Table 17.2). Nemaplex is an online tool and extensive website containing a lot of information, with an emphasis on terrestrial nematodes. It includes recommended colonizer-persister (*c-p*) assignments for the MI (Bongers, 1990), standardized assignments of trophic groups (Yeates *et al.*, 1993), and quantitative tools created by Neher and Darby (2006) and Ferris *et al.* (2001). It has a detailed section specifically dedicated to nematodes as environmental indicators in terrestrial environments. It is widely consulted, regularly updated, and can be used by prospective learners to distinguish terrestrial, free-living nematodes using comprehensive keys. Refinements to faunal analysis and correction of errors led to the creation of an enhanced, easy-to-use, new-generation tool: the Nematode Indicator Joint Analysis (NINJA) by Sieriebriennikov *et al.* (2014) (Table 17.2). This tool can also be employed by students and others to calculate NBIs based on the principles of computing the MI. A prerequisite for using the tool is the identification of free-living nematodes at least to the family level, with genus-level identification being preferable.

A truly comprehensive alphabetical list of all valid nematode genera and their trophic status, along with discussion of the various lines of evidence for trophic relationships, is available online (https://mapress.com/zt/issue/view/zootaxa.5114.1, accessed 12 September 2025). Another very valuable nematode database, Nemys (https://nemys.ugent.be/, accessed 27 August 2025), provides a comprehensive taxonomic overview of all described nematode species. Initially focused on free-living marine nematodes, it has since been expanded to include many terrestrial, freshwater and parasitic nematodes. H. Ferris also maintains a record of the most up-to-date classifications of terrestrial nematodes through the Nemaplex database (http://nemaplex.ucdavis.edu/, accessed 27 August 2025).

17.5 Possible Strategies to Integrate Environmental Nematology into Curricula of Formal and Non-Formal Education

The project 'Nematology Education in Sub-Sahara Africa (NEMEDUSSA)' is an example of an initiative addressing the need for training, capacity building and awareness raising in nematology. This was an EU Erasmus+ Capacity Building in

Higher Education project, running from 2021 to 2024, involving 16 partner institutes of which 14 are in sub-Saharan Africa. Although the project primarily focused on general nematology, the consortium developed educational materials for identifying free-living nematodes and their use as bioindicators.

In addition to investing in laboratory equipment for HEIs, staff members from the 14 African partner institutions participated in 'training for trainers' (ToTs) programmes at other partner institutes, including in soil nematology. The training prepared and equipped them to effectively impart their knowledge to students in programmes at their respective institutes. Investing in equipment and staff training transformed some of the partners' nematology laboratories into reference laboratories for research and training. Simultaneously, the partners developed educational material tailored to the African context to support the uptake of nematology in existing BSc and MSc course programmes, including trial runs in the participating universities. Other initiatives of this project raised societal awareness about nematodes through school visits, open houses, student clubs and short courses. A major outcome of this project has been the establishment of the Pan-African Nematology Network (PANEMA) focusing on training that brings together multiple stakeholders, networking and dissemination activities; this PANEMA network organizes monthly webinars and has organized three workshops in Kenya, South Africa and Benin. The project has brought greater attention to nematology in the participating HEIs as a discipline, particularly in the centres where trial runs and PANEMA workshops were conducted.

Since 2011, a long-term project, spanning more than 14 years, has been underway to train an ASEAN (Association of Southeast Asian Nations) Regional Diagnostic Network (ARDN) in nematology. The project is funded by the ASEAN–Australia–New Zealand Free Trade Area's Economic Cooperation Work Programme and the Mekong–Australia Partnership's Vietnam Enhanced Economic Engagement Strategy. The training involves all ten ASEAN countries and is being conducted mainly in ASEAN (Cambodia, Lao PDR, Myanmar, Thailand and Vietnam) as well as in Australia. This programme has benefited from the long-term engagement and is continuing. It is primarily focused on general and plant-parasitic nematology, but also includes components on terrestrial free-living nematodes, soil health and biocontrol. Universities from several ASEAN countries are participating in developing educational and promotional materials, 'train the trainer' sessions and practical fieldwork, with local staff and materials as crucial elements of the programme. The programme has developed and vastly increased the interest in nematology in the region as well as the local capacity and capability in all aspects of nematology.

The two strategic approaches discussed above are both important for developing environmental and biomonitoring nematology as integral parts of both formal and non-formal education and training. Some of the key requirements for these approaches have been the following.

- *Faculty expertise and facilities:* Institutions need well-trained academic and technical staff, along with adequate laboratory equipment, to effectively educate students and others.
- *Syllabus development:* A comprehensive syllabus for nematology training needs to be created to accommodate different education levels. Pre-existing model syllabuses, like the list of topics for an in-depth course on environmental nematode ecology in Table 17.3, can help educators save time and allow students to access course materials easily. Alternatively, modular educational resources, such as those developed for the Australia–ASEAN projects mentioned earlier, can also be considered.
- *Dissemination strategies:* These should target various stakeholders to raise awareness about the importance of educating, researching, and applying knowledge and experience gained in nematology.

Implementing these general strategies may help integrate environmental nematology into other programmes more effectively.

17.6 Challenges and Opportunities in Nematology Education

Nematology education faces several challenges, but strategic attention and innovative solutions are either apparent or available for investigation.

Table 17.3. Suggested topics for inclusion in a comprehensive course on environmental nematology. (Author's own table.)

Topic	Details
Nematode taxonomy	Classification systems, morphological and molecular-based taxonomy, tree of life
Nematode biology	Reproduction, life cycle, locomotion, interaction with biotic and abiotic environment
Nematodes and the soil/sediment food web	Internal food web, predator–prey relationships, parasitism, contribution to soil processes, energy pathways, mitigation of soil-borne diseases
Nematodes as bioindicators	Local soil/sediment quality assessment, biomonitoring, and other soil/sediment health bioindicator systems, trade-offs
Trophic group classification	Trophic groups, classification systems for terrestrial and aquatic species, taxonomic resolution, differences in opinion
Life-history groups	Historical evolution, relationship with the soil/aquatic environment and management
Nematode sampling	Hypothesis formulation, site selection, sampling strategies, corer selection, terrestrial–aquatic soils/sediments. Need for standardization
Nematode extraction	Sample storage, selection extraction method, extraction efficiency. Need for standardization
Nematode identification	Required taxonomic resolution, life stages, dauer larvae
Morphological identification and description	Requires experts in different groups of nematodes, identification literature, knowledge of morphology, morphometrics. Standardized guidelines for species descriptions, including consistency in terminology, measurements and presentation in taxonomic publications (see Mokievsky *et al.*, 2024)
Molecular identification	Extraction of DNA, matching methods to goal, primer selection, quantitative PCR (qPCR), Sanger sequencing, metabarcoding, assay standardization, reality check
Image analysis and artificial intelligence	Image databases, data quality and completeness, relationship between traditional taxonomy and computerized image recognition, trait analysis
Quantification and statistical analysis	Categories of diversity (taxonomic, functional, etc.)
Diversity indices/trophic index	Indices, index selection, effective number of species, rarefaction
Maturity indices	Maturity Index (MI), Maturity Index 2-5 (MI2-5), Sigma Maturity Index (ΣMI), colonizer–persister (*c-p*) triangle, trait-based models, non-linearity *c–p*1-5 classification
Pollution indices	*k*-Dominance curve; NemaSPEAR[%]-index (Nematode SPEcies At Risk), sentinel species, empirical models
Trophic group indices	Trophic groups, trophic diversity, Channel Index (CI), bacterivore:fungivore (B/F) ratio
Food web diagnostics	Enrichment Index (EI), Structure Index (SI), interpretation of Nematode Faunal Profile (NFP)
Metabolic footprints	Footprints, graphical interpretation of NFP footprint graph
Multivariate analysis	Principal components analysis (PCA), redundancy analysis (RDA), cluster analyses, network analyses, etc.
Data interpretation and reporting	Hypothesis testing, dissemination and translation of results to the public
Linking MI and food web diagnostics to soil functioning	General ecological theory, soil-based ecosystem services, online nematological analyses tools, Nematode Indicator Joint Analysis (NINJA) online tool, pitfalls in data processing
Reading and understanding an NFP figure	Test hypotheses related to the four quadrats of the NFP figure and functional footprint interpretation
Reading and understanding multivariate analysis figures	How to read biplot and triplot graphs of nematode communities and environmental factors

One of the main challenges is a depleted pool of expert taxonomists: several of the few experts on nematode taxonomy still need to be replaced after retirement, hampering knowledge transfer and jeopardizing the continuity of high-quality education in nematology.

Insufficient and limited financial resources allocated to environmental nematology research and education are also significant barriers to advancing this field, reducing scientific progress and public awareness. This creates a cycle where lack of funding limits progress, and limited progress fails to attract more funding. Persuading policy makers to allocate adequate resources is challenging due to the complexity and low visibility of nematodes.

The limited financial resources are connected to the challenge of attracting students to further their careers by studying free-living nematodes and their use as bioindicators. When opportunities are limited, it can discourage students. The lack of visibility can also be addressed by including NBIs more in curricula at the BSc and MSc levels, which can encourage more students to develop an interest.

In some cases, online tools are fragmented, which makes it difficult for users and policy makers to understand their impact on education. Individual nematologists' efforts also often lack visibility. Global coordination can unite these resources but recognizing contributors and reconciling differing methods and data pose challenges that need resources.

Exploring recent and future technological developments and innovations may assist in addressing the above challenges and training future students and nematologists in environmental nematology. Some approaches like the following may be integrated into both educators' toolkits and curricula.

- *The hybrid flexible or HyFlex teaching method:* This educational method integrates in-person and online learning, adding value and broadening geographic accessibility for educating students in environmental nematology. The availability of every class session and learning activity as a face-to-face session which is also online synchronously and/or asynchronously provides students with the option to choose how they want to participate and learn. It offers fairer opportunities for students with varying physical and emotional dis/abilities to access education.
- *Virtual simulation:* Visual aids such as photographs, figures and illustrations of nematodes and ecosystem processes can greatly enhance knowledge acquisition in environmental nematology. Thus, employing the concept of partial immersion within digital learning environments, such as computers, tablets or phones, could foster a realistic experience.
- *Modern digital microscopy:* Using state-of-the-art cameras and magnified optics, real-time projection of live nematode images offers an excellent opportunity for observation. It makes it easy to capture, save, measure and analyse the images. Observation from computer monitors or projection screens allows multiple people to observe the images simultaneously.
- *Artificial intelligence (AI):* The scientific landscape is currently undergoing a transformation with the integration of AI. In environmental nematology, AI can revolutionize data analysis, such as shown by Brito de Jesus *et al.* (2023) for marine nematodes, automate repetitive tasks and contribute to groundbreaking discoveries that were once inconceivable.
- *Molecular technologies:* Numerous advanced DNA sequencing technologies are becoming more accessible at a rapid pace; for example, new metabarcoding and sequencing methods are being developed to simultaneously identify nematode communities and other microbiota directly extracted from samples. This will require both the development and subsequent training in the analysis and interpretation of the data.
- *Harmonization of online databases and nematode-based indicator tools:* Environmental nematologists have access to a unique set of online resource databases and data analysis tools (Table 17.2). These resources could be pooled and their survival and maintenance guaranteed.

17.7 Conclusions

The role that nematodes can and do play as bioindicators has the potential to increase greatly due to the central position nematodes occupy within food webs, coupled with the global movement towards conserving our natural resources, maintaining healthy ecosystems and ensuring sustainable food production for a fast-growing global population. Nematologists will play a key role in assessing and reporting the status of aquatic and terrestrial ecosystems using nematodes as bioindicators, but this will only occur if nematology education is appropriately valued and managed. This will require effort to ensure that trained scientists, equipped with specialized infrastructure, can in turn train their successors. Education is vital for people to grow, discover and learn, and so the process of education should be continuous.

The ultimate key to all nematology education and training may be convincing governments, education and training managers, and other decision makers to include truly comprehensive nematology training in educational programmes. The review in this chapter has attempted to make the case for this. We, as authors, hope we have demonstrated that it is thoroughly worthwhile to ensure the transfer of knowledge and continuity of nematode bioindicator research and teaching at HEIs and research institutions, as well as to raise awareness among non-scientists about the value of NBIs in contributing to the monitoring and evaluation of ecosystems and environmental health.

Acknowledgements

This chapter would not have been possible without the information given by several nematology colleagues all over the world: respondents on IMaNema survey on nematology education in sub-Saharan Africa in 2021 and NEMEDUSSA partners; participants in the International Congress of Nematology (ICN) Nematology Educational Forum; Juvenil Enrique Cares, Vanessa da Silva Mattos, Giovanni Paiva dos Santos, Joey Genevieve Martinez, Manqiang Li and all other colleagues who contributed in one way or another.

References

Andrássy, I. (1976) *Evolution as a Basis for the Systemization of Nematodes*. Pitman, London.

Andrássy, I. (1984) *Klasse Nematoda (Ordnungen Monhysterida, Desmoscolecida, Araeolaimida, Chromadorida, Rhabditida)*. Bestimmungsbücher zur Bodenfauna Europas. Gustav Fischer, Berlin–Stuttgart, Germany

Austen, M.C. and McEvoy, A.J. (1997) The use of offshore meiobenthic communities in laboratory microcosm experiments: response to heavy metal contamination. *Journal of Experimental Marine Biology and Ecology* 211, 247–261. DOI: 10.1016/S0022-0981(96)02734-7.

Balsamo, M., Semprucci, F., Frontalini, F. and Coccioni, R. (2012) Meiofauna as a tool for marine ecosystem biomonitoring. In: Cruzado, A. (ed.) *Marine Ecosystems*. IntechOpen eBook, pp. 77–104. Available at: https://www.intechopen.com/books/1689 (accessed 27 August 2025).

Bloemers, G.F., Hodda, M., Lambshead, P.J.D.L., Lawton, J.H. and Wanless, F.R. (1997) The effects of forest disturbance on diversity of tropical soil nematodes. *Oecologia* 111, 575–582. DOI: 10.1007/s004420050274.

Bongers, T. (1990) The maturity index: an ecological measure of environmental disturbance based on nematode species composition. *Oecologia* 83, 14–19. DOI: 10.1007/BF00324627.

Bongers, T. and Ferris, H. (1999) Nematode community structure as a bioindicator in environmental monitoring. *Trends in Ecology and Evolution* 14, 224–228. DOI: 10.1016/S0169-5347(98)01583-3.

Bongers, T., de Goede R.G.M., Korthals, G.W. and Yeates, G.W. (1995) Proposed changes of c-p classification for nematodes. *Russian Journal of Nematology* 3, 61–62.

Brito de Jesus, S., Vieira, D., Gheller, P., Cunha, B.P., Gallucci, F. and Fonseca, G. (2023) Machine learning algorithms accurately identify free-living marine nematode species. *PeerJ* 9(11), e16216. DOI: 10.7717/peerj.16216.

Cortada, L., Dehennin, I., Bert, W. and Coyne, D. (2019) Integration of nematology as a training and research discipline in sub-Saharan Africa: progress and prospects. *Nematology* 22, 1–21. DOI: 10.1163/15685411-00003291.

Coull, B.C., Hicks, G.R.F. and Wells, J.B.J. (1981) Nematode/copepod ratios for monitoring pollution: a rebuttal. *Marine Pollution Bulletin* 12, 378–381.

De Goede, R.G.M., Bongers, T. and Ettema, C.H. (1993) Graphical presentation and interpretation of nematode community structure: c-p triangles. *Mededelingen van de Faculteit Landbouwwetenschappen Rijksuniversiteit Gent* 58, 743–750.

De Ley, P. and Bert, W. (2002) Video capture and editing as a tool for the storage, distribution, and illustration of morphological characters of nematodes. *Journal of Nematology* 34, 296–302.

Derycke, S, Tezerji, R.S. and Moens, T. (2012) Investigating the ecology and evolution of cryptic marine nematode species through quantitative real-time PCR of the ribosomal ITS region. *Molecular Ecology Resources* 12, 607–619.

Du Preez, G., Daneel, M., de Goede, R.G.M., Du Toit, M.J., Ferris, H. *et al.* (2022) Nematode-based indices in soil ecology: application, utility, and future directions. *Soil Biology and Biochemistry* 169, 108640. DOI: 10.1016/j.soilbio.2022.108640.

Ferris, H. (2010) Form and function: metabolic footprints of nematodes in the soil food web. *European Journal of Soil Biology* 46, 97–104. DOI: 10.1016/j.ejsobi.2010.01.003.

Ferris, H., Bongers, T. and de Goede, R.G.M. (2001) A framework for soil food web diagnostics: extension of the nematode faunal analysis concept. *Applied Soil Ecology* 18, 13–29. DOI: 10.1016/S0929-1393(01)00152-4.

Freckman, D.W. (1988) Bacterivorous nematodes and organic-matter decomposition. *Agriculture, Ecosystems & Environment* 24, 195–217. DOI: 10.1016/0167-8809(88)90066-7.

Gerlach, S.A. (1953) Die biozoenotische gliederung der Nematodenfauna an den deutschen Kuesten. *Zeitschrift fuer Morphologie und Oekologie der Tiere* 41, 411–512.

Goodey, J.B. (1963) *Soil and Freshwater Nematodes*. Methuen & Co., London.

Heip, C.H.R., Vincx, M. and Vranken, G. (1985) The ecology of marine nematodes. *Oceanography and Marine Biology* 23, 399–489.

Hodda, M. (2022a) Phylum Nematoda: a classification, catalogue and index of valid genera, with a census of valid species. *Zootaxa* 5114, 1–289. DOI: 10.11646/zootaxa.5114.1.1.

Hodda, M. (2022b) Phylum Nematoda: feeding habits for all valid genera using a new, universal scheme encompassing the entire phylum, with descriptions of morphological characteristics of the stoma, a key, discussion of evidence for trophic relationships and environmental affinities. *Zootaxa* 5114, 318–451. DOI: 10.11646/zootaxa.5114.1.3.

Hodda, M. (2022c) Phylum Nematoda: trends in species descriptions, the documentation of diversity, systematics, and the species concept. *Zootaxa* 5114, 290–317. DOI: 10.11646/zootaxa.5114.1.2.

Hodda, M. and Nicholas, W.L. (1986) Nematode diversity and industrial pollution in the Hunter River, NSW, Australia. *Marine Pollution Bulletin* 17, 251–255. DOI: 10.1016/0025-326X(86)90058-5.

Hodda, M., Bloemers, G.F., Lawton, J.H. and Lambshead, P.J.D. (1997) The effects of clearing and subsequent land-use on abundance and biomass of soil nematodes in tropical forest. *Pedobiologia* 41, 279–294. DOI: 10.1016/S0031-4056(24)00245-2.

Hodda, M., Peters, L. and Traunspurger, W. (2009) Nematode diversity in terrestrial, freshwater aquatic and marine systems. In: Wilson, M.J. and Kakouli-Duarte, T. (eds) *Nematodes as Environmental Indicators*. CAB International, Wallingford, UK, pp. 45–93. DOI: 10.1079/9781845933852.0045.

Höss, S., Claus, E., Von der Ohe, P.C., Brinke, M., Güde, H., Heininger, P. and Traunspurger, W. (2011) Nematode species at risk – a metric to assess pollution in soft sediments of freshwaters. *Environment International* 37, 940–949. DOI: 10.1016/j.envint.2011.03.013.

Höss, S., Sanders, D. and van Egmond, R. (2023) Determining the toxicity of organic compounds to the nematode *Caenorhabditis elegans* based on aqueous concentrations. *Environmental Science and Pollution Research* 30, 96290–96300. DOI: 10.1007/s11356-023-29193-2.

Kassam, A., Derpsch, R. and Friedrich, T. (2021) Development of conservation agriculture systems globally. *Journal of Agricultural Physics* 21, 10–51.

Korthals, G.W., de Goede, R.G.M., Kammenga, J.E. and Bongers, T. (1996) The maturity index as an instrument for risk assessment of soil pollution. In: Van Straalen, N.M. and Krivolutsky, D.A. (eds) *Bioindicator Systems for Soil Pollution*. Kluwer Academic Publishers, Dordrecht, The Netherlands, pp. 85–93. DOI: 10.1007/978-94-009-1752-1_8.

Lambshead, P.J.D. (1986) Sub-catastrophic sewage and industrial-waste contamination as revealed by marine nematode faunal analysis. *Marine Ecology Progress Series* 29, 247–260. DOI: 10.3354/meps029247.

Lambshead, P.J.D., Platt, H.M. and Shaw, K.M. (1983) The detection of differences among assemblages of marine benthic species based on an assessment of dominance and diversity. *Journal of Natural History* 17, 859–874. DOI: 10.1080/00222938300770671.

Lawton, J.H., Bignell, D.B., Bloemers, G.F., Eggleton, P. and Hodda, M. (1995) Carbon flux and diversity of nematodes and termites in Cameroon forest soils. *Biodiversity and Conservation* 5, 261–273. DOI: 10.1007/BF00055835.

McIntyre, A.D. (1971) Meiofauna and microfauna sampling. In: Holme, N.A. and McIntyre, A.D. (eds) *Methods for the Study of Marine Benthos*. Blackwell, Oxford, pp. 131–139.

Moens, T. and Vincx, M. (1998) On the cultivation of free-living marine and estuarine nematodes. *Helgolander Meeresuntersuchungen* 52, 115–139.

Mokievsky, V, Bezerra, T.N, Decraemer, W., Eisendle, U., Hodda, M. *et al.* (2024) Guidelines for species descriptions of free-living aquatic nematodes: characters, measurements and their presentation in taxonomic publications. *Zootaxa* 5543, 225–236.

Neher, D. (2001) Role of nematodes in soil health and their use as indicators. *Journal of Nematology* 33, 161–168.

Neher, D.A. (2010) Ecology of plant and free-living nematodes in natural and agricultural soil. *Annual Review of Phytopathology* 48, 371–394. DOI: 10.1146/annurev-phyto-073009-114439.

Neher, D.A. (2023) Moving up within the food web: protists, nematodes and other microfauna. In: Uphoff, N. and Thies, J. (eds) *Biological Approaches to Regenerative and Resilient Soil Systems*, 2nd edn. CRC Press, Boca Raton, Florida, pp. 157–168. DOI: 10.1201/9781003093718.

Neher, D.A. and Darby, B.J. (2006) Computation and application of nematode community indices: general guidelines. In: Eyualem, A., Traunspurger, W. and Andrássy, I. (eds) *Freshwater Nematodes: Ecology and Taxonomy*. CAB International, Wallingford, UK, pp. 211–222. DOI: 10.1079/9780851990095.0211.

Neher, D.A. and Darby, B.J. (2009) General community indices that can be used for analysis of nematode assemblages. In: Wilson, M.J. and Kakouli-Duarte, T. (eds) *Nematodes as Environmental Indicators*. CAB International, Wallingford, UK, pp. 107–103.

Peters, B.G. (1927) On the anatomy of the vinegar eelworm. *Journal of Helminthology* 5, 133–142. DOI: 10.1017/S0022149X00002261.

Platt, H.M. and Warwick, R.M. (1983) *Free-living Marine Nematodes. Part 1. British Enoplids: Pictorial Key to World Genera and Notes for the Identification of British Species*. Synopses of the British Fauna New Series No. 28. Cambridge University Press, London.

Platt, H.M. and Warwick, R.M. (1988) *Free-living Marine Nematodes. Part 2. British Chromadorids: Pictorial Key to World Genera and Notes for the Identification of British Species*. Synopses of the British Fauna New Series No. 38, Cambridge University Press, London.

Platt, H.M., Shaw, K.M., and Lambshead, P.J.D. (1984) Nematode species abundance patterns and their use in the detection of environmental perturbations. *Hydrobiologia* 118, 59–66. DOI: 10.1007/BF00031788.

Rafaelli, D.G. and Mason, C.F. (1981) Pollution monitoring with meiofauna, using the ratio of nematodes to copepods. *Marine Pollution Bulletin* 12, 158–163.

Riddle, D.L., Blumenthal, T., Meyer, B.J., Priess, J.R. (eds) (1997) *C. elegans II*, 2nd edn. Cold Spring Harbor Laboratory Press, Cold Spring Harbor, New York.

Schmidt-Rhaesa, A. (2013) *Handbook of Zoology, Vol. 2: Nematoda*. De Gruyter, Berlin. DOI: 10.1515/9783110274257.

Sieriebriennikov, B., Ferris, H. and de Goede, R.G.M. (2014) NINJA: an automated calculation system for nematode-based biological monitoring. *European Journal of Soil Biology* 61, 90–93. DOI: 10.1016/j.ejsobi.2014.02.004.

Sikora, R.A., Molendijk, L.P.G. and Desaeger, J. (2021) *Integrated Nematode Management and Crop Health: Future Challenges and Opportunities*. CAB International, Wallingford, UK. DOI: 10.1079/9781789247541.0001.

Tietjen, J.H. and Lee, J.J. (1975) Axenic culture and uptake of dissolved organic substances by the marine nematode *Rhabditis marina* Bastian. *Cahiers de Biologie Marine* 16, 685–693.

Traunspurger, W. (2021) *Ecology of Freshwater Nematodes*. CAB International, Wallingford, UK. DOI: 10.1079/9781789243635.0001.

van den Hoogen, J., Wardle, D.A., Traunspurger, W., de Goede, R.G.M., Adams, B.J. *et al.* (2020) A global database of soil nematode abundance and functional group composition. *Scientific Data* 7, 103. DOI: 10.1038/s41597-020-0437-3.

Warwick, R.M. (1981) The nematode–copepod ratio and its use in pollution ecology. *Marine Pollution Bulletin* 12, 329–333.

Warwick, R.M. (2019) Pollution: effects on marine communities. In: Cochran, J.K., Bokuniewicz, H.J. and Yager, P.L. (eds) *Encyclopedia of Ocean Sciences, Vol. 6: Ocean Interfaces & Human Impacts*, 3rd edn. Elsevier, Amsterdam, pp. 372–378. DOI: 10.1016/B978-0-12-409548-9.09052-7.

Warwick, R.M., Platt, H.M. and Somerfield, P.J. (1998) *Free-living Marine Nematodes. Part 3. Monhysterids: Pictorial Key to World Genera and Notes for the Identification of British Species*. Synopses of the British Fauna New Series No. 53. Cambridge University Press, London.

Wasilewska, L. (1979) The structure and function of soil nematode communities in natural ecosystems and agrocenoses. *Polish Ecological Studies* 5, 97–145.

Wasilewska, L. (1995a) Maturity and diversity of nematodes versus long-term succession after stress. *Nematologica* 41, 353.

Wasilewska, L. (1995b) Differences in development of soil nematode communities in single- and multi-species grass experimental treatments. *Applied Soil Ecology* 2, 53–64.

Wasilewska, L. (1997) Soil invertebrates as bioindicators, with special reference to soil- inhabiting nematodes. *Russian Journal of Nematology* 5, 113–126.

Wieser, W. (1953) Die Beziehung zwishen Mundhöhlengestalt, Ernährungsweise und Vorkommen bei freilebenden marinen Nematoden. *Arkiv für Zoologie* 4, 439–484.

Yeates, G.W. (1970) The diversity of soil nematode faunas. *Pedobiologia* 10, 104–107. DOI: 10.1016/S0031-4056(23)00401-8.

Yeates, G.W. (1987) *Resource Utilization by Nematodes: Notes for 8 Lectures to Nematology Department, Agricultural University, Wageningen, The Netherlands, November–December 1987*. NZ Soil Bureau Record No. 106. Department of Scientific and Industrial Research, New Zealand.

Yeates, G.W. (1994) Modification and qualification of the nematode maturity index. *Pedobiologia* 38, 97–101.

Yeates, G.W., Bongers, T., de Goede, R.G.M., Freckman, D.W. and Georgieva, S.S. (1993) Feeding habits in soil nematode families and genera – an outline for soil ecologists. *Journal of Nematology* 25, 315–331.

18 Commercial Aspects of the Use of Nematodes as Bioindicators

Simon J. Forster[1]*, Cécile P. Villenave[2], Camille Chauvin[2], Beatriz Calvo[3] and Marcus W. Trett[1]

[1]*Physalia Associates, Burwash Weald, UK;* [2]*ELISOL Environnement, Congénies, France;* [3]*Department of Pathology and Population Sciences, Royal Veterinary College, Hatfield, UK*

Abstract

This chapter discusses the commercial use of soil and aquatic nematodes for the assessment of prevailing environmental conditions for a range of client sectors, including manufacturing industries, agriculture, regulatory bodies/government authorities and the legal profession. A background to the development of the commercial nematology market and the establishment of professional nematologists in the commercial, rather than academic, realm is outlined. The ecological monitoring and assessment requirements of a range of commercial client groups are reviewed. Précis of standard and non-standard survey approaches and analytical protocols are presented, and the benefits and constraints of such studies are considered. The chapter concludes with the presentation of four representative aquatic and terrestrial case studies. These serve to demonstrate the commercial value of nematode bioindicator surveys to the clients in terms of both regulatory compliance and provision of cost-effective ecological and business management information.

18.1 Introduction

This chapter examines the background to the use of nematodes as bioindicators in commercial situations in two different environments: aquatic (marine, estuarine and fresh water) and terrestrial (soils). It highlights some of the key benefits that they can offer regulatory bodies, industry, agriculture and other sectors in the assessment of prevailing environmental conditions. The problems that can arise when undertaking work in this non-academic realm are examined, then the methods and protocols that are adopted for the nematode analyses and the standard and possibly less standard analyses that are undertaken to examine the resulting data are discussed. Representative case histories are also presented to illustrate some of the different bioindicator uses to which nematodes have been put.

In aquatic environments, the term 'meiofauna', coined initially by Mare (1942), is used to distinguish the size class from the larger macrofauna and the smaller microbial or nanofauna taxa. The meiofauna are defined loosely as organisms that pass through a 1 mm (1000 μm) mesh but are retained on a 42 μm mesh (Higgins and Theil, 1988). The diversity and abundance of meiofauna in aquatic sediment and terrestrial soil are extraordinary. Of the 33 recognised metazoan phyla, 22 are represented in the meiofauna of

*Email: simon.forster@physalia.uk

 Nematodes as Environmental Indicators: From Theory to Practice, 2nd Edition (T. Kakouli-Duarte *et al*. eds)
DOI: 10.1079/9781800624221.0018

aquatic sediments (Coull, 1988). Of the meiofauna phyla, the Nematoda is routinely the most abundant, regularly comprising up to 90% of all meiofauna (see Ridall and Ingels, 2021; Chapter 3, this volume). In soil ecosystems, the standard terminology and definitions vary from those of aquatic habitats. In soils, the term microfauna is used for animals of below 0.1 mm in length, mesofauna include animals between 0.1 and 2 mm, while macrofauna encompass animals between 2 and approximately 20 mm (Swift *et al.*, 1979). Nematodes as well as tardigrades and rotifers are included in the microfauna. However, unlike the aquatic meiofauna, the methods for extraction from soil of these three types of organisms differ and, consequently, they are not studied simultaneously. Soil microfauna analyses focus primarily on nematodes.

In both aquatic sediments and soils, millions of nematodes can occupy 1 m^2 of substrate and their communities can comprise an extraordinarily diverse array of species. The diversity and abundance of nematodes are discussed in further detail in Chapters 2 and 3 (this volume).

The ecological significance of meiofauna/microfauna in general and nematodes in particular was, until relatively recently, underestimated and this size class of animals was widely considered an ecological 'dead end' (see e.g. McIntyre and Murison, 1973). However, more recent studies have indicated that the meiofauna/microfauna performs important roles in a number of essential processes that sustain life on Earth; see Schratzberger and Ingels (2017) and Potapov (2022) for reviews of marine meiofauna and soil microfauna ecosystem function, respectively. The value of the nematofauna in respect of ecological services includes:

- Nutrient recycling and energy flow (Schratzberger and Ingels, 2017).
- Microbial activity regulation (Trap *et al.*, 2016; Schratzberger *et al.*, 2019).
- Food resource for 'higher' taxa (see e.g. Ptatscheck *et al.*, 2020).
- Mediation of habitat structural characteristics (see e.g. Murray *et al.*, 2002).
- Biological regulation (Steel and Ferris, 2016).

A detailed discussion of the role of nematodes in aquatic and terrestrial ecosystems is presented in Chapter 1 (this volume). As a consequence of their role in providing essential ecological services, any disturbance or change in environmental conditions that alters the composition of the nematofauna/meiofauna is likely to have consequences beyond the immediate confines of the sediment or soil in which it resides. Moreover, published field and laboratory studies, as well as experience gained from surveying and monitoring in a wide range of aquatic situations, showed that within the meiofauna (aquatic environment) and the microfauna (soil), it was the nematodes that provided the most serviceable information on prevailing environmental conditions (see e.g. Somerfield *et al.*, 1995; Neher, 2001; Muthumbi *et al.*, 2004; Fraschetti *et al.*, 2006; Gyedu-Adabio and Baird, 2006; Salamun *et al.*, 2012).

The abundance and diversity of the Nematoda, in conjunction with their important role in ecosystem processes, render this group the most appropriate representative of the meiofauna for use as bioindicators and in commercial biomonitoring of soil and sediment.

18.2 Background to the Use of Nematodes in Commercial Ecological Assessment

To understand how nematodes gained a commercial role as bioindicators, it is useful to appreciate some of the legal and regulatory factors that shaped the marketplace for ecological monitoring and assessment tools. Until the latter years of the 20th century, it was rare to find an environmental impact assessment (EIA) for a development that included reference to the natural environment in anything other than aesthetic terms. In Europe this began to change following the implementation of a European Community Directive (EC, 1985), which resulted in legal requirements in each of the member states for the inclusion of ecological components in environmental impact studies. The development of a legal framework paved the way for the evolution of *ecological* impact assessment (EcIA) as a field of study in its own right and, by raising the profile of potential hazards to the natural environment, led to an expansion in related areas such as ecological risk assessment (EcRA). Within a few years of the implementation of the Directive, demand for the services of fully trained, professional terrestrial and aquatic ecologists had increased significantly. It also brought about the emergence of professional bodies, such as the Institute of Ecology and Environmental Management (IEEM) (latterly the Chartered Institute of Ecology and Environmental Management (CIEEM)) based in the UK, that set standards and regulated the emerging profession of the 'consultant ecologist' (see IEEM, 2005, 2006).

In addition to baseline surveys, studies were needed to determine ecological effects of ongoing operations, and these investigations demanded the adoption of quantitative approaches. Legislation was produced to enforce environmental monitoring of certain industrial activities. For example, the European Community Titanium Dioxide Directive stated that any disposal of waste material 'must be accompanied by measures for the surveillance and monitoring of the environments concerned from a physical, chemical, biological and ecological point of view' (Council of the European Communities, 1982). Fully quantitative, field-based surveys remain the only practical means of assessing the spatial and temporal changes that underpin the detection and description of *actual* ecological effects, and these approaches are now used routinely at sites throughout the world.

18.2.1 Aquatic environments

With the introduction of environmental protection legislation, macrofaunal monitoring surveys (i.e. surveys based on assessment of the communities of larger aquatic invertebrates) were the norm (see, for example, EU Water Framework Directive: European Parliament and Council, 2000; EU Marine Strategy Framework Directive: European Parliament and Council, 2008). To a large extent they still are and, in several instances, quantitative macrofaunal surveys not only continue to form part of the operational requirements for manufacturing facilities as prescribed by the authorities, but their results also provide the basis for granting of wastewater discharge consents. These, in turn, help to determine the quantities of waste materials that can be released by the operators of discharge pipelines. Ultimately, it was the industries that were already obliged to monitor their ecological effects using a macrofaunal approach that created the demand for a move towards the use of integrated faunal monitoring systems – programmes that combined assessments of macro- and meiofauna (or in a few cases purely meiofauna), in parallel with physicochemical surveys. In South Africa, for example, the Department of Forestry, Fisheries and the Environment (DFFE) licences for industrial wastewater discharges at a number of sites specify the inclusion of meiofauna analyses in the annual ecological impact monitoring programmes that, ultimately, determine whether the permits to discharge are renewed (see Section 18.7.1, Case Study 1, below). The inclusion of the meiofauna size class in commercial environmental surveys led to the formal appearance of Nematoda as a functional component of ecological monitoring systems. A summary of the typical information collected and the survey stages for a comprehensive marine ecological monitoring survey is presented in Fig. 18.1.

What caused this change? The shift in emphasis arose from concerns over the natural patchiness and seasonal variability inherent in the communities of the larger invertebrate species (i.e. macrofauna). The natural variation in the assemblages of these invertebrates generated disquiet about potential false allegations of environmental harm that could arise in relation to point-source discharges as well as to diffuse inputs of materials entering aquatic habitats, such as leachates from landfills or terrestrial runoff from spill and leak incidents. In the light of this, a change to the higher-resolution monitoring ostensibly offered by meiofaunal assessments (see Platt and Warwick, 1980; Bongers and Ferris, 1999; Semprucci and Balsamo, 2012; Alves *et al.*, 2013) became not only a logical step forwards but also a commercially viable one. This was assisted by the fact that fewer and smaller-volume replicate samples appeared to be necessary to provide adequate descriptions of meiofaunal variance than for the macrofauna.

Materialization of meiofauna in the ecological monitoring toolbox for use commercially in aquatic environments occurred gradually between the late 1970s and mid-1980s; early patrons included pigment manufacturers, explosives factories, halogenated chemicals companies and pharmaceutical producers operating at sites around the world. Most of these were commercially sensitive studies and, hence, are not in

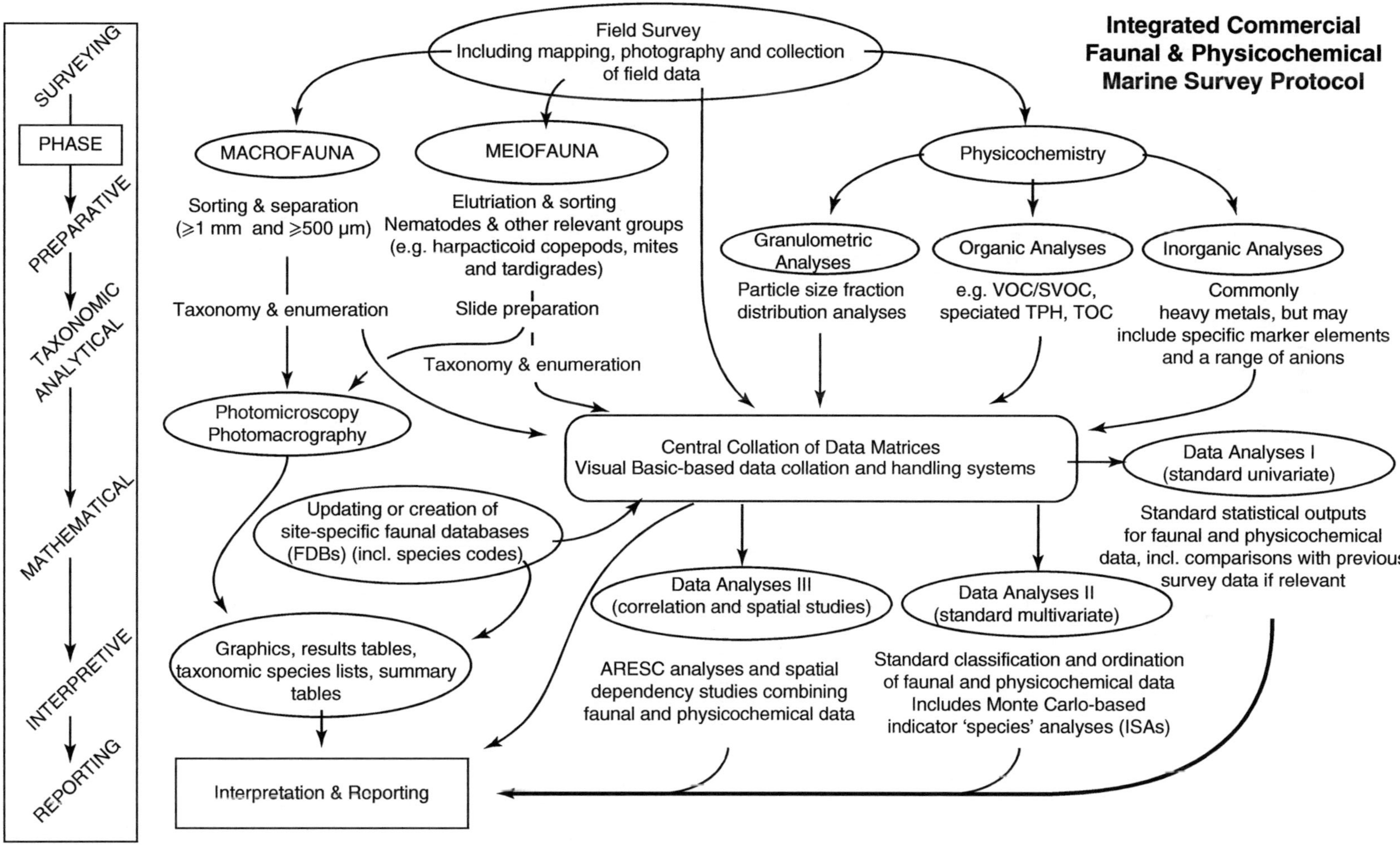

Fig. 18.1. Summary of the key operational stages in an integrated macrofaunal, meiofaunal and physicochemical marine monitoring survey undertaken by Physalia Associates for industry and regulators. VOC = volatile organic compound; SVOC = semi-volatile organic compound; TPH = total petroleum hydrocarbon; TOC = total organic carbon; ARESC = assessment and ranking of ecologically significant contaminants. (Author's own image.)

the public domain, although there were a few exceptions; see, for example, Newell *et al.* (1990a,b, 1991).

18.2.2 Soil environments

Water is subject to a European framework directive (EC, 2000) aimed at achieving a good ecological status for the waters of the European Union (EU). Similarly, air quality is controlled by EU directives (EC, 2008). However, soils do not benefit from such regulation. They are the great forgotten habitats of European law, despite their fundamental importance in the fight against climate change, the preservation of human health and biodiversity, and the protection of water and nature. These objectives are in addition to the need for good soil quality for agriculture, natural areas and, therefore, human health, as well as the requirement to reduce the mass of contaminated soils. The latest studies indicate that between 60% and 70% of soils in the EU are in poor health due to erosion, salinization, drought, desertification, pollution, biodiversity decline, organic matter depletion and compaction (European Commission, 2020). Intensive agriculture, with its aggressive practices towards soils and massive use of pesticides, contributes to the impoverishment and contamination of European soils. However, until now, there has been no regulatory obligation to maintain soil quality.

An attempt to establish a framework directive on soils started in 2006, but this was derailed by the abstention and rejection of several countries in 2014. However, in 2021, the European Commission presented a new soil strategy, which planned for the adoption of a framework directive in 2023, continuing from the one that failed to be adopted in 2014. The objective is simple: to achieve healthy soils. The European Commission defines healthy soil as 'a soil that is chemically, biologically, and physically sound, capable of continuously providing a wide range of ecosystem services' (European Commission, 2021). This includes acting as a carbon reservoir, supporting food and biomass production, serving as a source of raw materials, providing essential elements for biodiversity, and protecting groundwater through water absorption, storage and filtration.

In the agricultural sector, the recent move to sustainable farming practices has focused interest on the commercial importance of soil health and the economic value of increased productivity through nurturing a diverse and abundant soil fauna in general and nematodes in particular (Neher, 2001; Ockleford *et al.*, 2017; Lazarova *et al.*, 2021; Du Preez *et al.*, 2022). This has led to the development of soil quality surveys including characterization of soil biological functioning.

18.2.3 Potentially polluted or degraded soils

Another driver for the use of nematodes as bioindicators in applied ecological studies came from the EcRA domain. Single-species toxicity tests, including nematode-based screening of materials (see review by Höss, 2021; and Chapter 13, this volume), have been in use for over two decades. It is common to extrapolate from laboratory-based toxicity testing on single species (or artificially maintained assemblages) to predict effects of toxic materials in the natural environment and, hence, set effluent concentration discharge limits or sediment/soil contamination levels when issuing screening guidance (e.g. Buchman, 2008) or commercial works/discharge licences. However, unlike laboratory conditions, in the natural environment the permutations of conditions and features that may influence the effect of the stressor (e.g. a contaminant or disturbance) on the affected habitat/community are infinite and cannot be encapsulated sufficiently or with confidence by controlled laboratory tests. Arguments against this practice have been rehearsed by Chapman (2002a,b; Chapman *et al.*, 2003), who suggested that measurement of actual ecological effects *in situ* is a far more valid, but by no means simple, approach to determining and setting appropriate and precautionary soil, sediment and water guideline concentration values for contaminants.

Laboratory-based toxicity testing has an important role to play in investigations into the effects of disturbance and/or pollution on the faunal communities of aquatic sediments and soil and is discussed in detail in Chapter 15 (this volume). The current chapter, however, is concerned with the assessment of *in situ* environmental conditions and the use of nematode species and nematode communities to determine *actual* ecological effects that occur in the field and, as such, toxicity testing and laboratory-based experiments will not be considered further. Nevertheless, it is valuable to note that in studies concerning contaminated sites and soils, contract labs, such as ELISOL Environnement and Ecossa, intervene by conducting both types of analyses: these include studying the soil-extracted communities following the ISO 23611-4 standard, as well as performing ecotoxicological tests using *Caenorhabditis elegans* in accordance with ISO 10872 standard. Indeed, the information generated by these two types of methods can be complementary to comprehend the ecological state of a site.

18.3 Specific Commercial Needs Addressed by Nematodes as Bioindicators

Since the early 1980s commercial ecological assessments of the environment utilizing meiofaunal marine and freshwater nematodes and free-living soil nematodes have been employed in commercial contexts in a wide variety of situations and at sites worldwide.

In the aquatic environment, these have included baseline surveys and monitoring of industrial effluent from marine, estuarine and freshwater discharge pipelines, surveys of deep-sea industrial waste-dumping grounds, assessments of impacts of landfill and contaminated land leachates entering rivers and estuaries, baseline surveys at deep-sea mining sites, monitoring of oil-production facilities using nematode colonisation unit (NCU), impact assessments associated with gas-well blowouts, biodiversity assessments at protected sites, and risk assessment of chemicals that are associated with sediments.

In terrestrial environments, the majority of commercial nematode studies are dedicated to agricultural soils (e.g. for the assessment of soil quality changes over time following the adoption of practices that are more environmentally friendly). Beyond the agricultural context, baseline surveys and monitoring studies of soil nematofauna have been conducted in the vicinity of potentially polluting industries. These studies are typically focused on large areas (>10 hectares) and are funded by the manufacturers themselves or commissioned by engineering consulting firms. With regard to the restoration of degraded soils in urban and industrial brownfield sites, there is a new demand for evaluating the functionality of constructed soils in such areas in order to demonstrate and document the rehabilitation of these habitats.

In each of these scenarios, the studies addressed the highly specific needs of the customers, which included industry, local authorities, agricultural suppliers and agricultural advisers. By way of example, Table 18.1 presents a list of the most common, generic requirements of each of the customer sectors that the authors have worked with in the UK and France. Many, but not all, of these needs can and have been addressed in some way by nematode assessments and examples are touched on briefly in Table 18.1 below. Further details of specific case studies are presented later in this chapter (see Section 18.7).

The drive towards more sustainable *agriculture* over the last two decades and the need to reduce the inputs of pesticides and fertilizers for both economic and environmental reasons have highlighted the need for better management of 'soil health'. While the definition of soil quality and soil health has been debated (Kibblewhite *et al.*, 2008; Bünemann *et al.*, 2018; Harris *et al.*, 2022), the important role of soil biodiversity in general (Farfan and Wall, 2022) and nematodes in particular (Neher, 2001; Sánchez-Moreno, *et al.*, 2006; Neher and Powers, 2022) in soil function and land productivity is well documented. Consequently, surveying and assessing soil nematode communities has become an important tool in monitoring soil health and the management of agricultural land. These studies are commissioned primarily by agricultural advisors affiliated with organizations such as agricultural chambers, agricultural suppliers or agricultural cooperatives. Currently, due to the absence of regulatory obligations, few farmers themselves seek these studies directly. This is likely due to the perceived high cost of the analyses and the distant return on investment; such analyses are conducted within the framework of agricultural system sustainability rather than immediate profit.

Where novel biocides or genetically modified (GM) crops are developed by the *agrochemical*

Table 18.1. A list of the principal areas in which different customer sectors require inputs from ecological surveys and assessments that can, and in many cases have been, met by the use of nematodes as bioindicators. While the industrial sectors have historically represented the largest group of nematode-based assessment consumers, other sectors continue to grow steadily. (Author's own table.)

Customer sector	Needs	Comments
Industry (Chemical/petrochemical, explosives, pharmaceutical, power generation, construction, transport, mineral and aggregate extraction, offshore mining, paper, food processing and drinks manufacturing including some brewing companies, etc.)	Detection, description and delineation of *actual* ecological effects of operations on surrounding habitats	Part of planning permissions/licences granted for discharges to controlled waters and discharge consent compliance monitoring; for larger industries, this forms part of Environmental Management/Stewardship Systems
	Operations monitoring	Assessment of ecological impact of ongoing industrial activities and documentation of recovery following cessation of activities
	Environmental performance testing	Assessment of efficacy of treatment systems or mitigation measures
	Incident monitoring	Ecological assessments of effects following and arising from incidents (e.g. spills, leaks and 'loss of containment')
	Early warning and *in situ* monitoring systems	Biosentinel monitoring where risks perceived to be high
Regulatory bodies and local authorities	Baseline assessments	Often delegated to developers, forming part of planning applications for developments
	Post-development assessments and monitoring	Ostensibly as follow-up to above; *rarely enforced with developers*
	Compliance monitoring and certification	Part of their statutory role; work often subcontracted
	Status of contaminated land sites and prioritizing contaminated sites for clean-up and the targeting and design of remediation programmes	Now specific UK requirement to assess and rank contaminated land and to identify potential ecological receptors under Environmental Protection Act 1990 Part IIA
Water authorities and water companies	Abstraction licensing	Baseline assessments of habitats that will be exposed to potential stressors (contaminants/disturbance)
	Monitoring of wastewater treatment works	Periodic assessments (annual) of treatment efficacy and ecological effects
	Treatment failure/sewage release impact assessment	Assessment of effects of untreated sewage and nutrient enrichment on receiving environment
Legal profession and insurance industry	Evidence of any detectable impacts arising from incidents and their characterization	Description and 'quantification' of harm caused and production of 'expert witness' evidence
	Assessments for ascription of liability	Used to assess the scale of fines for damages

Remediation industry	Delineation of areas within sites requiring remediation	Landfills, contaminated land and former factory/brownfield sites; primarily terrestrial nematode surveys
	Assessment of success of the measures adopted	Certification of works undertaken for site closure/‘sign-off’
Agriculture	Assessment of soil health	Provide data for soil restoration/enhancement measures and assessment of soil management efficacy
	Risk assessment of genetically modified crops	Using nematode-based indices to assess the ecological risk of genetically modified organisms (GMOs) (e.g. *Bt*-maize) on non-target soil fauna
	Risk assessment of novel biocides	Using nematode-based indices to assess the ecological risk of agricultural biocides (nematicides, insecticides) on non-target soil fauna

industry, the effects on the wider fauna community and non-target species need to be assessed. Given the abundance and diversity of soil nematodes, this group provides the ideal tool by which to do so. Recent examples include studies where in-field nematodes were monitored over an extended period to assess potential effects of the product on free-living (non-target) nematodes. The data were used subsequently to support the registration of the product and/or GM crop.

For aquatic habitats, the *industrial sector* provides some of the principal customers for ecological monitoring surveys. There is usually a well-defined need to monitor effects of industrial operations on the environment which, in most cases, is set down in the planning permissions granted to individual factories and in the regularly reviewed licences to operate. Currently, it is rare that license-issuing *regulatory bodies* specifically stipulate the use of nematodes as monitoring tools. However, where sensitive ecosystems of high conservation value are vulnerable to the impacts of industrial activities, operation licences can require nematode monitoring to be undertaken. Such is the case in South Africa, where the DFFE continues to stipulate meiofauna (nematode) community monitoring of the receiving environments as a condition in licences to discharge wastewater to sea. At sites where macrofauna diversities and/or densities are low or suppressed, meiofauna communities may be the only bioindicator group by which the ecological impacts of activities can be assessed. Such is the case with many deep-sea mining sites and, in this context, monitoring effects using nematode communities has been recommended (see e.g. Ingels *et al.*, 2020; Pape *et al.*, 2021).

Where their cases relate to environmental issues, the requirements of *insurers and the legal profession* are usually highly specific. These most commonly centre on an incident that has resulted in environmental harm. The distinct advantages of nematode assessments in the detection, description and delineation of effects (see Chapters 4, 5, 6 and 7, this volume) are not lost on lawyers, who know that the sizes of areas affected determine the scale of fines.

In Europe, the *remediation industry* is growing rapidly as brownfield sites and contaminated land come under increasing pressures for redevelopment. In the UK alone, there are thought to be as many as 100,000 contaminated sites suitable for development (Property Investment Partners, 2023). In the UK there is now a requirement to undertake ecological assessments and to identify potentially affected ecological receptors (see Table 18.1). For the remediation companies, nematode assessments are already providing a flexible and sensitive means of identifying migration of contaminant materials from sites (see e.g. Trett and Thurgood, 2008) and to determine their effect and delineate their zone of influence. Once the remediation works have been undertaken, the same surveys are now furnishing information that is helping in the documentation of recovery and the assessment of the success of the measures that have been adopted (e.g. soil washing, the use of reactive barriers, *in situ* chemical treatment and removal, treatment of contaminated materials, natural attenuation).

18.4 Nematode Survey Protocols

Currently, for the majority of commercial surveys and monitoring programmes in aquatic environments and soils, taxonomy is based on morphological features examined under the microscope. Other taxonomic techniques have and are being developed, for example DNA barcoding, image analyses and artificial intelligence (AI) techniques. These are discussed in detail in Chapters 13 and 17 (this volume) and mentioned in Section 18.6 below. While these approaches have the potential to revolutionize nematode taxonomy, currently the techniques are either in the process of development (e.g. AI and image analysis techniques) or, in the case of DNA barcoding, require considerable pre-existing information regarding the species present (a DNA database of likely species occurrences) and does not produce information of quantitative abundance (Dell'Anno *et al.*, 2015). Therefore, for the time being at least, morphological taxonomy remains the 'go to' method for commercial nematode surveys.

18.4.1 Commercial marine, estuarine and freshwater nematode surveys

'Standard' survey

A summary of the laboratory protocols used for the processing of aquatic meiofaunal nematode samples is shown in Fig. 18.2. With some

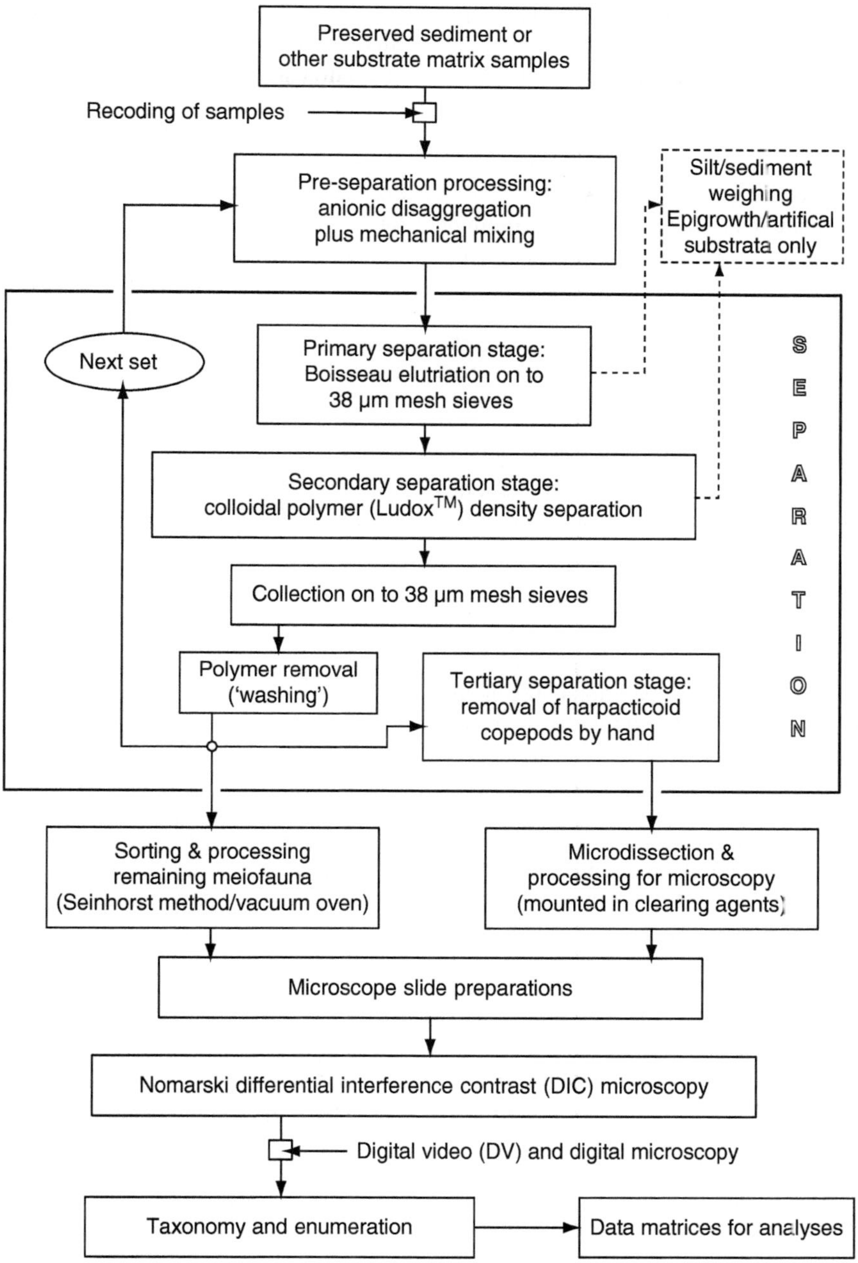

Fig. 18.2. The basic aquatic meiofauna processing protocol used and modified by staff at Physalia over many years. This hybrid, three-stage system is further modified to accommodate non-standard meiofaunal samples (e.g. epigrowth samples, artificial substrates collected from nematode colonization units). (Author's own image.)

modifications, this basic protocol has been used in aquatic studies since the early 1980s (see Newell *et al.*, 1990a,b). Special modifications of the illustrated protocol are used for different types of quantitative nematode samples. These include samples of intertidal algal epigrowth communities present on pier stanchions and river frontages and the artificial substrates from nematode colonisation unit (NCU) used in a range of natural and man-made habitats (see Section 18.7.2, Case Study 2, below). Note the use of 38 µm mesh sieves. These were adopted from the earliest surveys and prevent significant losses of small 'straight' adult and juvenile nematodes as opposed to coiled worms (Hooper, 1986).

All nematode morphological taxonomic work is carried out using Nomarski differential interference contrast (DIC) microscopes with slides made using the Seinhorst rapid glycerol technique (Seinhorst, 1959). This is used in preference to lactophenol, which, although a more rapid technique, is avoided on health grounds. Use of lactophenol (or a modified in-house 'phenol-free' formulation) is presently restricted to meiofaunal harpacticoid copepod preparations where the clearing of the cuticular structures is important for the copepod setotaxy.

In standard integrated macrofaunal–meiofaunal surveys undertaken in Northern Europe, the first 100 adult nematode specimens encountered in a random sample will be identified to species level (or to an operational taxonomic unit within a genus). Total counts are then made of all other nematodes in the sample and appropriate multiplication factors applied to determine the abundance per species per sample. In preliminary studies for specific surveys where, for example, post-development assessments will be undertaken or a future monitoring programme is planned, and in cases where a new procedure is being developed, such as the use of NCUs in a new location, the first 150 or even 200 nematodes specimens in a sample may be identified to provide a better understanding of the taxa present and their relative abundances.

Results from the taxonomic analyses include total abundances of nematode feeding-type groups which, for marine/outer estuarine species, are still based on Wieser's original classification (Wieser, 1953) rather than on more recent and potentially more valuable systems (namely Moens and Vincx, 1997). For freshwater and oligohaline habitats, feeding type classifications have corresponded with those described by Traunspurger (1997). These data, along with the final site-by-site results, are entered into standardized spreadsheets. With up to nearly 400 nematode taxa occurring in single 50-station monitoring surveys (e.g. Bass Strait, Tasmania), and cumulative totals of over 1000 taxa in the nematode faunal databases for some survey sites (e.g. Malaysia and South Africa), data entry and the cross-checking with microscope record sheets are time-consuming and arduous procedures. However, quality control and attention to detail are an extremely important part of the work and, just as chain-of-custody forms are required for handover of samples, faunal data sets are not accepted until signed off as checked and correct by the taxonomists.

Less conventional nematode sampling and assessment methods

In a limited number of circumstances, where the accurate assessment of conditions is required and where nematodes are the monitoring tool of choice, owing to either the paucity or even absence of macrofauna or to the nature of the stress under investigation, standard surveying protocols may be neither appropriate nor possible. Examples include:

- Assessment of the impacts on marginal habitats of increased shading caused by developments located beside watercourses. This is now a standard requirement of the Environment Agency in the UK for developments along built-up canals and estuaries (e.g. the Thames).
- Evaluation of the extent and degree of effects arising from marine and terrestrial spills that impinge on rocky shores at low or on falling tides, or assessment of the stability of patterns of ecological effects in the same habitats arising from domestic and industrial effluents that disperse along tidal rocky shores.
- Monitoring of prevailing water quality using submerged surfaces of a variety of structures such as barrages, pier stanchions, marine wind-farm turbines and mooring or channel-marking buoys.

In these situations, nematode communities can be readily sampled using quantitative epigrowth samples. Samples, comprising the algal and/or moss matrices that develop on submerged or on wetted and water-retentive substrates in intertidal habitats, can be collected from a known area, ranging from 25 cm × 25 cm down to 10 cm × 10 cm depending on the depth (thickness) of the epigrowth mats. Aspect, depth of immersion and, for intertidal habitats, degree of exposure to wave action and position within the littoral zone are clearly important factors and must be recorded accurately. These data can then be incorporated into multivariate correlation analyses (see Section 18.5 below). The laboratory separation protocols outlined above can be used with these samples, with the sediment present in the epigrowth matrices being removed during the elutriation and Ludox™ separation stages (see Fig. 18.2).

One step removed from the use of a natural, floral epigrowth matrix is the deployment of artificial matrices that are deployed strategically and become colonized by nematode assemblages. These artificial habitats have several benefits. Foremost among them is that the matrix, typically a coarse, inert fibre pad, can be standardized in size and structure, reducing the effects of variation in microhabitat. These can also be deployed in situations where natural, photosynthetic matrices (plant and algal epigrowth) could not occur (e.g. within cooling-water systems and water-abstraction culverts). This approach has been used in a range of projects where controlled sampling would otherwise have been difficult and where the role of nematodes as bioindicators was more important than characterization of effects on natural assemblages (see Section 18.7.2, Case Study 2, below). For sample processing, the standard protocol is modified to include two or three 'washing cycles' in which the colonist fauna is flushed from the pad using running water.

18.4.2 Commercial soil nematofauna surveys

Soil sampling

Nematode soil samples are collected using a small soil core or auger, typically in the 0–20 cm strata. Other depths can be chosen (e.g. 0–10 cm or 0–30 cm) according to the specific question of the commercial order. It is necessary to select areas which are homogeneous in terms of soil physicochemical properties, vegetation and soil use. The surface of the studied plots may vary depending on the context (agricultural soil, polluted soil, etc.), ranging from 100 m^2 to several hectares.

Several soil cores are collected and combined following a regular (grid) plan or according to a pattern (e.g. a zigzag pattern). The collected bulk sample is homogenized in a bucket, and the soil is gently mixed until the mass is uniform. The homogenization is then completed by the washing of the soil through an 8 mm aperture sieve to remove coarse organic material, roots and stones. This step of homogenization is essential to all the analyses that are based on it.

Approximately 500 g to 1 kg of fresh soil are taken from the bulk sample for extraction. At this stage, other subsamples from the bulk may be collected for soil physicochemical analyses and microorganisms analyses, as required. Soil samples for nematode analysis are stored at between 4 and 10°C prior to extraction. The storage period should be kept as short as possible, with a maximum of two weeks between sample collection and nematode extraction.

Characterization of nematode communities in soil samples

After sampling, the nematodes are extracted from the soil using the Oostenbrink wet funnel method. This involves three basic mechanisms for the separation of nematodes from soils: washing, sieving and active movement. The nematodes that are characterized quantitatively are free-living nematodes and plant-feeding nematodes (ectoparasitic plant-feeding nematodes and stages of endoparasitic nematodes that live in the soil).

The procedure is fully standardized in an International Organization for Standardization (ISO) standard dating from 2008 which was updated in 2022 (ISO 23611-4:2022, Soil Quality – Sampling of Soil Invertebrates – Part 4: Sampling, Extraction, and Identification of Soil Nematodes). The main steps are outlined below.

- A known weight of fresh soil (i.e. 300 g) is processed through Oostenbrink elutriators which have a constant extraction efficiency. An upward current of water separates the nematodes from soil particles and holds them in suspension while the heavier particles sink. Then the suspension of nematodes and small particles passes through three sieves with a mesh width of 50 μm.
- The catch is washed from the sieves on to a cotton-wool filter on a supporting sieve and is

placed in a Petri dish filled with tap water. After two days, the nematodes have separated themselves from the debris on the filter through their active downwards movement.

Alternatives to these two steps used to separate nematodes from soil particles are also proposed in the standard (ISO 23611-4:2022); some are suitable for extracting non-motile nematodes or eggs.

After extraction, the nematodes are counted in two aliquots of the suspension, then concentrated, preserved in a formalin solution and mounted on mass slides. At least 150 individuals are identified under a light microscope, based on morphological characteristics. Nematodes are identified to genus or family level (note: nematode populations in the soil are often dominated by juveniles).

The results of nematode counts and identifications are stored in an electronic database. The data are expressed in relation to the dry weight of soil. Taxa are assigned to feeding groups according to Yeates *et al.* (1993) and colonizer–persister (*c-p*) categories (Bongers, 1990). After this, the nematofaunal indices and proportions of feeding groups can be calculated along with standard diversity indices.

18.5 Nematode Survey: Data Analyses

18.5.1 Assessment of aquatic nematode assemblage data for commercial studies

Figure 18.1 shows the main stages in a marine invertebrate monitoring survey programme which integrates meiofauna, and hence nematodes, with analyses of groups of larger invertebrates. Meiofaunal core samples, in this particular case for quantitative assessments of nematodes and harpacticoid copepods, are collected at the same sites as samples for macrofauna (typically retained on either a 1 mm or 500 μm mesh sieve). In addition, samples are collected at each site for sediment physicochemical analyses. Granulometric samples provide data on the proportions of different particle-size fractions which, in aquatic (and terrestrial) habitats, prove to be important natural determinants of nematode community structures (see Section 18.7.1, Case Study 1, below). The samples for chemical analyses can include organic and inorganic contaminants, depending on the nature of the putative stress being investigated and the presence of any confounding influences (e.g. riverine inputs, neighbouring discharges and offshore moorings) and commonly occurring background and/or natural compounds and elements. In many cases, specific elements are also assayed for use in inter-element ratio analyses where they function as markers for materials of terrestrial and/or aquatic origin; for example, aluminium as a proxy for aluminosilicate of clay (see Ho *et al.*, 2012).

Marine survey areas comprise fixed grids of stations in which the sampling sites are, ideally, distributed in a 'herringbone' pattern across a feature of interest such as a point of discharge, a dredging area or a dumping ground. In some marine studies the survey grids may become modified over successive surveys due to observed patterns of net sediment transport and the associated movements of deposited materials on sea and estuary beds, or where natural features such as reefs or even wrecks are encountered and preclude collection of grab and core samples.

Ecological data are noisy, and, in this respect, nematode community data are no exception. However, some data are noisier than others and nematode data, thankfully, do not fall into this category, especially when their communities are constrained by adverse environmental conditions. The reasons for this have already been presented in this book (see Chapters 2, 3, 5 and 6, this volume) and are the key to understanding the value of nematodes as bioindicators. A range of community indices and approaches to the assessment of changes in nematode communities and the detection of environmental stress have been discussed in Chapters 4, 5 and 6 (this volume). However, there is no single community statistic or set of results from a mathematical analysis that gives access to a definitive appraisal of the effects of prevailing environmental conditions. To extract this information, we need to resort to a variety of techniques. In a commercial situation, this data mining process needs to be efficient and well-focused but should not prevent exploratory investigation or in any way blinker the final interpretation. This, coupled with judicious use of Occam's razor, leads to the most parsimonious and defensible analysis of nematode survey results and provides the basis for a robust ecological interpretation of prevailing conditions.

The analyses used in most commercial aquatic meiofaunal/nematode studies undertaken by the authors can be divided into two categories: *Level I* analyses, which encompass standard uni- and bivariate approaches; and *Level II* analyses that use multivariate community techniques. A third category (*Level III* analyses) is employed in more advanced studies where there may be a need to identify and rank measured environmental parameters, natural and anthropogenic, in terms of their relationships with nematode community structures or in specialized, bespoke analyses for a wide variety of non-standard survey situations. Examples of these three levels of analyses are presented in the sections below. Graphics summarizing the results of the analyses form an important part of the reporting process and strong emphasis is placed on relating the analytical outputs back to field situations (see e.g. Section 18.7.1, Case Study 1, below). This helps in the process of communicating the findings of surveys to audiences that are usually experts in areas other than ecology.

Level I analyses: standard univariate studies

Standard community description parameters are generated automatically when the data sets enter the system. For each site/sample, these include standard parameters such as species richness, densities and dominance/co-dominance. Diversity indices are also calculated, usually reciprocal infinite and finite Simpson's dominance indices (1 – D; Simpson, 1949; see also Hill, 1973; Chapter 4, this volume) are preferred as these are less influenced by the occurrence of rarer nematode taxa and, therefore, small variations in sample size (Whittaker, 1972). Finally, densities of individual nematode feeding types and ratios of Weiser's type 1B to type 2A ('non-selective' deposit feeders and detritivores to epistrate feeders) and their equivalents in freshwater systems (see Moens *et al.*, 2006) are calculated. Other indices have been developed and applied primarily to freshwater habitats. These include the NemaSPEAR[%]-index (Nematode Species At Risk) that uses the percentage abundance of pollution-sensitive nematode species (see e.g. Höss *et al.*, 2017; Schenk *et al.*, 2022) and the Maturity Index (MI) (Bongers, 1990). The latter is not part of the suite of standard parameters examined in the analyses. This may change if the MI can be modified to give higher resolution than family level and/or if the index and the ascription of *c-p* values can accommodate the revisions that are now occurring rapidly in the taxonomy of the Nematoda following the most recently published genomic studies (see e.g. De Ley *et al.*, 2006; Meldal *et al.*, 2007).

Overlaying plots of proportionally scaled symbols for the univariate parameters on to maps of the survey areas provides a rapid visual means of assessing the distribution of the data with respect to features of interest (e.g. factory discharges). When they are combined with appropriate parametric and non-parametric statistical tests, these are particularly valuable in detecting changes in time-series data derived from monitoring surveys.

Other Level I analytical approaches include the use of *k*-dominance curves. These are generated from the ranked dominance values and allow communities to be compared *en masse*. These can be valuable for coarse screening of the nematode data sets and also enable the validity of comparisons between their diversity indices to be determined (see Platt *et al.*, 1984).

Level II analyses: multivariate analysis

Patterns in the structures of the nematode assemblages are examined using transformed data matrices and a suite of both classification and ordination 'similarity' analysis methods (see Chapters 4, 5, 6 and 7, this volume). These techniques do not assume linear responses of species to environmental gradients and, hence, avoid the issues that can be encountered when using such techniques as principal components analysis and discriminant analysis. The application of both classification and ordination methods has distinct advantages. As these two multivariate techniques differ fundamentally in their underlying rationales, and hence the algorithms that they employ, the coherence of the clusters of 'similar' assemblages indicated in the classification studies can be examined in the light of the results of the ordination analyses. Where the results corroborate each other, the investigator can be confident that the observed patterns and structural relationship between the samples/communities are likely to be real and not artefacts of the analytical technique. Any deviations that occur between the results of the ordination and classification analyses are investigated more closely. It should come as no surprise that fewer inter-analysis anomalies occur with nematode data sets than with those derived from the macrofaunal groups.

Derived from the classification analysis, the key species that are statistically associated with each cluster in terms of their occurrence and abundance can be identified. This enables characteristic community structural types to be identified, mapped and linked spatially with habitat conditions, whether they be natural variables such as sediment coarseness or contamination status. In turn, this facilitates the delineation of areas of effect (or impact) relating to the anthropogenic disturbance and/or contamination.

Level III analyses: multivariate correlation analysis (and other techniques)

Multivariate correlation analyses are commonly included at this stage to assess the combined nematode community and sediment physicochemical data matrices (see Section 18.7.1, Case Study 1, below). While these approaches tend to focus on *statistically* significant relationships between community structures and the measured environmental factors, assessment and ranking of *ecologically* significant contaminants (ARESC) studies acknowledge the fact that lower-order relationships when multiplied up over many generations can lead to marked changes in community structures. This is particularly relevant when using nematodes as bioindicators where generation times (egg-to-egg) can be as low as three days (Vranken and Heip, 1983).

ARESC analysis is employed in aquatic meiofauna surveys to define the correlations between the measured environmental parameters and the structures of the nematode communities (see Trett *et al.*, 2000, 2009, 2012; Ellis *et al.*, 2002). The results of these analyses are summarized as ordination plots of the community structures with correlation lines for each of the measured environmental parameters superimposed on the ordination axes. Separately, tables presenting the correlation coefficients of individual parameters with each of the ordination axes are produced and the statistical significance of the correlations are calculated and ranked. Crucially, the results of these analyses enable: (i) the primary factors that are associated with the nematode community structures to be identified; and (ii) the associations of natural and anthropogenic influences to be distinguished (see Section 18.7.1, Case Study 1 and Fig. 18.3, below).

For experimental studies that consider the response of nematode communities to certain stressors over time (e.g. micro/mesocosm or field experiments), principal response curves (PRC) are a suitable multivariate tool to analyse the changes in species composition over time (Van den Brink and Ter Braak, 2009). This offers also the possibility to test for significant differences of nematode assemblages in different treatments (using Monte-Carlo permutation test). Examples for PRC studies on nematode communities can be found for microcosm experiments using single chemical compounds (e.g. Smit *et al.*, 2002; Brinke *et al.*, 2010; Höss *et al.*, 2020) or field studies comparing different crops (Höss *et al.*, 2015).

18.5.2 Assessment of soil nematode assemblage data for commercial studies

Nematofaunal parameters

The analyses of the nematofauna used as bioindicators to characterize the biological functioning of the soil are carried out either on individual plots or on sets of plots. Sets of plots are compared when the land being investigated has been subjected to different histories in terms of land use, management practices or risk of contamination, etc.

The analyses on individual plots mainly concern agricultural sites. However, studies on sets of plots are carried out in a number of different contexts:

- *Agriculture:* For groups of farmers implementing innovative agricultural practices, for example, or in experiments for the evaluation of products (e.g. agricultural biocides: nematicides, insecticides) newly developed by agricultural suppliers.
- *Polluted sites:* Assessment of the impacts of contamination on soil organisms.

1997				2004			
★Axis 1		Axis 2		★Axis 1		Axis 2	
Determinand	Pearson's *r*	Determinand	Pearson's *r*	Determinand	Pearson's *r*	Determinand	Pearson's *r*
■ Fe	−0.482 **	125–250 μm	0.798 **	125–250 μm	0.891 **	K	−0.625 **
> 2000 μm	−0.477 **	500–1000 μm	−0.706 **	500–1000 μm	−0.705 **	63–125 μm	−0.617 **
Ni	−0.473 **	1000–2000 μm	−0.632 **	1000–2000 μm	−0.693 **	Al	−0.542 **
Al	−0.460 **	63–125 μm	0.472 **	K	0.597 **	Mg	−0.531 **
Cu	−0.457 **	250–500 μm	0.463 **	> 2000 μm	−0.595 **	Ni	−0.524 **
TMBI	−0.452 **	□ Mn	−0.434 **	63–125 μm	0.581 **	Cu	−0.521 **
TOC%	−0.448 **	Ni	−0.343 **	Al	0.578 **	Na	−0.519 **
■ Pb	−0.444 **	■ Fe	−0.338 **	Ni	0.543 **	TOC%	−0.475 **
□ Cr	−0.415 **	Cu	−0.299 *	■ Fe	0.439 **	■ Fe	−0.469 **
□ V	−0.357 **	Al	−0.288 *	Mg	0.421 **	■ Pb	−0.418 **
□ Zn	−0.303 *	< 63 μm	0.255 *	Cu	0.360 **	500–1000 μm	0.405 **
500–1000 μm	0.234 *	□ Zn	−0.236 *	TMBI	0.297 *	Ca	0.301 *
□ Mn	−0.233 *			□ Mn	0.295 *	Sr	−0.285 *
				□ Zn	0.295 *		

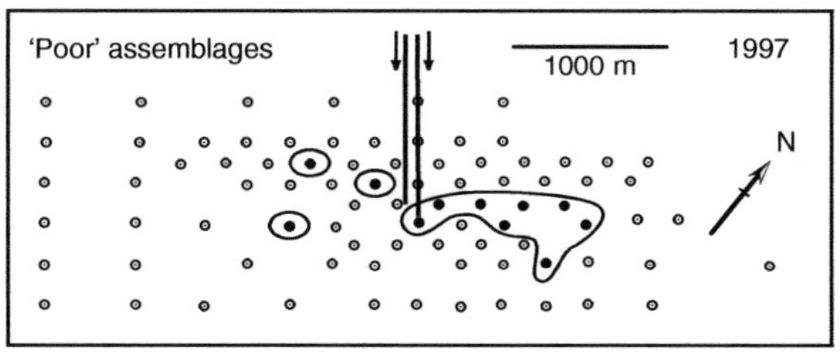

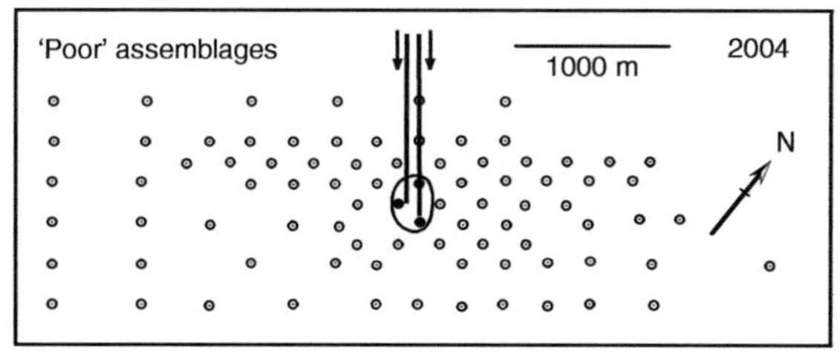

★ Axes accounting for the highest proportions of the observed variance in the meiofaunal nematode community structures

□ Elements closely associated with the industrial complex discharges ■ Elements associated with the industrial discharges and natural coastal riverine inputs

Statistical significance: $*P < 0.05$; $**P < 0.01$

Fig. 18.3. Comparison of the nematode multivariate correlation (ARESC) tables for monitoring surveys undertaken in 1997 and 2004 off the South African coast, showing the ranked, statistically significant environmental determinands and their correlation coefficients in relation to the nematode community structure axes. The maps illustrate the sites at which 'poor' nematode assemblages were recorded in the years shown in the tables above and are based on the same lower quartile criteria (namely depressed species richness, depressed diversity indices, elevated 1B:2A feeding type ratios in medium-to-coarse sand sediments and elevated dominance values). In both studies axis 1 accounts for the highest proportion of the variance observed in the nematode community structures. Note that between 1997 and 2004, the significance of effluent-associated elements declined and natural factors (primarily sediment particle-size fractions) increased in their significance, migrating from axis 2 in 1997 to axis 1 in 2004. These changes were tracked in annual monitoring surveys and paralleled the changes made by the factory complex in their effluent treatment systems (see text). ARESC = assessment and ranking of ecologically significant contaminants; TMBI = total metal burden index; TOC% = percentage of total organic carbon. (Author's own image.)

- *Follow-up of the rehabilitation of polluted sites:* Refunctionalization of soil and creation of fertile land.

The assessment and diagnosis of the soil condition is based on a set of nematode community parameters:

- *Abundances of the different nematode trophic groups:* Bacterial feeders (broken down into opportunistic bacterivores (*c–p*1) and other bacterivores), fungal feeders, omni-predators, facultative plant feeders, obligate plant feeders.
- *Nematofaunal indices:* Mainly Enrichment Index (EI), Structure Index (SI) and Nematode Channel Ratio (NCR).
- *Diversity indices:* Shannon index and number of taxa.

For agricultural plots, the abundances of the main genera of plant-parasitic nematodes can also be documented. These nematodes have the potential to cause crops damage and, therefore, are of interest to the customer. Even if these analyses are carried out in a soil stratum that is too superficial or at a taxonomic level with little resolution, the risk associated with the presence of these pests is identified and further investigations can be undertaken if necessary. When customers are not interested in characterizing the biological functioning of the soil but only in characterizing the risk on crops linked to plant-parasitic nematodes, more specific analyses can be carried out. These analyses, that search for a single taxon, or for a range of taxa, are commonly carried out by molecular biology methods.

Benchmarks

In order to diagnose the biological condition of soil in a specific plot, it is important to understand the variation ranges of nematofauna parameters based on the specific conditions of the study site. The abundance and composition of soil nematode communities are influenced by several factors, including:

- Land use, such as forests, meadows, dunes, agricultural fields, market gardens, vineyards, orchards, urban parks, etc.
- Physicochemical characteristics of the soil, including carbon and total nitrogen content, texture (sand, silt, clay), pH, limestone content, etc.
- Climatic conditions, such as latitude, longitude, altitude, annual precipitation, average temperature, etc.
- Management practices, including tillage, mineral fertilization, organic fertilization, pesticide use, forest management practices, etc.

Over a 20-year period of undertaking nematofauna analyses in various contexts, a database containing nematofaunal analyses (taxa abundance and synthetic nematofauna parameters) along with associated non-biological data (climatic conditions, soil characteristics, land use and practices) has been compiled by ELISOL Environnement. This has been used to assess the impact of different families of environmental influencing factors on the nematofauna. Only the analyses conducted in Europe were used and data from other areas (Africa, Asia, North and South America and overseas territories of France) were excluded. Similarly, greenhouse or tunnel studies were also excluded to focus on bulk soil analyses. With a total of 5000 nematofauna analyses available, statistical analyses were performed to determine the most influential environmental factors in explaining nematofauna parameters and to generate interpretation reference systems.

The results of these analyses allowed us to generate benchmarks based on land use, as this factor played a significant role in community structuring. While some soil parameters were also impactful, the available data were insufficient to generate references combining both land use and pedo-climatic conditions. Currently, we have developed standards for field crops, vineyards, market gardens, meadows and forests. These reference systems serve as an initial basis by which to assess soil conditions and can be further refined with additional data.

Once the statistical distribution of values for each nematofauna parameter in each soil-use category was known, and leveraging our nematology expertise, we transformed the biological variables into scores ranging from 0 to 10. A score of 0 represents the worst situation for the considered parameter; scores in excess of 6 represent a suitable situation and 10 indicates the best possible situation.

We applied this scoring method to six parameters, namely abundance of free-living nematodes (as an indicator of soil biological activity), EI, SI,

NCR, Shannon index and abundance of major plant-parasitic nematodes (as an estimate of parasitic pressure by plant-parasitic nematodes). The first five scores (excluding the plant-parasitic nematodes) were used to calculate an overall score and classify the soil's biological functioning into three categories: 'altered', 'to be monitored' and 'satisfactory'.

The generated reports for each soil sample provide results with varying levels of detail, including:

- overall assessment (altered, to be monitored, satisfactory);
- scores ranging from 0 to 10 for six parameters; and
- detailed figures showing the values of different nematofauna parameters.

These reports are well suited for agricultural situations. However, in other contexts, the expectations and presentation of nematofauna analysis results may vary. For example, in the context of built soils, users are interested in evaluating the functions performed by soil organisms. To meet this demand, we have created function scores based on the aggregation of different nematofauna parameters, including:

- biological fertility function;
- biological regulation function;
- biodiversity support function;
- dynamic function of carbon; and
- pressure on the roots.

Multivariate analyses

In studies involving a large number of samples/analyses, the differences in community structures can be investigated by applying multivariate analytical techniques in a manner similar to that described for the aquatic nematode analyses above. This enables the influences of the differing conditions (e.g. differing agricultural practices or metal contamination levels) on the nematode communities to be compared and the effects of environmental conditions/parameters to be identified.

18.6 Constraints Imposed on Commercial Nematode Surveys

While in an ideal world there are many things that we would wish to do in relation to studying, say, the impacts of an industrial pollution incident, we have to accept that in a practical, commercial situation these are not always possible to achieve. Indeed, in several cases there are some very real and serious constraints and to deny these would be disingenuous. For present purposes, we need to be aware of these constraints and how they affect the use of bioindicators, including nematodes and other organisms (microorganisms or macrofauna), for commercial applications. Experience shows that the problems fall into four categories. In each case it transpires that nematode assessments offer distinct benefits over using other biological groups that can help to minimize the effects of these limitations.

18.6.1 The 'control' conundrum

As is the case with all bioindicator groups, nematode communities can vary significantly from one site to another as each component nematode species has preferences for the microclimatic conditions of their habitat and specific physicochemical characteristics of the substrate. In order to assess the impact of a 'treatment' or 'disturbance' using a bioindicator, it is necessary to have a 'control' community against which the 'treatment' community can be compared. In soil nematode research, this is addressed by conducting one of two study types:

- *Diachronic studies*, where multiple measurements are taken over time, with T0 being conducted *before the application of the treatment*, i.e. 'before and after' comparisons.
- *Synchronic studies*, where different studied situations are compared simultaneously. In this case, the control is a location geographically close to the treatments and serves as a reference site.

While in studies investigating the effects of land management and/or treatments on soil nematofauna, pre-treatment baseline data are commonly available enabling diachronic studies to be established, the same cannot be said for studies investigating the effects of industry. It is rare to find established operations such as factories, refineries, power plants and even domestic wastewater treatment works that have pre-construction, baseline ecological data relating to their surrounding habitats and/or receiving environments for any of their waste discharges. If such data are available, it is even rarer to find that they encompass assemblages of soil or sediment nematodes. Consequently, an ecological assessment is forced to adopt the synchronic approach and examine spatial patterns in the structures of communities and to make comparisons with assemblages present at unaffected locations, either remote control sites or within a survey grid that extends beyond the zone of influence of the disturbance/contamination source.

The choice of the control site(s) is always a delicate one as the soil/substrate physicochemical characteristics and land use, in the case of terrestrial sites, have a considerable impact on nematode communities and the factors that influence the structures of the nematode communities are seemingly infinite. Simplistic rules for control site selection, as used for other study subjects, are not usually applicable to the nematofauna. For example, during studies on soil carbon, it is customary to consider that the reference value to which an agricultural plot can be compared (its potential carbon content) is the nearest forest located in the same pedo-climatic context. Such a comparison does not make sense for soil nematode communities since taxon assemblages are highly dependent on soil use.

In commercial terrestrial nematode studies, particularly for evaluating the biological status of soil, the investigators are interested in characterizing individual plots. Often, no suitable control data are available. To overcome this, the development of interpretation frameworks, as described in Section 18.6, has been instigated. The results obtained in the evaluated plot are compared to a set of values that have been defined for the same context of the plot.

In freshwater scenarios, when assessing sediment quality in the absence of suitable control sites, the NemaSPEAR[%]-index can provide an appropriate interpretation framework. In such cases, predetermined NemaSPEAR[%] value thresholds relating to the EU Water Framework Directive sediment quality classifications are utilized to indicate sediment quality: i.e. high (>54%), good (30–54%), moderate (20–30%), poor (10–20%) and bad (<10%) ecological status (see Höss *et al.*, 2017).

In marine situations, where no interpretation frameworks are available and the ecological effects of industrial contaminants is to be investigated, nematode assessments have a clear advantage over those that employ other bioindicator groups. The complexity of their communities makes statistically rigorous assessment of relationships between physicochemical parameters and community structure possible. This approach can and usually does incorporate bioreactive materials specific to the relevant industrial operation, which, in the absence of the baseline data, can establish the likely extent of the effects that have been caused by applying the multivariate analytical approaches outlined in Section 18.5 above.

18.6.2 Financial constraints

Restricted funding of regulatory bodies and competing budgetary demands in industry and agriculture can preclude collection and analyses of large numbers of bioindicator samples, which, in turn, can affect the size of survey areas and/or sampling sufficiency and the numbers of control sites that can be examined. This leads to direct and obvious consequences and, overall, will compromise the ability to discriminate between anthropogenic effects and natural variation in communities. Again, nematode-based assessments can partially offset this problem. As discussed in Chapter 2 (this volume), assemblages of nematodes encapsulate higher spatial and temporal information than communities of larger invertebrates and, at the same time, fewer, smaller samples may be required to achieve sampling sufficiency.

18.6.3 Time constraints

For surveys to be of value to customers there are several situations where results are needed as soon after the fieldwork as possible. For example, rapid results may be required where incidents of loss of containment (spills and leaks) are being assessed, where the surveys form part of compliance monitoring and are needed for the issue of operating consents, or where soil health assessments are required for land management purposes (such as ecological rehabilitation, ecological engineering). A year or six months later is not necessarily what insurers, lawyers, regulators and factory environmental managers or farmers want to hear. Again, nematode surveys help in this area. Often, fewer nematode samples are required to achieve statistically robust data and analysing species-rich nematode data sets is rewarded more rapidly with mathematically coherent results.

18.6.4 Expertise constraints

There is a severe, worldwide shortage of nematode taxonomists and those who are active in the field are, like the present authors, getting older! Taxonomy is rarely taught in universities, and most taxonomists learn their skills during postgraduate studies or as part of their continued professional development in the workplace. Furthermore, there is an unhelpful and, in our opinion, untrue perception that nematode taxonomy is more challenging than that for other faunal groups. In aquatic environments where identification is done at the species level, this problem of competent taxonomists is very significant. For terrestrial diagnostics, such as those described in Section 18.5, the analyses can be based on a less demanding level of identification, i.e. the family or the genus level, because what is exploited are the functional characteristics of the groups of nematodes and not the presence of key species or community species composition structures.

Whether undertaking taxonomy to species, genus or family level, identification using microscopy is a very time-consuming process. To assist in this and to reduce the requirement for high-level morphological taxonomic expertise, alternative 'shortcut' methods have been developed by applying molecular technology. While DNA 'barcode' taxonomy provides a valuable alternative approach (see ISA, 2011; Schenk *et al.*, 2020; Chapter 15, this volume), the requirement for a DNA database for species 'likely to be encountered' makes such an approach untenable for small- and medium-scale aquatic surveys. This is particularly pertinent at sites beyond Northern Europe and North America where the majority of nematode species are yet to be described, let alone DNA-profiled. Furthermore, the time-consuming process of nematode extraction is still required for effective DNA extraction (Griffiths *et al.*, 2018) and this type of analysis does not determine the abundances of the nematode populations but only provides the 'presence and absence' of the taxa (Geisen *et al.*, 2018). Despite these current shortcomings, with further development, molecular taxonomic techniques have the potential to radically change the approach to nematode surveys and make the use of nematodes as bioindicators much more widely available. Other methods based on the use of AI (use of neural networks, for example) are also promising (Thevenoux *et al.*, 2021; Lu *et al.*, 2021; Qing *et al.*, 2022) for automatization of the identification of nematodes.

18.7 Examples of Commercial Studies Using Nematodes in Aquatic and Terrestrial Environments

18.7.1 Case Study 1: Long-term monitoring of liquid wastes discharges to coastal waters

Customers: Operators on a coastal industrial complex in South Africa

The coastal industrial complex supports over 20 manufacturing plants which include paint and dye factories, a pigment manufacturer and several specialist chemical producers. Treated and untreated liquid wastes from these factories are pumped to sea via two long outfalls, which discharge at a distance of between 1.5 and 2 km offshore in approximately 30 m depth of water. The seabed of the highly active coastal zone system is characterized by medium to coarse sands and patches of sandstone reef. Principal effluent components of ecological concern were heavy metals with the primary source being an acid-iron effluent characterized by high concentrations of iron, manganese, zinc, chromium, vanadium and arsenic. Since the late 1980s, annual sea surveys have been undertaken examining macro- and meiofaunal communities in a 6 km × 2 km survey area located across the ends of the outfalls and parallel to the shoreline. This comprised a grid of 50 sampling stations.

From the earliest days, with over 300 taxa being documented per survey round, the nematode data provided higher-resolution information regarding the impacts of the effluent than the less diverse, more patchy macrofauna communities (see Newell *et al.*, 1990b). The distributions of nematode community parameters such as species richness, diversity indices and community dominance/co-dominance values as well as the community multivariate analyses indicated the existence of extensive effects on the seabed (see Fig. 18.3, 1997 data and map). This corresponded closely to the occurrence of elevated concentrations of heavy metals in the benthic sediment.

With a high-resolution monitoring system in place, attention was focused on the causes of the observed impacts and steps were taken to reduce these. By the late 1990s, multivariate correlation studies using data from more comprehensive sediment physicochemical analyses were being incorporated into the surveys specifically for use with benthic nematode community data. Over the following seven years, significant capital expenditure was made by the owners of the factory complex to address the effects at sea. The year-on-year results of the meiofaunal nematode and physicochemistry correlation analyses were used to track and document changes at sea and to indicate the key effluent-related elements/compounds that were most associated with the benthic impacts.

By 2004, the results of the multivariate correlation (ARESC) analyses showed that natural factors (primarily the proportions of different sediment particle-size fractions as well as marker elements for materials of terrestrial origin) had become primary correlates with the structures of the nematode communities. Using the same criteria as those employed in 1997, the zone of residual impacts in 2004 is illustrated in Fig. 18.3 (see 2004 data and map). These lie within the mixing ('boil') zone for the discharges (computed as a function of volume of effluent emitted and mean sea depth at the point of discharge) and this is now recognized by the regulatory authorities (Department of Water Affairs and Forestry) as an 'acceptable footprint'.

In 2016, the factory responsible for the acid-iron effluent and the associated heavy metal burden ceased production and the acid-iron effluent was removed from the wastewater discharge. Despite this, subsequent surveys have demonstrated that the heavy metal contamination of the sediment in the vicinity of the outfall has persisted. This contamination was associated with 'iron-staining' of the sand, caused by the precipitation of iron(iii) hydroxide from the neutralized acidic effluent. Iron(iii) hydroxide flocc has the propensity to 'strip' heavy metal irons from water (see Pang *et al.*, 2009) *and* to adhere to sand particle surfaces. It was likely that the persistent metal contamination of the sediment related to the bound metals within the iron(iii) hydroxide staining. The analyses of the nematode community data since the factory closure in 2016 have demonstrated that the residual sediment contamination does not affect the benthic nematode communities and is, in terms of ecological effect, not bioaccessible. Table 18.2 presents the 2021 multivariate correlation analyses data for ARESC axis 1 and 2 (accounting for over 80% of the nematode community data variance). The only measured parameters that were correlated with the structures of the nematode communities were natural variables (sediment particle-size fractions and background/ naturally occurring metals).

18.7.2 Case Study 2: Bespoke *in situ* monitoring system – Caspian Sea

Customer: Oil industry and service company

The shallow north Caspian Sea is part of an extensive and internationally important RAMSAR-designated site and supports numerous endemic species, some of which are listed as critically endangered. Accordingly, the relatively recent discovery of exploitable oil reserves beneath the sea (late 1990s), set against significant changes in world politics in relation to traditional Middle East oil supplies and subsequent increases in oil demand, has placed considerable pressures on these habitats and their species, as surrounding countries permit the construction of offshore production facilities. In recognition of the ecological importance of the shallow-water habitats, monitoring and assessment programmes were agreed between international oil companies and several of the governments concerned. Clearly these programmes needed to be sensitive enough to detect any subtle, long-term degradation relating to fugitive materials arising from the construction and operation of the production platforms, as well as the acute effects of

Table 18.2. Multivariate correlation (ARESC) analysis results for the 2021 nematode communities collected during the effluent discharge marine monitoring benthic survey off the South African coast. Parameters highlighted in green were statistically significantly correlated ($P < 0.05$) with the structures of the nematode communities. The effluent-related metals are highlighted in red. ARESC = assessment and ranking of ecologically significant contaminants; TOC = total organic carbon. (Author's own table.)

	Axis 1			Axis 2	
Parameter	*r*	*P* value	Parameter	*r*	*P* value
125–250 µm	0.83093	0.00000	250–500 µm	−0.59359	0.01795
500–1000 µm	−0.75810	0.00001	1–2 mm	0.56700	0.02012
K	0.68681	0.00018	TOC	0.46950	0.15535
Mg	0.67648	0.00021	125–250 µm	−0.22768	1
Al	0.61014	0.00209	500–1000 µm	0.34621	1
63–125 µm	0.55483	0.00949	K	−0.16729	1
Sr	0.52410	0.01848	Mg	0.00580	1
Na	0.50135	0.02830	Al	0.03291	1
Cu	0.45443	0.07121	63–125 µm	0.14239	1
Se	0.43271	0.09903	Sr	0.22891	1
1–2 mm	−0.40772	0.14375	Na	0.15587	1
<63 µm	0.37734	0.22276	Cu	−0.08988	1
Ni	0.36792	0.23732	Se	0.01126	1
Co	0.36393	0.23732	<63 µm	−0.20359	1
As	0.27610	0.71176	Ni	−0.28803	1
>2 mm	−0.27904	0.71176	Co	−0.05520	1
V	−0.26539	0.76245	As	−0.01204	1
250–500 µm	0.03582	1	>2 mm	0.28688	1
TOC	0.22831	1	V	−0.04273	1
Ca	0.17430	1	Ca	−0.03213	1
Fe	0.20705	1	Fe	0.17213	1
Mn	0.00700	1	Mn	0.09560	1
Zn	0.04760	1	Zn	−0.01067	1
Pb	0.03054	1	Pb	−0.15216	1
Cr	−0.12634	1	Cr	−0.17557	1
Sn	−0.10525	1	Sn	0.03838	1
Cd	0.20797	1	Cd	−0.01956	1

loss-of-containment incidents and the impacts of clean-up (see Green and Trett, 1989).

In some quarters, the agreements to monitor were based on a belief that traditional benthic macroinvertebrate surveys would meet all requirements. The commencement of baseline sampling soon dispelled this notion. Offshore, the sediments in the oligohaline waters (salinity *c.*1.2%) support a comparatively depauperate macroinvertebrate fauna in which larvae of salt-tolerant chironomids (Diptera: Nematocera) are seasonally the dominant faunal group along with the small polychaete worm, *Manayunkia caspica* (see Green, 1968). This poverty of the macrofauna may reflect the fact that the shallow northern sea (minimum depth *c.*2.5 m) freezes over in December and may not thaw until late March by which time ice drift can also have caused significant physical impacts on the seabed.

From earlier studies that we had undertaken (1997/98), it was clear that the North Caspian meiofauna was considerably richer than the macrofauna and in excess of 100 nematode taxa have now been recorded from this region. To overcome the problems of ice and those of restrictions placed on the collection of samples from the nature reserve, a monitoring system based on NCUs was used. These comprise inert, 9 cm × 11 cm, coarse fibre pads, which are held in large mesh, heavy-duty bags, similar to those used in marine oyster farms. Between 2003 and 2005, the colonization units were suspended at three different depths around the perimeter of artificial islands that were created to support land-based types of drilling rigs. After a period of just eight weeks' immersion, densities of up to 20,000 nematodes per pad were recorded, equivalent to over 2 million individuals per m^2 of the artificial substrate. Although up to 47 taxa were recorded in a sampling round, at most representatives of 12 nematode species were present in any one single pad sample. Representatives of all the major nematode trophic groups were present in the sample assemblages. However, 'non-selective' deposit feeders (Weiser type 1B species, e.g. Axonolaimidae and Xyalidae) predominated in samples from the lower water column while selective epistrate feeders and diatomivorous species (Weiser type 2A species, e.g. Chromadoridae and Microlaimidae) were dominant in those from the upper water column.

Throughout the duration of the monitoring study, no spill or contamination incidents occurred, and chemical stresses were not detected by these 'biosentinel' units. Water depth and physical disturbance caused by wind and wave action in the upper water column appeared to be the major determinants of the structures of the assemblages that colonize the pads.

This novel approach to aquatic nematode bioindicator monitoring using naturally colonized, artificial habitats has proved to be ideally suited for situations where sensitive ecological monitoring is required at sites where natural substrate either cannot be sampled or does not exist. For example, NCUs have been deployed within power station cooling-water systems where there was a requirement for the efficacy of biocide treatment regimes to be monitored. The results are quantitative and easily repeatable. The survey equipment is minimal and deployment of NCUs is a very simple process that can be undertaken at any location with suitable access to the waterbody in question, usually without the need for a boat.

18.7.3 Case Study 3: Development of green spaces around a sports infrastructure

Customer: Local authority

The study commenced one year after the completion of a sports infrastructure construction and the establishment of associated green spaces. An ecological landscape park was created within a 20-hectare perimeter surrounding the infrastructure to compensate for the destruction of habitats of two protected species that were originally present on the site and that required preservation. The two species were a lizard and a snake, and their preservation guided the choice of developments to recreate their habitats as well as habitats for a diverse range of species native to the region. During the site development and the creation of green spaces, deep silts and topsoil from elsewhere were used to reconstruct the soils. The same constructed soil was utilized throughout the entire site area.

To assess biodiversity and its evolution in the initial years following construction, naturalistic monitoring of the compensation measures was conducted, including ecological, faunal and floristic inventories. The objective was to enhance the landscape project surrounding the infrastructure and extend its educational focus to the ecological environment. Monitoring activities (flora, amphibians, reptiles, insects, nematofauna, chiroptera, avifauna) were carried out for six consecutive years in the spring, except for the initial sampling which took place in the autumn.

Four main measures were implemented to promote biodiversity in unsealed areas:

- The establishment of a network of wetlands around the perimeter to serve as habitats for amphibians and as a food-production area for insectivores.
- Wooded areas along a part of the perimeter designed as bird refuges.
- An ecological corridor along the entire south and west perimeter consisting of continuous lawns serving as preferred habitats for reptiles.
- Gabion areas to further enhance habitats for reptiles.

The objective of the nematofauna study was to characterize the site based on the biological functioning of its soils in different ecological zones that were designed and implemented. For the soil study, we selected the five major ecosystems established on the site and sampled three zones within each major ecosystem (three repetitions) in the landscaped park. These five ecosystems within the site enclosure were supplemented by a control zone located outside the site, but in close proximity, with undisturbed soils maintained as grassland (control). The five ecosystems were: dry ditch, underground corridor, meadow corridor, wet ditch, bed under bark and control. Figure 18.4 illustrates the evolution of certain nematofauna parameters over a six-year period following the installation of green spaces and vegetation development.

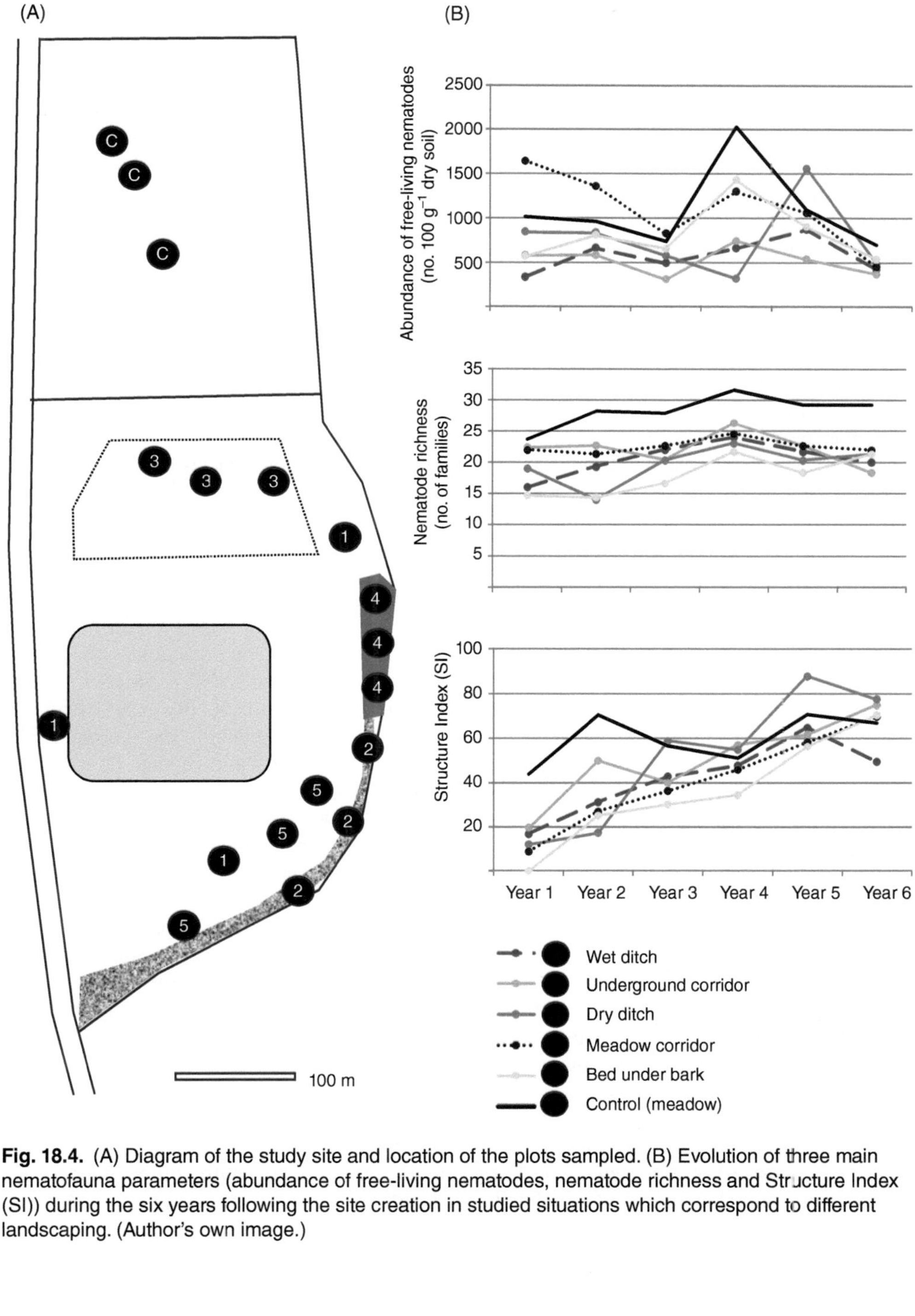

Fig. 18.4. (A) Diagram of the study site and location of the plots sampled. (B) Evolution of three main nematofauna parameters (abundance of free-living nematodes, nematode richness and Structure Index (SI)) during the six years following the site creation in studied situations which correspond to different landscaping. (Author's own image.)

Annual variations observed were mostly due to the effect of climate, as the green spaces followed the same tendency/pattern as the control. A slight increase in the abundance of free-living nematodes was observed in green spaces over time, indicating the stability of resources for these soil organisms. However, this abundance was generally lower than that in the control zone across the majority of green spaces. The abundance of phytophagous nematodes increased more rapidly compared to free-living nematodes in relation to the establishment of vegetation (results not shown). Taxonomic diversity showed minimal growth across ecosystems and remained lower than that of the control. On the other hand, functional diversity of soil nematodes, measured by the SI, increased significantly and reached a level similar to that of the control, indicating an improvement in soil functionality.

Biological activity demonstrated a positive and significant evolution over time, reflecting more intense and complex soil biological activity in the studied green spaces. Throughout the study, an improvement in the biological activity of the soil was observed in various green spaces created around the infrastructure, driven by the establishment of functional diversity. After six years, it became evident that the soils had reached a certain level of maturity in the soil food web and a stabilized level of biological activity close to that of the control zone.

18.7.4 Case Study 4: Evolution of the soil biological state in plots of a group of farmers changing their practices

Customer: Agricultural cooperative

This study was carried out to assess the effect of the introduction of cover crops on the biological functioning of the soil of cereal-production plots. Seventeen farmers belonging to the same agricultural cooperative had set up different types of cover crops merged into their rotation. The project consisted of four steps: (i) training on soil nematofaunal analysis and sampling for farmers and technicians of the cooperative; (ii) soil sampling in 2019, before the establishment of cover crops; (iii) soil sampling in 2022, after three consecutive years of setting up cover crops; and (iv) reporting of the results to farmers and technicians of the cooperative.

In 2019, four plots had a 'satisfactory' level of soil biological functioning evaluated by the analyses of soil nematofauna, 12 plots had a 'to be monitored' level and one plot had an 'altered' level (see Fig. 18.5). This latter plot had a low abundance of free-living nematodes, a low Shannon diversity index (H′) and a low SI. The 12 plots 'to be monitored' had medium to high free-living nematode abundance levels but medium to low diversity and SI levels. Based on the practice surveys, it was agreed that these results were likely related to the long periods of bare soil between two cereal crops (>3 months); a situation that could be remedied by the introduction of cover crops.

Between 2019 and 2022, six plots improved from 'to be monitored' to 'satisfactory' level, nine plots did not evolve (two of which maintained 'satisfactory' level) and two plots deteriorated from 'satisfactory' to 'to be monitored' level (see Fig. 18.5). In the case of the two plots that had regressed, the farmers had failed to set up cover crops for no-till techniques. This failure resulted in a soil compaction, a reduction in the abundance of free-living nematodes and an unsatisfactory level for cereal production. In the case of improved plots, this was essentially linked to an improvement in the SI promoted by multi-species covers and regular litter inputs. This improvement allows most plots to move closer to the food web diagnostic target area. For the seven 'altered' and 'to be monitored' plots that demonstrated no change in biological function status, the addition of cover crops in the rotation was not enough to improve the soil biological functioning. Other agroecological levers needed to be implemented at these sites, such as variation of the intensity of tillage, the inputs of organic matter and mineral fertilizers, the length of rotations and the use of pesticides.

These results provided an opportunity for discussions between cooperative technicians and farmers to improve production practices and communicate with other cooperative members.

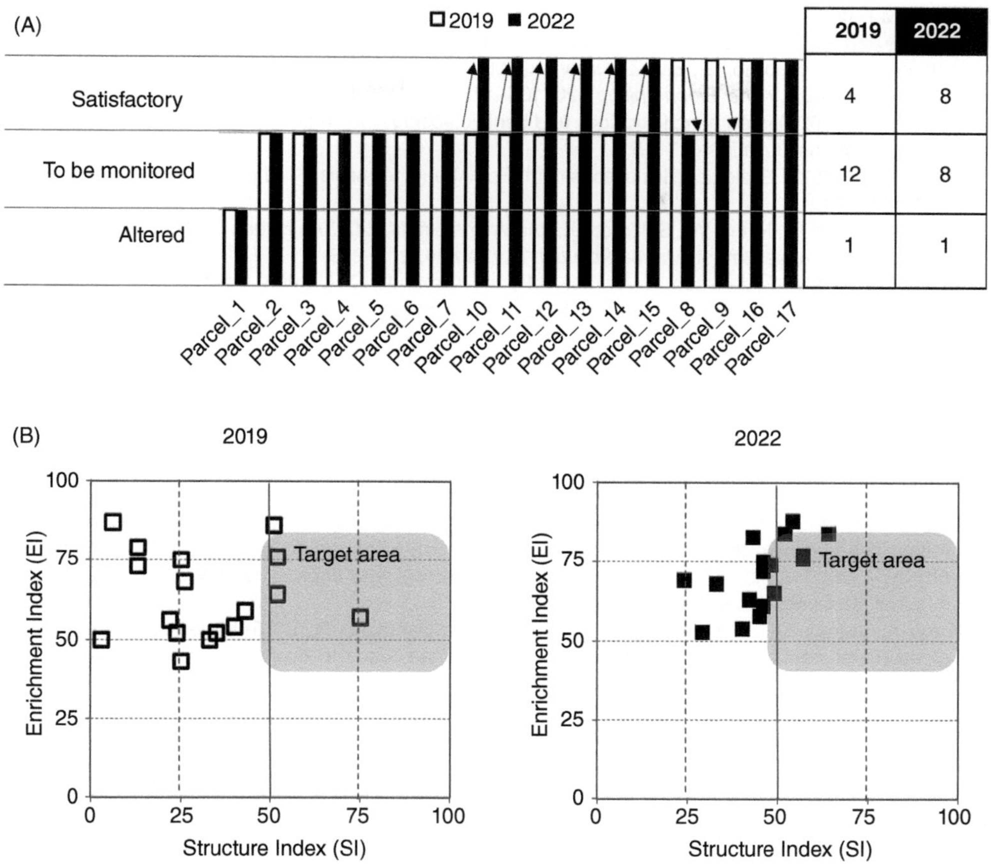

Fig. 18.5. (A) Diagram of the number of parcels per soil functioning class evaluated in 2019 and in 2022, three years after the establishment of intercultural plant cover. 'Altered' = at least three nematode parameters indicate a biological dysfunction; 'to be monitored' = one or two parameter(s) indicate(s) a biological dysfunction; 'satisfactory' = no nematode parameter indicates a biological dysfunction. Five nematode parameters are taken into account to calculate this score: abundance of free-living nematodes; Enrichment Index (EI); Structure Index (SI); Nematode Channel Ratio (NCR); and Shannon diversity index (H'). (B) Evolution of the soil food web diagnosis between 2019 and 2022 of the 17 parcels followed during the project. The 'Target area' indicates a satisfactory soil food web diagnosis for crops including a moderate EI and a moderate to high SI. (Author's own image.)

References

Alves, A.S., Adãob, H., Ferreroc, T.J., Marquesa, L.C., Costad, M.J. and Patrício, J. (2013) Benthic meiofauna as indicator of ecological changes in estuarine ecosystems: the use of nematodes in ecological quality assessment. *Ecological Indicators* 24, 462–475.

Bongers, T. (1990) The maturity index; an ecological measure of environmental disturbance based on nematode species composition. *Oecologia* 83, 14–19.

Bongers, T. and Ferris, H. (1999) Nematode community structure as a bioindicator in environmental monitoring. *Trends in Ecology & Evolution* 14(6), 224–228.

Brinke, M., Höss, S., Fink, G., Ternes, T.A., Heininger, P. and Traunspurger, W. (2010) Assessing effects of the pharmaceutical ivermectin on meiobenthic communities using freshwater microcosms. *Aquatic Toxicology* 99, 126–137.

Buchman, M.F. (2008) *NOAA Screening Quick Reference Tables*. NOAA OR&R Report 08-1. Office of Response and Restoration Division, National Oceanic and Atmospheric Administration, Seattle, Washington.

Bünemann, E.K., Bongiorno, G., Bai, Z., Creamer, R.E., De Deyn, G. *et al.* (2018) Soil quality – a critical review. *Soil Biology and Biochemistry* 120, 105–125.

Chapman, P.M. (2002a) Integrating toxicology and benthic ecology: putting the 'eco' back into ecotoxicology. *Marine Pollution Bulletin* 44, 7–15.

Chapman, P.M. (2002b) Ecological risk assessment (EcRA) and hormesis. *Science of The Total Environment* 288, 131–140.

Chapman, P.M., Ho, K., Muns, W., Solomon, K. and Weinstein, M.P. (2002) Issues in sediment toxicity and ecological risk assessment. *Marine Pollution Bulletin* 44, 271–278.

Chapman, P.M., Wang, F., Janssen, C., Goulet, R.R. and Kamunde, C.N. (2003) Conducting ecological risk assessments of inorganic metals and metalloids – current status. *Human and Ecological Risk Assessment* 9, 641–697.

Coull, B.C. (1988) Ecology of marine meiofauna. In: Higgins, R.P. and Theil, H. (eds) *Introduction to the Study of Meiofauna*. Smithsonian Institute Press, London, pp. 18–36.

Council of the European Communities (1982) Council Directive 82/883/EEC of 3 December 1982 on procedures for the surveillance and monitoring of environments concerned by waste from the titanium dioxide industry. *Official Journal of the European Communities* L378, 1–14. Available at: https://eur-lex.europa.eu/eli/dir/1982/883/oj/eng (accessed 15 January 2025).

De Ley, P., Decraemer, W. and Eyualem-Abebe (2006) Introduction: summary of present knowledge and research addressing the ecology and taxonomy of freshwater nematodes. In: Eyualem-Abebe, Andrássy, I and Traunsperger, W. (eds) *Freshwater Nematodes: Ecology and Taxonomy*. CAB International, Wallingford, UK, pp. 3–30.

Dell'Anno, A., Carugati, L., Corinaldesi, C., Riccioni, G. and Danovaro, R. (2015) Unveiling the biodiversity of deep-sea nematodes through metabarcoding: are we ready to bypass the classical taxonomy? *PLoS One* 10(12), e0144928. DOI: 10.1371/journal.pone.0144928.

Du Preez, G., Daneel, M., De Goede, R., Du Toit, M.J., Ferris, H. *et al.* (2022) Nematode-based indices in soil ecology: application, utility, and future directions. *Soil Biology and Biochemistry* 169, 108640.

EC (European Community) (1985) Council Directive of 27 June 1985 on the assessment of the effects of certain private and public projects (85/337/EEC). *Official Journal of the European Communities* LR175/40, 5 July 1985.

EC (European Community) (2008) Directive 2008/50/EC of the European Parliament and of the Council of 21 May 2008 on ambient air quality and cleaner air for Europe. Available at: https://eur-lex.europa.eu/legal-content/EN/ALL/?uri=CELEX%3A32008L0050 (accessed 15 January 2025).

Ellis, R.J., Best, J.G., Fry, J.C., Morgan, P., Neish, B. *et al.* (2002) Similarity of microbial and meiofaunal community analyses for mapping ecological effects of heavy-metal contamination in soil. *FEMS Microbiology Ecology* 40, 113–122.

European Commission (2020) *Caring for Soil is Caring for Life – Ensure 75% of Soils are Healthy by 2030 for Healthy Food, People, Nature and Climate*. Publications Office of the European Union, Luxembourg. Available at: https://data.europa.eu/doi/10.2777/918775 (accessed 15 January 2025).

European Commission (2021) *EU Soil Strategy for 2030. Reaping the Benefits of Healthy Soils for People, Food, Nature and Climate. Communication from the Commission to the European Parliament, the Council, the European Economic and Social Committee and the Committee of the Regions*. Brussels, 17 November 2021, COM(2021) 699 final. Available at: https://eur-lex.europa.eu/legal-content/EN/TXT/?uri=CELEX%3A52021DC0699 (accessed 15 January 2025).

European Parliament and Council (2000) Directive 2000/60/EC establishing a framework for Community action in the field of water policy. *Official Journal of the European Communities* L327, 1–73. Available at: https://eur-lex.europa.eu/eli/dir/2000/60/oj (accessed 15 January 2025).

European Parliament and Council (2008) Directive 2008/56/EC establishing a framework for community action in the field of marine environmental policy (Marine Strategy Framework Directive). *Official Journal of the European Union* L164, 19–40. Available at: https://eur-lex.europa.eu/eli/dir/2008/56/oj/eng (accessed 15 January 2025).

Farfan, M.A. and Wall, D.H. (2022) Biodiversity. In: Elias, S.A. (ed.) *Reference Module in Earth Systems and Environmental Sciences*. Elsevier.

Fraschetti, S., Gambi, C., Giangrande, A., Musco, L., Terlizzi, A. and Danovaro, R. (2006) Structural and functional response of meiofauna rocky assemblages to sewage pollution. *Marine Pollution Bulletin* 52, 540–548.

Geisen, S., Snoek, L.B., ten Hooven, F.C., Duyts, H., Kostenko, O. *et al.* (2018) Integrating quantitative morphological and qualitative molecular methods to analyse soil nematode community responses to plant range expansion. *Methods in Ecology and Evolution* 9(6), 1366–1378.

Geyedo-Adabio, T.K. and Baird, D. (2006) Response of meiofauna and nematode communities to increased levels of contaminants in a laboratory microcosm experiment. *Ecotoxicology and Environmental. Safety* 63, 443–450.

Green, J. (1968) *The Biology of Estuarine Animals*. Sidgwick and Jackson, London.

Green, J. and Trett, M.W. (1989) *The Fate and Effects of Oils in Freshwater*. Springer, Dordrecht, The Netherlands.

Griffiths, B.S., de Groot, G.A., Laros, I., Stone, D. and Geisen, S. (2018) The need for standardisation: exemplified by a description of the diversity, community structure and ecological indices of soil nematodes. *Ecological Indicators* 87, 43–46.

Harris, J.A., Evans, D.L. and Mooney, S.J. (2022) A new theory for soil health. *European Journal of Soil Science* 72(4), e13292.

Higgins, R.P. and Theil, H. (1988) Prospectus. In: Higgins, R.P. and Theil, H. (eds) *Introduction to the Study of Meiofauna*. Smithsonian Institute Press, London, pp. 11–13.

Hill, M.O. (1973) Diversity and evenness, a unifying notion and its consequences. *Ecology* 54, 427–432.

Ho, H.H., Swennen, R, Cappuyns, V, Vassilieva, E. and Van Tran, T. (2012) Necessity of normalization to aluminum to assess the contamination by heavy metals and arsenic in sediments near Haiphong Harbor, Vietnam. *Journal of Asian Earth Sciences* 56, 229–239.

Hooper, D.J. (1986) Extraction of free-living stages from soils. In: Southey, J.F. (ed.) *Laboratory Methods for Work with Soil and Plant-Parasitic Nematodes*. Her Majesty's Stationery Office, London, pp. 5–30.

Höss, S. (2021) Single- and multi-species toxicity testing with nematodes. In: Traunspurger, W. (ed.) *Ecology of Freshwater Nematodes*. CAB International, Wallingford, UK, pp. 298–322. DOI: 10.1079/9781789243635.0010.

Höss, S., Reiff, N., Ottermanns, R., Pagel-Wieder, S., Dohrmann, A.B. *et al.* (2015) Risk assessment of the cultivation of a stacked *Bt*-maize variety (MON89034 × MON88017) for nematode communities. *Soil Biology and Biochemistry* 91, 109–118.

Höss, S., Heininger, P., Claus, E., Möhlenkamp, C., Brinke, M. and Traunspurger, W. (2017) Validating the NemaSPEAR[%]-index for assessing sediment quality regarding chemical-induced effects on benthic communities in rivers. *Ecological Indicators* 73, 52–60.

Höss, S., Roessink, I., Brock, T.C.M. and Traunspurger, W. (2020) Response of a nematode community to the fungicide fludioxonil in sediments of outdoor freshwater microcosms compared to a single species toxicity test. *Science of The Total Environment* 710, 135627.

Höss, S., Reiff, N., Traunspurger, W. and Helder, J. (2021) On the balance between practical relevance and standardization – testing the effects of zinc and pyrene on native nematode communities in soil microcosms. *Science of The Total Environment* 788, 147742. DOI: 10.1016/j.scitotenv.2021.147742.

IEEM (Institute of Ecology and Environmental Management) (2005) *Code of Professional Conduct*. IEEM, Winchester, UK .

IEEM (Institute of Ecology and Environmental Management) (2006) *Guidance on Survey Methodology*. IEEM, Winchester, UK.

Ingels, J., Vanreusel, A., Pape, E., Pasotti, F., Macheriotou, L. *et al.* (2020) Ecological variables for deep-ocean monitoring must include microbiota and meiofauna for effective conservation. *Nature Ecology and Evolution* 5, 27–29. DOI: 10.1038/s41559-020-01335-6.

ISA (International Seabed Authority) (2011) *Marine Benthic Nematode Molecular Protocol Handbook (Nematode Barcoding)*. ISA Technical Study No. 7. ISA, Jamaica.

Kibblewhite, M.G., Ritz, K. and Swift, M.J. (2008) Soil health in agricultural systems. *Philosophical Transactions of the Royal Society B: Biological Sciences* 363(1492), 685–701.

Lazarova, S., Coyne, D., Rodríguez, M.G., Peteira, B. and Ciancio, A. (2021) Functional diversity of soil nematodes in relation to the impact of agriculture – a review. *Diversity* 13(2), 64.

Lu, X., Wang, Y., Fung, S. and Qing, X. (2021) I-Nema: a biological image dataset for nematode recognition. *arXiv:2103.08335v1* [q-Bio.Qm], 15 March. Available at: http://arxiv.org/abs/2103.08335 (accessed 15 January 2025).

Mare, M.F. (1942) A study of a marine benthic community with special reference to micro-organisms. *Journal of the Marine Biological Association of the United Kingdom* 25, 517–554.

Meldal, B.H.M., Debenham, M.J., De Ley, P., Tandingan De Ley, I., Vanfleteren, J.R. *et al.* (2007) An improved molecular phylogeny of the Nematoda with special emphasis on marine taxa. *Molecular Phylogenetics and Evolution* 42, 622–636.

Moens, T. and Vincx, M. (1997) Observations of the feeding ecology of estuarine nematodes. *Journal of the Marine Biological Association of the United Kingdom* 72, 211–227.

Moens, T., Traunsperger, W. and Bergtold, M. (2006) Feeding ecology of free-living benthic nematodes. In: Eyualem-Abebe, Andrássy, I and Traunsperger, W. (eds) *Freshwater Nematodes: Ecology and Taxonomy*. CAB International, Wallingford, UK, pp. 103–131.

Murray, J.M., Meadows, A. and Meadows, P.S. (2002) Biogeomorphological implications of microscale interactions between sediment geotechnics and marine benthos: a review. *Geomorphology* 47(1), 15–30.

Muthumbi, A.W., Vanreusel, A., Duineveld, G., Soetaert, K. and Vincx, M. (2004) Nematode community structure along the continental slope off the Kenyan coast, western Indian Ocean. *International Review of Hydrobiology* 89, 188–205.

Neher, D.A. (2001) Role of nematodes in soil health and their use as indicators. *Journal of Nematology* 33(4), 161–168.

Neher, D.A. and Powers, T.O. (2022) Nematodes. In: Elias, S.A. (ed.) *Reference Module in Earth Systems and Environmental Sciences.* Elsevier.

Newell, P.F., Newell, R.C. and Trett, M.W. (1990a) Environmental impacts of an acid-iron effluent on macrobenthic and meiobenthic assemblages of the St Lawrence River. *Science of The Total Environment* 97/98, 771–781.

Newell, R.C., Newell, P.F. and Trett, M.W. (1990b) Assessment of the impact of liquid waste on benthic invertebrate assemblages. *Science of The Total Environment* 97/98, 885–867.

Newell, R.C., Maughan, D.W., Trett, M.W., Newell, P.F. and Seiderer, L.J. (1991) Modifications of benthic community structure in response to acid-iron waste discharge. *Marine Pollution Bulletin* 22, 112–118.

Ockleford, C., Adriaanse, P., Berny, P., Brock, T., Duquesne, S. *et al.* (2017) Scientific Opinion addressing the state of the science on risk assessment of plant protection products for in-soil organisms. *EFSA Journal* 15(2), e4690. DOI: 10.2903/j. efsa.2017.4690.

Pang, F., Teng, S., Teng, T.T. and Kadir, O.A. (2009) Heavy metals removal by hydroxide precipitation and coagulation-flocculation methods from aqueous solutions. *Water Quality Research Journal of Canada* 44(2), 174–182.

Pape, E., Bezerra, T.N., Gheerardyn, H., Buydens, M., Kieswetter, A. and Vanreusel, A. (2021) Potential impacts of polymetallic nodule removal on deep-sea meiofauna. *Scientific Reports* 11, 19996. DOI: 10.1038/s41598-021-99441-3.

Platt, H.M. and Warwick, R.M. (1980) The significance of free-living nematodes to the littoral ecostystem. In: Price, J.H., Irvine, D.E.G. and Farnham, W.F. (eds) *The Shore Environment, Vol. 2: Ecosystems*. Academic Press, London, pp. 729–759.

Platt, H.M., Shaw, K.M. and Lambshead, P.J.D. (1984) Nematode species abundance patterns and their use in the detection of environmental perturbations. *Hydrobiologia* 118, 59–61.

Potapov A.M. (2022) Multifunctionality of belowground food webs: resource, size and spatial energy channels. *Biological Reviews* 97, 1691–1711.

Property Investment Partners (2023) Contaminated Land Development – Guide for Property Developers. Available at: https://investmentproperty.co.uk/property-investment-resources/contaminated-land-development (accessed 17 May 2023).

Ptatscheck, C., Brüchner-Hüttemann, H., Kreuzinger-Janik, H., Weber, S. and Traunspurger, W. (2020) Are meiofauna a standard meal for macroinvertebrates and juvenile fish? *Hydrobiologica* 847, 2755–2778.

Qing, X., Wang, Y., Lu, X., Li, H., Wang, X. *et al.* (2022) NemaRec: a deep learning-based web application for nematode image identification and ecological indices calculation. *European Journal of Soil Biology* 110, 103408. DOI: 10.1016/j.ejsobi.2022.103408.

Ridall, A. and Ingels, J. (2021) Suitability of free-living marine nematodes as bioindicators: status and future considerations. *Frontiers in Marine Science* 8, 685327. DOI: 10.3389/fmars.2021.685327.

Salamun, P., Renco, M., Kucanova, E., Brazova, T., Papajova, I. *et al.* (2012) Nematodes as bioindicators of soil degradation due to heavy metals. *Ecotoxicology* 21, 2319–2330.

Sánchez-Moreno, S., Minoshima, H., Ferris, H. and Jackson, L.E. (2006) Linking soil properties and nematode community composition: effects of soil management on soil food webs. *Nematology* 8, 703–715.

Schenk, J., Höss, S., Brinke, M., Kleinbölting, N., Brüchner-Hüttemann, H. and Traunspurger, W. (2020) Nematodes as bioindicators of polluted sediments using metabarcoding and microscopic taxonomy. *Environment International* 143, 105922. DOI: 10.1016/j.envint.2020.105922.

Schenk, J., Hoss, H, Kleinbolting, N. and Traunspurger, W. (2022) Suitability of molecular taxonomy for assessing polluted sediments using the NemaSPEAR[%] index. *Ecological Indicators* 137, 108761. DOI: 10.1016/j.ecolind.2022.108761.

Schratzberger, M. and Ingels, J. (2017) Meiofauna matters: the roles of meiofauna in benthic ecosystems. *Journal of Experimental Marine Biology and Ecology* 502, 12–25.

Schratzberger, M., Holterman, M., van Oevelen, D. and Helder, J. (2019) A worm's world: ecological flexibility pays off for free-living nematodes in sediments and soils. *BioScience* 69, 867–876.

Seinhorst, J.W. (1959) A rapid method for the transfer of nematodes to anhydrous glycerine. *Nematologica* 4, 67–69.

Semprucci, F. and Balsamo, M. (2012) Free-living marine nematodes as bioindicators: past, present and future perspectives. *Environmental Research Journal* 6(1), 17–35.

Simpson, E.H. (1949) Measurement of diversity. *Nature* 163, 688.

Smit, C.E., Schouten, A.J., Van Den Brink, P.J., van Esbroek, M.L. and Posthuma, L. (2002) Effects of zinc contamination on a natural nematode community in outdoor soil mesocosms. *Archives of Environmental Contamination and Toxicology* 42(2), 205–216. DOI: 10.1007/s00244-001-0029-y.

Somerfield, P.J., Rees, H.L. and Warwick, R.M. (1995) Interrelationships in community structure between marine meiofauna and macrofauna in relation to dredgings disposals. *Marine Ecology Progress Series* 127, 103–112.

Steel, H. and Ferris, H. (2016) Soil nematode assemblages indicate the potential for biological regulation of pest species. *Acta Oecologica* 73, 87–96.

Swift, M.J., Heal, O.W. and Anderson, J.M. (1979) *Decomposition in Terrestrial Ecosystems*. Blackwell Scientific, Oxford.

Thevenoux, R., Van Linh, L.E., Villessèche, H., Buisson, A., Beurton-Aimar, M. *et al.* (2021) Image based species identification of *Globodera* quarantine nematodes using computer vision and deep learning. *Computers and Electronics in Agriculture* 186, 106058. DOI: 10.1016/j.compag.2021.106058.

Trap, J., Bonkowski, M., Plassard, C., Villenave, C. and Blanchart, E. (2016) Ecological importance of soil bacterivores for ecosystem functions. *Plant and Soil* 398, 1–24. DOI: 10.1007/s11104-015-2671-6.

Traunspurger, W. (1997) Bathymetric, seasonal and vertical distribution of feeding-types of nematodes in an oligotrophic lake. *Vie Milieu* 47, 1–7.

Trett, M.W. and Thurgood, R. (2008) Assessment and monitoring of actual ecological effects; nematodes in the service of industry and regulators. In: *Proceedings of the 10th International UFZ–Deltares–TNO Conference on Soil–Water Systems (CONSOIL 2008)*. UFZ, Leipzig, Germany, pp. 82–91.

Trett, M.W., Calvo Urbano, B., Forster, S.J., Hutchinson, J.D., Feil, R.L. *et al.* (2000) Terrestrial meiofauna and contaminated land assessment. *Environmental Science and Technology* 34, 1594–1602.

Trett, M.W., Urbano, B.C., Forster, S.J. and Trett, S.P. (2009) Commercial aspects of the use of nematodes as bioindicators. In: Wilson, M.J. and Kakouli-Douarte, T. (eds) *Nematodes as Environmental Indicators*. CAB International, Wallingford, UK, pp. 275–314.

Trett, M.W., Moss, R., Calvo, B. and Wilkinson, G. (2012) Carbon disulphide – contaminant or biogenic compound? A new perspective. In: Morrison, R.D. and O'Sullivan, G. (eds) *Environmental Forensics*. Royal Society of Chemistry, Cambridge, pp. 232–243.

Van den Brink, P.J. and Ter Braak, C.J.F. (2009) Principal response curves: analysis of time-dependent multivariate responses of biological community to stress. *Environmental Toxicology and Chemistry* 18(2), 138–148. DOI: 10.1002/etc.5620180207.

Vranken, G. and Heip, C. (1983) The productivity of marine nematodes. *Ophelia* 26, 429–442.

Whittaker, R.H. (1972) Evolution and measurement of species diversity. *Taxon* 21, 213–251.

Wieser, W. (1953) Die Beziehung zwischen Munholengestalt, Ernährungsweise und Vorkommen beifreilebenden marinen Nematoden. *Arkiv für Zoologie* 4, 439–484.

Yeates, G.W., Bongers, T., de Goede, R.G.M., Freckman, D.W. and Georgieva, S.S. (1993) Feeding habits in soil nematode families and genera – an outline for soil ecologists. *Journal of Nematology* 25, 315–331.

Index

Page numbers in **bold** type refer to figures; Page numbers in *italic* type refer to tables

abundance 98–101, *99*, **100**, **101**
Acrobeloides 125, 129, 195
'active spiking' method 265
acute-to-chronic ratios (ACR = LC50/EC50) 212
adaptive response 226
additive effect 240
advanced citriculture production (ACPS) 125–126
Agaonidae wasps 69
agglomerative clustering 85
Aglaophyton major 66
agricultural/agriculture
  habitats 195
  management systems 125–126
  manure 234
  regeneration 300
  soils **193**, 277
  systems 10, 194–195, 276
agrochemicals 132–134
agroecosystem 125, 183, 185, 187, 195
Aichi Targets 278
algae 3, 208
alien plants, invasions of 137
  plant invasions in different habitats 138
  time since initial invasion 137
  see also plants
alpha diversity 45–46, 78, 83, 198
alpine nematode 73, 170–171
alternative nematode extraction technique 265, 267
American Society for Testing and Materials (ASTM) 206, 217
Anaplectus 140
Andrássy formula 100
anhydrobiosis 168, 172–174
animal-parasitic nematodes (APNs) 67, 293
animal parasitism 68
anise 133
annulation 254, 255
Antarctic
  deserts 122, 168
  nematodes 168–170
anthelmintic-resistant nematodes 12
aphelenchid nematodes 13
Aphelenchoides ritzemabosi 72, 172
Aphelenchus avenae 13
apple replant disease (ARD) 125
aquatic ecosystems 22, 39, 160, 300
  bioindication with marine nematodes 111–114, *112*, **113**
  in situ bioindication 106–108, *107*, **108**
  management of 107
  NemaSPEAR[%]-index 108–111, **109**, **110**, **111**
  pollution of 105–106
  quality of waterbodies 108
aquatic habitats
  freshwater 155–158
  marine 150–155
aquatic nematodes **41**, 107, 324–326
  biodiversity 285
  environments 92, 106, 107, 283–285, *284*, 312, 314–316
  food webs 14
  macrophytes 19
  microbial communities 20
  organisms 159
  sediments 313
  see also aquatic ecosystems

Arctic and alpine regions 170–171
arthropods 9, 12, 172, 281
artificial intelligence (AI) 257, 307
ASEAN Regional Diagnostic Network (ARDN) 305
assemblages 47, 52
 abundance of 23
 assessment of 23
 cluster analysis 84–85, **85**
 community 82–84, **84**
 composition 154–155
 description of 77–78
 multivariate techniques 84
 ordination techniques 85–88, **86**, **87**
 structure of 6
 taxonomic and functional breadth of 46
 univariate identity-independent indices 78–82, *79–80*, **81**
Association of Southeast Asian Nations (ASEAN) 305
ATP levels 232
atrazine 234, 239, 281
automated testing approaches 210
average linkage method 85

Bacillus thuringiensis 216
bacteria/bacterial
 co-evolution 69–71
 defence signals 20
 overgrazing of 20
 productivity 52
bacterial-feeding nematodes 4, 5, 7, 9, 15, 16, 97, 98, 133, 211
 abundance of 21
 excretions of 10
 feeding category 93
bacterivores 8, 52, 120, 122, 123
bacterivorous nematodes 51
Baermann
 extraction system 254
 funnels 42, 253–254
 tray separation 254
Bamako Convention 274
Basal Index (BI) *99*
Bathing Water Directive 283
bats, catastrophic consequences on 73
behavioural assay 234
Belgian trial 134
beneficial nematodes 4, 293, 299, 300
benthic nematodes communities **48**, 49, 159
benzo[a]pyrene (BaP) 234
benzyl butyl phthalate (BBP) 265
beta diversity 78, 83–84, **84**
bioaccumulation studies 209
bioavailability 207, 224, 239–240, 303, 314
biobased fertilizers 134
biodiversity 20, 42, 73, 119, 258, 278–280
 competition-derived 45
 data spanning 45
 evaluation 282
 invertebrate 73
 loss 278–283
 representation of 43
 soil assessment 281–283
 species sensitivity distribution models and nematodes 280–281
 see also diversity
biofilms
 colonization 47
 matrices 52
 recolonization process 47
 sampling nematodes in 42
biogeochemistry 49, 151
bioindicators 126, 292–293, *294–297*, 299, 300, 304, 312–313
 appropriate 105–106
 commercial ecological assessment 313–317
 of ecosystem health 298
 specific commercial needs 317–320
biological diversity 37, 73, 202, 278, 298
biological interactions, nematode diversity
 freshwater nematodes 50–51, **51**
 marine nematodes 49–50
 terrestrial nematodes 51–52
biological processes 171, 222, 227–228, *231*, 239
biomass 98–101, *99*, **100**, **101**
biome, characteristics **193**
biosensors, *Caenorhabditis elegans* 224
 detection limit and sensitivity of 224, *225*
 dose–response of 224–226, **226**
 energy balance perturbation 232–234
 of ethanol exposure 234–235
 of heat-shock response 229
 of humic substances 235–236
 microarrays 227–228
 profiling by transcriptomics 226–229, *228*
 RNA-Seq 228–229
 strains 240, 241
 variability in response of 226
biosphere 21
BIOSORT dispense 241
biotic
 groups 194
 interactions 68, **68**
bioturbation 5, 18–19, 40, 124
Brillouin's maximum diversity 82
Bromus tectorum 137–138
Bt toxin 216
Bursaphelenchus xylophilus 69, 72, 73

cadmium 135–136, 224, 229, 231, **231**, 265
Caenorhabditis elegans 2, 171, **208**, 209, 222, 298

automated testing approaches 210
biosensors *see* biosensors
description of 222–223
metabolic perturbations in 234
molecular and genetic testing 210
multiple transgenic 240
passive dosing 209–210
reactive oxygen species (ROS) 210
sensitivity of 211
strain **226**
toxicity tests with 216–217
calcium 171–172
Cannon River basin waters 240
canonical correspondence analysis (CCA) 86, **86**, 88, **159**
carbendazim 281
carbon 8–9, 18, 123–124
carbon dioxide 276
carbon storage 276–277
Central Integrated Pest Management Centres (CIPMCs) 277
centroid method 85
Cephalobidae 23, 51, 194, 195, 198
Channel Index (CI) *99*, 133
Chartered Institute of Ecology and Environmental Management (CIEEM) 314
Chemical Effects in Biological Systems (CEBS) 223
chemicals
authorization of 208
disinfection 133
nematicides 22
pollution 108, 156, 157, 160, 187
chemoattractant 35
chemoautotrophic bacteria 18
Chromadorea 68, 211
Chromadoria 69
Chromadorina 47, 48
chromium 135, 136, 331
chrysene 152
Cichorium intybus 137
circular bioeconomy 133
Clarion-Clipperton Fracture Zone 153–154
climate change 138–139, 174–175, **193**
effects of 72
elevated CO_2 139
and environmental restoration 182–186, **183**, 186–188, **186**
interactions of global change factors 140
marine nematode response to 154–155
precipitation pattern 140
warming 139
climate regulation 120, 123–124, 262
cluster analysis 84–85, **85**
clustered regularly interspaced short palindromic repeats (CRISPR) 224
coastal
sediments 18, 111, 112
waters, liquid wastes discharges to 331–332
Cobbs sieving 254
cold deserts 168–169
colonizer–persister (c-p) values 92–94, **93**, 304
commercial ecological assessment 313–314
aquatic environments 314–316
potentially polluted/degraded soils 316–317
soil environments 316
commercial nematode surveys 320, 329
control conundrum 329–330
examples in aquatic and terrestrial environments 331
expertise constraints 331
financial constraints 330
time constraints 330–331
Common Agricultural Policy (CAP) 276
community 87–88, **197**, 261–262, **262**
basic food web status 263
classical composition 88
ecosystem status 263–264
heterogeneity 262–263
patterns 40
structure 45–46
taxonomic and functional structure of 52
see also community assemblage
community assemblage 88, 92
beta diversity 83–84, **84**
ecological succession 83
graphic representations 85, **85**
species co-occurrence patterns 82–83
Composite Footprint 100
contamination 23, 105, 109–110, 112, 114, 136–137, 151–152, 156–157, 214, 216, 316, 326, 330, 332
degrees of 110
diesel 17
gradient 157
heavy metals 151
Tisza River 216
see also pollution
controlled baseline studies 41
Convention on Biological Diversity 27, 202, 278
Convention on International Trade in Endangered Species (CITES) 278
Convention on the Conservation of European Wildlife and Natural Habitats 78
Convention on Wetlands (Ramsar Convention) 78
convolutional neural networks (CNNs) 257
co-occurrence analysis **81**, 112
COPAS (Complex Object Parametric Analyser and Sorter) 241
copper 96, 123, 135, 151, 157, 231, 240, 281
cosmopolitanism 40, 67
c-p assignments 94, 304
*c-p*1 bacterial feeders 97

crop/cropping 12, 22
 primary causes of losses 6
 productivity 1
crust formation 45
cryptic
 diversity 2
 species 39, 67, 154
cryptobiotic species 73
cuticular microvilli 8
cyst nematodes 4–6, 16, 20, 69, 257
cytochromes 237

Database for Annotation, Visualization and Integrated Discovery (DAVID) 227
decomposition 6–8, 13, 18–19, 38, 92, 98, 122–124, 126, 132, 133, 169, 173, 184, 187, 263, 276, 293
deep-sea
 mining sites 151, 153–154, 317, 320
 nematodes 43, 67
 sediments 21, 22, 171
deep-subsurface sediments 171
Deladenus siricidicola 12–13
De Man formulae 91
Department of Agriculture, Forestry and Fisheries (DAFF) 278
Department of Forestry, Fisheries and the Environment (DFFE) 314
denaturing gradient gel electrophoresis (DGGE) 260
detoxification, metal 231–232, 239
differential interference contrast (DIC) microscopes 321, 322
differentially expressed circular RNAs (DEcircRNAs) 232
Diplolaimelloides 20
direct gradient analysis 85, 86
discriminant analysis 85, 325
disease
 causing organisms 121
 management 121
dissemination strategies 305
distribution ecology 169, 173
diversity 6, 78, *79–80*, 187, 196
 area within 40, **41**
 bacteria co-evolution 69–71
 definition of 38–40
 drivers of *53–54*
 comparable 52
 evidence on 55
 functional 39, 98–101, *99*, **100**, **101**
 global 43, 72–73
 indices *80*, 92
 as invasive species 71–72
 latitudinal effects on 45
 molecular methods 42–43
 of nematodes 293
 number of samples/individuals 40–41
 nutrients on 49
 plant parasitism 68–69, **68**
 predicting 43
 reduction of 55
 sampling and extraction techniques affect recovery 42
 and speciation 66–68
 time span 41–42
 use of 78
 variety of 82
 water depth on 45
 weighted indicators 101
divisive clustering 85
DNA 260
 barcoding 20
 extraction 258, 260
 microarrays 223, 227, 234, 237, 241
 sequences 227
Dorylaimida **3**, 8, 14, 51, 87, 93, 94
downregulation (genes) 237
Drinking Water Directive 283
drought 159–160, **160**
dynamic energy budget modelling 232, 265

ecological
 impact assessment (EcIA) 313
 knowledge 77
 monitoring systems 314
 quality status (EQS) 108, **113**, *113*, 283
 restoration, nematodes as indicators of 186–188, **186**
 risk assessment (EcRA) 313
 services 313
 succession 19–20, 83, 88
 theory 40
Ecological Society of America 201
Ecossa 317
ecosystem 39, 119
 abundances 5, 17–18
 aquatic bacterial community diversity 13
 body structure 2, **3**
 coastal sand dune 16
 description of 1–2
 diversity of 5, 77
 economic crop loss due to plant-feeding 14
 effects of 5–7
 as environmental indicators 22–23
 as environmental modifiers 19–20
 excretion by bacterial-feeding soil 9–10
 feeding 3, 10–12, **11**
 food resources of 7–8
 functions 9, 14–16, **15**, **16**, **17**, 67, 276, 283
 history 2–3
 indicator value 2
 of lakes and oceans 18–19

management 22, 77
microbial feeding in soils and sediments 8–9
organizational structure of 5
parasites of vertebrates 12
physical environments 4
plant-feeding 10
populations governed by environmental modulation of top-down regulation 16–17
predation and omnivory 14
properties 293
regulation of populations of invertebrates 12–13
response to stimuli 4–5
services 96
species diversity within vertebrate hosts 19
status 263
structure and function 10
suppression of mycelial growth 13–14, *13*
synoptic integration to system, landscape and biosphere levels 21
ecotoxicity testing
C. elegans 216–217
chemical uptake routes 213–214
description of 206
invertebrate species 211–212
single substances or chemical mixtures 209–210
vs. species 211
standard test organisms 211–212, **213**
suitability of 207–209, **207**, **208**
toxicity 206–207, *208*, 212–215, **215**
ecotoxicogenomics in *Caenorhabditis elegans* 223, 229
energy balance perturbation 232–234
ethanol exposure 234–235
heat-shock response 229
humic substances 235–236
metal exposure *223*, 229–232
nanoparticle exposure 237–239
phthalate exposure 237
polychlorinated biphenyl exposure 236–237, **236**, **238**
steroid hormone response 234
ecotoxicological methods, classification of **207**
ectotrophic mycorrhizal fungi **13**
education
approaches to 300
challenges and opportunities in 305–307
dedicated hands-on short courses 303–304
and environmental nematology 292–293, *294–297*, 298–300
existing 300
formal educational programmes 300, 303
formal programmes 300, 303
online platforms as tools 304
subjects in existing BSc and MSc programmes 303, **303**
and training 293
Eel nematode 298
eigenanalysis-based methods 85–86
electrical conductivity **158**
ELISOL Environnement 317, 328
elutriator extraction 254
Enchytraeus crypticus 211, 281
endocrine-disrupting chemicals (EDCs) 213, 234
endoparasitic nematodes 254, 353
energy balance 232
Enoplea 68, 211
Enoploides 16, 17
enrichment 98
indicators of 97
nutrient 151–152
opportunists 8, 96–97
organic 151–152
Enrichment, Structure and Basal indices 97, 299, 328
Enrichment Index (EI) values 132, 299
entomopathogenic nematodes (EPNs) 6, 9, 94, 125, 126
environmental DNA (eDNA) 283
environmental indicators 6, 22–23, 273, 274, 285, 303, 304
environmental modifiers
influence ecological succession in grasslands 19–20
predispose hosts to organisms 20
sediment agitation 19
succession of aquatic microbial communities 20
environmental policy 278–280
in agriculture 276–277
and biodiversity loss 278–283
description of 273–274
on pesticide application 277–278
on soil and water pollution 274–275
soil biodiversity assessment 281–283
species sensitivity distribution models and nematodes 280–281
environment/environmental 258
change, sensitivity and responsiveness to 93
comprehensive course on *306*
curricula of formal and non-formal education 304–305
disturbance 2
heterogeneity 200
impact assessment (EIA) 313
physical and chemical properties of 43
policies in aquatic environments 283–285, *284*
stressors 167, 171
variables 86
see also environmental policy
Environmental Protection Act 275
Environmental Protection Agency 275
'enzyme sharing' concept 19
essential biodiversity variables 282

estuarine quality paradox 112
ethanol 234–235
eukaryote organisms 231
European Climate Law 276
European Community Titanium Dioxide Directive 314
European Food Safety Authority 105, 277
European Green Deal 274
European Parliament 276, 314
European Union Biodiversity Strategy 2030 280
European Union Marine Strategy Framework Directive 105, 314
European Union Water Framework Directive (EU WFD) 155, 316
European Water Framework Directive (WFD) 274
European water-related policies *284*
EU Soil Strategy for 2030 274
eutrophication 158–159, **159**
EXCALIBUR project 282
extraction 42
- description of 253
- efficiency 254–255
- methods 254
- sample pretreatment 254
- selection of 253–254
- standardization 255
- technique 253
- time 254
- tools 253, **253**
- *see also* sampling

extrapolation 81, 211, 214, 255
extreme environments
- adaptation to 171
- Arctic and alpine regions 170–171
- cold deserts 168–169
- description of 167–168
- freeze tolerance 172–173
- heat tolerance and thermophily 173
- hot deserts 169
- pH 171–172
- ultraviolet (UV) radiation 174
- *see also* environment/environmental

Fallopia japonica 138
false discovery rate (FDR) analysis 227
fauna/faunal 21, 97, 283
- distribution patterns of 21
- monitoring systems 314
- profile **98**

feeding
- activities of nematodes 10
- groups 22–23
- stage 9
- strategies 8
- types 7–8, 9

Fergusonina 69
fibroblast growth factor (FGF) genes 239
Filenchus 122
Filenchus misellus 13
Five-Year Plan EFC Memo 277
flotation/centrifugation (FC) technique 267
food
- availability 49
- production 134
- resources 7–8, **11**
- security 72

food web 38–39, 52, 283
- assimilation 120
- complexity 78
- functioning 263
- salt marsh benthic 17–18

forest ecosystems 122, 184, 198
Fram to Fork Strategy 276
free-living nematodes (FLNs) 121, 252, 293, 299
freeze tolerance 172–173
freshwater 7, 51, 67, 92, 283, 299–300
- biofilms 52
- communities 52–55, 111
- ecosystems 38, 52, 106, 156, 235
- environments 43
- latitudinal patterns 44
- nematodes 44–48, 50–51, 115, 160, 254, 299–300, 317
- sediments 108, 214–215, **215**
- stress on 106
- substrate texture and food availability 48–49, **48**
- water depth/altitude on nematode diversity 45–46
- *see also* aquatic nematodes

freshwater habitats 46–47, 155–156
- contaminated samples across river basins 156–157, **156**
- drought and temperature 159–160, **160**
- eutrophication 158–159, **159**
- pollution gradient approach 157–158, **158**

functional diversity 39, 98–101, *99*, **100**, **101**, 150
Functional Genomics Data (FGED) Society 227
fungal/fungi 208
- feeders 142
- feeding nematodes 10, 15, 16, 196

fungivores 120, 123, 126, 194, 198

gamma diversity 6, 45, 78
Gammaproteobacteria 18
Gantt chart 251
gene expression *228*, 234
- database **228**
- software tool **228**

Gene Expression Omnibus (GEO) 227
gene ontology (GO) 227

gene responses, identification and classification of 242
genetic
  modification 222
  testing 210
  variation 42–43
genetically modified (GM) crops 317, 320
genetically modified plants 216
genome function 222–223
genome mapping technologies 232
genomic tools 222
global
  biodiversity 73, 175, 200, 280
  change 72–73, 144, 174, 182, 186
  nematode diversity 38
  ocean deoxygenation 155
Global Soil Biodiversity Initiative 202, 282
Globodera 72
Glomus etunicatus 13
Glomus margarita 13
glutathione *S*-transferase (GST) genes 229
Gobiosoma bosc 17–18
grain size structure 48
granular activated carbon (GAC) 239–240
grazing/grazers 9–10, 17, 20
  animals 12
  on fungal mycelia 13
  rates 13
green fluorescent protein (GFP) 210, 224
Green House Gass (GHG) 276–277
green spaces, development of 334–336, **335**
groundwater 17, 22, 157–158, 207, 283, 316
Group on Earth Observations Biodiversity Observation Network (GEOBON) 282

habitat disturbance on nematode diversity
  freshwater nematodes 46–47
  marine nematodes 46
  terrestrial nematodes 47
Halicephalobus mephisto 171
harmonization of online databases and nematode-based indicator tools 307
hazardous wastes 274
heat fixation 255
heat-induced changes 160
heat-shock proteins (HSPs) 229
heat-shock response *230, 233*
heat tolerance 173
heavy metals contamination 151
Heracleum sosnowskyi 138
herbicide prometryne 124
herbivorous nematodes 194
hierarchical clustering 85
higher education institutions (HEIs) 300, 305
high-throughput technologies 222–223
homogeneous soils 6
horizontal gene transfer 69
host-parasite interactions 69
hot deserts 169
hybrid flexible teaching method 307
hydrophobic chemicals 209–210, 265

ice nucleation activity (INA) 172
identity-independent indices 78–82, *79–80*, **81**
immotile nematodes 253
immunosenescence 237
Indian Council for Agricultural Research (ICAR) 277
indices
  Enrichment, Structure and Basal 97
  identity-independent 78–82, *79–80*, **81**
  nematode-based indices (NBIs) *see* nematode-based indices (NBIs)
  univariate identity-independent 78–82, *79–80*, **81**
indicator value 2, 23, 140
indirect gradient analysis 85
inferential hypothesis-testing 84
inhibitory effect 240
in situ
  bioindication 106–108, *107*, **108**
  monitoring system 332–334
Institute of Ecology and Environmental Management (IEEM) 314
integrated nematode management (INM) 277
integrated pest management (IPM) 277
Intergovernmental Panel on Climate Change (IPCC) 182, 185, 276
Intergovernmental Science-Policy Platform on Biodiversity and Ecosystem Services and Protected Areas (IPBES) 278
Intergovernmental Technical Panel on Soils 202, 276
International Congress of Nematology (ICN) 308
International Master of Science in Agro- and Environmental Nematology (IMaNema) 299–301
International Organization for Standardization (ISO) 206, 324
intervention values 275
invasive species 46, 69, 71–72
invertebrates 68
  complex life cycles with 4
  regulation of 13
  species 210–212
in vivo toxicity tests 207
Irish trials 134
irrigation 15, 16

K-dominance curve 298–299, 325
K-strategist 39, 136, 281

laboratory-based toxicity testing 316
lactophenol 322

landscape 21
lakes
- littoral zones of 45–46, 47
- nematodes of 18–19
- sediments 47

Land Use and Coverage Area frame Survey Soil (LUCAS Soil) 282
land uses, ecosystem functioning and bioindication 122–123
- bioindicators 126
- carbon and climate regulation 123–124
- description of 119–120
- disease and pest regulation 125–126
- intensification 195
- land-use change and anthropogenic activities **186**
- nutrient cycling 122–123
- practices 120
- soil ecosystems 120–122, **121**
- water regulation and purification 124–125

latitudinal patterns of nematode diversity
- freshwater 44
- marine 43–44
- terrestrial 44–45

large particle flow cytometry 210
Laxus 18
lethality data 212, *225*
limno-nematode diversity 44
lipid metabolism 232
Litoditis marina 39, 67, 281
littoral macrophytes 18
LUCAS soil survey 282
Ludox 42, 322

machine learning methods 258
macrobenthic invertebrates 155
macrofauna/macrofaunal 49–50, **50**, 312–313, 314, **315**
macroscopic organisms 40
marine habitats 150–151, 195
- bioindicators across 112, 114
- climate change 154–155
- deep-sea mining 153–154
- extreme environmental conditions 155
- hydrocarbon and heavy metal pollution 151
- microplastics 152
- organic enrichment 151–152
- trawling 152–153

marine nematodes 7, 40, 49, 50, 67, 283, 298, 299–300
- across marine habitats 112–114, *112*, **113**
- biofilms 52
- bioindication with 111–114, *112*, **113**
- coastal systems 153
- communities 46, 52–55
- ecosystems 38, 50, 51, 52, 150–151
- habitat disturbance on diversity 46
- latitudinal patterns of diversity 43–44
- nematode assemblages 43
- organisms 66
- principles of 111–112
- sediments 14
- substrate texture and food availability 47–48
- systems 21, 111–112
- water depth/altitude on diversity 45

masculinization 258
Maturity Index (MI) 92, 94, 108, 125, 133, 187, 252, 298
Maturity Index 2-5 (MI2-5) 96
maturity index family
- calculation and use of 94, *95*
- maturity index (MI) 94, *95*
- maturity index 2-5 (MI2-5) *95*, 96
- plant-parasitic index (PPI) 94–96
- Sigma Maturity Index (ΣMI) *95*, 96
- Sigma Maturity Index 2-5 (ΣMI2-5) *95*, 96

McMurdo Dry Valleys (MDVs) 168
meiobenthic nematodes 18
meiofauna/meiofaunal **50**, 154, 312–313, 314, **315**
meiofaunal colonization units (MCUs) 333–334
meta-analysis 136–137
metallothioneins (MTs) 210, 224, 231–232
metal pollution 232
methane 170, 276
MIAME (Minimum Information About a Microarray Experiment) 227
microalgae 8, 52
Microarray and Gene Expression Data (MGED) Society 227
microbes
- communities of 1–2
- nematode grazing on 5, 9
- rich habitats 208

microbial
- biomass 15
- communities 133, 193
- dispersion 121, 125
- diversity 51
- feeding nematodes 5, 14–16, 51
- grazing 120, 121

microbiota 50, **50**
microcosm toxicity testing 207
microelements 125, 135–137, 144
microfauna 313
microinvertebrates 2, 3
microorganisms 170, 197, 200
microphytobenthos biomass (MPB) 17, 18, 50
microplastics 152, 281
mineral-associated organic matter (MAOMC) 123
mineral fertilizers 22, 276
mining sites 122–123, 141, 241, 317, 320
MINSEQE (Minimum Information About a Next-generation Sequencing Experiment) 227

modern digital microscopy 307
molecular
- biomarkers 223
- functions *227*, *231*
- methods 259, 260
- techniques 2
- technologies 307
- testing 210

molecular identification of nematodes
- flow chart structuring **259**
- methods 258–261, **259**

monitored macrozoobenthos (MZB) 156
Monte Carlo permutation tests 87
morphological nematode identification 41, 200, 255, 256–257, **256**
- flow chart structuring **256**
- killing and fixing 255
- machine learning 257–258
- slide preparation and microscope selection 255–256

multi-species test systems (MSTSs) 267
multi-species toxicity testing *266*, 267
multivariate analysis 84, 325, 329
multivariate correlation analysis 326
multivariate techniques 84
- cluster analysis 84–85, **85**
- ordination techniques 85–88, **86**, **87**

municipal wastewater treatment plant (WWTP) 215

nanoparticles (NPs) 237
nanopesticides 281
nanoplastics 234, 281
nanoplastics (particles) 232
nano-remediation 232
National Biodiversity Strategy and Action Plan 280
National Center for Biotechnology Information (NCBI) 227, 260
National Environmental Management and Biodiversity Act 280
National Mission on Agricultural Extension and Technology 277
natural ecosystems 1, 6, 22, 24, 51, 94, 96, 187
natural organic matter (NOM) 235
nature protection 140–141
- assessment of soil biodiversity and function 142
- ecosystem restoration after intensive land use 141–142

necromass 123
necromeny 6
neem cake 133
Nemaplex 24, 94, 96, 100, 257, 306
NemaSPEAR[%]-index 108–111, **109**, **110**, **111**, 112, 156, 157, 158, 159, 263, 330
- based on molecular taxonomy 110–111
- correlation of **109**
- linear regression **111**
- with macrozoobenthos-based indices 110
- principles of 108–109
- within reference sites 110
- validation of 109–110, **109**

nematode-based indices (NBIs) 263, 300, *301–302*, 303
- abundance, biomass and functional diversity 5, 98–101, *99*, **100**, **101**
- as bioindicators 91–92
- calculating 96
- development and application of 92
- ecosystem status 92–96
- flow chart structuring 262, **262**
- soil food web condition 96–98, *97*, **98**
- toolbox 263

Nematode Channel Ratio (NCR) 137, 328
nematode colonization units (NCUs) *321*
Nematode Indicator Joint Analysis (NINJA) online tool 263, 304
nematode metabolic footprints (NMFs) 92, 100, **100**, 101
- Bacterivore 101
- Fungivore 101
- Omnivore 101
- Predator 101

nematodes
- abundances 14, 45, 52
- beta and gamma diversity of 45
- bioindicator potential of 23
- biological control agents of 13–14
- characteristics of 22
- combined influences of 1–2
- distribution 40, 45
- diversity 38–43
- effects on resources 5–7
- extraction methods 42
- fungal biocontrol of 14
- head and pharyngeal regions of 2, **3**
- herbivores 12
- intra-host speciation of 19
- morphological and morphometric identification of 7
- parasitic nature of 6
- parasitism 12
- populations 45
- population regulation of invertebrates 13
- recovery, from sampling and extraction 2
- responses of **108**
- survey protocols 45, 320–323
- suspension 255
- techniques for extracting 23
- temporal stability of 14
- trophic groups **121**
- vertical distribution 42

nematofaunal parameters 26–328
Nematology Educational Forum 308
Nematology Education in Sub-Sahara Africa (NEMEDUSSA) 304

Nemys 304
neurotoxicants/neuro toxicity 212
Nitrate Directive 274
Nitrous oxide 276
NOM Fuchskuhle 235
non-governmental organizations (NGOs) 277
non-metric multidimensional scaling (NMDS) 86, **110**
non-predatory nematodes 51, **51**
non-probability sampling 252
Northern blotting 227, 235
Northern blot analysis 235
nuclear hormone receptors (NHRs) 234
nutrient
- cycling 122–123, 126
- enrichment 151–152

Occam's razor 324–325
Oceans Act 283
oceans, nematodes of 18–19
oil degradation 124
oligochaete 7, 16, 18, 214
olive mill wastewater (OMW) 133
'omic' data 242
omnivores 14, 120, 126, 142, 212
Oncholaimus campylocercoides 18
online databases, harmonization of 307
Oostenbrink elutriation 42, 255
operational taxonomic unit (OTU) frequencies 141
optimal sampling strategy 250
ordination techniques 85–88, **86**, **87**
organic
- debris 4, 18, 265
- enrichment 151–152
- farming practices 15
- molecules 8–9, 235
- pollution 106–107, 152, 157

overgrazing 9, 20
oxidative stress 231–232, 235, 241
oxygen-consuming microbes 49–50
oxygen minimum zones (OMZs) 155

PacBio platforms 260
Palaeonema phyticum 66
Pan-African Nematology Network (PANEMA) 305
Panagrellus 140, 298
parasitism 12
- evolutionary history of 68
- plant 68–69, **68**

Paratylenchus 138
passive dosing (PD) 209–210, 265–267
Pellioditis marina 298
persistent organic pollutants (POPs) 224, 274
perturbed soil, restoring nematode communities in 186–188, **186**
pesticides 22, 106, 132, 209, 212, 213, 214, 275–278
pest regulation 120, 125–126
pH 171–172
Phasmarhabditis 13
phenanthrene 151, 210
phenotypic plasticity 172–173
phyla 3, 68, 70, 312–313
phylum nematoda 36, 68
physicochemical marine monitoring survey **315**
pig manure (PM) 134
Pinus ponderosa 13
plant parasites 91, 96, 261–262
plant-parasitic assemblage 94
plant-parasitic index (PPI) 94–96
plant-parasitic nematodes (PPNs) 4, 14, 19, 67–68, 122, 125, 126, 293, 303
plants
- based alternatives 132–133
- communities, climate-driven shifts in 73
- composition, effects of nematode feeding 10–12
- diversity 49
- ecology 83
- feeding nematodes 4, 10, 12, 14, 20, 22
- host-dependent 96
- invasions 137
- parasitism 68–69, **68**
- photosynthesis 3
- rehabilitation 123
- resistance and defence mechanisms of 121
- rhizosphere 4
- viruses 10

plastic particles 232
Plectidae 23
Plectus aquatilis 171
plough system 101, **101**
Pochonia chlamydosporia 14
polar ordination (PO) 86
pollution 42, 157
- advantages and disadvantages associated with use of nematodes in *107*
- anthropogenic 106, 216
- of aquatic ecosystems 105–106
- artificial 135
- contaminated areas 275
- gradient approach 157–158, **158**
- -induced expression 223
- monitoring and assessment 274–275
- resistant nematodes 112
- trend-based 112

Pollution Load Index (PLI) 112
Pollution Prevention Act 275
polychlorinated biphenyls (PCBs) 236, **236**, 237, **238**
polychlorinated insecticide toxaphene 211
polycyclic aromatic hydrocarbons (PAHs) 112, 209–212, 267
polyploid apomictic species 69

polystyrene 210, 213, 239
polyvinyl chloride (PVC) 152, 237
port sediments **114**
Postgraduate International Nematology Course (PINC) 299
potassium 171–172
potato cyst nematodes 6–7
poultry litter ash (PLA) 134
PPP (plant protection products) Regulation 277–278, 280
Pratylenchus coffeae 69
predators 14, 126, 212
prey nematodes 17
principal components analysis (PCA) **86**
principal coordinates analysis (PCoA) 86
principal response curves (PRC) 87, 88
progesterone 234
propagules 71, 187
proteolysis 231, 232
Protocol for the Assessment of Sustainable Soil Management 276
Pseudomonas syringae 20
Pseudokirchneriella subcapitata 281
Punctodora 48

quantitative PCR (qPCR) 231, 258–259, 260
quarantine pests 72

rDNA 94 *see also* DNA
REACH (Registration, Evaluation, Authorisation and Restriction of Chemicals) 209, 280
reactive oxygen species (ROS) 210
recovered nitrogen from manure (RENURE) materials 134
redundancy analysis (RDA) 86, **215**
Registration, Evaluation, Authorisation, and Restriction of Chemicals (REACH) 280
remediation industry 320
ReNu2Farm project 134
RENURE treatments 135
reporter gene expression 241
resilience 38, 52, 77, 101, 154, 175, 184–185, 232, 277, 283, 293
resource farming 5, 9
Rhabditis marina 18
risk assessments 209, 211, 214, 216, 239, 241, 264, 275, 280, 281, 283, 313, 317
river
  basins 156–157, **156**, 240
  ecosystems 41, 46
RNA interference (RNAi) 231, 237, **238**, 239
RNA sequencing (RNA-Seq) 223, 228–229, 234, 239, 241
Romanomermis culicivorax 12–13
root-feeding nematodes 16, **17**, 19–20

salinization 45, 182, 194–196, 316
salt marsh benthic food web 17–18
sample-based rarefaction curves **160**
sampling
  collect and process 252–253
  cores per sample 252
  extraction technique and recovery 42
  flow chart structuring 251, **251**
  processing 200, 261, 323
  research questions and hypotheses 250–251
  site and sample location selection 251–252, **251**
Schistonchus 69
seabed mining 153–154
Secretariat of the Convention on Biological Diversity 278
sediments 48, 106, *208*, 214
  agitation 19
  aquatic 42
  ecology 300
  sampling depth for 42
  and soil toxicity testing 207, **208**, 216, 217
  testing 207
Seinhorst rapid glycerol technique 322
selenium 135, 136
semi-arid grasslands 140, 187, 196
serpentine receptor class epsilon (SRE) 236
shallow-water studies 153
Shannon diversity index 82, 101, 137, 328
Sigma Maturity Index (ΣMI) 96
Sigma Maturity Index 2-5 (ΣMI2-5) 96
signal transduction 229
simple random sampling 251
single-species toxicity testing 214, 264
  non-standardized test systems 264–267, *266*
  standardized test systems 264, *265*
Smart Soil Organism Detector 254
soil 52, 214, 216, 300
  biodiversity 44, 142, 200, 201, 281–283
  biological state in plots 336–337, **337**
  biota 16, 98, 120
  carbon and nitrogen 194
  carbon assessment protocols 277
  carbon sequestration (SCS) 276
  C cycling 123, 124
  characteristics 192
  chemical properties 170
  community 16
  C stabilization 124, 126
  disturbance 14
  diversity of 196, 202
  ecological restoration on 187
  ecosystems 120–122, **121**, 235, 262, **262**, 263, 313
  environments 96–97, 316
  fauna 10
  food web 14–15, 21, 91, 96–98, *97*, **98**

soil (*continued*)
functioning 101, 120
habitats 201
invertebrates 20
management **197**
microorganisms 16
moisture 42
nematode assemblages in 21
nitrate availability **15**
organisms 194–195, 199
pollution 137, 274–275
properties **193**
remediation 275
restoration 187
sampling depth for 42
texture 49
toxicity 207, *208*
Soil Biodiversity Observation Network (Soil BON) 282
Soil BON Foodweb 282
SOILGUARD project 282
SoildiverAgro project 282
soil nematode assemblage data for commercial studies
benchmarks 328–329
multivariate analyses 329
parameters 326–328
soil nematode communities 141, 183–186, 267
anthropogenic impacts on 194–196
(sub)continental scale 196–197, **197**
drivers of 192–194, **193**
elevated atmospheric CO_2 183–184
global scale 198
interactive effects of 186
large- and local-scale studies 198–199, **199**
methods to synchronize 199–200
precipitation and drought 185–186
scientific values and societal relevance of 201–202
warming 184–185
soil nematodes 4, 124–125, 196
abundance 170
diversity 49, 170, 197
virtual loss of 195
soil organic matter (SOM) 276
Solanum elaeagnifolium 137, 138
Solidago gigantea 138
solid waste 215–216
South Africa
coastal industrial complex in 331–332
Environment Protection Act 1986 275
Integrated Coastal Management Act, 2008 275
Waste Act, 2008 275
Water Act 1974 275
Water (Prevention and Control of Pollution) Cess Act 1977 275
soyabean crops 47, 257
SPEAR (SPEcies At Risk)-index 106, 156, 157
species
area curves 81
complex 39
co-occurrence patterns 82
differentiation 2
rarefaction curves **81**
replacement indices 83–84, **84**
sensitivity distributions (SSDs) 212, **212**, 280–281
specific sensitivities 106–107
Spray Service Providers (SSP) initiative 278
Stilbonema 18
Strategic Plan for Biodiversity 278
Strengthening and Modernization of Pest Management Approach (SMPMA) 277
stress
induced metabolic perturbation 232
induced protein denaturation 229
response biology 222
responses *233*, 240
responsive gene 229
specific gene expression 223
Structure Index (SI) *99*
sublethal stress levels **225**
Sub Mission on Plant Protection and Plant Quarantine (SMPPQ) 277
substrate texture and food availability
freshwater nematodes 48–49, **48**
marine nematodes 47–48
terrestrial nematodes 49
see also sediments; soil
Suillus granulatus 13
survey protocols, nematode 320
communities in soil samples 323–324
sampling and assessment methods 322–323
'standard' survey 320–322
syllabus development 305
symbiotic bacteria 8
synergistic effect 185, 186, 240
synthetic pesticides 22
system 21

taxonomic
abilities 262
classifications 92–94, **93**
nematode community 257
precision 78
resolution 78
Taxonomic Distinctness Index 156, **156**
temperature 159–160, **160**
terminal restriction fragment length polymorphism (T-RFLP) 260
terrestrial habitats
agrochemicals and plant-based alternatives 132–133
climate change 138–140
invasions of alien plants 137–138

microelement effects on nematode communities 135–137
nature protection 140–142
nematode assemblage in *143*, 144
recycling-derived and bio-based fertilizers 134–135
terrestrial nematodes 67
communities 52–55
diversity **44**
ecosystems 38, 39–40, 92, 123, 300
environments 43, 92, 298, 304, 331
habitat disturbance on diversity 47
latitudinal patterns of diversity 44–45
substrate texture and food availability 49
water depth/altitude on diversity 46
testosterone 234
thermophily 173
Thornenematidae 93–94
Tisza River contamination 216
tomato crop
phenology **15**
yield **16**
total organic carbon (TOC) 112
toxicant with the test organism (TKTD modelling) 265
toxicity 212, 264
contaminated field samples, testing of 214–216, **215**
of contaminated soils 240
ecotoxicology, testing in 206–207, *208*
end points 212–213
flow chart structuring *264*
multi-species testing 267
single-species testing 264–267, *265*, *266*
standards for *265*
test protocol 206, 216–217, 264
toxicokinetics and toxicodynamics (TKTD) 264
toxic potential of the measured organic chemicals (maximal TEC-Q) *219*, 215
toxic response 226
'training for trainers' (ToTs) programmes 305
trait-based indices 112
transcriptional regulation *236*
transcriptional response 223
transcriptomics, profiling by 226–227, *228*
transgenic biosensor 222, 224
transgenic strains 224, *225*, *230*, 232, *233*, 239
applicability of 241
in environmental monitoring and toxicology 241
trawling, marine nematode response to 152–153
tree-related microhabitats (TreM) 49
trend-based pollution 112
tricarboxylic acid (TCA) cycle 232, 234
tundra regions 198
Turbatrix aceti 298
Tylenchidae 96, 256
Tylencholaimus 122, 141
Tylencholaimus parvus 13

UDP-glucuronosyltransferase (UGT) genes 229
UK Environmental Bill 280
ultraviolet (UV) radiation 174
UN 2030 Agenda for Sustainable Development 274, 277
UN Conventions 202, 276, 277
United Nations Environment Programme (UNEP) 278
Univariate studies 325
up-regulation (genes) 235
urbanization 198

vegetation
diversity **193**
succession 16
vertebrates 12, 68
hosts 4, 19
intestinal parasites of 8
vertical distributions 42, 169
virtual simulation 307
vitellogenin genes 234

Ward's method 85
wastewater 215–216, 239
assessment 320
discharge 314
toxicity of 208
treatment plant (WWTP) 215
waterbodies 108, 283
ecological status in 106
NemaSPEAR[%]-index 108–111, **109**, **110**, **111**
quality of 108
water depth/altitude on nematode diversity
freshwater 45–46
interface 46–47
marine 45
pollution 157, 274–275
terrestrial 46
Western blot experiment 234
Whitehead trays 42
White Paper on the Conservation and Sustainable use of Biodiversity 280
wire worms 20
World Bank 275
World Health Organization (WHO) 278
World Heritage Convention 278

Zero Liquid Discharge Strategy 275
zinc 109, 123, 136, 144, 187, 224, 231, 235, 331
Zullini's study 157